Molecules at Play in Cancer

Molecules at Play in Cancer

Editor

Dumitru A. Iacobas

MDPI • Basel • Beijing • Wuhan • Barcelona • Belgrade • Manchester • Tokyo • Cluj • Tianjin

Editor
Dumitru A. Iacobas
Undergraduate Medical Academy
Prairie View A&M University
Prairie View
United States

Editorial Office
MDPI
St. Alban-Anlage 66
4052 Basel, Switzerland

This is a reprint of articles from the Special Issue published online in the open access journal *Current Issues in Molecular Biology* (ISSN 1467-3045) (available at: www.mdpi.com/journal/cimb/special_issues/molecules_cancer).

For citation purposes, cite each article independently as indicated on the article page online and as indicated below:

LastName, A.A.; LastName, B.B.; LastName, C.C. Article Title. *Journal Name* **Year**, *Volume Number*, Page Range.

ISBN 978-3-0365-7303-8 (Hbk)
ISBN 978-3-0365-7302-1 (PDF)

Contents

About the Editor . . . ix

Preface to "Molecules at Play in Cancer" . . . xi

Dumitru Andrei Iacobas
Molecules at Play in Cancer
Reprinted from: *Curr. Issues Mol. Biol.* **2023**, *45*, 2182-2185, doi:10.3390/cimb45030140 . . . 1

Nae Takizawa, Susumu Tanaka, Koshiro Nishimoto, Yuki Sugiura, Makoto Suematsu and Chisato Ohe et al.
Familial Hyperaldosteronism Type 3 with a Rapidly Growing Adrenal Tumor: An In Situ Aldosterone Imaging Study
Reprinted from: *Curr. Issues Mol. Biol.* **2021**, *44*, 128-138, doi:10.3390/cimb44010010 . . . 5

Katja Seipel, Naomi Porret, Gertrud Wiedemann, Barbara Jeker, Vera Ulrike Bacher and Thomas Pabst
sBCMA Plasma Level Dynamics and Anti-BCMA CAR-T-Cell Treatment in Relapsed Multiple Myeloma
Reprinted from: *Curr. Issues Mol. Biol.* **2022**, *44*, 1463-1471, doi:10.3390/cimb44040098 . . . 17

Piotr Łacina, Aleksandra Butrym, Diana Frontkiewicz, Grzegorz Mazur and Katarzyna Bogunia-Kubik
Soluble CD147 (BSG) as a Prognostic Marker in Multiple Myeloma
Reprinted from: *Curr. Issues Mol. Biol.* **2022**, *44*, 350-359, doi:10.3390/cimb44010026 . . . 27

Elena Monti, Alessandro Mancini, Emanuela Marras and Marzia Bruna Gariboldi
Targeting Mitochondrial ROS Production to Reverse the Epithelial-Mesenchymal Transition in Breast Cancer Cells
Reprinted from: *Curr. Issues Mol. Biol.* **2022**, *44*, 5277-5293, doi:10.3390/cimb44110359 . . . 37

Erika Maldonado-Rodríguez, Marisa Hernández-Barrales, Adrián Reyes-López, Susana Godina-González, Perla I. Gallegos-Flores and Edgar L. Esparza-Ibarra et al.
Presence of Human Papillomavirus DNA in Malignant Neoplasia and Non-Malignant Breast Disease
Reprinted from: *Curr. Issues Mol. Biol.* **2022**, *44*, 3648-3665, doi:10.3390/cimb44080250 . . . 55

Lyudmila V. Bel'skaya and Elena A. Sarf
Salivaomics of Different Molecular Biological Subtypes of Breast Cancer
Reprinted from: *Curr. Issues Mol. Biol.* **2022**, *44*, 3053-3074, doi:10.3390/cimb44070211 . . . 73

Zhewen Zhang, Juan Yi, Bei Xie, Jing Chen, Xueyan Zhang and Li Wang et al.
Parkin, as a Regulator, Participates in Arsenic Trioxide-Triggered Mitophagy in HeLa Cells
Reprinted from: *Curr. Issues Mol. Biol.* **2022**, *44*, 2759-2771, doi:10.3390/cimb44060189 . . . 95

Wallax Augusto Silva Ferreira and Edivaldo Herculano Correa de Oliveira
Expression of *GOT2* Is Epigenetically Regulated by DNA Methylation and Correlates with Immune Infiltrates in Clear-Cell Renal Cell Carcinoma
Reprinted from: *Curr. Issues Mol. Biol.* **2022**, *44*, 2472-2489, doi:10.3390/cimb44060169 . . . 109

Hiroyuki Tsuchiya, Ririko Shinonaga, Hiromi Sakaguchi, Yutaka Kitagawa and Kenji Yoshida
NEAT1–SOD2 Axis Confers Sorafenib and Lenvatinib Resistance by Activating AKT in Liver Cancer Cell Lines
Reprinted from: *Curr. Issues Mol. Biol.* **2023**, *45*, 1073-1085, doi:10.3390/cimb45020071 . . . 127

Athanasios Armakolas, Vasiliki Dimopoulou, Adrianos Nezos, George Stamatakis, Martina Samiotaki and George Panayotou et al.
Cellular, Molecular and Proteomic Characteristics of Early Hepatocellular Carcinoma
Reprinted from: *Curr. Issues Mol. Biol.* **2022**, *44*, 4714-4734, doi:10.3390/cimb44100322 . . . . . . **141**

Mouadh Barbirou, Amanda Miller, Yariswamy Manjunath, Arturo B. Ramirez, Nolan G. Ericson and Kevin F. Staveley-O'Carroll et al.
Single Circulating-Tumor-Cell-Targeted Sequencing to Identify Somatic Variants in Liquid Biopsies in Non-Small-Cell Lung Cancer Patients
Reprinted from: *Curr. Issues Mol. Biol.* **2022**, *44*, 750-763, doi:10.3390/cimb44020052 . . . . **163**

Takuma Hayashi, Kenji Sano and Ikuo Konishi
Possibility of SARS-CoV-2 Infection in the Metastatic Microenvironment of Cancer
Reprinted from: *Curr. Issues Mol. Biol.* **2022**, *44*, 233-241, doi:10.3390/cimb44010017 . . . . **177**

Tania Sultana, Umair Jan, Hyunsu Lee, Hyejin Lee and Jeong Ik Lee
Exceptional Repositioning of Dog Dewormer: Fenbendazole Fever
Reprinted from: *Curr. Issues Mol. Biol.* **2022**, *44*, 4977-4986, doi:10.3390/cimb44100338 . . . . . . **187**

Gábor Somlyai, Lajos I. Nagy, László G. Puskás, András Papp, Beáta Z. Kovács and István Fórizs et al.
Deuterium Content of the Organic Compounds in Food Has an Impact on Tumor Growth in Mice
Reprinted from: *Curr. Issues Mol. Biol.* **2022**, *45*, 66-77, doi:10.3390/cimb45010005 . . . . **197**

Wade R. Gutierrez, Jeffrey D. Rytlewski, Amanda Scherer, Grace A. Roughton, Nina C. Carnevale and Krisha Y. Vyas et al.
Loss of Nf1 and Ink4a/Arf Are Associated with Sex-Dependent Growth Differences in a Mouse Model of Embryonal Rhabdomyosarcoma
Reprinted from: *Curr. Issues Mol. Biol.* **2023**, *45*, 1218-1232, doi:10.3390/cimb45020080 . . . . . . **209**

Hye-Ran Kim, Choong Won Seo, Sang Jun Han, Jae-Ho Lee and Jongwan Kim
Zinc Finger E-Box Binding Homeobox 2 as a Prognostic Biomarker in Various Cancers and Its Correlation with Infiltrating Immune Cells in Ovarian Cancer
Reprinted from: *Curr. Issues Mol. Biol.* **2022**, *44*, 1203-1214, doi:10.3390/cimb44030079 . . . . . . **225**

Hyeji Jeon, Su Min Seo, Tae Wan Kim, Jaesung Ryu, Hyejeong Kong and Si Hyeong Jang et al.
Circulating Exosomal miR-1290 for Diagnosis of Epithelial Ovarian Cancer
Reprinted from: *Curr. Issues Mol. Biol.* **2022**, *44*, 288-300, doi:10.3390/cimb44010021 . . . . **237**

Sanda Iacobas and Dumitru Andrei Iacobas
Personalized 3-Gene Panel for Prostate Cancer Target Therapy
Reprinted from: *Curr. Issues Mol. Biol.* **2022**, *44*, 360-382, doi:10.3390/cimb44010027 . . . . **251**

Wenjia Liu, Nanjiao Ying, Xin Rao and Xiaodong Chen
MiR-942-3p as a Potential Prognostic Marker of Gastric Cancer Associated with AR and MAPK/ERK Signaling Pathway
Reprinted from: *Curr. Issues Mol. Biol.* **2022**, *44*, 3835-3848, doi:10.3390/cimb44090263 . . . . . . **275**

Se-Il Go, Gyung Hyuck Ko, Won Sup Lee, Jeong-Hee Lee, Sang-Ho Jeong and Young-Joon Lee et al.
Cyclin D1 Serves as a Poor Prognostic Biomarker in Stage I Gastric Cancer
Reprinted from: *Curr. Issues Mol. Biol.* **2022**, *44*, 1395-1406, doi:10.3390/cimb44030093 . . . . . . **289**

Wen-Jung Chen, Wen-Wei Sung, Chia-Ying Yu, Yu-Ze Luan, Ya-Chuan Chang and Sung-Lang Chen et al.
PNU-74654 Suppresses TNFR1/IKB Alpha/p65 Signaling and Induces Cell Death in Testicular Cancer
Reprinted from: *Curr. Issues Mol. Biol.* **2022**, *44*, 222-232, doi:10.3390/cimb44010016 . . . . **301**

Roberta B. Andrade, Giovanna C. Cavalcante, Marcos A. T. Amador, Fabiano Cordeiro Moreira, André S. Khayat and Paulo P. Assumpção et al.
The Search for Cancer Biomarkers: Assessing the Distribution of INDEL Markers in Different Genetic Ancestries
Reprinted from: *Curr. Issues Mol. Biol.* **2022**, *44*, 2275-2286, doi:10.3390/cimb44050154 . . . . . . **313**

Eric di Luccio, Satoru Kaifuchi, Nobuaki Kondo, Ryota Chijimatsu, Andrea Vecchione and Takaaki Hirotsu et al.
Nematode-Applied Technology for Human Tumor Microenvironment Research and Development
Reprinted from: *Curr. Issues Mol. Biol.* **2022**, *44*, 988-997, doi:10.3390/cimb44020065 . . . . **325**

José R. Almeida, Bruno Mendes, Marcelo Lancellotti, Gilberto C. Franchi, Óscar Passos and Maria J. Ramos et al.
Lessons from a Single Amino Acid Substitution: Anticancer and Antibacterial Properties of Two Phospholipase A_2-Derived Peptides
Reprinted from: *Curr. Issues Mol. Biol.* **2021**, *44*, 46-62, doi:10.3390/cimb44010004 . . . . **335**

About the Editor

Dumitru A. Iacobas

Born and educated in Romania (PhD in Physics—Biophysics, awarded by the University of Bucharest), Dr. Iacobas is currently a Research Professor and Director of the Personalized Genomics Laboratory within the Undergraduate Medical Academy of the Prairie View A&M University. In addition to optimizing gene expression wet protocols, he introduced the Genomic Fabric Paradigm, which considers the transcriptome as a multi-dimensional mathematical object subjected to dynamic sets of transcription control and gene expression inter-coordination. Moreover, he developed the Gene Master Regulator approach, which identifies the most legitimate gene targets for personalized anti-cancer therapy.

Preface to "Molecules at Play in Cancer"

The spectacular evolution of molecular and genomic technologies, supported by advanced mathematical and bioinformatics analytical tools, allows for both a deep understanding of the intimate mechanisms of cancerization and the exploration of novel anti-cancer therapeutic avenues. With these in mind, I proposed to the MDPI journal *Current Issues in Molecular Biology* (*CIMB*) a Special Issue dedicated to advances in cancer molecular etiology and treatment.

"Molecules at Play in Cancer"is a collection of 24 articles presenting recent findings about molecules with significant roles either in the malignancy, development, and proliferation of cancer cells or in their selective killing within an affected organ. This reprint covers a wide range of tissues affected by cancer: adrenal glands (Takizawa et al.), blood (Seipel et al.; Łacina et al.), breasts (Monti et al.; Maldonado-Rodríguez et al.; Bel'Skaya and Sarf), cervix (Zhang et al.), kidneys (Fereira et al.), liver (Tsuchiya et al.; Armakolas et al.), lungs (Barbirou et al.; Hayashi et al.; Sultana et al.; Somlyai et al.), muscles (Gutierez et al.), ovaries (Kim et al; Jeon et al.), prostate (Iacobas and Iacobas), stomach (Liu et al.; Go et al.), and testicles (Chen et al). The influence of favoring factors such as sex (Gutierez et al.) and genetic ancestry (Andrade et al.) are also discussed. In addition to the traditional mouse models of human cancers (Somlyai et al.; Gutierez et al.), nematodes, such as *Caenorhaditis elegans* (di Lucio et al.), were very useful in detecting cancer in urine samples owing to their superior smelling capabilities.

Some authors proposed either general (Almeida et al.; Somlyai et al.; Sultana et al.) or personalized (Iacobas & Iacobas) molecular solutions for anticancer therapy, with some interesting findings including anti-bacterial (Almeida et al.) and dewormer (Sultana et al.) drugs being repurposed for cancer treatment and some personalized solutions being based on the unique combination of Gene Master Regulators of the cancer nodules in the patient tumor (Iacobas & Iacobas).

By purpose and content, this Special Issue is addressed to the vast number of life science researchers and health care workers (doctors, physician assistants, and nurses). Our initiative was honored by 24 groups totaling 166 scientists from: Asia (China, Japan, and Korea), Europe (Greece, Hungary, Italy, Poland, Portugal, Russia, Switzerland, and the U.K.), North America (Mexico and the U.S.A.), and South America (Brazil and Ecuador). The number of involved researchers and the geographic distribution of their academic institutions indicate the great interest of the worldwide scientific community in understanding the molecular causes of cancer and the efforts to fight these lethal diseases.

Acknowledgments: this reprint would not have been possible without the excellent administrative work of Ms. Norah Tang, Managing Editor of *CIMB*, and the rigorous evaluations provided by the invited reviewers.'

Dumitru A. Iacobas
Editor

current issues in molecular biology

MDPI

Editorial

Molecules at Play in Cancer

Dumitru Andrei Iacobas

Personalized Genomics Laboratory, Undergraduate Medical Academy, Prairie View A&M University, Prairie View, TX 77446, USA; daiacobas@pvamu.edu

Despite its wide range of incidence, cancer can spontaneously occur in any part of the body and invade regions other than the originally affected tissue. Extensive research conducted by thousands academic institutions all over the world and huge industry investments in developing diagnostic bioassays and therapeutic agents spanning decades are yet to produce valuable results for a better understanding of cancer formation and the design of effective treatments. From genes to transcripts, proteins (enzymes included) and metabolites, a long list of molecular factors are believed to be responsible for triggering the transformation of normal cells into cancer cells, many of these factors being also considered as potential actionable molecules for targeted therapies.

For these reasons, the most significant recent studies on molecular-based diagnostics and the molecular therapy of several types of cancer are collated in this Special Issue of *CIMB*, entitled "Molecules at Play in Cancer". The 24 chapters present the contributions of 166 researchers from 16 countries: Brazil (14), China (12), Ecuador (2), Greece (10), Hungary (8), Italy (4), Japan (25), Korea (29), Mexico (11), Poland (5), Portugal (2), Russia (2), Switzerland (7), the UK (1) and the USA (27). The articles reported new molecular findings regarding the following types of cancer: blood [1,2], breast [3–5], kidney [6], liver [7,8], lung [9–12], ovary [10,13,14], prostate [15], soft tissue [16], gastric [17,18], testicular [19], cervical [20,21], and adrenal gland [22].

However, beyond the relevance of these studies for understanding the roles of various molecular factors in cancer development and anti-cancer therapies, I would like to discuss the types of investigated biological specimens that provided experimental evidence for the authors' conclusions. The types of the specimens were: (1) fresh tissues from cancer patients and healthy counterparts; (2) formalin-fixed, paraffin-embedded (FFPE), long-term-stored human tissues; (3) standard and genetically engineered human cancer cell lines; (4) tissues from diseased and healthy animal models; (5) cancer nodules and surrounding normal tissue from the same tumor of each patient.

Due to large numbers, comparing either fresh or FFPE samples from cancer and healthy humans has the advantage of statistical significance, especially when the investigated populations are homogeneous regarding race, sex, age group and other important cancer-favoring factors. When profiled in-house, experiments on human tissues require the approval of the local Institutional Review Board (IRB), which often limits the spectrum of the experimental approaches. The use of either cell lines or animal models, even with the required Institutional Animal Care and Use Committee (IACUC) approval, has the advantage of allowing the genetic engineering and testing of various treatments.

Most authors of this Special Issue compared fresh tissue samples from cancer patients with those obtained from healthy counterparts in order to identify novel biomarkers or simply test the predictive value of already established biomarkers. Thus, basigin, a membrane-bound glycoprotein, was identified by Łacina et al. [2] as a multiple myeloma biomarker by comparing the expression of the encoding gene in the peripheral blood from 62 patients with multiple myeloma with that of 25 healthy donors. The studied cancer patients were sometimes subdivided into groups based on the etiology of their disease. For instance, Armakolas et al. [8] compared the proteome profiles, expression of

Citation: Iacobas, D.A. Molecules at Play in Cancer. *Curr. Issues Mol. Biol.* **2023**, *45*, 2182–2185. https://doi.org/10.3390/cimb45030140

Received: 27 February 2023
Accepted: 2 March 2023
Published: 7 March 2023

selected miRNAs and abundance of circulating tumor cells among a group of 56 people with advanced and 33 people with early hepatocellular carcinoma, 28 with cirrhosis and 5 healthy controls. Barbirou et al. sequenced circulating tumor cells to identify somatic variants in people with non-small-cell lung cancer [9]. An interesting study by Bel'skaya et al. [5] compared the saliva "omic" characteristics of 487 females with breast cancer with those of 298 healthy controls. Seipel et al. proved that anti B-cell maturation antigen CAR-T cell treatment is efficient in four out of five cases of patients with relapsed multiple myeloma [1].

The significance of infection with human papillomavirus (HPV) in the development of breast cancer was tested by Maldonado-Rodríguez et al. [4] through quantifying the presence of the HPV DNA in 116 formalin-fixed breast tissues of 59 malignant neoplasms, 5 in situ neoplasms, 1 borderline neoplasm and 20 benign neoplasms. The study concluded that virus presence is not a sufficient condition for developing cancer. However, as shown by Hayashi et al. [10], the mortality of infectious diseases such as COVID-19 is higher for women with a pulmonary metastatic niche caused by ovarian adenocarcinoma.

Both fresh and formalin-fixed tissues were used by Go et al. [18] to assess the association of a high expression of epidermal growth factor receptor and cyclin D1 with gastric cancer, and by Jeon et al. [14] to test the association of circulating exosomal miR-1290 with epithelial ovarian cancer. This strategy increases the statistical relevance of studies, while providing a direct evaluation of how much formalin fixation and long-term storage affects the expression levels of the biomarkers.

Other authors used human cancer cell lines to elucidate the responsible molecular mechanisms. Thus, HeLa cells were used by Zhang et al. [21] to test the role of Parkin as synergistic mediator of mitophagy in dysfunctional mitochondria. The testicular teratocarcinoma NCCIT and NTERA2 lines helped Chen et al. [19] to evaluate the therapeutic role of PNU-74654, and prostate cancer LNCaP and DU145 lines were used by Iacobas and Iacobas [15] to identify gene master regulators.

For several studies, genomic data were downloaded from The Cancer Genome Atlas (TCGA). Liu et al. [17] used these data to determine the prognostic value of miR-942-3p for the gastric cancer, Ferreira et al. [6] to determine the value of *GOT2* in clear–cell renal-cell carcinoma, and Kim et al. [13] to determine that of zinc finger E-box binding homeobox 2 in ovarian cancer.

Although most of these studies were carried out on human samples, two groups, Somlyai et al. [12] and Gutierez et al. [16], used mouse models, while Almeida et al. [23] tested the proposed solutions on bacteria. A comprehensive review by Lucio et al. [24] presents the major accomplishments in cancer research that have been possible by the use of the nematode *Caenorhaditis elegans*.

The interesting articles of this Special Issue reveal the potential anti-cancer benefits of using deuterium-depleted water (Somlyai et al., [12]) or even the broad-spectrum antiparasitic activity of the dog dewormer Fenbendazole (Sultana et al. [11]), which is sometimes self-administrated by desperate Korean patients.

Two articles indicated that race and sex should be considered when discussing the relevance of cancer biomarkers. Thus, an epidemiological study by Andrade et al. [20] identified the genetic heterogeneity among subpopulations with different ancestry in Brazil, and sex differences were analyzed in a mouse model of embryonal rhabdomyosarcoma by Gutierez et al. [16].

Nevertheless, in addition to race, sex and age, combinations of cancer risk factors such as medical history, diet, habits and exposure to stress, toxins and radiation make each human a dynamic unique subject. Therefore, is it really possible to identify biomarkers characterizing all patients with a particular form of cancer? Although preferred by the pharma industry for economic reasons, are "fit-for-all" treatments really effective for everyone? Tumor heterogeneity further complicates the characterization of cancer subtypes and requires complex approaches to destroy most of the primary cancer clones at once. Moreover, both the uniqueness of favoring factors for each person and tumor heterogeneity question the validity of meta-analyses that compare the genomes and/or transcriptomes of subpopulations of cancer-stricken and healthy individuals. Instead, we should refer the genomes and/or transcriptomes of cancer cells to those of the normal cells within the tissue of the same person, with an emphasis on how to better personalize treatment to fit patient own characteristics [15].

Conflicts of Interest: The author declares no conflict of interest.

References

1. Seipel, K.; Porret, N.; Wiedemann, G.; Jeker, B.; Bacher, V.U.; Pabst, T. sBCMA Plasma Level Dynamics and Anti-BCMA CAR-T-Cell Treatment in Relapsed Multiple Myeloma. *Curr. Issues Mol. Biol.* **2022**, *44*, 1463–1471. [CrossRef] [PubMed]
2. Łacina, P.; Butrym, A.; Frontkiewicz, D.; Mazur, G.; Bogunia-Kubik, K. Soluble CD147 (BSG) as a Prognostic Marker in Multiple Myeloma. *Curr. Issues Mol. Biol.* **2022**, *44*, 350–359. [CrossRef] [PubMed]
3. Monti, E.; Mancini, A.; Marras, E.; Gariboldi, M.B. Targeting Mitochondrial ROS Production to Reverse the Epithelial-Mesenchymal Transition in Breast Cancer Cells. *Curr. Issues Mol. Biol.* **2022**, *44*, 5277–5293. [CrossRef]
4. Maldonado-Rodríguez, E.; Hernández-Barrales, M.; Reyes-López, A.; Godina-González, S.; Gallegos-Flores, P.I.; Esparza-Ibarra, E.L.; González-Curiel, I.E.; Aguayo-Rojas, J.; López-Saucedo, A.; Mendoza-Almanza, G.; et al. Presence of Human Papillomavirus DNA in Malignant Neoplasia and Non-Malignant Breast Disease. *Curr. Issues Mol. Biol.* **2022**, *44*, 3648–3665. [CrossRef]
5. Bel'Skaya, L.V.; Sarf, E.A. «Salivaomics» of Different Molecular Biological Subtypes of Breast Cancer. *Curr. Issues Mol. Biol.* **2022**, *44*, 3053–3074. [CrossRef] [PubMed]
6. Ferreira, W.A.S.; de Oliveira, E.H.C. Expression of *GOT2* Is Epigenetically Regulated by DNA Methylation and Correlates with Immune Infiltrates in Clear-Cell Renal Cell Carcinoma. *Curr. Issues Mol. Biol.* **2022**, *44*, 2472–2489. [CrossRef]
7. Tsuchiya, H.; Shinonaga, R.; Sakaguchi, H.; Kitagawa, Y.; Yoshida, K. NEAT1–SOD2 Axis Confers Sorafenib and Lenvatinib Resistance by Activating AKT in Liver Cancer Cell Lines. *Curr. Issues Mol. Biol.* **2023**, *45*, 1073–1085. [CrossRef]
8. Armakolas, A.; Dimopoulou, V.; Nezos, A.; Stamatakis, G.; Samiotaki, M.; Panayotou, G.; Tampaki, M.; Stathaki, M.; Dourakis, S.; Koskinas, J. Cellular, Molecular and Proteomic Characteristics of Early Hepatocellular Carcinoma. *Curr. Issues Mol. Biol.* **2022**, *44*, 4714–4734. [CrossRef]
9. Barbirou, M.; Miller, A.; Manjunath, Y.; Ramirez, A.B.; Ericson, N.G.; Staveley-O'Carroll, K.F.; Mitchem, J.B.; Warren, W.C.; Chaudhuri, A.A.; Huang, Y.; et al. Single Circulating-Tumor-Cell-Targeted Sequencing to Identify Somatic Variants in Liquid Biopsies in Non-Small-Cell Lung Cancer Patients. *Curr. Issues Mol. Biol.* **2022**, *44*, 750–763. [CrossRef]
10. Hayashi, T.; Sano, K.; Konishi, I. Possibility of SARS-CoV-2 Infection in the Metastatic Microenvironment of Cancer. *Curr. Issues Mol. Biol.* **2022**, *44*, 233–241. [CrossRef]
11. Sultana, T.; Jan, U.; Lee, H.; Lee, H.; Lee, J.I. Exceptional Repositioning of Dog Dewormer: Fenbendazole Fever. *Curr. Issues Mol. Biol.* **2022**, *44*, 4977–4986. [CrossRef]
12. Somlyai, G.; Nagy, L.I.; Puskás, L.G.; Papp, A.; Kovács, B.Z.; Fórizs, I.; Czuppon, G.; Somlyai, I. Deuterium Content of the Organic Compounds in Food Has an Impact on Tumor Growth in Mice. *Curr. Issues Mol. Biol.* **2023**, *45*, 66–77. [CrossRef] [PubMed]
13. Kim, H.-R.; Seo, C.W.; Han, S.J.; Lee, J.-H.; Kim, J. Zinc Finger E-Box Binding Homeobox 2 as a Prognostic Biomarker in Various Cancers and Its Correlation with Infiltrating Immune Cells in Ovarian Cancer. *Curr. Issues Mol. Biol.* **2022**, *44*, 1203–1214. [CrossRef] [PubMed]
14. Jeon, H.; Seo, S.M.; Kim, T.W.; Ryu, J.; Kong, H.; Jang, S.H.; Jang, Y.S.; Kim, K.S.; Kim, J.H.; Ryu, S.; et al. Circulating Exosomal miR-1290 for Diagnosis of Epithelial Ovarian Cancer. *Curr. Issues Mol. Biol.* **2022**, *44*, 288–300. [CrossRef]
15. Iacobas, S.; Iacobas, D.A. Personalized 3-Gene Panel for Prostate Cancer Target Therapy. *Curr. Issues Mol. Biol.* **2022**, *44*, 360–382. [CrossRef] [PubMed]
16. Gutierrez, W.R.; Rytlewski, J.D.; Scherer, A.; Roughton, G.A.; Carnevale, N.C.; Vyas, K.Y.; McGivney, G.R.; Brockman, Q.R.; Knepper-Adrian, V.; Dodd, R.D. Loss of Nf1 and Ink4a/Arf Are Associated with Sex-Dependent Growth Differences in a Mouse Model of Embryonal Rhabdomyosarcoma. *Curr. Issues Mol. Biol.* **2023**, *45*, 1218–1232. [CrossRef] [PubMed]
17. Liu, W.; Ying, N.; Rao, X.; Chen, X. MiR-942-3p as a Potential Prognostic Marker of Gastric Cancer Associated with AR and MAPK/ERK Signaling Pathway. *Curr. Issues Mol. Biol.* **2022**, *44*, 3835–3848. [CrossRef]
18. Go, S.-I.; Ko, G.H.; Lee, W.S.; Lee, J.-H.; Jeong, S.-H.; Lee, Y.-J.; Hong, S.C.; Ha, W.S. Cyclin D1 Serves as a Poor Prognostic Biomarker in Stage I Gastric Cancer. *Curr. Issues Mol. Biol.* **2022**, *44*, 1395–1406. [CrossRef]
19. Chen, W.-J.; Sung, W.-W.; Yu, C.-Y.; Luan, Y.-Z.; Chang, Y.-C.; Chen, S.-L.; Lee, T.-H. PNU-74654 Suppresses TNFR1/IKB Alpha/p65 Signaling and Induces Cell Death in Testicular Cancer. *Curr. Issues Mol. Biol.* **2022**, *44*, 222–232. [CrossRef]
20. Andrade, R.B.; Cavalcante, G.C.; Amador, M.A.T.; Moreira, F.C.; Khayat, A.S.; Assumpção, P.P.; Ribeiro-dos-Santos, Â.; Santos, N.P.C.; Santos, S. The Search for Cancer Biomarkers: Assessing the Distribution of INDEL Markers in Different Genetic Ancestries. *Curr. Issues Mol. Biol.* **2022**, *44*, 2275–2286. [CrossRef]
21. Zhang, Z.; Yi, J.; Xie, B.; Chen, J.; Zhang, X.; Wang, L.; Wang, J.; Hou, J.; Wei, H. Parkin, as a Regulator, Participates in Arsenic Trioxide-Triggered Mitophagy in HeLa Cells. *Curr. Issues Mol. Biol.* **2022**, *44*, 2759–2771. [CrossRef] [PubMed]

22. Takizawa, N.; Tanaka, S.; Nishimoto, K.; Sugiura, Y.; Suematsu, M.; Ohe, C.; Ohsugi, H.; Mizuno, Y.; Mukai, K.; Seki, T.; et al. Familial Hyperaldosteronism Type 3 with a Rapidly Growing Adrenal Tumor: An In Situ Aldosterone Imaging Study. *Curr. Issues Mol. Biol.* **2022**, *44*, 128–138. [CrossRef]
23. Almeida, J.R.; Mendes, B.; Lancellotti, M.; Franchi, G.C., Jr.; Passos, Ó.; Ramos, M.J.; Fernandes, P.A.; Alves, C.; Vale, N.; Gomes, P.; et al. Lessons from a Single Amino Acid Substitution: Anticancer and Antibacterial Properties of Two Phospholipase A_2-Derived Peptides. *Curr. Issues Mol. Biol.* **2022**, *44*, 46–62. [CrossRef] [PubMed]
24. di Luccio, E.; Kaifuchi, S.; Kondo, N.; Chijimatsu, R.; Vecchione, A.; Hirotsu, T.; Ishii, H. Nematode-Applied Technology for Human Tumor Microenvironment Research and Development. *Curr. Issues Mol. Biol.* **2022**, *44*, 988–997. [CrossRef] [PubMed]

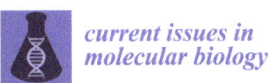

current issues in molecular biology

MDPI

Case Report

Familial Hyperaldosteronism Type 3 with a Rapidly Growing Adrenal Tumor: An In Situ Aldosterone Imaging Study

Nae Takizawa [1,†], Susumu Tanaka [2,†], Koshiro Nishimoto [3,*,†], Yuki Sugiura [4], Makoto Suematsu [4], Chisato Ohe [5], Haruyuki Ohsugi [1], Yosuke Mizuno [6], Kuniaki Mukai [4], Tsugio Seki [7], Kenji Oki [8], Celso E. Gomez-Sanchez [9] and Tadashi Matsuda [1]

1 Department of Urology and Andrology, Kansai Medical University, Osaka 573-1191, Japan; yamamotn@hirakata.kmu.ac.jp (N.T.); haosugi@gmail.com (H.O.); matsudat@takii.kmu.ac.jp (T.M.)
2 Department of Anatomy, Kansai Medical University, Osaka 573-1191, Japan; tanakass@hirakata.kmu.ac.jp
3 Department of Uro-Oncology, Saitama Medical University, International Medical Center 1397-1 Yamane, Hidaka 350-1298, Japan
4 Departments of Biochemistry, School of Medicine, Keio University, Tokyo 160-8582, Japan; yuki.sgi@gmail.com (Y.S.); gasbiology@keio.jp (M.S.); k-mukai@keio.jp (K.M.)
5 Department of Pathology, Kansai Medical University, Osaka 573-1191, Japan; ohec@hirakata.kmu.ac.jp
6 Biomedical Research Center, Division of Morphological Science, Saitama Medical University, Saitama 350-0495, Japan; mizuno@saitama-med.ac.jp
7 Department of Medical Education, California University of Science and Medicine, Colton, CA 92324, USA; SekiT@cusm.org
8 Department of Molecular and Internal Medicine, Graduate School of Biomedical and Health Sciences, Hiroshima University, Hiroshima 734-8551, Japan; kenjioki@hiroshima-u.ac.jp
9 Medical Service, G.V. (Sonny) Montgomery VA Medical Center, Department of Pharmacology and Toxicology, Medicine, University of Mississippi Medical Center, Jackson, MS 39216, USA; cgomez-sanchez@umc.edu
* Correspondence: kn7961@5931.saitama-med.ac.jp; Tel.: +81-(0)4-2984-4111
† These authors contributed equally to this work.

Citation: Takizawa, N.; Tanaka, S.; Nishimoto, K.; Sugiura, Y.; Suematsu, M.; Ohe, C.; Ohsugi, H.; Mizuno, Y.; Mukai, K.; Seki, T.; et al. Familial Hyperaldosteronism Type 3 with a Rapidly Growing Adrenal Tumor: An In Situ Aldosterone Imaging Study. *Curr. Issues Mol. Biol.* **2022**, *44*, 128–138. https://doi.org/10.3390/cimb44010010

Academic Editor: Dumitru A. Iacobas

Received: 22 November 2021
Accepted: 21 December 2021
Published: 28 December 2021

Abstract: Primary aldosteronism is most often caused by aldosterone-producing adenoma (APA) and bi-lateral adrenal hyperplasia. Most APAs are caused by somatic mutations of various ion channels and pumps, the most common being the inward-rectifying potassium channel *KCNJ5*. Germ line mutations of *KCNJ5* cause familial hyperaldosteronism type 3 (FH3), which is associated with severe hyperaldosteronism and hypertension. We present an unusual case of FH3 in a young woman, first diagnosed with primary aldosteronism at the age of 6 years, with bilateral adrenal hyperplasia, who underwent unilateral adrenalectomy (left adrenal) to alleviate hyperaldosteronism. However, her hyperaldosteronism persisted. At the age of 26 years, tomography of the remaining adrenal revealed two different adrenal tumors, one of which grew substantially in 4 months; therefore, the adrenal gland was removed. A comprehensive histological, immunohistochemical, and molecular evaluation of various sections of the adrenal gland and in situ visualization of aldosterone, using matrix-assisted laser desorption/ionization imaging mass spectrometry, was performed. Aldosterone synthase (CYP11B2) immunoreactivity was observed in the tumors and adrenal gland. The larger tumor also harbored a somatic β-catenin activating mutation. Aldosterone visualized in situ was only found in the subcapsular regions of the adrenal and not in the tumors. Collectively, this case of FH3 presented unusual tumor development and histological/molecular findings.

Keywords: familial hyperaldosteronism type 3; *KCNJ5*; adrenal tumor; β-catenin; MALDI-IMS; CYP11B2

1. Introduction

Primary aldosteronism (PA) is caused by excessive and autonomous secretion of aldosterone and is classified with aldosterone-producing adenoma (APA), bilateral idiopathic hyperaldosteronism (IHA), unilateral hyperplasia, or aldosterone-producing carcinoma.

Somatic mutations in ion channel/pump genes, including the inwardly rectifying subfamily J, member 5 potassium channel (*KCNJ5*), have been identified in a significant percentage of APAs (APA-associated mutations) [1]. *KCNJ5* mutations cause a loss in specificity of the channel's selectivity filter for potassium. This leads to sodium leakage into the cells, causing depolarization of the membrane potential; this results in increased calcium influx into adrenocortical cells, causing autonomous aldosterone production [1]. There are four types of familial hyperaldosteronism (FH1–FH4) [2]. FH3 is caused by a germline mutation of *KCNJ5* that leads to adrenal hyperplasia with a marked increase in the secretion of aldosterone [1]. We and others have recently reported cases of non-familial juvenile PA due to mosaicism of somatic *KCNJ5*-mutated and non-mutated cells [3,4], in which the mutated cells/tissues were hyperplastic.

We previously described an immunohistochemistry protocol for aldosterone synthase (CYP11B2) that distinguishes CYP11B2 from the cortisol-synthesizing enzyme steroid 11β-hydroxylase (CYP11B1) [5]. Using CYP11B2 staining, putative aldosterone-producing cells were visualized in the zona glomerulosa of normal adrenals from infants and adults [5–7], as well as several PA lesions [4,5,8]. However, aldosterone biosynthesis requires a cascade of steroidogenic enzymes, and the presence of CYP11B2 alone is not sufficient for the synthesis of aldosterone. To visualize the aldosterone localization in adrenal sections, we recently developed a protocol for the in situ detection of aldosterone using state-of-the-art matrix-assisted laser desorption/ionization imaging mass spectrometry (MALDI-IMS) [9]. Since the steroid hormones, including aldosterone, are released into the blood stream immediately after production, i.e., there is no intracellular storage of steroid hormones, the detection of aldosterone in cells using MALDI-IMS indicates that those cells are actively producing aldosterone. In the present study, we describe an FH3 case with results of comprehensive molecular and CYP11B2 immunohistochemical analyses and correlated them with aldosterone localization in adrenal tissue.

2. Case

We present the case of a 27-year-old Japanese female with a history of severe juvenile PA. She was diagnosed with PA due to bilateral adrenal hyperplasia following adrenal vein sampling at the age of six years and treated with spironolactone and potassium supplementation with moderate control of her blood pressure [10]. However, when she was 15 years old, her serum creatinine level increased to 2.04 mg/dL (normal range: 0.4–1 mg/dL) due to severe hypertension and persistent high plasma aldosterone concentration (PAC: 2511 pg/mL (normal range: 35.7–240 pg/mL)). Since multiple adrenal vein catheterization attempts failed, and computed tomography (CT) indicated that her left adrenal gland (pink arrowhead in Figure 1A) was more hyperplastic than the right, she underwent left adrenalectomy, expecting to alleviate hyperaldosteronism. However, although lower than before, PAC remained elevated (1280 pg/mL) after surgery. At the age of 21 years, she developed end-stage chronic renal failure, thereby requiring intermittent hemodialysis. When she was 26 years old, a CT detected two adrenal tumors (22 × 17 mm and 10 × 6 mm) in her right adrenal gland (red and blue arrowheads in Figure 1B, respectively). Four months later, the larger tumor grew further (28 × 25 mm), and the smaller tumor remained unchanged (Figure 1C), suggesting that the larger tumor might be an adrenocortical carcinoma. She underwent a laparoscopic right adrenalectomy with removal of the intact gland in toto. Her PAC fell into the normal range (83 pg/mL) while on replacement with prednisolone for bilateral adrenalectomy.

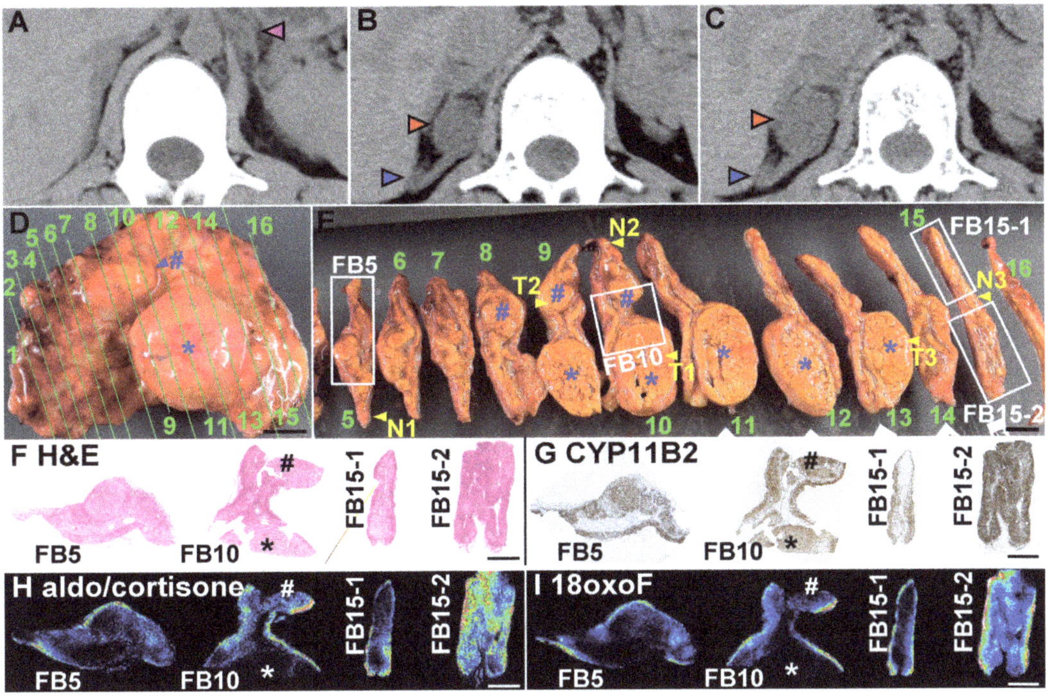

Figure 1. CT and histological findings of the case. (**A**) CT findings at 15 years of age. The left adrenal gland was removed after CT examination. (**B**) CT findings at 26 years of age. Red and blue arrowheads indicate the larger and smaller adrenal tumors in the right adrenal gland, respectively. (**C**) CT findings 4 months after the CT shown in panel B. The larger tumor significantly enlarged in 4 months. (**D**) Macroscopic findings of the extracted right adrenal. The larger (*) and the smaller (#) tumors presumably corresponded to the large (red arrowhead) and small (blue arrowhead) tumors in panels (**B**) and (**C**), respectively. The adrenal was cut into 16 pieces at the green lines. (**E**) Cut surfaces of the extracted adrenal. The green numbers in panel (**D**) correspond to the numbers in panel (**E**). The cut surface numbers in panels (**D**,**E**) correspond to those in parentheses in Supplementary Figure S1, which shows the sections after formaldehyde fixation. Frozen tissue blocks, in an optimal cutting temperature compound, were prepared from 4 portions, indicated by white frames (FB5, FB10, FB15-1, and FB15-2). Flash frozen tissues were also taken from 3 non-tumor portions (N1–N3) and 3 tumor portions (T1–T3). (**F**–**I**) Hematoxylin and eosin staining, immunohistochemistry for CYP11B2, MALDI-imaging of aldosterone and cortisone (aldo/cortisone), and that of 18-oxocortisol (18oxoF), respectively, of frozen tissues.

3. Materials and Methods

3.1. DNA and RNA Isolation from Flash Frozen Tissues, Blood, and Hair Root

Using the AllPrep DNA/RNA Mini Kit (catalog#: 80204, Qiagen, Valencia, CA, USA) and ISOHAIR (catalog#: 315-3403, NIPPON GENE CO., LTD., Tokyo, Japan), genomic DNA and RNA (DNA/RNA) #86, 87, 88, 89, 90, and 91 were prepared from N1, N2, N3, T1, T2, and T3, respectively, as previously reported [11]. DNA #92, 93, 155, and 156 were isolated from the patient's blood, mother's blood, patient's hair root, and father's blood, respectively, according to the manufacturer's instruction.

3.2. Whole Exome Sequencing

We performed whole exome sequencing of genomic DNA samples from T1, N1, and blood (Bl), which was carried out at RIKEN GENESIS CO., LTD. (Tokyo, Japan), as follows.

DNA was sheared into approximately 200 bp fragments and used to construct a library for multiplexed paired-end sequencing with the SureSelectXT Reagent Kit (catalog#: G9641B, Agilent Technologies, Santa Clara, CA, USA). The constructed library was hybridized to biotinylated cRNA baits from the SureSelectXT Human All Exon V6 Kit (catalog#: 5190–8865, Agilent Technologies, Santa Clara, CA, USA) for target enrichment. Targeted sequences were purified with magnetic beads, amplified, and sequenced on an Illumina HiSeq 2500 platform in paired-end 101 bp configuration.

The raw sequence read data of the three samples passed the quality checks in FASTQC (https://www.bioinformatics.babraham.ac.uk/projects/fastqc/ (accessed on 1 February 2017)). Read trimming via base quality was performed using Trimmomatic [12]. Read alignment was performed with the Burrows–Wheeler Aligner [13] (version 0.7.15-r1140). hs37d5 was used as the reference human genome. PCR duplicate reads were removed using Picard (version 2.9.0-1-gf5b9f50-SNAPSHOT, https://broadinstitute.github.io/picard/ (accessed on 1 February 2017)). Non-mappable reads were removed using SAMtools (version 1.3.1) [14]. After filtering out those reads, we applied the Genome Analysis Toolkit [15] (GATK version 3.5-0-g36282e4) base quality score recalibration and performed SNP and INDEL discovery (HaplotypeCaller). Finally, we identified 350, 346, and 343 variants in samples T1, N1, and Bl, respectively, and the variants were annotated using ANNOVAR (version 2016Feb1) [16]. As expected, the *KCNJ5* (p.G151R) mutation was identified in these three samples.

Variants that passed quality control were prioritized according to the following strategies. We only retained variants predicted to modify protein function; these included the nonsense, splice site, coding indel, and missense variants. We removed variants with minor allele frequencies >0.4% for the ESP6500 (ESP6500siv2_all provided by ANNOVAR) database, >0.4% for each population of the Exome Aggregation Consortium (exac03 provided by ANNOVAR), >0.4% in HGVD (containing genetic variations determined by exome sequencing of 1208 individuals in Japan) [17], and >0.4% in 2KJPN (whole-genome sequences of 2049 Japanese healthy individuals and construction of a highly accurate Japanese population reference panel). After removing these variants, we focused on variants identified only in the tumor sample. Variants that appeared to be mapping artifacts, and were too common in in-house controls, were also excluded from further analyses. Consequently, several somatic mutations were found in sample T1. Sanger sequencing of these genes confirmed mutations in catenin β 1 (*CTNNB1*), centromere protein E (*CENPE*), leucine zipper- and EF-hand-containing transmembrane protein 2 (*LETM2*), and ALG10 Alpha-1,2-Glucosyltransferase B (*ALG10B*) in T1 and T3, but not in T2, N1–N3, and Bl, suggesting that these genes might be associated with the rapid growth of the larger tumor. Among these genes, it is well known that mutation in *CTNNB1* is associated with tumor growth in adrenocortical carcinoma via the constitutively activated nuclear β catenin protein [18].

3.3. Microarray Analyses

Microarray analyses of T1–T3 (RNA#89–91, respectively) and N1–N3 (RNA#86 – 88) were performed using the Human Clariom™ S Array and GeneChip WT PLUS Reagent Kit (Thermo Fisher Scientific, #902916 and 902280) [7]. N1 was presumably contaminated with cells from the adrenal medulla, because a few genes known to be expressed in the adrenal medulla were highly expressed in N1 samples (e.g., tyrosine hydroxylase). Genes that exhibited a fold change of 1.3 or more in T1 and T3, as compared to N2 and N3, were used for pathway analysis using the Kyoto Encyclopedia of Genes and Genomes Database. Six pathways were significantly identified as upregulated, as follows: "Protein digestion and absorption" ($p = 0.0016$), "Renin-angiotensin system" ($p = 0.0023$), "Adipocytokine signaling pathway" ($p = 0.0058$), "Cell cycle" ($p = 0.0091$), "Pancreatic secretion" ($p = 0.0171$), and "p53 signaling pathway" ($p = 0.0339$). A similar analysis, using the WikiPathway Database, revealed three up-regulated pathways, as follows: "Retinoblastoma Gene in Cancer" ($p = 0.0003$), "Splicing factor NOVA regulated synaptic proteins" ($p = 0.0091$), and

"Deregulation of Rab and Rab Effector Genes in Bladder Cancer" ($p = 0.0114$). Upstream steroidogenic enzymes for aldosterone synthesis (i.e., cytochrome p450 family 11 subfamily A member 1 (CYP11A1), 3-β-hydroxysteroid dehydrogenase (HSD3B2), and 21-hydroxylase (CYP21A1)) did not exhibit variation in expression between samples, suggesting that the localization of aldosterone shown by MALDI-IMS (mainly in the subcapsular area but not in the tumors) was not due to a lack of the upstream steroidogenic enzyme expression, but other unidentified reasons. Overall, the status of gene variants and gene expression was consistent with the clinical course of the case.

3.4. *Confirmation of CYP11B2 Expression*

We compared the expression levels of *CYP11B2* mRNA between this case and archived APA cases, previously adrenalectomized in the Kansai Medical University (APA#7–APA#26), using qRT-PCR for *CYP11B2* (Supplementary Table S1). *CYP11B2*-expression levels in the tumors of cases APA#7, 18, 19, and 26 were lower than those in paired adjacent adrenal tissues, suggesting incorrect sampling or sampling from non-APA tumors; therefore, these samples were removed from the following analyses. Two tumors (tumors 1 and 2) were sampled from APA#9, but the *CYP11B2* expression of tumor 2 was lower than that of adjacent normal; therefore, tumor 2, but not tumor 1, was also removed from the following analyses. Normal adrenal APA#23 (APA#23N) showed the lowest *CYP11B2* expression level, and a fold difference of each sample (16 pairs) over APA#23N was calculated. Sanger sequencing of these cases for *KCNJ5* revealed that 10 cases harbored *KCNJ5* mutations (62.5%, p.G151R [$n = 6$], and p.L168R [$n = 3$], p.L168Hfs*93 [$n = 1$]) in samples from tumors, but not in their paired adjacent adrenals. An average fold change of APA samples with *KCNJ5* mutation (513,214.9 ± 290,452.9 [mean ± S.D.]) was similar to that without *KCNJ5* mutation (708,321.1 ± 297,156.8, $p = 0.319$, unpaired Student's t-test using ΔΔCt values). As expected from the results of CYP11B2 immunohistochemistry, *CYP11B2* expression levels were not different between non-tumor (816,286.5 ± 428,233.7-fold) and tumor portions (539,737.3 ± 336,381.8-fold) of the case ($p = 0.393$, unpaired Student's t-test using ΔΔCt values). *CYP11B2* expression levels in the case (T1–T3 and N1–N3, 582,237 [interquartile range: 448,734–977,189]-fold) and APA (574,401 (328,933–817,145)-fold) were significantly higher than that of the paired adjacent normal adrenals (966 (62–11,986)-fold), and those in the case and APA were similar (Supplementary Figure S6). Consequently, whole enlarged adrenal in the case expressed high levels of CYP11B2 in mRNA and protein as APAs did.

4. Result

4.1. *Analyses of the Surgically Removed Adrenal Gland*

Comprehensive pathological and molecular analyses were approved by the institutional review boards. Immediately after surgery, the adrenal gland was cut into 16 pieces, as shown in Figure 1D,E. The adrenal gland had two apparent tumors (# and * in Figure 1D,E) and many smaller nodules. Flash frozen tissues were also taken from three non-tumor portions (N1–N3 in Figure 1E), two portions from the larger tumor (T1 and T3), and one portion from the smaller tumor (T2). Frozen blocks, embedded in optimal cutting temperature compound, were prepared from four portions (FB5, FB10, FB15-1, and FB15-2 in Figure 1E), as previously reported [7,9]. The remaining adrenal tissues were fixed with 10% formalin (Supplementary Figure S1) and used for formalin-fixed paraffin-embedded (FFPE) blocks for regular pathological diagnosis. Sanger sequencing of *KCNJ5* was performed, as previously reported [11], and a de novo *KCNJ5* mutation (p.G151R) was detected in genomic DNA from N1–N3 (DNA #86–88, respectively), T1–T3 (#89–92, respectively), as well as her blood (Blood, #92) and hair root (#155), but not in blood samples from her mother (#93) and father (#156) (see "DNA and RNA isolation from flash frozen tissues, blood, and hair root" in the Materials and Methods, Supplementary Figure S2, and details of DNA samples shown in Supplementary Table S1). Histological analyses, using the frozen blocks (Figures 1F and S3) and FFPE tissues (Supplementary Figure S4), were performed. Microscopically, most of the tumor cells in the larger and smaller tumors (* and

in Figure 1D–E, respectively) were composed of compact cells and lipid-rich cells but did not fulfill the criteria for adrenocortical carcinoma (Weiss score: 2; Ki-67 proliferation index: 4.0%), leading to the diagnosis of an adrenal adenoma. The adjacent non-tumor portion lost typical adrenocortical zonation in hematoxylin and eosin staining and revealed many cells that contained lipid vacuoles (Figure 1F, Supplementary Figures S3 and S4). CYP11B2 immunohistochemistry confirmed that the tumors were APAs, as CYP11B2 was expressed throughout the tumors (* and # in Figure 1G) [5]. The cortex of the adjacent non-tumor portion had many CYP11B2-positive cells with irregular arrangement, similar to the adrenal cortices of the previously removed left adrenal gland (Supplementary Figure S5) and as in a previously reported FH3 case [8].

4.2. Production of Aldosterone in FH3 Adrenal

We performed MALDI-IMS, using SolariX attached with Fourier transform ion cyclotron resonance mass spectrometry (Bruker Daltonics, Billerica, MA, USA), to demonstrate in situ aldosterone production throughout the adrenal, as we previously reported [9]. Aldosterone and cortisone, which share identical mass-to-charge ratio values (m/z), were identified mainly in the subcapsular areas of non-tumor adrenal gland, but not in tumors (Figure 1H), irrespective of strong CYP11B2 expression throughout the non-tumor adrenal gland and adrenal tumors (Figure 1G). The hybrid steroid 18-oxo-cortisol, a steroid marker of aldosterone-producing cells [9], was similarly detected in the subcapsular area only (Figure 1I). To determine the CYP11B2 mRNA levels in various areas of the adrenal, including the APA and adjacent adrenal cortices to the APA, we performed quantitative real-time polymerase chain reaction (qRT-PCR), as previously reported [4,7,11,19]. We confirmed that the CYP11B2 expression levels were not significantly different between the tumor and non-tumor portions (T1 – T3 and N1 – N3, respectively) (ΔCT in Supplementary Table S1, $p = 0.303$, Student's *t*-test). In addition, there was no significant difference in the expression level of CYP1B2 between the case, i.e., the average of T1–T3 and N1–N3, and unrelated archived cases of sporadic APA ($n = 16$, Supplementary Figure S6, data of the archived APA cases are shown in Supplementary Table S1). Immunohistochemistry of KCNJ5 was performed in frozen sections obtained from a normal adrenal gland of a renal cell carcinoma patient (left in Supplementary Figure S7A) and FB10 (right), as previously reported [19]. KCNJ5 was detected only in the subcapsular area of the normal adrenal tissue, as previously reported (Supplementary Figure S7B) [19]; whereas, in this case, it was found throughout the adrenal cortex and tumors (Supplementary Figure S7C,D), suggesting that the *KCNJ5* mutation induced KCNJ5 and CYP11B2 co-expression throughout the adrenal cortex and tumors in the patient. Irrespective of high levels of CYP11B2 and KCNJ5, the tumors produced much lower levels of aldosterone and 18oxoF than the non-tumor portion, as shown in Figure 1H,I.

Whole exome sequencing confirmed a germline mutation of *KCNJ5* in T1, N1, and the patient's blood (Materials and Methods). Several somatic mutations were identified in the larger tumor, T1, but not in N1 and blood, which included β-catenin (*CTNNB1*, c.134C > A, p.S45Y), *ADAM17*, *CENPE*, *COL12A1*, *LETM2*, *ALG10B*, and *SRCAP*. Among these mutations, the *CTNNB1* mutation presumably caused rapid tumor growth. Immunohistochemistry confirmed nuclear CTNNB1 expression in the tumor but not in the non-tumor portions, suggesting activation of CTNNB1 in the tumor (Supplementary Figure S8). Microarray analyses of T1–T3 (RNA#89 – 91, respectively) and N1–N3 (RNA#86 – 88, respectively) were performed, as previously reported [7], and confirmed that genes of the cell proliferation pathway were upregulated in T1 and T3 (* in Supplementary Table S2). Except for hydroxysteroid 17-β-dehydrogenase 14 (HSD17B14), which is not associated with aldosterone synthesis, expression of steroidogenic enzymes did not differ between the aldosterone-negative tumors (T1–T3) and non-tumor portions (N1–N3), suggesting that aldosterone production in the non-tumor subcapsular area was controlled by other factor(s) than the steroidogenic enzymes, including CYP11B2.

4.3. Cellular Progression in Non-Tumor and Tumor Portions of the Case

The non-tumor adrenal gland was hyperplastic (Figure 2A,B) and harbored mitotic cells (yellow arrowhead in Figure 2C). To assess the cell cycle progression status of the adrenal cells of the patient, we compared the Ki-67 index [20] between the non-tumor portions, the larger tumor (* in Figures 1 and 2), smaller tumor (# in Figure 1), and archived sporadic APAs and their adjacent adrenal sections ("adjacent") in cases APA #8, 10–17, and 20–25 ($n = 15$) in Supplementary Table S1. It is noteworthy that APA #9 was removed from the analysis because the case harbored a non-APA tumor (sample name: KS-APA_9_T2, Supplementary Table S1). Non-tumor portions were analyzed using two parts each from FFPE blocks #4, 10, and 14 ($n = 6$, Supplementary Figure S4). The larger tumor was analyzed using two parts each from FFPE block #9, 10, and 14 ($n = 6$). The smaller tumor was analyzed using four parts from FFPE block #8 ($n = 4$). Upon comparing these five groups (Kruskal–Wallis one-way analysis of variance on ranks, followed by post hoc comparison with Dunn's methods), we found that the larger tumor (3.20 [interquartile range: 2.84–3.70] unit) had a higher index than the APAs (0.47 [0.39–0.77] unit, $p = 0.001$) and their adjacent adrenal tissue (0.45 [0.36–0.70] unit, $p < 0.001$) (Figure 3). Interestingly, the non-tumor portion of the patient showed a higher Ki-67 index (1.09 [0.94–1.48] unit) than the adjacent adrenal tissue in sporadic APAs ($p = 0.047$). These results suggest that chronic stimulation from the mutated *KCNJ5* channel and/or high aldosterone concentration around the cells might be associated with increased cell cycle progression and/or second hit mutations in genes, including *CTNNB1*.

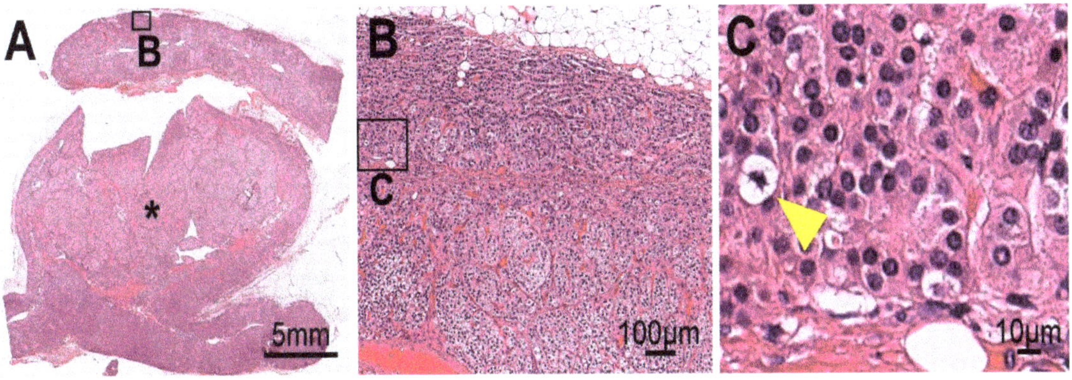

Figure 2. Adrenal histology of the case. The non-tumor adrenal portions were hyperplastic (panels (**A**) and (**B**)) and harbored mitotic cells (yellow arrowhead in panel (**C**)). *: the larger tumor (* in Figure 1).

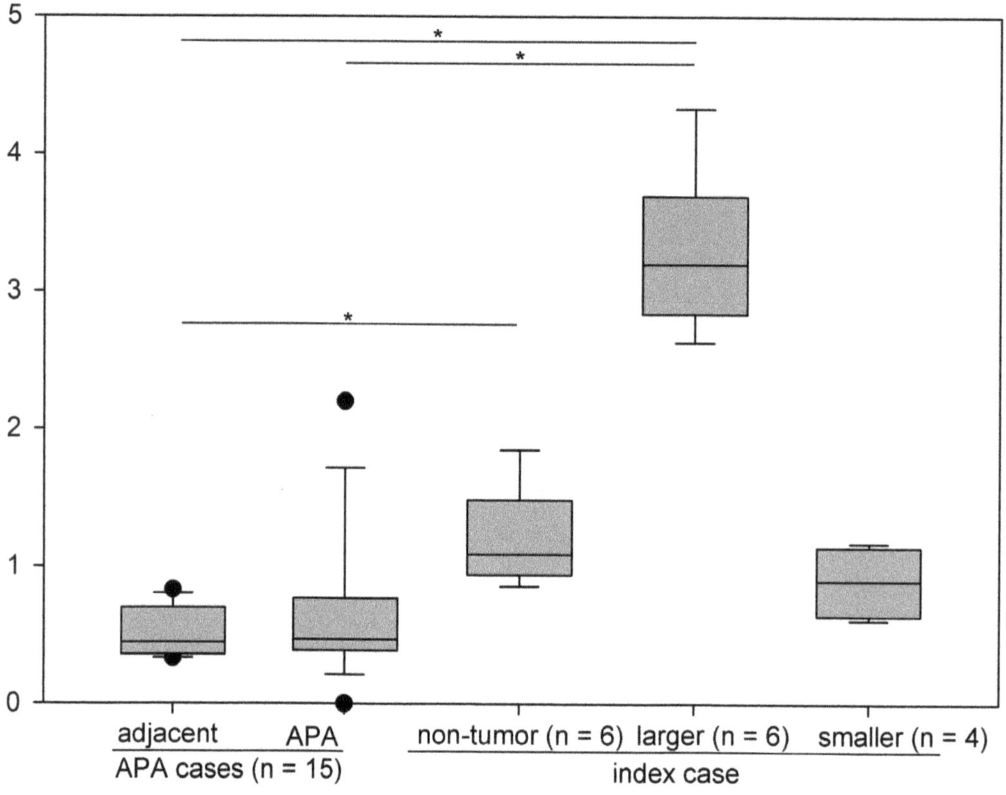

Figure 3. Comparison of Ki-67 index among APA cases (n = 15 each) and index case. * $p < 0.05$.

5. Discussion

This is a case of FH3 with unusual tumor development and histological/molecular findings. The patient was initially diagnosed with PA at the age of 6 years, and her adrenals were removed at the age of 15 years (left adrenal, to alleviate hyperaldosteronism) and at the age of 27 years (right adrenal, due to an enlarging tumor on CT). Clinical data of this patient from birth to 14 years old, including those of kidney biopsy at 10 years old, were described in a preceding article [10], and those from the age of 14 years are provided in Supplementary Table S3. Comprehensive pathological and molecular analyses of the removed right adrenal and blood resulted revealed FH3. The findings can be summarized as follows: (i) an abnormal cellular arrangement of CYP11B2-positive cells throughout the adrenal gland, similar to observed in a previously reported FH3 case [8]; (ii) limited localization of aldosterone production, primarily in the non-tumorous sections, with widespread and very strong CYP11B2 expression in the tumor areas; (iii) rapid tumor growth, which may represent an early stage of adrenocortical carcinoma, caused presumably by second hit mutation of *CTNNB1* (p.S45Y) in *KCNJ5* mutated cells of the larger tumor. The presence of a β-catenin mutation causing constitutive activation of adrenal cell growth has been shown to induce adrenal hyperplasia and adrenal cancer development in mice [21].

There were several peculiar aspects in this case. The pathological findings were remarkably similar to those of previously reported cases [22,23], but with the development of adrenal tumors. The *KCNJ5* mutation, absent in the parents of the patient, represents a de novo mutation. In this case, the larger adrenal tumor also had a second mutation of *CTNNB1*, which led to accelerated cellular growth. Furthermore, even though the patient underwent a bilateral adrenalectomy, serum aldosterone was still detectable and within the normal range, suggesting the presence of extra-adrenal adrenocortical cells harboring the *KCNJ5* mutation, causing aldosterone production [24].

While CYP11B2 staining has been used as a marker for aldosterone production, in the current case, in situ imaging of aldosterone revealed that aldosterone was only found in the non-tumorous areas of the distorted adrenal, which is an additional peculiar finding. Microarray and qRT-PCR studies revealed that all steroidogenic enzymes responsible for aldosterone synthesis were present in sufficient quantities to lead to the increased production of aldosterone in both the tumorous and non-tumorous areas. However, aldosterone production did not occur in the tumorous areas. It has been reported that *KCNJ5* mutations result in depolarization of the adrenal cells with stimulation of calcium mobilization and calmodulin phosphorylation, resulting in transcriptional induction of steroidogenic enzymes [1]. The fact that the tumors exhibited increased levels of steroidogenic enzymes suggests that there is an additional deficiency (or deficiencies) in the steps required for aldosterone biosynthesis in this case. For the synthesis of aldosterone, in addition to transcription of steroidogenic enzymes, cholesterol transporters to mitochondria, such as steroidogenic acute regulatory protein (StAR), are also needed. In addition, StAR must be phosphorylated to exert its action. We speculate that phosphorylation of StAR was repressed, as microarray analysis revealed that StAR expression was not downregulated (Supplementary Table S2).

Although some authors have reported that a treatment with mineralocorticoid receptor antagonist controls hyperaldosteronism well in FH3, most cases in FH3 needed bilateral adrenalectomy for management of severe hyperaldosteronism [1,23]. We had held off a decision of bilateral adrenalectomy because we preferred to avoid lifelong glucocorticoid replacement therapy and the risk of adrenal crisis from bilateral adrenalectomy, if possible. However, that led to the chronic kidney disease and, ultimately, maintenance dialysis at an early age. Notably, this case suggested that chronic stimulation from *KCNJ5* mutation and/or severe hyperaldosteronism contribute to tumorigenic transformation. Therefore, we believe that FH3 patients should undergo bilateral adrenalectomy as soon as possible before irreversible organ damages occur, if mineralocorticoid receptor antagonists cannot control patients' blood pressure.

6. Conclusions

This is an interesting case of a de novo germline mutation of the *KCNJ5* gene that resulted in severe hyperaldosteronism with serious target organ damage, resulting in end-stage renal disease. Steroidogenic and molecular studies of the adrenal gland demonstrated discrepancies that need further investigation.

Supplementary Materials: The following are available online at https://www.mdpi.com/article/10.3390/cimb44010010/s1, Figure S1: The sections after formaldehyde fixation, Figure S2: Pedigree of the index patient and the results of Sanger sequencing, Figure S3: The original microscopic imaging data, Figure S4: HE and IHC of FFPE tissue sections, Figure S5: HE and IHC of FFPE tissue sections of previously resected adrenal gland, Figure S6: CYP11B2 expression, Figure S7: *KCNJ5* immunostaining, Figure S8: CTNNB1 immunostaining, Table S1: Clinical data, qPCR data, and *KCNJ5* mutation analysis of the index case, the parents of the index case, and other APA cases, Table S2: Microarray data, Supplementary Table S3: clinical course from 15 years old.

Author Contributions: Conceptualization, N.T., S.T., K.N., T.S., C.E.G.-S. and T.M.; formal analysis, N.T., S.T., K.N., Y.S., M.S, Y.M., K.M. and K.O.; investigation, N.T., S.T., K.N., Y.S., M.S, C.O., H.O., Y.M., K.M., T.S., K.O., C.E.G.-S. and T.M.; resources, N.T., S.T., K.N., Y.S., M.S, C.O., H.O., Y.M., K.M., K.O., C.E.G.-S. and T.M.; data curation, N.T., S.T., K.N., Y.S., M.S., C.O., Y.M., K.M. and K.O.; writing—original draft preparation, N.T., S.T., K.N., Y.S., C.O., K.M. and K.O.; writing—review and editing, N.T., S.T., K.N., Y.S., M.S., C.O., H.O., K.M., T.S., K.O., C.E.G.-S. and T.M.; visualization, N.T., S.T., K.N., Y.S., C.O., K.M. and K.O.; supervision, K.N., M.S, T.S. and T.M.; project administration, K.N., M.S, T.S. and T.M.; funding acquisition, K.N., M.S. and C.E.G.-S. All authors have read and agreed to the published version of the manuscript.

Funding: K.N. was supported by JSPS KAKENHI, grant number 18K09205. The infrastructure of imaging metabolomics was partly supported by a grant from Shimadzu Corporation. Please note, this corporation did not contribute to the study design and data analyses; thus, there are no conflicts of interests to declare. K.N. was also partly supported by a grant from the Ministry of Health, Labor, and Welfare, Japan (20FC1020). Dr. Celso E. Gomez-Sanchez was supported by the R01 HL144847 grant, from the National Heart, Lung, and Blood Institute, the 1U54GM115428 grant from the National Institute of General Medical Sciences, and the BX004681 grant from the Department of Veteran Affairs.

Institutional Review Board Statement: The study was conducted according to the guidelines of the Declaration of Helsinki. In addition, comprehensive molecular analyses were performed under approval of the institutional review boards of Saitama Medical University International Medical Center (approval #18-308), Keio University School of Medicine (#20090018), and Kansai Medical University (#2016902).

Informed Consent Statement: Informed consent was obtained from all subjects involved in the study.

Data Availability Statement: The data presented in this study are available on request from the corresponding author. The data are not publicly available due to privacy and ethical concerns.

Acknowledgments: The authors are grateful to the patient and her parents for their cooperation and participation in this study. The authors would like to thank the Department of Pathology and Internal Medicine II at Kansai Medical University Hospital for their contribution to the diagnosis, and Masakazu Kohda (Intractable Disease Research Center, Graduate School of Medicine, Juntendo University, Tokyo, Japan) for the contribution to analyze the sequence data. This study was also supported in part by Shimadzu Corporation, Kyoto, Japan.

Conflicts of Interest: The authors declare no conflict of interest.

References

1. Choi, M.; Scholl, U.I.; Yue, P.; Björklund, P.; Zhao, B.; Nelson-Williams, C.; Ji, W.; Cho, Y.; Patel, A.; Men, C.J.; et al. K^+ channel mutations in adrenal aldosterone-producing adenomas and hereditary hypertension. *Science* **2011**, *331*, 768–772. [CrossRef]
2. Perez-Rivas, L.G.; Williams, T.A.; Reincke, M. Inherited Forms of Primary Hyperaldosteronism: New Genes, New Phenotypes and Proposition of a New Classification. *Exp. Clin. Endocrinol. Diabetes* **2019**, *127*, 93–99. [CrossRef]
3. Maria, A.G.; Suzuki, M.; Berthon, A.; Kamilaris, C.; Demidowich, A.; Lack, J.; Zilbermint, M.; Hannah-Shmouni, F.; Faucz, F.R.; Stratakis, C.A. Mosaicism for *KCNJ5* causing early-onset primary aldosteronism due to bilateral adrenocortical hyperplasia. *Am. J. Hypertens.* **2020**, *33*, 124–130. [CrossRef]
4. Tamura, A.; Nishimoto, K.; Seki, T.; Matsuzawa, Y.; Saito, J.; Omura, M.; Gomez-Sanchez, C.E.; Makita, K.; Matsui, S.; Moriya, N.; et al. Somatic *KCNJ5* mutation occurring early in adrenal development may cause a novel form of juvenile primary aldosteronism. *Mol. Cell. Endocrinol.* **2017**, *441*, 134–139. [CrossRef] [PubMed]
5. Nishimoto, K.; Nakagawa, K.; Li, D.; Kosaka, T.; Oya, M.; Mikami, S.; Shibata, H.; Itoh, H.; Mitani, F.; Yamazaki, T.; et al. Adrenocortical zonation in humans under normal and pathological conditions. *J. Clin. Endocrinol. Metab.* **2010**, *95*, 2296–2305. [CrossRef] [PubMed]
6. Nishimoto, K.; Seki, T.; Hayashi, Y.; Mikami, S.; Al-Eyd, G.; Nakagawa, K.; Morita, S.; Kosaka, T.; Oya, M.; Mitani, F.; et al. Human Adrenocortical Remodeling Leading to Aldosterone-Producing Cell Cluster Generation. *Int. J. Endocrinol.* **2016**, *2016*, 7834356. [CrossRef]
7. Nishimoto, K.; Tomlins, S.A.; Kuick, R.; Cani, A.K.; Giordano, T.J.; Hovelson, D.H.; Liu, C.J.; Sanjanwala, A.R.; Edwards, M.A.; Gomez-Sanchez, C.E.; et al. Aldosterone-stimulating somatic gene mutations are common in normal adrenal glands. *Proc. Natl. Acad. Sci. USA* **2015**, *112*, E4591–E4599. [CrossRef] [PubMed]
8. Gomez-Sanchez, C.E.; Qi, X.; Gomez-Sanchez, E.P.; Sasano, H.; Bohlen, M.O.; Wisgerhof, M. Disordered zonal and cellular CYP11B2 enzyme expression in familial hyperaldosteronism type 3. *Mol. Cell. Endocrinol.* **2017**, *439*, 74–80. [CrossRef]
9. Sugiura, Y.; Takeo, E.; Shimma, S.; Yokota, M.; Higashi, T.; Seki, T.; Mizuno, Y.; Oya, M.; Kosaka, T.; Omura, M.; et al. Aldosterone and 18-Oxocortisol Coaccumulation in Aldosterone-Producing Lesions. *Hypertension* **2018**, *72*, 1345–1354. [CrossRef] [PubMed]
10. Takaya, J.; Isozaki, Y.; Hirose, Y.; Higashino, H.; Noda, Y.; Kobayashi, Y. Long-term follow-up of a girl with primary aldosteronism: Effect of potassium supplement. *Acta Paediatr.* **2005**, *94*, 1336–1338. [CrossRef] [PubMed]
11. Nishimoto, K.; Seki, T.; Kurihara, I.; Yokota, K.; Omura, M.; Nishikawa, T.; Shibata, H.; Kosaka, T.; Oya, M.; Suematsu, M.; et al. Case Report: Nodule Development from Subcapsular Aldosterone-Producing Cell Clusters Causes Hyperaldosteronism. *J. Clin. Endocrinol. Metab.* **2016**, *101*, 6–9. [CrossRef]
12. Bolger, A.M.; Lohse, M.; Usadel, B. Trimmomatic: A flexible trimmer for Illumina sequence data. *Bioinformatics* **2014**, *30*, 2114–2120. [CrossRef]
13. Li, H.; Durbin, R. Fast and accurate short read alignment with Burrows-Wheeler transform. *Bioinformatics* **2009**, *25*, 1754–1760. [CrossRef]

14. Li, H.; Handsaker, B.; Wysoker, A.; Fennell, T.; Ruan, J.; Homer, N.; Marth, G.; Abecasis, G.; Durbin, R. The Sequence Alignment/Map format and SAMtools. *Bioinformatics* **2009**, *25*, 2078–2079. [CrossRef] [PubMed]
15. McKenna, A.; Hanna, M.; Banks, E.; Sivachenko, A.; Cibulskis, K.; Kernytsky, A.; Garimella, K.; Altshuler, D.; Gabriel, S.; Daly, M.; et al. The Genome Analysis Toolkit: A MapReduce framework for analyzing next-generation DNA sequencing data. *Genome Res.* **2010**, *20*, 1297–1303. [CrossRef] [PubMed]
16. Wang, K.; Li, M.; Hakonarson, H. ANNOVAR: Functional annotation of genetic variants from high-throughput sequencing data. *Nucleic Acids Res.* **2010**, *38*, e164. [CrossRef]
17. Higasa, K.; Miyake, N.; Yoshimura, J.; Okamura, K.; Niihori, T.; Saitsu, H.; Doi, K.; Shimizu, M.; Nakabayashi, K.; Aoki, Y.; et al. Human genetic variation database, a reference database of genetic variations in the Japanese population. *J. Hum. Genet.* **2016**, *61*, 547–553. [CrossRef] [PubMed]
18. Else, T.; Kim, A.C.; Sabolch, A.; Raymond, V.M.; Kandathil, A.; Caoili, E.M.; Jolly, S.; Miller, B.S.; Giordano, T.J.; Hammer, G.D. Adrenocortical carcinoma. *Endocr. Rev.* **2014**, *35*, 282–326. [CrossRef] [PubMed]
19. Chen, A.X.; Nishimoto, K.; Nanba, K.; Rainey, W.E. Potassium channels related to primary aldosteronism: Expression similarities and differences between human and rat adrenals. *Mol. Cell. Endocrinol.* **2015**, *417*, 141–148. [CrossRef]
20. Klöppel, G.; La Rosa, S. Correction to: Ki67 labeling index: Assessment and prognostic role in gastroenteropancreatic neuroendocrine neoplasms. *Virchows Arch. Int. J. Pathol.* **2018**, *472*, 515. [CrossRef]
21. Berthon, A.; Sahut-Barnola, I.; Lambert-Langlais, S.; de Joussineau, C.; Damon-Soubeyrand, C.; Louiset, E.; Taketo, M.M.; Tissier, F.; Bertherat, J.; Lefrancois-Martinez, A.M.; et al. Constitutive beta-catenin activation induces adrenal hyperplasia and promotes adrenal cancer development. *Hum. Mol. Genet.* **2010**, *19*, 1561–1576. [CrossRef] [PubMed]
22. Geller, D.S.; Zhang, J.; Wisgerhof, M.V.; Shackleton, C.; Kashgarian, M.; Lifton, R.P. A novel form of human mendelian hypertension featuring nonglucocorticoid-remediable aldosteronism. *J. Clin. Endocrinol. Metab.* **2008**, *93*, 3117–3123. [CrossRef] [PubMed]
23. Scholl, U.I.; Nelson-Williams, C.; Yue, P.; Grekin, R.; Wyatt, R.J.; Dillon, M.J.; Couch, R.; Hammer, L.K.; Harley, F.L.; Farhi, A.; et al. Hypertension with or without adrenal hyperplasia due to different inherited mutations in the potassium channel *KCNJ5*. *Proc. Natl. Acad. Sci. USA* **2012**, *109*, 2533–2538. [CrossRef] [PubMed]
24. Abdelhamid, S.; Muller-Lobeck, H.; Pahl, S.; Remberger, K.; Bonhof, J.A.; Walb, D.; Rockel, A. Prevalence of adrenal and extra-adrenal Conn syndrome in hypertensive patients. *Arch. Intern. Med.* **1996**, *156*, 1190–1195. [CrossRef]

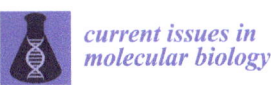

current issues in molecular biology

Article

sBCMA Plasma Level Dynamics and Anti-BCMA CAR-T-Cell Treatment in Relapsed Multiple Myeloma

Katja Seipel [1,*], Naomi Porret [2], Gertrud Wiedemann [2], Barbara Jeker [3], Vera Ulrike Bacher [2] and Thomas Pabst [3,*]

1 Department for Biomedical Research, University of Bern, 2008 Bern, Switzerland
2 Department of Hematology and Central Hematology Laboratory, Inselspital, Bern University Hospital, University of Bern, 3010 Bern, Switzerland; NaomiAzur.Porret@insel.ch (N.P.); gertrud.wiedemann@insel.ch (G.W.); veraulrike.bacher@insel.ch (V.U.B.)
3 Department of Medical Oncology, Bern University Hospital, 3010 Bern, Switzerland; Barbara.Jeker@insel.ch
* Correspondence: katja.seipel@dbmr.unibe.ch (K.S.); Thomas.Pabst@insel.ch (T.P.)

Abstract: BACKGROUND: Novel chimeric antigen receptor T-cells (CAR-T) target the B-cell maturation antigen (BCMA) expressed on multiple myeloma cells. Assays monitoring CAR-T cell expansion and treatment response are being implemented in clinical routine. METHODS: Plasma levels of soluble BCMA (sBCMA) and anti-BCMA CAR-T cell copy numbers were monitored in the blood, following CAR-T cell infusion in patients with relapsed multiple myeloma. sBCMA peptide concentration was determined in the plasma, applying a human BCMA/TNFRS17 ELISA. ddPCR was performed using probes targeting the intracellular signaling domains 4-1BB und CD3zeta of the anti-BCMA CAR-T construct. RESULTS: We report responses in the first five patients who received anti-BCMA CAR- T cell therapy at our center. Four patients achieved a complete remission (CR) in the bone marrow one month after CAR-T infusion, with three patients achieving stringent CR, determined by flow cytometry techniques. Anti-BCMA CAR-T cells were detectable in the peripheral blood for up to 300 days, with copy numbers peaking 7 to 14 days post-infusion. sBCMA plasma levels started declining one to ten days post infusion, reaching minimal levels 30 to 60 days post infusion, before rebounding to normal levels. CONCLUSIONS: Our data confirm a favorable response to treatment in four of the first five patients receiving anti-BCMA CAR-T at our hospital. Anti-BCMA CAR-T cell expansion seems to peak in the peripheral blood in a similar pattern compared to the CAR-T cell products already approved for lymphoma treatment. sBCMA plasma level may be a valid biomarker in assessing response to BCMA-targeting therapies in myeloma patients.

Keywords: B-cell maturation antigen (BCMA); soluble BCMA (sBCMA); multiple myeloma (MM); anti-BCMA CAR-T cell therapy

Citation: Seipel, K.; Porret, N.; Wiedemann, G.; Jeker, B.; Bacher, V.U.; Pabst, T. sBCMA Plasma Level Dynamics and Anti-BCMA CAR-T-Cell Treatment in Relapsed Multiple Myeloma. *Curr. Issues Mol. Biol.* **2022**, *44*, 1463–1471. https://doi.org/10.3390/cimb44040098

Academic Editor: Dumitru A. Iacobas

Received: 7 February 2022
Accepted: 22 March 2022
Published: 24 March 2022

1. Introduction

Multiple myeloma (MM) is a neoplasm of clonal plasma cells. MM is considered treatable, but incurable. Remission may be induced by treatment options involving steroids, chemotherapy, antibodies, targeted compounds, and autologous stem cell transplantation (ASCT). Advancements in treatments, including the introduction of immunomodulatory drugs (IMID), proteasome inhibitors (PI), and monoclonal antibodies have prolonged survival to a five-year survival rate of about 50% [1–4]. New therapies targeting BCMA are currently being investigated in clinical trials and will be incorporated into routine use, with the aim of further improving outcome rates.

The B-cell maturation antigen (BCMA, TNFRSF17) is a cell surface receptor of the TNF receptor superfamily preferentially expressed in malignant and normal plasma cells and mature B lymphocytes, specifically binding to TNFSF13B/TALL-1/BAFF (B-cell activating factor) and to various TRAF family members, and, thus, transducing signals for cell survival and proliferation [5,6]. Proteolytic shedding of the BCMA receptor reduces its cell-surface expression and the ligand-mediated survival of B cell subsets. This shedding is mediated by protease γ-secretase, and is partially dependent on ligand binding and receptor interactions. Shed receptors of soluble BCMA (sBCMA) may serve as biomarkers for auto-immunity and lymphoma [7]. sBCMA is shed from plasma cells, myeloma cells, and plasmacytoid dendritic cells. sBCMA peptide concentrations are elevated in MM plasma (median 500 ng/mL, range 100–1700 ng/mL) compared to healthy donors (median 40 ng/mL, range 10–80 ng/mL), with no difference between newly diagnosed patients and those with relapsed disease [8–10]. A remarkable decline in sBCMA levels was observed in patients with good responses to BCMA-targeted immunotherapy, suggesting sBCMA as a new biomarker for monitoring response to MM therapy [11]. In preclinical studies, anti-BCMA-CAR-T therapy showed low antigen-independent signaling and potent in vitro killing of myeloma tumor cells, across a range of BCMA expression levels, as well as sustained elimination of tumors and 100% survival, after single-dose administration in a mouse model of human multiple myeloma [12].

Idecaptagene-vicleucel (bb2121) is an anti-BCMA chimeric antigen receptor T-cell therapy for the treatment of multiple myeloma, approved by the FDA in 2021 for the treatment of adults with relapsed or refractory multiple myeloma, who have received at least three lines of anti-CD38/PI/IMID treatment. In clinical studies the median progression-free survival was 11.8 months in phase I [13] and 8.8 months in phase 2 trials [14]. In the absence of CAR-T therapy, RR/MM patients had a median PFS of 6.6 months and OS of 13.5 months [15]. Response to bb2121 is heterogeneous and often transient, possibly due to the presence of sBCMA in the plasma [7,9,11]. Serial sBCMA concentrations may decline more significantly in hematologic responders (PR/CR/sCR) than in non-responders (SD/PD) before day 28 post infusion [16].

The assays monitoring CAR-T expansion and assessing response to this novel therapeutic option await implementation into clinical routine. Here, we evaluate the first five patients with relapsed MM treated with anti-BCMA CAR-T cell therapy at the Inselspital Bern, Switzerland.

2. Materials and Methods

In the work described here, we aimed to monitor response in anti-BCMA CAR T-cell recipients, by establishing a panel of laboratory assessments. We evaluated whether introducing routine laboratory parameters might facilitate timely identification of response and trigger therapeutic interventions. We established a digital droplet PCR (ddPCR) assay for CAR-T−specific T-cell receptor (TCR) measurement from peripheral blood (PB) and introduced sBCMA assessments at consecutive time points before and after CAR-T cell infusion.

2.1. Patients

The first five patients with relapsed/refractory multiple myeloma (RR/MM) receiving anti-BCMA CAR-T therapy at Bern University Hospital, Switzerland, were included in this study. They received their individual anti-BCMA CAR-T cell infusion between May and September 2021 after several lines of prior therapy. All patients gave written informed consent, and the study was approved by the local ethics committee of Bern, Switzerland (No. 2018-00628).

2.2. Lymphocyte Apheresis, Lymphocyte Depletion Chemotherapy, and CAR-T Infusion

Lymphocyte collections with the Spectra Optia (Terumo BCT) device were performed using the continuous mononuclear cell collection (CMNC) procedure. Peripheral CD3+ cell counts of blood and lymphocyte products were analyzed by multi-parameter flow cytometry (BD FACSCanto II). Five patients received idecabtagene vicleucel (ide-cel; previously bb2121). Ide-cel was manufactured following leukapheresis and then infused at dose levels aiming at 4.5E + 08 CAR+ T cells after 2 days interval, following lympho-depletion with 3 days of fludarabine 30 mg/m^2 + cyclophosphamide 300 mg/m^2.

2.3. Response Criteria

The criteria for assessing response were according to the International Myeloma Working Group [16,17]. Negative immuno-fixation on the serum and urine and disappearance of any soft tissue plasmacytomas and <5% plasma cells in bone marrow was required for CR response. In addition, normal free light chain (FLC) ratio and absence of clonal cells in the bone marrow by immunohistochemistry or immunofluorescence was required for stringent CR.

2.4. Establishment of ddPCR for CAR-T Quantification

In analogy to our previous study on CD19-CAR-T cell specific digital PCR [17], we designed a specific ddPCR assay to quantify sequences of the intracellular domain of the bb2121 CAR-T construct, using serial recipients' peripheral blood samples. We designed primers and probes for the CAR-T transgenes to target the intracellular junction sequence between the effector (4-1BB) and co-stimulatory (CD3z) domains, similarly to Milone [18]. The procedure for the selection of the PCR primer sequences for quantitative assessment of the bb2121 CAR-T construct was performed following the strategy previously reported by Raje et al., with the primer sequences published there [12]. The reference gene was RPP30 (ribonuclease P protein subunit 30) [19].

2.5. Determination of sBCMA Plasma Levels in the Peripheral Blood

The Human BCMA/TNFRSF17 ELISA Kit (EH41RB, Thermo Fisher Scientific, Waltham, MA, USA) is a solid-phase sandwich enzyme-linked immunosorbent assay (ELISA) designed to detect and quantify the level of human BCMA in cell culture supernatants, plasma, and serum. ELISA assay was done according to the manufacturer's instructions using BCMA protein standard. sBCMA levels were represented as the mean of triplicate samples for each specimen.

3. Results

We evaluated the first five myeloma patients treated with bb2121 CAR-T cell therapy at a single academic center (Table 1). After CAR-T cell administration, patients can develop specific acute toxicities, including cytokine release syndrome (CRS) or immune effector cell-associated neurotoxicity syndrome (ICANS) [20], with up to 40% of patients requiring ICU admission. Three of the five patients in our cohort developed CRS grade 1 or 2, and two patients had ICANS grade 1 or 2. Response determination entailed measurement of standard myeloma blood parameters every three weeks. Four patients achieved a complete remission (CR) in the bone marrow one month after CAR-T infusion, with three patients achieving stringent CR, as determined by flow cytometry techniques. bb2121 CAR-T DNA was detectable in peripheral blood for up to 300 days, with copy numbers peaking 7 to 14 days post infusion (Figure 1A). sBCMA plasma levels started dropping 1–10 days post infusion and reached minimal levels 60 to 80 days post infusion (Figure 1B). In three patients, sBCMA levels started to rebound immediately after the CAR-T levels declined. In one patient, high levels of sBCMA were consistently detected after the initial reduction, and this patient died of progression 100 days post CAR-T cell infusion.

3.1. Patients Characteristics

Five patients with relapsed/refractory multiple myeloma were included in this study (Table 1). They received anti-BCMA CAR-T cells after two to five lines of prior therapy. Response determination entailed detection of standard myeloma blood parameters every three weeks, according to the international myeloma working group [21]. The CAR-T treatment resulted in complete remission (CR) in four patients with CAR-T peak levels exceeding 2E + 05 copies/μg gDNA and sBCMA plasma concentration dropping to minimal levels (<10 ng/mL) within 90 days post infusion.

Table 1. R/R MM Patient characteristics.

ID	C-80	C-85	C-86	C-89	C-91	Average (Range)
age at diagnosis (years)	63	58	55	69	45	58 (45–69)
stage R-ISS	III	II	II	na	II	
plasma cell infiltration (%)	85	50	80	90	60	73 (50–90)
FISH	t(11;14)	t(4;14)	t(11;14)/+1q	na	t(4;14)/+1q	
IgA (g/L)		22	50	46		
light chain kappa (mg/L)		23		6	223	
light chain lambda (mg/L)	13,400		382	604	11	
lines of prior therapy	3	2	5	2	3	3 (2–5)
prior ASCT	1	1	1	0	1	0.8 (0–1)
number of relapses	3	2	2	2	0	2 (0–3)
time to CART (years)	7	6	6	1	1	4.2 (1–7)
CAR-T peak (copies/µg gDNA)	3.2E + 05	2.0E + 05	3.0E + 05	6.0E + 05	1.3E + 04	2.9E + 05
CAR-T peak day post infusion	7	7	13	9	14	10 (7–13)
sBCMA pre-infusion (ng/mL)	115	100	120	280	200	163 (100–280)
sBCMA 4 weeks post infusion	10	10	50	60	120	50 (10–120)
sBCMA 8 weeks post infusion	5	5	12	15	115	33 (5–115)
status, 4 weeks post infusion	sCR	sCR	sCR	CR	SD	
status, 8 weeks post infusion	sCR	sCR	sCR	CR	PD	
status, 6 months post infusion	sCR	sCR	sCR	relapse	deceased	

Abbreviations: relapsed/refractory (R/R); multiple myeloma (MM); autologous stem cell transplant (ASCT); complete remission (CR); stable disease (SD); progressive disease (PD).

Patient 1 (C-80), first diagnosed with multiple myeloma, stage III (R-ISS) received three lines of prior therapy and autologous stem cell transplant (ASCT). After the third relapse, anti-BCMA CAR-T treatment resulted in stringent complete remission four weeks after CAR-T cell infusion. CAR-T copy number in the peripheral blood peaked on day 7 at high levels (>3E + 05 copies/µg gDNA), followed by decline to minimal levels after six months, then remained at low levels (100 copies/µg gDNA) until end of observation. sBCMA plasma levels, initially at 115 ng/mL, started dropping immediately after infusion, and reached minimal levels (5 ng/mL) eleven weeks after infusion. sBCMA plasma levels started to rise to normal levels 4 months after infusion, and 10 months after infusion the response status was still sCR.

Patient 2 (C-85), first diagnosed with multiple myeloma, stage II (R-ISS), had received two lines of prior therapy and ASCT. After second relapse, anti-BCMA CAR-T treatment resulted in stringent complete remission four weeks after CAR-T cell infusion. CAR-T copy number in the peripheral blood peaked on day 7 at intermediate levels (2E + 05 copies/µg gDNA) and subsided below the detection limit after five months. sBCMA plasma levels, initially at 100 ng/mL, started dropping immediately after CAR-T cell infusion, and reached minimal levels (10 ng/mL) four weeks after infusion, persisting below normal levels for nine months when the response status was still sCR.

Patient 3 (C-86), first diagnosed with multiple myeloma, stage II (R-ISS) received five lines of prior therapy and ASCT. After second relapse, anti-BCMA CAR-T treatment resulted in complete remission four weeks after infusion. CAR-T copy number in the peripheral blood peaked on day 13 at high levels (3.2E + 05 copies/µg gDNA) and was still at 128 copies/µg gDNA after six months. sBCMA plasma levels, initially at 120 ng/mL, started dropping 10 days after infusion, and reached minimal levels (10 ng/mL) two months after CAR-T cell infusion. sBCMA plasma levels started to rise above normal levels (rebound) three months after infusion. Eight months after CAR-T infusion sBCMA plasma concentration had reached pretreatment levels, while the response status was still CR.

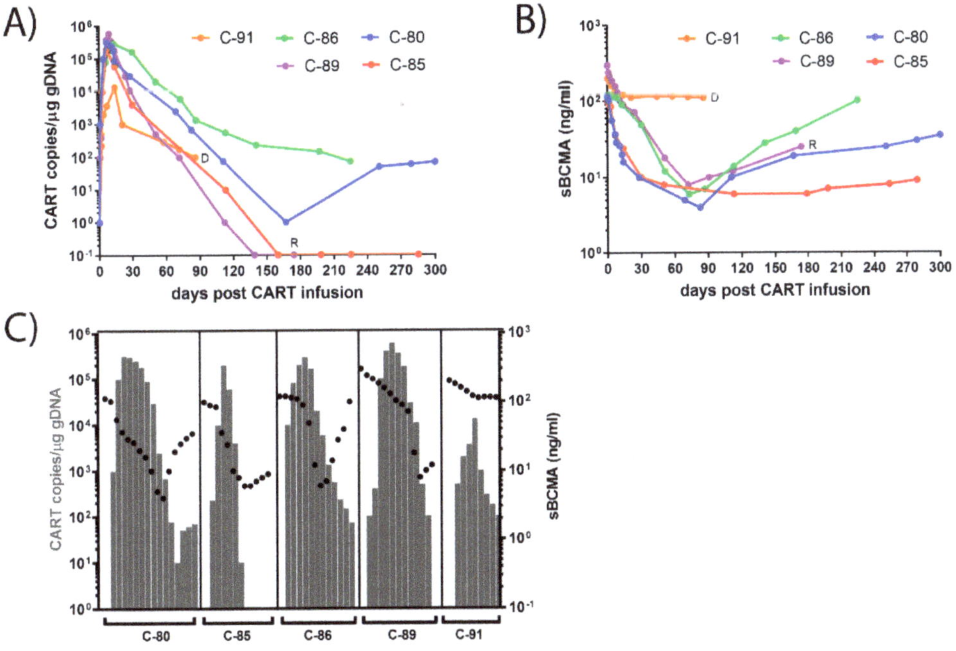

Figure 1. Dynamics of CAR-T and sBCMA plasma levels in the peripheral blood of patients. (**A**) Dynamics of CAR-T copies/µg genomic DNA in the peripheral blood. Rapid rise to high peak levels (>1.9E + 05 copies/µg), followed by decline to undetectable levels (C-85, C-89) or persistence at low levels (C-80, C-86). Rapid rise to mediocre peak levels (1.3E + 04 copies/µg) and decline to low levels (C-91). The concentration of the CAR-T copies/µg gDNA is given on a logarithmic scale on the y-axis. (**B**) Dynamics of sBCMA concentrations in the plasma. Rapid decline to minimal levels (C-80, C-85, C-86, C-89), followed by rebound to normal levels (C-80, C-86). Persistence of elevated levels (C-91). Response determination at the end of the study entailed three sCR (C-80, C-85, C-86), one relapsed (C-89, R), one deceased (C-91, D). The concentration of the sBCMA is given on a logarithmic scale on the y-axis. (**C**) Dynamics of CART copies/µg gDNA (grey bars) and sBCMA (ng/mL) in the plasma (black dots).

Patient 4 (C-89) had received two lines of prior therapy, but was not eligible for ASCT. After second relapse anti-BCMA CAR-T, therapy resulted in complete remission four weeks after infusion. CAR-T copy number in the peripheral blood peaked on day 9 at high levels (6E + 05 copies/µg gDNA) and subsided below the detection limit after five months. sBCMA plasma levels, initially at 280 ng/mL, started dropping immediately after infusion and reached minimal levels two months after CAR-T cell infusion. Six months after CAR-T infusion, the response determination indicated a relapse.

Patient 5 (C-91), first diagnosed with multiple myeloma, stage II (R-ISS), had received two lines of prior therapy and ASCT. Anti-BCMA CAR-T treatment resulted in stable disease (SD) four weeks after infusion. CAR-T copy number in the peripheral blood peaked on day 14 at low levels (1.3E + 04 copies/µg gDNA) and was at 100 copies/µg gDNA after three months when disease was progressive (PD). sBCMA plasma levels, initially at 200 ng/mL, started dropping immediately after infusion; however, they soon reached a plateau at 120 ng/mL. The disease progressed, and the patient died 100 days after CAR-T infusion.

3.2. Dynamics of CAR-T Concentration in the Peripheral Blood

We determined the dynamics of CAR-T cell presence in the peripheral blood by evaluating the copy number, as well as persistence after infusion, similarly to the analysis done in the phase I and phase II studies [13,14]. Expansion of CAR-T cells peaked above 2–E + 05 copies/µg gDNA in the four patients with good responses, and at 1.3E + 04 copies/µg DNA) in the one patient with partial response (Figure 1A). Three patterns of dynamics were observed: 1. Rapid increase to high peak levels (>1.9E + 05 copies/µg), followed by decline to undetectable levels (C-85, C-89). 2. Rapid increase to high peak levels, followed by decline with persistence at low levels (C-80, C-86). 3. Rapid increase to mediocre peak levels (1.3E + 04 copies/µg), followed by decline with persistence at low levels (C-91). CAR-T cells were present for less than six months after infusion in two patients (C-85, C-89), and over six months in two patients (C-80, C-86).

3.3. Dynamics of sBCMA Plasma Levels

Proteolytic shedding of the BCMA receptor reduces its cell-surface expression and ligand-mediated survival of B cell subsets. This shedding is mediated by protease γ-secretase, and is partially dependent on ligand binding and receptor interactions. sBCMA plasma levels are elevated in MM sera (100–300 ng/mL) compared to healthy donors (10–80 ng/mL), with no difference between newly diagnosed patients and those with relapsed disease [8–10]. A remarkable decrease in sBCMA level was previously observed in patients with good responses to BCMA-targeted immunotherapy, suggesting sBCMA as a suitable biomarker to assess response to MM therapy [11]. Here, we assessed soluble BCMA as a serum-based universal marker of myeloma burden, and correlated sBCMA levels with response duration. sBCMA plasma levels started dropping 1–10 days post infusion, and minimal levels (<10 ng/mL) were reached 60 to 80 days post-infusion (Figure 1B,C). After the initial decline, sBCMA levels started to rise again (rebound), reaching normal levels in three cases, and rising to pretreatment levels in one patient eight months after CAR-T cell infusion. Four patterns of dynamics of sBCMA levels were observed: 1. Rapid decline to minimal levels, followed by rebound to normal levels (C-80, C-89). 2. Rapid decline to minimal levels with persistence (C-85). 3. Rapid decline to minimal levels, followed by rebound to normal levels and rise to pretreatment levels (C-86). 4. Persistently high sBCMA plasma levels (C-91). Shedding of BCMA may reduce the efficacy of ide-cel CAR-T treatment, since the target is no longer present on the target cell, and sBCMA can act as a decoy that neutralizes the compound (Figure 2).

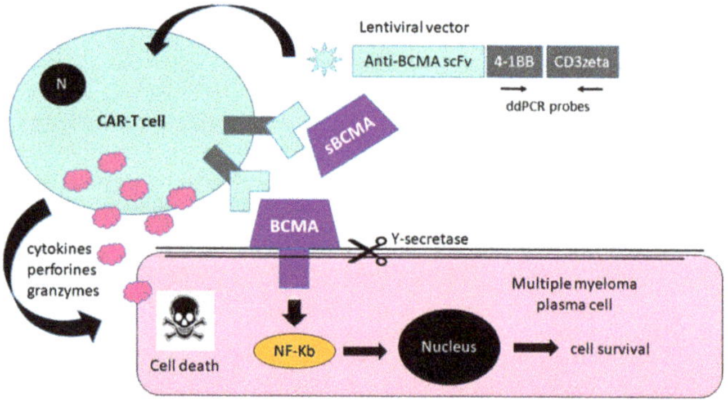

Figure 2. Anti-BCMA CAR-T cell treatment. CAR-T cells target the BCMA protein present on the surface of the myeloma cells and release cytokines, perforines, and granzymes. BCMA shedding is partially dependent on ligand binding and receptor interactions. sBCMA may act as a decoy blocking the CAR-TCR; thus, leading to reduced efficacy of anti-BCMA-CAR-T therapy.

4. Discussion

Four of the first five patients receiving anti-BCMA CAR-T cells at the University Hospital, Bern, demonstrated favorable response to anti-BCMA CAR-T treatment. Complete remission (CR) in the bone marrow was prevalent in four patients (80%) one month post CAR-T infusion, with stringent CR (sCR) present in three patients (60%), as determined by flow cytometry techniques. This is in accordance with results from a phase II study [14], where, at the target dose of 4.50E + 08 CAR-T cells, a response was observed in 44 of 54 patients (81%), and a complete response or better was observed in 21 of 54 patients (39%). Anti-BCMA CAR-T cell expansion appeared to peak in the peripheral blood in multiple myeloma patients, with a similar pattern compared to CAR-T cell products already approved for commercial use in lymphomas [22,23].

Responses to anti-BCMA CAR-T treatment in this small cohort were found to associate with peak expansion by qPCR (>2E + 05 copies/μg DNA for CR vs. 1E + 04 copies/μg for PR/SD), similarly to data reported by Cohen et al., 2019 [16], who described a significant association of response with peak expansion by qPCR (median 7E + 05 copies/μg DNA for ≥PR vs. 6E + 03 copies/μg for <PR), as well as with persistence over the first 28 days. CAR-T cells were detectable for less than six months in two patients, and for up to 10 months in two patients, at low levels. A similar duration was described in the cellular kinetic analysis of the phase II trial where CAR-T cells were present in 59% of patients at 6 months and in 36% at 12 months after infusion [14]. The low level presence of CAR-T cells (100 copies/μg gDNA) may be insufficient to guard against myeloma recurrence. In addition, several months after infusion, the CAR-T cells may become functionally compromised and the myeloma cells may become resistant to the CAR-T cells.

With respect to sBCMA levels, a clinical reference value has not been defined. However, sBCMA peptide concentrations are elevated in MM plasma (100–1000 ng/mL) compared to healthy donors (10–80 ng/mL) [14,22,23]. In four of the five MM patients in our study, sBCMA concentrations in the plasma, at 100–300 ng/mL before CAR-T infusion, declined to minimal levels below 10 ng/mL within 90 days post anti-BCMA CAR-T infusion, before rebounding to normal levels. Higher baseline sBCMA of 300 ng/mL was detected in two patients, one without clinical response and one with good initial clinical response but early relapse, indicating that higher baseline sBCMA may affect therapeutic efficacy, possibly by blocking the chimeric T cell receptors of the effector cells.

sBCMA levels above normal levels after the decline of CAR-T cells may indicate the impending end of remission. sBCMA plasma concentrations of 100 ng/mL can represent normal levels in one patient, and indicate recurrence of myeloma in another patient. sBCMA plasma levels may be easily determined together with standard remission controls on a monthly basis and can serve, not only as response monitoring tool, but also as an early predictor of response.

A remarkable decrease in sBCMA level was observed in the four patients with good responses to bb2121 therapy. With this limited number of patients, definite conclusions on the utility of sBCMA as a response monitoring marker are not possible. However, the study may present a proof of principle, suggesting that sBCMA may represent a promising biomarker of response to BCMA-targeted immunotherapy. sBCMA concentrations were found to decline more significantly after CAR-T cell infusions in responders (CR/sCR) than in non-responders (SD). sBCMA was detectable in all five patients at the end of the study, suggesting that BCMA antigen loss is not a prevalent mechanism of escape from ide-cel therapy. sBCMA concentrations remained at low levels in long-term responders, confirming sBCMA plasma levels as a useful adjunctive biomarker for assessing myeloma disease response and progression. Larger studies are required to evaluate the prognostic significance of sBCMA plasma levels for their potential as biomarker of response and predictor of response to BCMA-targeting therapies in relapsed multiple myeloma patients.

Identification of the observed CAR-T copy number kinetics suggests investigation of whether the pattern 'rapid rise to high peak levels, followed by decline, with persistence at low levels' may be associated with superior efficacy of CAR-T treatment. Identification of the observed sBCMA plasma level dynamics suggests further investigation of whether the pattern 'rapid decline to minimal levels < 10 ng/mL', as well as 'duration at minimal levels', may be associated with superior efficacy of CAR-T treatment. These questions can be addressed once the therapy is approved and more MM patients can be admitted to BCMA CAR-T therapy.

Author Contributions: Conceptualization, T.P.; investigation, K.S., G.W., B.J. and N.P.; writing—original draft preparation, K.S.; writing—review and editing, K.S., V.U.B. and T.P.; visualization, K.S.; funding acquisition, T.P. All authors have read and agreed to the published version of the manuscript.

Funding: This work was supported by a grant from the Swiss National Science Foundation (SNF) #310030_127509 to TP.

Institutional Review Board Statement: The study was conducted according to the guidelines of the Declaration of Helsinki, and approved by decisions of the local ethics committee of Bern, Switzerland, decision number #221/15.

Informed Consent Statement: Informed consent was obtained from all subjects involved in the study according to the Declaration of Helsinki. General consent form available.

Data Availability Statement: Data is contained within the article.

Conflicts of Interest: The authors declare no conflict of interest.

References

1. Palumbo, A.; Anderson, K. Multiple Myeloma. *N. Engl. J. Med.* **2011**, *364*, 1046–1060. [CrossRef]
2. Rajkumar, S.V. Treatment of Multiple Myeloma. *Nat. Rev. Clin. Oncol.* **2011**, *8*, 479–491. [CrossRef]
3. Kumar, S.K.; Dispenzieri, A.; Lacy, M.Q.; Gertz, M.A.; Buadi, F.K.; Pandey, S.; Kapoor, P.; Dingli, D.; Hayman, S.R.; Leung, N.; et al. Continued Improvement in Survival in Multiple Myeloma: Changes in Early Mortality and Outcomes in Older Patients. *Leukemia* **2014**, *28*, 1122–1128. [CrossRef]
4. Goldschmidt, H.; Ashcroft, J.; Szabo, Z.; Garderet, L. Navigating the Treatment Landscape in Multiple Myeloma: Which Combinations to Use and When? *Ann. Hematol.* **2019**, *98*, 1–18. [CrossRef]
5. Shu, H.B.; Johnson, H. B Cell Maturation Protein Is a Receptor for the Tumor Necrosis Factor Family Member TALL-1. *Proc. Natl. Acad. Sci. USA* **2000**, *97*, 9156–9161. [CrossRef]
6. Shah, N.; Chari, A.; Scott, E.; Mezzi, K.; Usmani, S.Z. B-Cell Maturation Antigen (BCMA) in Multiple Myeloma: Rationale for Targeting and Current Therapeutic Approaches. *Leukemia* **2020**, *34*, 985–1005. [CrossRef]
7. Kumar, S.; Paiva, B.; Anderson, K.C.; Durie, B.; Landgren, O.; Moreau, P.; Munshi, N.; Lonial, S.; Bladé, J.; Mateos, M.-V.; et al. International Myeloma Working Group Consensus Criteria for Response and Minimal Residual Disease Assessment in Multiple Myeloma. *Lancet Oncol.* **2016**, *17*, e328–e346. [CrossRef]
8. Mathys, A.; Bacher, U.; Banz, Y.; Legros, M.; Mansouri Taleghani, B.; Novak, U.; Pabst, T. Outcome of Patients with Mantle Cell Lymphoma after Autologous Stem Cell Transplantation in the Pre-CAR T-Cell Era. *Hematol. Oncol.* **2021**. [CrossRef]
9. Nydegger, A.; Novak, U.; Kronig, M.-N.; Legros, M.; Zeerleder, S.; Banz, Y.; Bacher, U.; Pabst, T. Transformed Lymphoma Is Associated with a Favorable Response to CAR-T-Cell Treatment in DLBCL Patients. *Cancers* **2021**, *13*, 6073. [CrossRef]
10. Meinl, E.; Krumbholz, M. Endogenous Soluble Receptors SBCMA and STACI: Biomarker, Immunoregulator and Hurdle for Therapy in Multiple Myeloma. *Curr. Opin. Immunol.* **2021**, *71*, 117–123. [CrossRef]
11. Ali, S.A.; Shi, V.; Maric, I.; Wang, M.; Stroncek, D.F.; Rose, J.J.; Brudno, J.N.; Stetler-Stevenson, M.; Feldman, S.A.; Hansen, B.G.; et al. T Cells Expressing an Anti-B-Cell Maturation Antigen Chimeric Antigen Receptor Cause Remissions of Multiple Myeloma. *Blood* **2016**, *128*, 1688–1700. [CrossRef]
12. Friedman, K.M.; Garrett, T.E.; Evans, J.W.; Horton, H.M.; Latimer, H.J.; Seidel, S.L.; Horvath, C.J.; Morgan, R.A. Effective Targeting of Multiple B-Cell Maturation Antigen-Expressing Hematological Malignances by Anti-B-Cell Maturation Antigen Chimeric Antigen Receptor T Cells. *Hum. Gene Ther.* **2018**, *29*, 585–601. [CrossRef]
13. Raje, N.; Berdeja, J.; Lin, Y.; Siegel, D.; Jagannath, S.; Madduri, D.; Liedtke, M.; Rosenblatt, J.; Maus, M.V.; Turka, A.; et al. Anti-BCMA CAR T-Cell Therapy Bb2121 in Relapsed or Refractory Multiple Myeloma. *N. Engl. J. Med.* **2019**, *380*, 1726–1737. [CrossRef]
14. Munshi, N.C.; Anderson, L.D.; Shah, N.; Madduri, D.; Berdeja, J.; Lonial, S.; Raje, N.; Lin, Y.; Siegel, D.; Oriol, A.; et al. Idecabtagene Vicleucel in Relapsed and Refractory Multiple Myeloma. *N. Engl. J. Med.* **2021**, *384*, 705–716. [CrossRef]
15. Brechbühl, S.; Bacher, U.; Jeker, B.; Pabst, T. Real-World Outcome in the Pre-CAR-T Era of Myeloma Patients Qualifying for CAR-T Cell Therapy. *Mediterr. J. Hematol. Infect. Dis.* **2021**, *13*, e2021012. [CrossRef]
16. Cohen, A.D.; Garfall, A.L.; Stadtmauer, E.A.; Melenhorst, J.J.; Lacey, S.F.; Lancaster, E.; Vogl, D.T.; Weiss, B.M.; Dengel, K.; Nelson, A.; et al. B Cell Maturation Antigen–Specific CAR T Cells Are Clinically Active in Multiple Myeloma. *J. Clin. Investig.* **2019**, *129*, 2210–2221. [CrossRef]

17. Pabst, T.; Joncourt, R.; Shumilov, E.; Heini, A.; Wiedemann, G.; Legros, M.; Seipel, K.; Schild, C.; Jalowiec, K.; Mansouri Taleghani, B.; et al. Analysis of IL-6 Serum Levels and CAR T Cell-Specific Digital PCR in the Context of Cytokine Release Syndrome. *Exp. Hematol.* **2020**, *88*, 7–14.e3. [CrossRef]
18. Milone, M.C.; Fish, J.D.; Carpenito, C.; Carroll, R.G.; Binder, G.K.; Teachey, D.; Samanta, M.; Lakhal, M.; Gloss, B.; Danet-Desnoyers, G.; et al. Chimeric Receptors Containing CD137 Signal Transduction Domains Mediate Enhanced Survival of T Cells and Increased Antileukemic Efficacy In Vivo. *Mol. Ther.* **2009**, *17*, 1453–1464. [CrossRef]
19. Härmälä, S.K.; Butcher, R.; Roberts, C.H. Copy Number Variation Analysis by Droplet Digital PCR. *Methods Mol. Biol.* **2017**, *1654*, 135–149. [CrossRef]
20. Messmer, A.S.; Que, Y.-A.; Schankin, C.; Banz, Y.; Bacher, U.; Novak, U.; Pabst, T. CAR T-Cell Therapy and Critical Care: A Survival Guide for Medical Emergency Teams. *Wien. Klin. Wochenschr.* **2021**, *133*, 1318–1325. [CrossRef]
21. Durie, B.G.M.; Harousseau, J.-L.; Miguel, J.S.; Bladé, J.; Barlogie, B.; Anderson, K.; Gertz, M.; Dimopoulos, M.; Westin, J.; Sonneveld, P.; et al. International Uniform Response Criteria for Multiple Myeloma. *Leukemia* **2006**, *20*, 1467–1473. [CrossRef] [PubMed]
22. Lee, L.; Bounds, D.; Paterson, J.; Herledan, G.; Sully, K.; Seestaller-Wehr, L.M.; Fieles, W.E.; Tunstead, J.; McCahon, L.; Germaschewski, F.M.; et al. Evaluation of B Cell Maturation Antigen as a Target for Antibody Drug Conjugate Mediated Cytotoxicity in Multiple Myeloma. *Br. J. Haematol.* **2016**, *174*, 911–922. [CrossRef] [PubMed]
23. Ghermezi, M.; Li, M.; Vardanyan, S.; Harutyunyan, N.M.; Gottlieb, J.; Berenson, A.; Spektor, T.M.; Andreu-Vieyra, C.; Petraki, S.; Sanchez, E.; et al. Serum B-Cell Maturation Antigen: A Novel Biomarker to Predict Outcomes for Multiple Myeloma Patients. *Haematologica* **2017**, *102*, 785–795. [CrossRef] [PubMed]

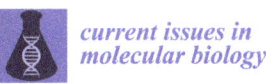

current issues in
molecular biology

MDPI

Article

Soluble CD147 (BSG) as a Prognostic Marker in Multiple Myeloma

Piotr Łacina [1,*], Aleksandra Butrym [2], Diana Frontkiewicz [3], Grzegorz Mazur [4] and Katarzyna Bogunia-Kubik [1]

1 Laboratory of Clinical Immunogenetics and Pharmacogenetics, Hirszfeld Institute of Immunology and Experimental Therapy, Polish Academy of Sciences, 53-114 Wrocław, Poland; katarzyna.bogunia-kubik@hirszfeld.pl
2 Department of Cancer Prevention and Therapy, Wroclaw Medical University, 50-556 Wrocław, Poland; aleksandra.butrym@umw.edu.pl
3 Department of Haematology, Sokołowski Specialist Hospital, 58-309 Wałbrzych, Poland; d.frontkiewicz@gmail.com
4 Department of Internal, Occupational Diseases, Hypertension and Clinical Oncology, Wroclaw Medical University, 50-556 Wrocław, Poland; grzegorz.mazur@umw.edu.pl
* Correspondence: piotr.lacina@hirszfeld.pl; Tel.: +48-713-709-960 (ext. 236)

Abstract: CD147 (basigin, BSG) is a membrane-bound glycoprotein involved in energy metabolism that plays a role in cancer cell survival. Its soluble form is a promising marker of some diseases, but it is otherwise poorly studied. CD147 is overexpressed in multiple myeloma (MM) and is known to affect MM progression, while its genetic variants are associated with MM survival. In the present study, we aimed to assess serum soluble CD147 (sCD147) expression as a potential marker in MM. We found that sCD147 level was higher in MM patients compared to healthy individuals. It was also higher in patients with more advanced disease (ISS III) compared to both patients with less advanced MM and healthy individuals, while its level was observed to drop after positive response to treatment. Patients with high sCD147 were characterized by worse progression-free survival. sCD147 level did not directly correlate with bone marrow CD147 mRNA expression. In conclusion, this study suggests that serum sCD147 may be a prognostic marker in MM.

Keywords: BSG; basigin; CD147; MM; multiple myeloma; survival

Citation: Łacina, P.; Butrym, A.; Frontkiewicz, D.; Mazur, G.; Bogunia-Kubik, K. Soluble CD147 (BSG) as a Prognostic Marker in Multiple Myeloma. *Curr. Issues Mol. Biol.* **2022**, *44*, 350–359. https://doi.org/10.3390/cimb44010026

Academic Editor: Dumitru A. Iacobas

Received: 25 November 2021
Accepted: 10 January 2022
Published: 14 January 2022

1. Introduction

Multiple myeloma (MM) is an incurable bone marrow malignancy associated with the presence of atypical plasma cells and with occurrence of end organ damage. It is the second most common haematological malignancy and accounts for 2% of all cancer cases [1,2]. MM is proceeded by monoclonal gammopathy of undetermined significance (MGUS), which has a 1% chance of progressing to malignant MM [3]. While novel therapeutic options, such as autologous stem cell transplantation and immunomodulatory drugs, led in the last two decades to a significant increase in overall survival, mean survival of MM patients is still relatively low at approximately 5 years [4,5].

CD147, also known as basigin (BSG) and extracellular matrix metalloproteinase inducer (EMMPRIN), is a heavily glycosylated member of the Ig superfamily. It is encoded by the *BSG* gene located on chromosome 19p13.3 and is ubiquitously expressed on various types of cells [6,7]. It is primarily described as a transmembrane protein. Its main isoform, basigin-2, is composed of a longer extracellular domain including a signal sequence and two Ig domains, as well as of shorter transmembrane and cytoplasmic domains [8,9]. CD147 can also be found in body fluids in the form of soluble CD147 (sCD147). sCD147 can be secreted by cells as a full-length protein released with microvesicles [10]. Alternatively, the extracellular domain can be cleaved and released through one of two pathways, one involving matrix metalloproteinases (MMPs), and one involving ADAM12 [11,12]. CD147 is a multifunctional protein involved in many cellular pathways [9,13]. Additionally, recent reports indicate that it may function as an alternative entry receptor for the SARS-CoV-2 virus associated with the COVID-19 pandemic [14,15].

CD147 is overexpressed in many cancers and is known to promote cancer progression [16,17]. It was found to be a marker of risk, poor prognosis, overall and progression-free survival, as well as chemotherapy resistance [18,19]. Although many CD147-associated pathways may be responsible for this, it was shown that interaction between CD147 and monocarboxylate transporters (MCTs) contributes most to this pro-tumour effect [20]. MCTs are membrane-bound transporters of monocarboxylates such as lactic acid [21]. They are important components of energy metabolism, as lactic acid is a by-product of glycolysis which needs to be removed from the cell to avoid dangerous decrease in cytosolic pH [22]. Most cancer cells rely primarily on glycolysis for energy production, a phenomenon known as the Warburg effect, making proper functioning of MCTs crucial for them [23].

CD147 functions as a chaperone of MCT1 (also known as SLC16A1) and its downregulation is known to be detrimental to proper lactate transport and tumour survival [24–26]. CD147 is also known to be involved with expression of other proteins such as MMPs, and vascular endothelial growth factor (VEGF), while its own expression can be controlled by various other factors, e.g., receptor activator for nuclear Factor κ B ligand (RANKL) [27–29]. Its ability to induce VEGF expression also contributes to cancer development and progression, as VEGF is an important pro-angiogenic factor [29].

Like in other cancers, CD147 was shown to be overexpressed in MM and to be associated with MM progression [30]. Likewise, MCT1 and MCT4 were also overexpressed in MM patients, although only MCT1 was indispensable for continued MM cell proliferation [24]. Genetic variants of both CD147 and MCT1 were found to influence survival in MM patients [31], and CD147 is known to be involved in response to MM treatment [32,33]. Myeloma cells were observed to exhibit increased lactate transport, whereas CD147 gene expression was found to correlate with key regulators of glycolysis and the Warburg effect, further substantiating the pro-myeloma effect of CD147 [24,34].

Soluble CD147 is thought to support cancer proliferation by interacting with membrane-bound CD147 [35]. It was shown to be an easily detectable biomarker in some diseases [36–38], although its role in haematological malignancies is poorly studied. We recently showed thatsCD147 is overexpressed in acute myeloid leukaemia (AML) patients compared to healthy individuals and that high CD147 is associated with worse overall survival [39].

In the present study, we aimed to determine whether serum sCD147 could be used as a potential prognostic marker in MM. Furthermore, we wanted to establish if sCD147 level correlated with the proangiogenic factor VEGF and mRNA expression of CD147 in the bone marrow.

2. Materials and Methods

2.1. Patients and Controls

The study included 62 newly diagnosed MM patients and 25 healthy blood donors serving as the control group. Both groups were nearly equally divided into men and women (the ratio of females was 30/59 and 12/25, respectively). The study was approved by the Wroclaw Medical University Bioethical Committee (ethical approval code: 369/2019). According to International Staging System (ISS) stratification, 21.4% of patients were in stage I, 33.9% were in stage II, and 44.6% were in stage III. Most patients were administered either the bortezomib, melphalan, prednisone (VMP); 35.2%, or the bortezomib, thalidomide, dexamethasone (VTD); 29.6% regimen as first line therapy. Further clinical data of patients analysed in the study are included in Table 1.

Table 1. Characteristics of MM patients included in the study.

Data	Median and Range (n = 59)
Age	70 (43–88)
white blood cell count (G/L)	6.7 (2.4–20.9)
haemoglobin (g/dL)	10.1 (5.6–14.7)
total protein (g/dL)	8.4 (5.1–15.9)
albumin (g/dL)	3.8 (1.7–5.0)
lactate dehydrogenase (U/L)	225 (100–1595)
β2-microglobulin (mg/L)	4.4 (1.3–78.5)
C-reactive protein (mg/L)	10.2 (0.7–102.8)
creatinine (mg/dL)	1.00 (0.36–8.87)
calcium (mg/dL)	9.4 (7.4–14.0)

2.2. ELISA Analysis of Serum Samples

Peripheral blood from 62 MM patients and 25 healthy individuals was collected, allowed to clot, and subsequently centrifuged for 15 min at 1000× g. Serum samples were then collected, aliquoted, and kept at −70 °C until further use. Serum samples were used for measurements of sCD147 and VEGF concentrations, which were performed using the Human EMMPRIN/CD147 Quantikine ELISA Kit and Human VEGF Quantikine ELISA Kit (R&D Systems, Inc., Minneapolis, MN, USA) according to manufacturer's instructions. All samples were run in duplicate. Subsequently, absorbance was measured in a Sunrise microplate reader with Magellan analysis software (Tecan Trading AG, Männedorf, Switzerland).

2.3. RNA Isolation and Gene Expression Analysis

Bone marrow aspirates from a group of 29 MM patients and 3 non-MM patients (working as a PCR control group) were collected, and mononuclear cells were isolated by Lymphodex (inno-train Diagnostik GmbH, Kronberg im Taunus, Germany) density-gradient centrifugation. Total RNA was then extracted using the RNeasy Plus Mini Kit (QIAGEN, Hilden, Germany) according to manufacturer's protocol. RNA purity and integrity were verified on the DeNovix DS-11 specrophotometer (DeNovix Inc., Wilmington, DE, USA), and by gel electrophoresis. A total of 2000 ng of isolated RNA was used for reverse transcription into cDNA using the High-Capacity cDNA Reverse Transcriptase kit (Applied Biosystems, Waltham, MA, USA) and RNase Inhibitor (Applied Biosystems, Waltham, MA, USA) was added to the reaction mix. The reaction was performed in a SimpliAmp Thermal Cycler (Applied Biosystems, Waltham, MA, USA) according to manufacturer's instructions. The resulting cDNA was stored at −70 °C.

CD147 (BSG), MCT1 (SLC16A1), MCT4 (SLC16A3), and VEGF gene expression were measured using quantitative real-time PCR, and the raw expression data were normalized to β-actin (ACTB), which was used as a reference gene. TaqMan Gene Expression assays specific to each gene of interest, as well as TaqMan Gene Expression Master Mix (Applied Biosystems, Waltham, MA, USA) were used for the experiment. TaqMan Gene Expression probes (Applied Biosystems, Waltham, MA, USA) used were: Hs00936295_m1 (CD147), Hs01560299_m1 (MCT1), Hs00358829_m1 (MCT4), Hs00900055_m1 (VEGF), and Hs01060665_g1 (ACTB). Samples were run in duplicate. The reactions were performed on LightCycler 480 II (Roche Diagnostics, Rotkreuz, Switzerland) and according to manufacturer's instructions. Relative expression was then calculated using the $2^{-\Delta\Delta Ct}$ method.

2.4. Statistical Analysis

The non-parametric Mann–Whitney U test was used for comparison between serum sCD147 and clinical parameters such as white blood cell count, haemoglobin, total protein, albumin, lactate dehydrogenase, β2-microglobulin, creatinine or C-reactive protein. Spearman's coefficient was used to assess correlations with serum sCD147, serum

VEGF, CD147/MCT1/MCT4/VEGF gene expression, and clinical parameters. Overall and progression-free survival were analysed using the Kaplan–Meier curves and Gehan–Breslow–Wilcoxon test, as well as multivariate Cox proportional hazards model, and the non-parametric Wilcoxon signed-rank test was employed to compare sCD147 level before and after response to treatment. These analyses were performed with the Real Statistics Resource Pack for Microsoft Excel 2013 (version 15.0.5023.1000, Microsoft, Redmond, WA, USA), RStudio (RStudio, PBC, Boston, MA, USA), and GraphPad Prism (version 8.0.1, GraphPad Software, San Diego, CA, USA). p-values < 0.05 were considered statistically significant, while those between 0.05 and 0.10 were indicative of a trend.

3. Results

3.1. Serum Soluble CD147 Is Increased in MM Patients

We analysed expression of sCD147 in serum of a group of MM patients (n = 62) and in the control group of healthy individuals (n = 25). The median sCD147 value was 4441.78 pg/mL (interquartile range: 3435.10–5798.96 pg/mL) in patients, and 3894.45 pg/mL (interquartile range: 2903.50–4544.15 pg/mL) in the control group. We found sCD147 to be significantly higher in serum of MM patients compared to the control group (p = 0.016, Figure 1).

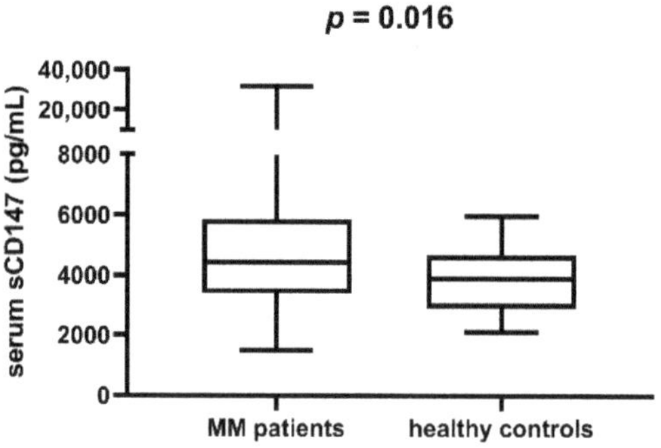

Figure 1. Serum sCD147 in multiple myeloma (MM) patients and the control group of healthy individuals. sCD147 is higher in the former group (p = 0.016).

3.2. Serum Soluble CD147 Is Associated with More Advanced Disease and Worse Survival

We analysed sCD147 expression in the context of some of the clinical parameters of MM. sCD147 was higher in patients in the more advanced stage III (mean: 5202.93 pg/mL, interquartile range: 3909.71–7986.53 pg/mL), compared to patients in stages I-II (mean: 3782.70, interquartile range: 3352.70–5196.10 pg/mL), according to the International Staging System (ISS) criteria (p = 0.012, Figure 2). sCD147 expression in stage III patients was also significantly higher than that of healthy individuals (p = 0.001).

Regarding other clinical parameters, we observed that sCD147 correlated with β2-microglobulin level (R = 0.279, p = 0.033, Figure 3A) and creatinine (R = 0.429, p = 0.001, Figure 3B). However, no associations with either white blood cell count, haemoglobin, total protein, albumin, lactate dehydrogenase, or C-reactive protein were observed. Additionally, we compared serum levels of sCD147 and of the proangiogenic factor VEGF. However, we did not find any correlation between them (R = −0.010, p = 0.941).

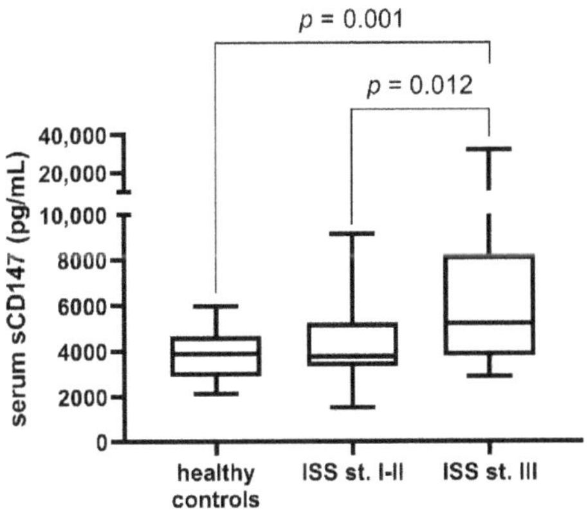

Figure 2. Serum sCD147 in patients in stages I-II and in patients in stage III of the International Staging System, or ISS. Healthy controls are also included. Patients in the more advanced stage III are characterized by higher sCD147 levels than both patients in the stages I-II (p = 0.014), and healthy controls (p = 0.002). Concurrently, healthy individuals did not significantly differ from patients in stages I-II (p = 0.279).

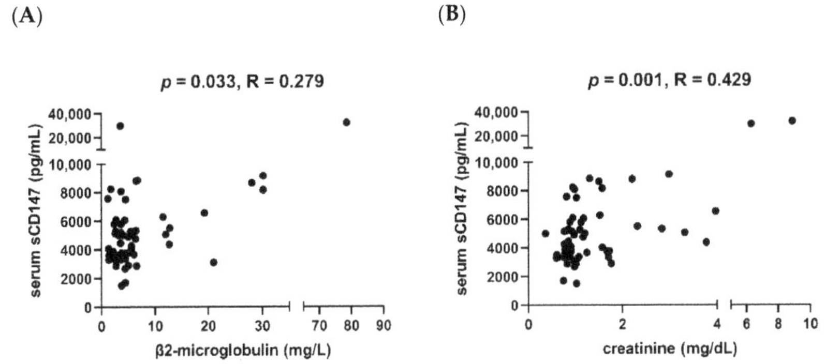

Figure 3. Correlation between serum soluble CD147 and two clinical parameters in MM patients—β2-microglobulin (**A**) and creatinine (**B**). Serum soluble CD147 is characterized by a weak-moderate correlation with these two parameters.

In the next step, we used Kaplan–Meier curves to analyse the difference in overall (OS) and progression-free survival (PFS) between patients with high and low sCD147. High sCD147 was defined in this and further analyses as being above the upper quartile in our study group (5798.96 pg/mL), and low sCD147 was below the upper quartile. While no difference was observed in OS, we observed that patients with high sCD147 were characterized by shorter PFS than patients with low sCD147 (p = 0.046, Figure 4).

Additionally, we constructed a Cox proportional hazards model including sCD147 status (low/high) and adjusting for age, β2-microglobulin level, creatinine level, ISS stage, and therapy (use of immunomodulatory drugs). This analysis confirmed high sCD147 to be an independent marker of adverse PFS (p = 0.038).

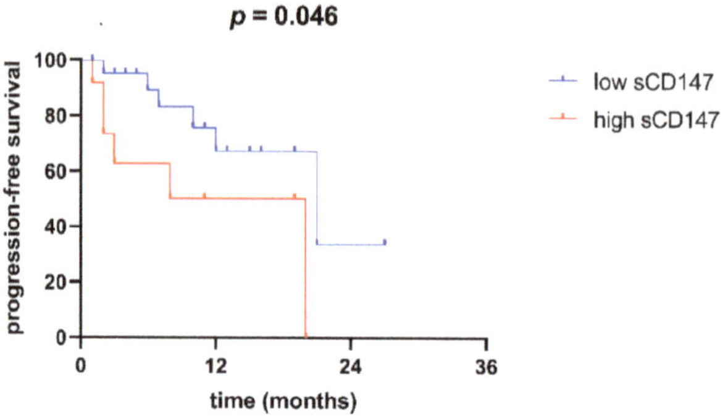

Figure 4. Progression-free survival of patients with high serum sCD147 (above the upper quartile, or 5798.96 pg/mL) and in patients with low serum sCD147 (below the upper quartile). Patients with high sCD147 are characterized by a more adverse PFS ($p = 0.046$).

3.3. *Serum Soluble CD147 Levels Drop in Response to Treatment*

In a subsection of MM patients ($n = 10$), we compared sCD147 levels after positive response to treatment (very good partial response or better) to those at diagnosis. We found sCD147 levels to significantly decrease in response to treatment in most patients ($p = 0.025$, Figure 5).

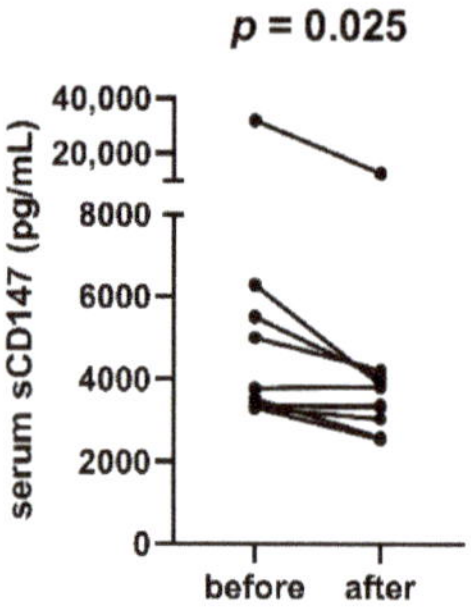

Figure 5. Serum sCD147 levels in multiple myeloma patients before and after positive response to treatment. sCD147 levels are decreased in most patients after remission ($p = 0.025$).

3.4. *Serum Soluble CD147 Does Not Correlate with BSG mRNA Expression Levels*

We measured relative mRNA *BSG* expression in bone marrow samples of a subgroup ($n = 29$) of MM patients. sCD147 level in serum did not correlate with mRNA CD147 expression in those patients (R = 0.027, $p = 0.896$). In addition to BSG, we also measured mRNA expression of various other genes associated with BSG in MM. We found that, as expected, BSG mRNA expression correlated strongly with expression of VEGF (R = 0.420, $p = 0.023$), MCT1/SLC16A1 (R = 0.811, $p < 0.001$), but not with MCT4/SLC16A3 (R = 0.061, $p = 0.755$). While no statistically significant associations between mRNA BSG expression and clinical parameters or survival were observed, we found a trend towards lower BSG expression in patients with positive (very good, partial, or better) response to treatment ($p = 0.093$).

4. Discussion

A growing body of evidence points to CD147 being a major factor in cancer progression and survival [16–19]. This may be a result of its multifunctional nature and involvement in a multitude of regulatory pathways that include cell migration, proliferation, and angiogenesis. Another pathway associated with tumour progression and promoted by

CD147 is epithelial–mesenchymal transition (EMT) [40,41]. Furthermore, recent research suggests that control of lactate transport through MCTs is a major function of CD147 in the context of cancer [20]. As CD147 was recently shown to function as an auxiliary receptor for SARS-CoV-2 infection [14,15], a good knowledge of its expression patterns may be of importance in the context of the COVID-19 pandemic.

Various studies suggests that CD147 is involved in development, progression, and response to treatment of MM [24,30,42]. CD147 is also implicated in response to treatment, particularly treatment involving immunomodulatory drugs [32,33]. In our previous studies, we showed that some genetic variants of CD147 and MCT1 affect survival of MM patients [31]. Soluble CD147 is a form of CD147 found in all body fluids and an easily measured potential biomarker, which we recently showed to be associated with survival in AML patients [39]. However, little is known about sCD147 in MM patients.

The main goal of the present study was to establish if sCD147 could be a prognostic marker in MM. Earlier study showed that serum/plasma sCD147 is elevated in many diseases, including some cancers [36–38,43,44] and our earlier studies show that it is elevated in AML [39]. Here, we observed that sCD147 level is significantly higher in MM patients compared to healthy individuals. Furthermore, sCD147 is higher in patients with more advanced disease than in both patients with less advanced MM, and healthy individuals. Additionally, we showed that sCD147 dropped in most analysed patients as a result of achieving remission, regardless of treatment regimen. High sCD147 also predicted shorter PFS, independently of variables such as age, ISS stage, or treatment regimen (whether immunomodulatory drugs were used or not).

All this evidence suggests that sCD147 is a potential prognostic marker associated with MM in general, and more particularly with MM progression and survival. Our results resemble those found in a study on breast cancer, which showed higher sCD147 in patients with primary breast cancer compared to benign diseases, as well as in patients with advanced cancer compared to early stage disease [37]. Similarly, sCD147 was shown to be elevated in patients with hepatocellular carcinoma (HCC), and to correlate with HCC tumour size and worse survival [36,45].

In our previous study, we likewise demonstrated that sCD147 is higher in AML patients and correlates with various AML clinical parameters and survival [39]. However, data on sCD147 expression in other cancers are scarce. Given that sCD147 is abundant in serum and plasma, but also in body fluids such as saliva and urine, it can be easily measured by protein detection methods such as ELISA. Therefore, it appears to be an interesting candidate for a potential biomarker not just in MM, but in cancer in general.

The role of sCD147 in multiple myeloma, and in cancer in general, is not very clear. CD147-containing microvesicles were shown to be internalized by myeloma cells and to increase their proliferation [42]. sCD147 is known to dimerize with membrane-bound CD147 in the tumour microenvironment, to stimulate production of more CD147 as well as various other proteins enhancing tumour survival [35,46]. However, we did not observe sCD147 to correlate with levels of the pro-angiogenic factor VEGF, even though there was a correlation between VEGF and CD147 expression on the mRNA level. This suggests that sCD147 might not be involved in BSG-dependent induction of VEGF expression and angiogenesis [29]. However, it is worth noting that VEGF regulation is very complex and involves many factors and pathways [47]. This includes post-transcriptional regulation processes, which means that its mRNA expression may not necessarily correspond to actual protein expression [48]. Additionally, we did not find the sCD147 level to correlate with CD147 mRNA expression. However, mRNA expression of both VEGF and MCT1 (SLC16A1) did correlate with CD147 mRNA level, as expected [29,33]. This may be due to the fact that CD147 is primarily expressed on the cell surface. Only a part of it is secreted, and this secretion is dependent on three specialised secretion pathways (one involving microvesicles, and two involving shedding of the CD147 extracellular domain) [10–12]. These results suggest that sCD147 secretion may not be the main way for CD147 to exert its pro-tumour effect, although sCD147 still appears to be an interesting biomarker in multiple myeloma.

In conclusion, we showed that sCD147 may be a prognostic marker in MM, although our results may require confirmation on a larger cohort of patients. sCD147 appears to be a promising marker of cancer.

Author Contributions: Conceptualization, P.Ł. and K.B.-K.; methodology, P.Ł., D.F. and K.B.-K.; formal analysis, P.Ł. and K.B.-K.; investigation, P.Ł. and K.B.-K.; resources, P.Ł., A.B., G.M. and K.B.-K.; data curation, P.Ł., A.B. and D.F.; writing—original draft preparation, P.Ł., D.F. and K.B.-K.; writing—review and editing, P.Ł., A.B. and G.M.; visualization, P.Ł.; supervision, G.M. and K.B.-K.; project administration, P.Ł. and K.B.-K.; funding acquisition, P.Ł. and K.B.-K. All authors have read and agreed to the published version of the manuscript.

Funding: This study was funded by the National Science Centre (Poland), grant number 2018/29/N/NZ5/02022.

Institutional Review Board Statement: The study was conducted according to the guidelines of the Declaration of Helsinki, and approved by the Institutional Ethics Committee of Wroclaw Medical University (protocol code: 368/2019, 25 April 2019).

Informed Consent Statement: Informed consent was obtained from all subjects involved in the study.

Data Availability Statement: The data presented in this study are available on request from the corresponding author. The data are not publicly available due to privacy or ethical restrictions.

Acknowledgments: The authors would like to thank Katarzyna Gębura and Monika-Chaszczewska-Markowska for their assistance.

Conflicts of Interest: The authors declare no conflict of interest.

References

1. Raab, M.S.; Podar, K.; Breitkreutz, I.; Richardson, P.G.; Anderson, K.C. Multiple myeloma. *Lancet* **2009**, *374*, 324–339. [CrossRef]
2. Palumbo, A.; Anderson, K. Multiple myeloma. *N. Engl. J. Med.* **2011**, *364*, 1046–1060. [CrossRef]
3. Kyle, R.A.; Rajkumar, S.V. Monoclonal gammopathies of undetermined significance. *Rev. Clin. Exp. Hematol.* **2002**, *6*, 225–252. [CrossRef]
4. Kumar, S.K.; Rajkumar, S.V.; Dispenzieri, A.; Lacy, M.Q.; Hayman, S.R.; Buadi, F.K.; Zeldenrust, S.R.; Dingli, D.; Russell, S.J.; Lust, J.A.; et al. Improved survival in multiple myeloma and the impact of novel therapies. *Blood* **2008**, *111*, 2516–2520. [CrossRef] [PubMed]
5. Munker, R.; Shi, R.; Nair, B.; Devarakonda, S.; Cotelingam, J.D.; McLarty, J.; Mills, G.M.; Glass, J. The Shreveport Myeloma Experience: Survival, Risk Factors and Other Malignancies in the Age of Stem Cell Transplantation. *Acta Haematol.* **2016**, *135*, 146–155. [CrossRef]
6. Iacono, K.T.; Brown, A.L.; Greene, M.I.; Saouaf, S.J. CD147 immunoglobulin superfamily receptor function and role in pathology. *Exp. Mol. Pathol.* **2007**, *83*, 283–295. [CrossRef] [PubMed]
7. Kaname, T.; Miyauchi, T.; Kuwano, A.; Matsuda, Y.; Muramatsu, T.; Kajii, T. Mapping basigin (BSG), a member of the immunoglobulin superfamily, to 19p13.3. *Cytogenet. Cell Genet.* **1993**, *64*, 195–197. [CrossRef]
8. Muramatsu, T.; Miyauchi, T. Basigin (CD147): A multifunctional transmembrane protein involved in reproduction, neural function, inflammation and tumor invasion. *Histol. Histopathol.* **2003**, *18*, 981–987. [CrossRef]
9. Landras, A.; Reger de Moura, C.; Jouenne, F.; Lebbe, C.; Menashi, S.; Mourah, S. CD147 Is a Promising Target of Tumor Progression and a Prognostic Biomarker. *Cancers* **2019**, *11*, 1803. [CrossRef]
10. Sidhu, S.S.; Mengistab, A.T.; Tauscher, A.N.; LaVail, J.; Basbaum, C. The microvesicle as a vehicle for EMMPRIN in tumor-stromal interactions. *Oncogene* **2004**, *23*, 956–963. [CrossRef] [PubMed]
11. Egawa, N.; Koshikawa, N.; Tomari, T.; Nabeshima, K.; Isobe, T.; Seiki, M. Membrane type 1 matrix metalloproteinase (MT1-MMP/MMP-14) cleaves and releases a 22-kDa extracellular matrix metalloproteinase inducer (EMMPRIN) fragment from tumor cells. *J. Biol. Chem.* **2006**, *281*, 37576–37585. [CrossRef]
12. Albrechtsen, R.; Wewer Albrechtsen, N.J.; Gnosa, S.; Schwarz, J.; Dyrskjøt, L.; Kveiborg, M. Identification of ADAM12 as a Novel Basigin Sheddase. *Int. J. Mol. Sci.* **2019**, *20*, 1957. [CrossRef] [PubMed]
13. Grass, G.D.; Toole, B.P. How, with whom and when: An overview of CD147-mediated regulatory networks influencing matrix metalloproteinase activity. *Biosci. Rep.* **2015**, *36*, e00283. [CrossRef] [PubMed]
14. Wang, K.; Chen, W.; Zhang, Z.; Deng, Y.; Lian, J.Q.; Du, P.; Wei, D.; Zhang, Y.; Sun, X.X.; Gong, L.; et al. CD147-spike protein is a novel route for SARS-CoV-2 infection to host cells. *Signal Transduct. Target. Ther.* **2020**, *5*, 283. [CrossRef] [PubMed]

15. Xu, C.; Wang, A.; Geng, K.; Honnen, W.; Wang, X.; Bruiners, N.; Singh, S.; Ferrara, F.; D'Angelo, S.; Bradbury, A.; et al. Human Immunodeficiency Viruses Pseudotyped with SARS-CoV-2 Spike Proteins Infect a Broad Spectrum of Human Cell Lines through Multiple Entry Mechanisms. *Viruses* **2021**, *13*, 953. [CrossRef]
16. Riethdorf, S.; Reimers, N.; Assmann, V.; Kornfeld, J.W.; Terracciano, L.; Sauter, G.; Pantel, K. High incidence of EMMPRIN expression in human tumors. *Int. J. Cancer* **2006**, *119*, 1800–1810. [CrossRef]
17. Nabeshima, K.; Iwasaki, H.; Koga, K.; Hojo, H.; Suzumiya, J.; Kikuchi, M. Emmprin (basigin/CD147): Matrix metalloproteinase modulator and multifunctional cell recognition molecule that plays a critical role in cancer progression. *Pathol. Int.* **2006**, *56*, 359–367. [CrossRef]
18. Li, Y.; Xu, J.; Chen, L.; Zhong, W.D.; Zhang, Z.; Mi, L.; Zhang, Y.; Liao, C.G.; Bian, H.J.; Jiang, J.L.; et al. HAb18G (CD147), a cancer-associated biomarker and its role in cancer detection. *Histopathology* **2009**, *54*, 677–687. [CrossRef]
19. Xin, X.; Zeng, X.; Gu, H.; Li, M.; Tan, H.; Jin, Z.; Hua, T.; Shi, R.; Wang, H. CD147/EMMPRIN overexpression and prognosis in cancer: A systematic review and meta-analysis. *Sci. Rep.* **2016**, *6*, 32804. [CrossRef]
20. Marchiq, I.; Albrengues, J.; Granja, S.; Gaggioli, C.; Pouysségur, J.; Simon, M.P. Knock out of the BASIGIN/CD147 chaperone of lactate/H+ symporters disproves its pro-tumour action via extracellular matrix metalloproteases (MMPs) induction. *Oncotarget* **2015**, *6*, 24636–24648. [CrossRef]
21. Halestrap, A.P.; Meredith, D. The SLC16 gene family-from monocarboxylate transporters (MCTs) to aromatic amino acid transporters and beyond. *Pflugers Arch.* **2004**, *447*, 619–628. [CrossRef] [PubMed]
22. Kennedy, K.M.; Dewhirst, M.W. Tumor metabolism of lactate: The influence and therapeutic potential for MCT and CD147 regulation. *Future Oncol.* **2010**, *6*, 127–148. [CrossRef] [PubMed]
23. Vander Heiden, M.G.; Cantley, L.C.; Thompson, C.B. Understanding the Warburg effect: The metabolic requirements of cell proliferation. *Science* **2009**, *324*, 1029–1033. [CrossRef]
24. Walters, D.K.; Arendt, B.K.; Jelinek, D.F. CD147 regulates the expression of MCT1 and lactate export in multiple myeloma cells. *Cell Cycle* **2013**, *12*, 3175–3183. [CrossRef]
25. Granja, S.; Marchiq, I.; Le Floch, R.; Moura, C.S.; Baltazar, F.; Pouysségur, J. Disruption of BASIGIN decreases lactic acid export and sensitizes non-small cell lung cancer to biguanides independently of the LKB1 status. *Oncotarget* **2015**, *6*, 6708–6721. [CrossRef] [PubMed]
26. Huang, P.; Chang, S.; Jiang, X.; Su, J.; Dong, C.; Liu, X.; Yuan, Z.; Zhang, Z.; Liao, H. RNA interference targeting CD147 inhibits the proliferation, invasiveness, and metastatic activity of thyroid carcinoma cells by down-regulating glycolysis. *Int. J. Clin. Exp. Pathol.* **2015**, *8*, 309–318.
27. Knutti, N.; Huber, O.; Friedrich, K. CD147 (EMMPRIN) controls malignant properties of breast cancer cells by interdependent signaling of Wnt and JAK/STAT pathways. *Mol. Cell. Biochem.* **2019**, *451*, 197–209. [CrossRef]
28. Rucci, N.; Millimaggi, D.; Mari, M.; Del Fattore, A.; Bologna, M.; Teti, A.; Angelucci, A.; Dolo, V. Receptor activator of NF-kappaB ligand enhances breast cancer-induced osteolytic lesions through upregulation of extracellular matrix metalloproteinase inducer/CD147. *Cancer Res.* **2010**, *70*, 6150–6160. [CrossRef] [PubMed]
29. Tang, Y.; Nakada, M.T.; Kesavan, P.; McCabe, F.; Millar, H.; Rafferty, P.; Bugelski, P.; Yan, L. Extracellular matrix metalloproteinase inducer stimulates tumor angiogenesis by elevating vascular endothelial cell growth factor and matrix metalloproteinases. *Cancer Res.* **2005**, *65*, 3193–3199. [CrossRef] [PubMed]
30. Arendt, B.K.; Walters, D.K.; Wu, X.; Tschumper, R.C.; Huddleston, P.M.; Henderson, K.J.; Dispenzieri, A.; Jelinek, D.F. Increased expression of extracellular matrix metalloproteinase inducer (CD147) in multiple myeloma: Role in regulation of myeloma cell proliferation. *Leukemia* **2012**, *26*, 2286–2296. [CrossRef]
31. Łacina, P.; Butrym, A.; Mazur, G.; Bogunia-Kubik, K. BSG and MCT1 Genetic Variants Influence Survival in Multiple Myeloma Patients. *Genes* **2018**, *9*, 226. [CrossRef] [PubMed]
32. Eichner, R.; Heider, M.; Fernández-Sáiz, V.; van Bebber, F.; Garz, A.K.; Lemeer, S.; Rudelius, M.; Targosz, B.S.; Jacobs, L.; Knorn, A.M.; et al. Immunomodulatory drugs disrupt the cereblon-CD147-MCT1 axis to exert antitumor activity and teratogenicity. *Nat. Med.* **2016**, *22*, 735–743. [CrossRef] [PubMed]
33. Bolomsky, A.; Hübl, W.; Spada, S.; Müldür, E.; Schlangen, K.; Heintel, D.; Rocci, A.; Weißmann, A.; Fritz, V.; Willheim, M.; et al. IKAROS expression in distinct bone marrow cell populations as a candidate biomarker for outcome with lenalidomide-dexamethasone therapy in multiple myeloma. *Am. J. Hematol.* **2017**, *92*, 269–278. [CrossRef] [PubMed]
34. Panchabhai, S.; Schlam, I.; Sebastian, S.; Fonseca, R. PKM2 and other key regulators of Warburg effect positively correlate with CD147 (EMMPRIN) gene expression and predict survival in multiple myeloma. *Leukemia* **2017**, *31*, 991–994. [CrossRef]
35. Knutti, N.; Kuepper, M.; Friedrich, K. Soluble extracellular matrix metalloproteinase inducer (EMMPRIN, EMN) regulates cancer-related cellular functions by homotypic interactions with surface CD147. *FEBS J.* **2015**, *282*, 4187–4200. [CrossRef]
36. Lee, A.; Rode, A.; Nicoll, A.; Maczurek, A.E.; Lim, L.; Lim, S.; Angus, P.; Kronborg, I.; Arachchi, N.; Gorelik, A.; et al. Circulating CD147 predicts mortality in advanced hepatocellular carcinoma. *J. Gastroenterol. Hepatol.* **2016**, *31*, 459–466. [CrossRef]
37. Kuang, Y.H.; Liu, Y.J.; Tang, L.L.; Wang, S.M.; Yan, G.J.; Liao, L.Q. Plasma soluble cluster of differentiation 147 levels are increased in breast cancer patients and associated with lymph node metastasis and chemoresistance. *Hong Kong Med. J.* **2018**, *24*, 252–260. [CrossRef] [PubMed]
38. Rurali, E.; Perrucci, G.L.; Gaetano, R.; Pini, A.; Moschetta, D.; Gentilini, D.; Nigro, P.; Pompilio, G. Soluble EMMPRIN levels discriminate aortic ectasia in Marfan syndrome patients. *Theranostics* **2019**, *9*, 2224–2234. [CrossRef]

39. Łacina, P.; Butrym, A.; Turlej, E.; Stachowicz-Suhs, M.; Wietrzyk, J.; Mazur, G.; Bogunia-Kubik, K. BSG (CD147) serum level and genetic variants are associated with overall survival in acute myeloid leukaemia. *J. Clin. Med.* **2022**, *11*, 332. [CrossRef]
40. Xu, T.; Zhou, M.; Peng, L.; Kong, S.; Miao, R.; Shi, Y.; Sheng, H.; Li, L. Upregulation of CD147 promotes cell invasion, epithelial-to-mesenchymal transition and activates MAPK/ERK signaling pathway in colorectal cancer. *Int. J. Clin. Exp. Pathol.* **2014**, *7*, 7432–7441. [PubMed]
41. Fang, F.; Li, Q.; Wu, M.; Nie, C.; Xu, H.; Wang, L. CD147 promotes epithelial-mesenchymal transition of prostate cancer cells via the Wnt/β-catenin pathway. *Exp. Ther. Med.* **2020**, *20*, 3154–3160. [CrossRef] [PubMed]
42. Arendt, B.K.; Walters, D.K.; Wu, X.; Tschumper, R.C.; Jelinek, D.F. Multiple myeloma dell-derived microvesicles are enriched in CD147 expression and enhance tumor cell proliferation. *Oncotarget* **2014**, *5*, 5686–5699. [CrossRef]
43. Sun, H.; Wen, W.; Zhao, M.; Yan, X.; Zhang, L.; Jiao, X.; Yang, Y.; Fang, F.; Qin, Y.; Zhang, M.; et al. EMMPRIN: A potential biomarker for predicting the presence of obstructive sleep apnea. *Clin. Chim. Acta* **2020**, *510*, 317–322. [CrossRef] [PubMed]
44. Amezcua-Guerra, L.M.; Ortega-Springall, M.F.; Guerrero-Ponce, A.E.; Vega-Memije, M.E.; Springall, R. Interleukin-17A enhances the production of CD147/extracellular matrix metalloproteinase inducer by monocytes from patients with psoriasis. *Eur. Rev. Med. Pharmacol. Sci.* **2020**, *24*, 10601–10604. [CrossRef]
45. Wu, J.; Hao, Z.W.; Zhao, Y.X.; Yang, X.M.; Tang, H.; Zhang, X.; Song, F.; Sun, X.X.; Wang, B.; Nan, G.; et al. Full-length soluble CD147 promotes MMP-2 expression and is a potential serological marker in detection of hepatocellular carcinoma. *J. Transl. Med.* **2014**, *12*, 190. [CrossRef]
46. Tang, Y.; Kesavan, P.; Nakada, M.T.; Yan, L. Tumor-stroma interaction: Positive feedback regulation of extracellular matrix metalloproteinase inducer (EMMPRIN) expression and matrix metalloproteinase-dependent generation of soluble EMMPRIN. *Mol. Cancer Res.* **2004**, *2*, 73–80.
47. Pagès, G.; Pouysségur, J. Transcriptional regulation of the Vascular Endothelial Growth Factor gene-A concert of activating factors. *Cardiovasc. Res.* **2005**, *65*, 564–573. [CrossRef] [PubMed]
48. Arcondéguy, T.; Lacazette, E.; Millevoi, S.; Prats, H.; Touriol, C. VEGF-A mRNA processing, stability and translation: A paradigm for intricate regulation of gene expression at the post-transcriptional level. *Nucleic Acids Res.* **2013**, *41*, 7997–8010. [CrossRef]

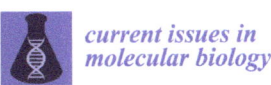

current issues in molecular biology

MDPI

Article

Targeting Mitochondrial ROS Production to Reverse the Epithelial-Mesenchymal Transition in Breast Cancer Cells

Elena Monti [1,†], Alessandro Mancini [2,3,†], Emanuela Marras [1] and Marzia Bruna Gariboldi [1,*]

[1] Department of Biotechnology and Life Sciences (DBSV), University of Insubria, Via J.H. Dunant 3, 21100 Varese, Italy
[2] Department of Translational Medical Sciences, University of Campania "Luigi Vanvitelli", 80138 Naples, Italy
[3] BioUp Sagl, 6900 Lugano, Switzerland
* Correspondence: marzia.gariboldi@uninsubria.it; Tel.: +39-0331-339418; Fax: +39-0331-339459
† These authors contributed equally to this work.

Citation: Monti, E.; Mancini, A.; Marras, E.; Gariboldi, M.B. Targeting Mitochondrial ROS Production to Reverse the Epithelial-Mesenchymal Transition in Breast Cancer Cells. *Curr. Issues Mol. Biol.* **2022**, *44*, 5277–5293. https://doi.org/10.3390/cimb44110359

Academic Editor: Dumitru A. Iacobas

Received: 10 October 2022
Accepted: 26 October 2022
Published: 29 October 2022

Abstract: Experimental evidence implicates reactive oxygen species (ROS) generation in the hypoxic stabilization of hypoxia-inducible factor (HIF)-1α and in the subsequent expression of promoters of tumor invasiveness and metastatic spread. However, the role played by mitochondrial ROS in hypoxia-induced Epithelial-Mesenchymal Transition (EMT) activation is still unclear. This study was aimed at testing the hypothesis that the inhibition of hypoxia-induced mitochondrial ROS production, mainly at the mitochondrial Complex III UQCRB site, could result in the reversion of EMT, in addition to decreased HIF-1α stabilization. The role of hypoxia-induced ROS increase in HIF-1α stabilization and the ability of antioxidants, some of which directly targeting mitochondrial Complex III, to block ROS production and HIF-1α stabilization and prevent changes in EMT markers were assessed by evaluating ROS, HIF-1α and EMT markers on breast cancer cells, following 48 h treatment with the antioxidants. The specific role of UQCRB in hypoxia-induced EMT was also evaluated by silencing its expression through RNA interference and by assessing the effects of its downregulation on ROS production, HIF-1α levels, and EMT markers. Our results confirm the pivotal role of UQCRB in hypoxic signaling inducing EMT. Thus, UQCRB might be a new therapeutic target for the development of drugs able to reverse EMT by blocking mitochondrial ROS production.

Keywords: EMT; UQCRB; ROS; hypoxia; HIF-1α

1. Introduction

The Epithelial-Mesenchymal Transition (EMT) is a biological process characterized by typical biochemical changes of epithelial cells that allow the acquisition of a mesenchymal phenotype and by the loss of cell-cell interactions and apicobasal polarity [1]. These changes lead to increased invasiveness, resistance to stress and to apoptotic cell death, and increased production of extracellular matrix proteins [1–3]. During the process, specific transcriptional modulators reduce the expression of the membrane glycoprotein E-cadherin, leading to the destabilization of adherent junctions and promoting cell mobility while increasing the synthesis of the mesenchymal protein vimentin, which modifies the composition of the intermediate filaments of the cytoskeleton, favoring invasiveness. The activation of the EMT program depends on the synergy between extracellular signals originating from the tumor microenvironment and the genomic and epigenomic structure of the cells. A number of extracellular ligands, including transforming growth factor β (TGF β), as well as microenvironmental cues, such as a reduction in oxygen partial pressure (frequently encountered in solid tumors), have been shown to activate and maintain EMT [2–4]. Under hypoxic conditions, EMT is induced as part of a pleiotropic adaptive response that is largely controlled by transcription factors belonging to the Hypoxia-Inducible Factor (HIF) family. HIFs are heterodimeric transcription factors consisting of a common β subunit that is constitutively present in cells, and of an inducible α subunit, which is specific for each family member. Under normoxic conditions, the latter is continuously synthesized and degraded following hydroxylation of two critical proline residues (P402 and P564) by specific prolyl hydroxylases (PHD1—3), which allows the interaction with the Von Hippel-Lindau (VHL) E3 ubiquitin ligase and subsequent protein degradation [5,6].

PHD activity is oxygen-dependent, and also requires Fe (II) and oxoglutarate as cofactors. Thus, under hypoxic conditions, the inducible α subunit is stabilized and the heterodimeric transcription factor accumulates in the nucleus, binds to specific sequences in gene promoters, and modifies the expression of several target genes, including those encoding for transcriptional modulators involved in EMT (Slug, Snail, Twist, Zeb) [7–9].

A number of studies published in the late 1990s/early 2000s have shown that intracellular generation of reactive oxygen species (ROS) is implicated in the hypoxic stabilization of HIF-1α, providing a possible mechanism through which ROS may promote tumor invasiveness and metastatic spread. These studies emphasize the role played by the mitochondrial electron transport chain, and more specifically of complex III, as a source for HIF-1α-stabilizing ROS during hypoxia [10–13]. Mitochondrial complex III, also called Ubiquinol-cytochrome c oxidoreductase or cytochrome bc1 complex, transfers electrons from ubiquinol to cytochrome c and couples this redox reaction to the translocation of protons from the matrix to the intermembrane space through a mechanism known as "Q cycle" [14–18]. During this cycle unstable ubisemiquinone intermediates are formed that can transfer their unpaired electron to oxygen, leading to superoxide formation; mitochondrial superoxide dismutase (SOD) then converts superoxide to H_2O_2, which inhibits PHD activity by direct attack and/or by decreasing Fe (II) availability, thus causing HIF-1α stabilization. Subsequent studies have established the ubisemiquinone formed at the Qo site of complex III as the crucial electron source for hypoxic ROS production. Accordingly, specific Qo inhibitors, such as stigmatellin, are able to counteract hypoxic ROS production and HIF-1α stabilization, as well as some HIF-1-dependent cellular responses such as VEGF release and angiogenesis [19]. In contrast, Qi site inhibitors, such as antimycin A, increase superoxide production by complex III and are therefore ineffective in preventing hypoxic adaptations [14,20].

Recently, the ubiquinol-cytochrome c reductase binding protein (UQCRB), a subunit of mitochondrial Complex III, has been found to play a pivotal role in hypoxic mitochondrial ROS generation, thereby emerging as an important modulator of tumor angiogenesis by the upregulation of both hypoxic signaling. Interestingly, the small molecule Terpestacin and its derivatives, which specifically bind UQCRB, inhibit the angiogenic response to pro-angiogenic stimuli, such as hypoxia [21–23], making UQCRB a potential new therapeutic target for antiangiogenic drug development.

While the role of mitochondrial ROS in hypoxia-induced angiogenesis has been successfully established, their implication in EMT activation is still unclear. The present study was aimed at testing the hypothesis that the inhibition of hypoxia-induced mitochondrial ROS production could result in the reversion of EMT, in addition to decreased HIF-1α stabilization, thus providing a rationale for the development of drugs able to counteract tumor progression induced by the EMT by blocking mitochondrial ROS production. We first evaluated the role of hypoxia-induced ROS increase in HIF-1α stabilization, in a triple-negative breast cancer cell line that has often been used in the study of EMT, namely MDA-MB468. Secondly, we assessed the ability of a number of antioxidants, mostly targeting mitochondrial Complex III and including Terpestacin, to block ROS production and HIF-1α stabilization and prevent changes in EMT markers.

Finally, the specific role of UQCRB in hypoxia-induced EMT was evaluated by silencing its expression through RNA interference and assessing the effects of its downregulation on ROS production, HIF-1α levels and EMT markers.

2. Materials and Methods

2.1. Chemicals

Myxothiazol (Myxo), mito-TEMPO (mitoTP), standard chemicals, and cell culture reagents were purchased from Euroclone s.r.l. (Milan, Italy), unless otherwise indicated. Terpestacin was kindly provided by Dr. HoJeong Kwon (Yonsei University, South Korea).

2.2. Cell Lines, Drug Treatment and Hypoxia Induction

The breast cancer cell lines MCF7, MDA-MB468, and MDA-MB231 were originally obtained from American Type Culture Collection (Rockville, MD, USA) and recently authenticated by morphological inspection, growth curve analysis, and short tandem repeat profiling. MCF7 cells are hormone responsive, whereas the other two cell lines derive from triple negative breast carcinomas (TNBCs), a particularly aggressive and untreatable subset of breast cancers [24]. The 293FT cell line (Invitrogen, Milan, Italy) was used to produce replication-incompetent lentiviral particles. Cells were maintained in RPMI1640 (MCF7 and MDA-MB231; ECM2001L) and DMEM (MDA-MB468 and 293FT; ECB7501L) medium supplemented with 10% fetal bovine serum, 1% glutamine (ECB3000D), 1% antibiotic mixture, 1% sodium pyruvate (S8636) and 1% non-essential amino acids (ECB3054D) and incubated at 37 °C in a humidified 5% CO_2 atmosphere. Cells were cultured for less than 10 weeks (18–20 passages) and were regularly checked for Mycoplasma (Molecular Biology Reagent Set *Mycoplasma* species, Euroclone, UK).

For all experiments, unless otherwise indicated, cells were seeded, treated 24 h later with Myxo (T5580, 1 μM), mitoTP (SML0737, 100 and 200 μM) or Terpestacin (25 μM) in serum-free medium and incubated for 48 h at 37 °C in normoxia (O_2 21%) or hypoxia (O_2 1%). Drug concentrations were based on literature data [22,25–27]. Hypoxia was induced by placing the cells for the indicated times inside a modular incubator chamber (Billups Rothenberg Inc., Del Mar, CA, USA) flushed with a mixture of 1% O_2, 5% CO_2, and 94% N_2 at 37 °C.

2.3. Scratch Wound Healing Assay

Cell migration was evaluated on the studied cell line by the scratch wound healing assay. Briefly, 7×10^5 cells/well were grown (approximately to confluence) in 6-well plates, and a scratch was produced in the cell monolayers using a 100-μL pipette tip. The culture medium was then replaced by serum-free medium and the plates were incubated under normoxic or hypoxic conditions. Pictures of the scratch wound were taken immediately following scratch formation (0) and after 48 h incubation, through a camera connected to an Olympus IX81 microscope. The percentages of the open scratch wound were evaluated by the TScratch software.

2.4. Evaluation of ROS Levels

Intracellular ROS generation was evaluated using 2,7-dichlorodihydrofluorescein diacetate (DCFH-DA, D6883) as a probe. At the end of 48 h treatment with mitoTP (100 and 200 μM), Myxo (1 μM), or Ter (25 μM), under both normoxic or hypoxic conditions, the cells were detached, washed, and re-suspended (10^6 cells/mL) in PBS containing 10 μM DCFH-DA; after 45 min incubation in the dark at 37 °C, cell samples were then analyzed with a FACSCalibur flow cytometer (Becton Dickinson Mountain View, CA, USA) and data were processed using CellQuestPro software (Becton Dickinson- version 6.0). DCFH-DA could not be used to analyze ROS production in infected cells as they also express GFP (Green Fluorescent Protein, see below); thus, for these experiments, DCFH-DA was replaced by dydroethidine (D7008, HE) [28]. Briefly, cells were detached, washed and resuspended (10^6 cells/mL) in HE (25 μM in PBS); following 30′ incubation at 37 °C in the dark, samples were analyzed as described [28,29]. Intracellular ROS generation was quantitated in arbitrary units based on the mean fluorescence intensity (MFI).

2.5. Western Blot Analysis

Western blot analysis was carried out to detect the expression of HIF-1α, E-cadherin, N-cadherin, vimentin, cytokeratin-19, Snail, Slug, and UQCRB in whole cell lysates, following normoxic or hypoxic incubation and/or drug treatment. For whole lysates, cells were lysed in a buffer containing NaCl 120 mM, NaF 25 mM, EDTA 5 mM, EGTA 6 mM, sodium pyrophosphate 25 mM in TBS 20 mM pH 7.4, PMSF 2 mM, Na_3VO_4 1 mM, phenylarsine oxide 1 mM, 1% (*v*/*v*) NP-40 and 10% Protease Inhibitor Cocktail. Protein concentration was determined by the BCA assay (23235, Thermo Fisher, Monza (MI), Italy); 100 μg of protein per sample was loaded onto polyacrylamide gels (8% or 12%) and separated under denaturing conditions. The protein bands were then transferred onto Hybond-P membranes (GE10600023, Sigma Aldrich, Milan, Italy) and Western blot analysis was performed by standard techniques, with mouse monoclonal anti-human HIF-1α (610958, BD Bioscience), E-cadherin (5085) and N-cadherin (13A9, Thermo Fisher, Monza (MI), Italy) antibodies; mouse monoclonal anti-human cytokeratin 19 (A53B, Thermo Fisher, Monza (MI), Italy) antibody; rabbit polyclonal anti-human vimentin, Snail and Slug antibodies (ABCAM). Equal loading of the samples was verified by re-probing the blots with an anti-mouse-β-tubulin antibody (SC-5274, Santa Cruz Biotechnology, Segrate (MI), Italy). Protein bands were visualized using a peroxidase-conjugated anti-mouse secondary antibody and the Westar Supernova Substrate (XLS-3, Cyanagen, Bologna, Italy).

2.6. Construction of Lentiviral Vectors

Lentiviral particles were generated using a second-generation transient expression system, composed of (i) the pCMV R8.74 packaging construct, (ii) the pMD2.G envelope expression construct, and (iii) the pLVTHM/GFP transfer vector, for silencing of UQCRB expression by RNA interference. The pLVTHM/GFP contains a green fluorescent protein (GFP) cDNA under the transcriptional control of an intronless human elongation factor 1-α (EF1-α-short) promoter. All constructs were kindly provided by Dr. Didier Trono (School of Life Sciences, Swiss Institute of Technology, Lausanne, Switzerland; www.epfl.ch/labs/tronolab/ (accessed on 25 May 2005)). The transfer vector pLVTHM/shUQCRB/GFP or pLVTHM/shScrambledUQCRB/GFP was generated as follows. A sense strand of 19 nucleotides specific for UQCRB, preceded by overhangs specific for MluI cloning, was designed to be followed by a short loop sequence (TTCAAGAGA), by the reverse complement of the sense strand, with five terminal thymidines to act as a RNA polymerase III transcriptional stop signal, and by a sticky sequence specific for ClaI. The forward oligonucleotide (5′-GCGTCCCCGACAGGATGTTTCGCATTATTCAAGAGATAATGCGAAACATCCTGTCTT TTTGGAAAT-3′ corresponding to the 300–319 nucleotides of UQCRB (NCBI Reference Sequence: NM_001199975.3), was annealed with a complementary reverse oligonucleotide (3′-AGGGGCTGTCCTACAAAGCGTAATAAGTTCTCTATTACGCTTTGTAGGACAGAAAA ACCTTTAGC-5′) in annealing buffer (100 mM potassium acetate, 30 mM HEPES pH7.4, 2 mM magnesium acetate). Annealed oligos were phosphorylated with T4 Polynucleotide Kinase (New England Biolabs Ltd.) and cloned into the MluI-ClaI sites of a pLVTHM lentiviral vector. The transfer vector pLVTHM/shScrambledUQCRB/GFP was constructed in the same way (5′-GCGTCCCCAGTGTTGGACGATTATCACTTCAAGAGAGTGATAATCGTC CAACACTTTTTGGAAAT-3′ sequence obtained from www.genscript.com (accessed on 15 January 2016)).

2.7. Generation of Lentiviral Particles and Target Cell Infection

Lentiviral particles pseudotyped through the VSV envelope glycoprotein were produced by co-transfecting 5×10^6 293FT cells with 40 μg of total plasmid DNA: the (i) pCMVDR8.74, (ii) pMD2.G, and (iii) pLVTHM/shUQCRB/GFP or pLVTHM/scrambled UQCRB/GFP vectors, with the calcium phosphate precipitation method, as previously described [30]. Transduction experiments were performed in a medium containing 4μg/mL polybrene (Sigma-Aldrich, Milan, Italy). Viral titration was performed by flow cytometer-counting GFP-expressing NIH3T3 cells 48 h after infection. For in vitro shRNA-UQCRB silencing, 60% confluent MDA-MB468 cells were infected for 4 h with 10 MOI lentiviral vectors; the particle-containing medium was then replaced with fresh medium and the cells were incubated at 37 °C for 48 h.

2.8. Evaluation of Cell Viability

Cell viability was evaluated through the dye exclusion assay following treatment and/or under normoxic and hypoxic conditions. Briefly, 2×10^5 cells were seeded onto 6-well plates and 24 h later were treated and incubated for 48 h in the presence of 21% or 1% O_2. Cells were then detached and counted using a Burker hemocytometer, following Trypan blue staining.

2.9. Statistical Analysis

Statistical analysis of the data obtained from flow cytometric studies, densitometric analysis of western blot results, and Scratch Wound Healing assay were performed by means of one- or two-way ANOVA, with Bonferroni's test for multiple comparisons, using GraphPad PRISMsoftware (version 4.03).

3. Results

3.1. Characterization of Breast Cancer Cell Lines

Three breast cancer cell lines were initially considered for the present study, namely MCF-7, MDA-MB468, and MDA-MB231, representing three distinct subtypes of the disease [24,31]. MCF7 cells belong to the luminal A subtype, expressing hormone receptors and exhibiting the least aggressive behavior of all breast cancer subtypes. In contrast, both MDA-MB468 and MDA-MB231 derive from triple-negative breast cancers, lacking ER-α, PR, and HER-2 expression, and they belong to the basal A and basal B groups, respectively, based on GE profiling. To find the ideal breast cancer cellular model to test the hypothesis that the inhibition of hypoxia-induced mitochondrial ROS production could result in the reversion of EMT, in addition to decreased HIF-1α stabilization, the three cell lines were characterized concerning their migratory capacity and ability to increase ROS and HIF-1α protein levels and to exhibit changes in EMT markers in response to hypoxia. For this purpose, ROS, HIF-1α, E-cadherin, and vimentin levels and cell migration were evaluated following 48 h incubation of the cells under normoxic (O_2 21%) or hypoxic (O_2 1%) conditions. Figure 1 shows that in MDA-MB468 cells 48 h hypoxia-induced ROS levels increase, as indicated by a shift to the right on the fluorescein peak. Fluorescence intensity was quantitated in arbitrary units based on Median Fluorescence Intensity (MFI—Figure 1a). Differences in HIF-1α, E-cadherin, and vimentin expression were also observed following 48 h incubation in hypoxia. In particular, the E-cadherin-vimentin switch, typical of the EMT, was present, along with the expected increase in HIF-1α levels (Figure 1d). Moreover, this cell line showed intrinsic migratory capacity, both in normoxia and hypoxia, as indicated in Table 1, reporting the percentages of the open scratch wound.

Table 1. Percentage of open scratch wound in MDA-MB468, MDA-MB231 and MCF7 cells following 48 h incubation under normoxic (O_2 21%) and hypoxic conditions (O_2 1%) (mean S.E. of 3 independent experiments; * $p < 0.05$ vs. 0 h; *** $p < 0.001$ vs. 0 h and O_2 21%).

	MDA-MB468		MDA-MB231		MCF7	
Time (h)	O_2 21%	O_2 1%	O_2 21%	O_2 1%	O_2 21%	O_2 1%
0	80.0 ± 2.0	75.0 ± 1.0	78.0 ± 2.0	78.0 ± 3.0	76.5 ± 1.5	78.0 ± 3.0
48	54.5 ± 0.5 *	36.5 ± 4.0 ***	59.0 ± 1.5 *	46.5 ± 1.5 ***	74.0 ± 1.0	74.5 ± 2.5

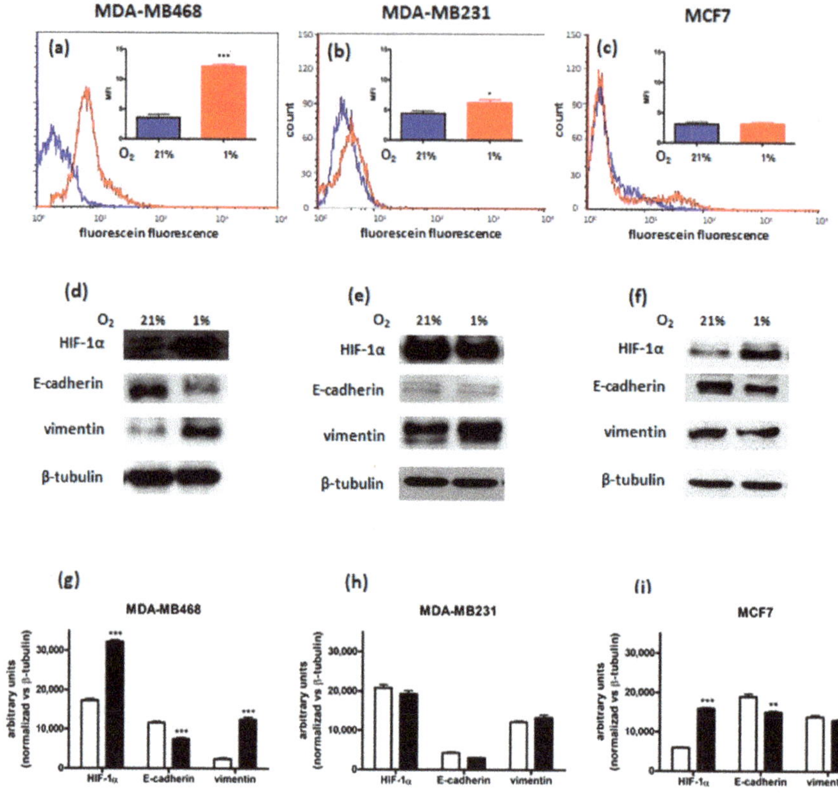

Figure 1. MDA−MB468, MDA−MB231 and MCF7 cells response to 48 h hypoxia (O_2 1%). (**a**–**c**): ROS levels (blue: O_2 21%; red: O_2 1%). Fluorescence intensity was quantitated based on the Median Fluorescence Intensity (MFI) and results obtained in 3 independent experiments are reported in the graph (*** $p < 0.001$ and * $p < 0.05$ vs. normoxia). (**d**–**f**): HIF−1α, E−cadherin and vimentin protein levels in MDA−MB468 (**d**), MDA-MB231 (**e**), and MCF7 (**f**) cell lines (representative western blot analysis out of 3 independent experiments with similar results). Densitometric analysis (**g**–**i**) were performed on all western blot experiments (*** $p < 0.001$ and ** $p < 0.01$ vs. normoxia).

A different pattern was observed in the other triple-negative breast cancer cell line characterized, namely MDA-MB231. In this cell line, under hypoxic conditions, migratory capacity and only a slight increase in ROS levels were observed (Table 1 and Figure 1b, respectively). However, the same EMT pattern was present both under normoxic and hypoxic conditions (Figure 1e), thus making MDA-MB231 cells unsuitable for our purposes. In contrast, MCF7 cells, representing a less aggressive breast cancer subtype (i.e., responsive to estrogens and progesterone), did not show either migratory activity or significant alterations in both ROS production and EMT-related protein levels following exposure to hypoxia (Table 1 and Figure 1c,f, respectively).

Therefore, only MDA-MB468 cells responded to hypoxic stimuli by increasing ROS and HIF-1α levels, as well as by activating EMT, as indicated by the decrease in E-cadherin and the increase in vimentin levels, and cell motility, in agreement with previously reported data [32–34]. In MDA-MB468 cells, hypoxia also induced an increase of the two E-cadherin transcription repressors, Snail and Slug (Figure S1), confirming the switch from epithelial to mesenchymal phenotype as reported by other authors [35]. However, the results obtained in this first part of the study suggest that such responses are highly cell/tumor type-dependent.

Based on these results, MDA-MB468 were chosen for further investigations.

3.2. *Effects of the Inhibition of ROS Production on Hypoxia-Induced Responses*

To confirm the role of hypoxia-induced ROS increase in EMT induction, the effects of the inhibition of ROS generation in hypoxic MDA-MB468 cells were evaluated following treatment with antioxidants/ROS-scavengers or inhibitors of the mitochondrial chain Complex III. In particular, the effects of Mito-Tempo (MitoTP), a mitochondria-specific antioxidant, were evaluated. Furthermore, the effects of two specific inhibitors of mitochondrial chain Complex III, the antibiotic Myxothiazol (Myxo), which targets the mitochondrial complex III at the Qo site, close to the heme group b566 [20,36] and Terpestacin (Ter), which inhibits the Mitochondrial Complex III by specifically binding the UQCRB (ubiquinol-cytochrome c reductase binding protein) subunit [37] were also evaluated.

3.2.1. Decrease of ROS Levels Using MitoTP

MitoTP is a small molecule that acts specifically as a mitochondrial ROS scavenger and was shown to inhibit cell migration inhibitor [38,39]. In our model, MitoTP significantly reduced ROS levels, in a dose-dependent manner (Figure 2A), and induced a parallel decrease in HIF-1α protein levels in hypoxic MDA-MB468 cells. These effects were associated with a decrease in vimentin and with an increase in E-cadherin levels (Figure 2B,C). Furthermore, MitoTP showed the same effect on cell survival in both normoxia and hypoxia (Figure S2).

3.2.2. Decrease of ROS Levels Using Inhibitors of Mitochondrial Complex III

Myxothiazol (Myxo) interacts with the Qo site, blocking electron transfer within complex III, thereby preventing ROS production, and inhibiting HIF-1α stabilization and HIF transcriptional activity in hypoxia. In agreement with these reports, we observed a significant decrease in hypoxia-induced ROS production following 48 h treatment with Myxo, as shown in Figure 3A. In MDA-MB468 cells Myxo was also able to significantly decrease hypoxia-induced HIF-1α and vimentin protein levels and increase hypoxia-decreased E-cadherin levels (Figure 3B,C). Furthermore, Myxo was shown to reduce the increase of Slug, N-cadherin, and Snail protein levels induced by 48 h incubation of the cells in hypoxia (Figure S3). A reduction in cell viability was observed following 48 h treatment; however, the same extent of viability reduction was induced by Myxo also in hypoxia (Figure S4).

In addition, in MDA-MB468 cells Ter significantly reduced the hypoxia-induced ROS production (Figure 4A). Furthermore, Ter was able to prevent hypoxia-induced HIF-1α stabilization, resulting in the reversion of the mesenchymal phenotype. For this set of experiments, CK19 was used as an epithelial marker instead of E-cadherin; Ter was found to prevent both the decrease in CK19 and the increase in vimentin levels induced by hypoxia (Figure 4B,C). Furthermore, treatment with Ter only produced a modest decrease in viability in normoxic MDA-MB468 cells, which is also observed under hypoxic conditions (Figure S5).

3.3. *Effects of the UQCRB Silencing*

Finally, the effects of UQCRB downregulation on ROS production and HIF-1α and EMT protein levels were evaluated on MDA-MB468/pLVTHM, MDA-MB468/scrambledUQCRB and MDA-MB468/shUQCRB, obtained by infection of MDA-MB468 cells, as indicated in the Materials and Methods section, following 48 h incubation in the presence of 21% or 1% O_2. The effects of the UQCRB silencing on cell migration and cell viability were also evaluated.

The results from the western blot analysis of whole cell lysates showed that following the infection of MDA-MB468 cells with shUQCRB, the UQCRB protein levels were significantly reduced in both normoxic and hypoxic MDA-MB468/shUQCRB cells, indicating successful silencing of the protein (Figure 5A,B). As hypothesized, a significant reduction in hypoxia-induced ROS increase was observed in MDA-MB468/shUQCRB cells, compared to MDA-MB468/pLVTHM and MDA-MB468/scrambledUQCRB cell lines (Figures 5C and S6).

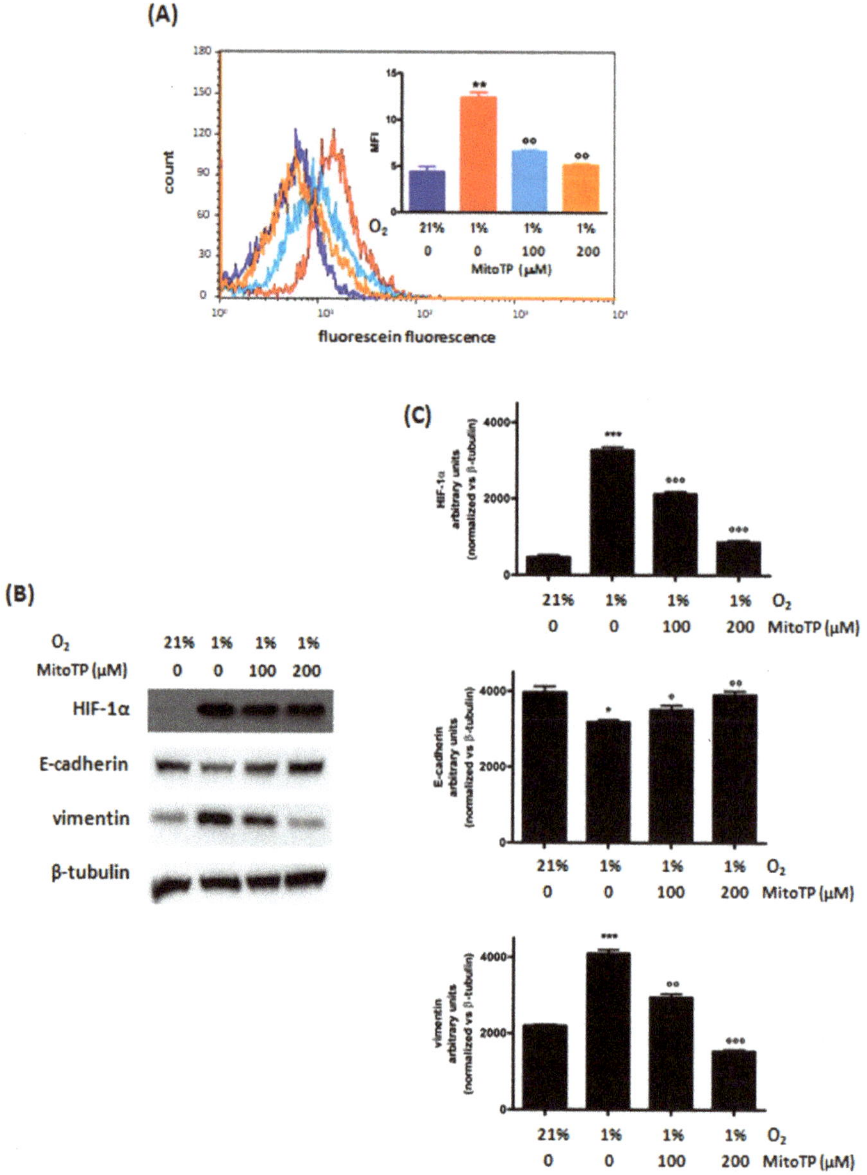

Figure 2. Effects of 48 h treatment with MitoTP (100 and 200 µM and incubation in normoxic (O_2 21%) or hypoxic (O_2 1%) conditions in MDA−MB468 cells: ROS (**A**) and HIF−1α, E−cadherin and vimentin (**B**) levels. The figure shows a representative flow cytometric and Western blot analysis out of 3 independent experiments with similar results. For flow cytometric analysis of ROS levels (blue line: normoxic control; red line: hypoxic control; light blue line: hypoxic MitoTP 100 µM treated cells; orange line: hypoxic MitoTP 200 µM treated cells) fluorescence intensity was quantitated based on the Median Fluorescence Intensity (MFI) and results obtained in 3 independent experiments are reported in the graph. Densitometric analysis (**C**) was performed on all experiments (* $p < 0.05$, ** $p < 0.01$ and *** $p < 0.001$ vs. normoxic control; $^{\circ}$ $p < 0.05$, $^{\circ\circ}$ $p < 0.01$, $^{\circ\circ\circ}$ $p < 0.001$ vs. hypoxic control).

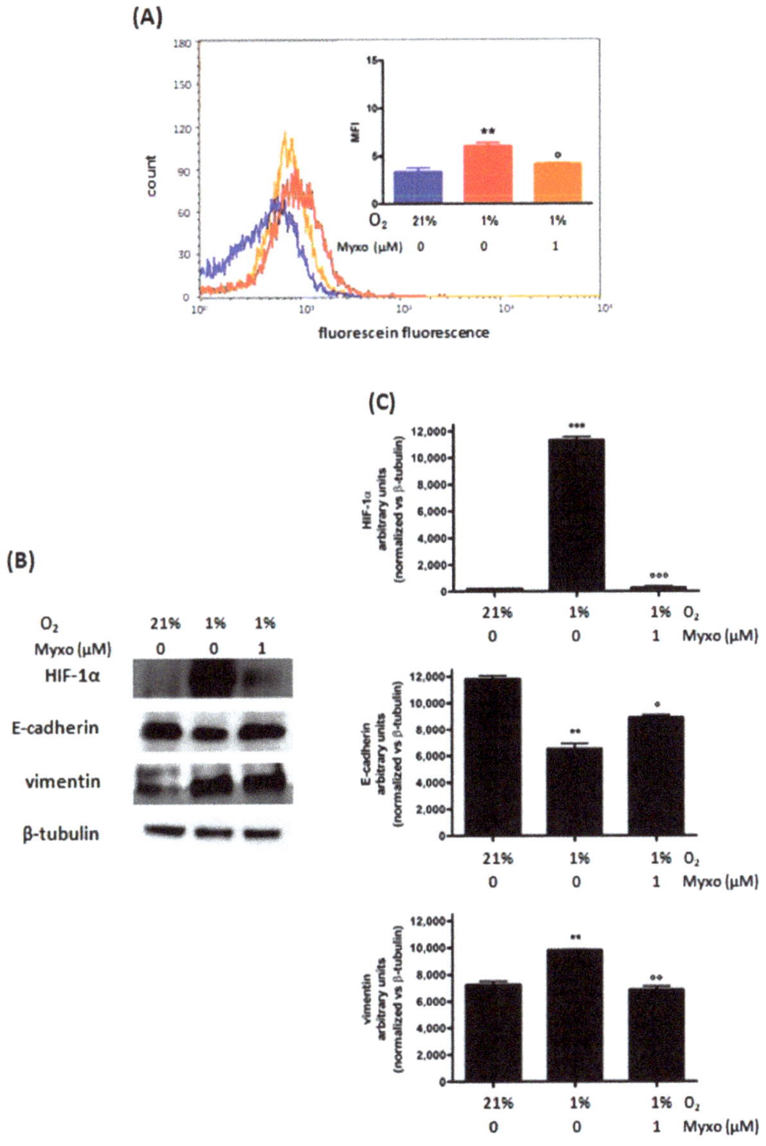

Figure 3. Effects of 48 h treatment with Myxo treatment (1 μM) and incubation in normoxic (O_2 21%) or hypoxic (O_2 1%) conditions in MDA−MB468 cells: ROS (**A**) and HIF−1α, E−cadherin and vimentin (**B**) levels. The figure shows representative flow cytometric and Western blot analyses out of 3 independent experiments with similar results. For flow cytometric analysis of ROS levels (**A**); blue line: normoxic control; red line: hypoxic control; orange line: hypoxic Myxo treated cells). The fluorescence intensity was quantitated based on the Median Fluorescence Intensity (MFI) and results obtained in 3 independent experiments are reported in the graph. Densitometric analysis (**C**) was performed on all experiments (*** $p < 0.001$ and ** $p < 0.01$ vs. normoxic control; °°° $p < 0.001$, °° $p < 0.01$ and ° $p < 0.05$ vs. hypoxic control).

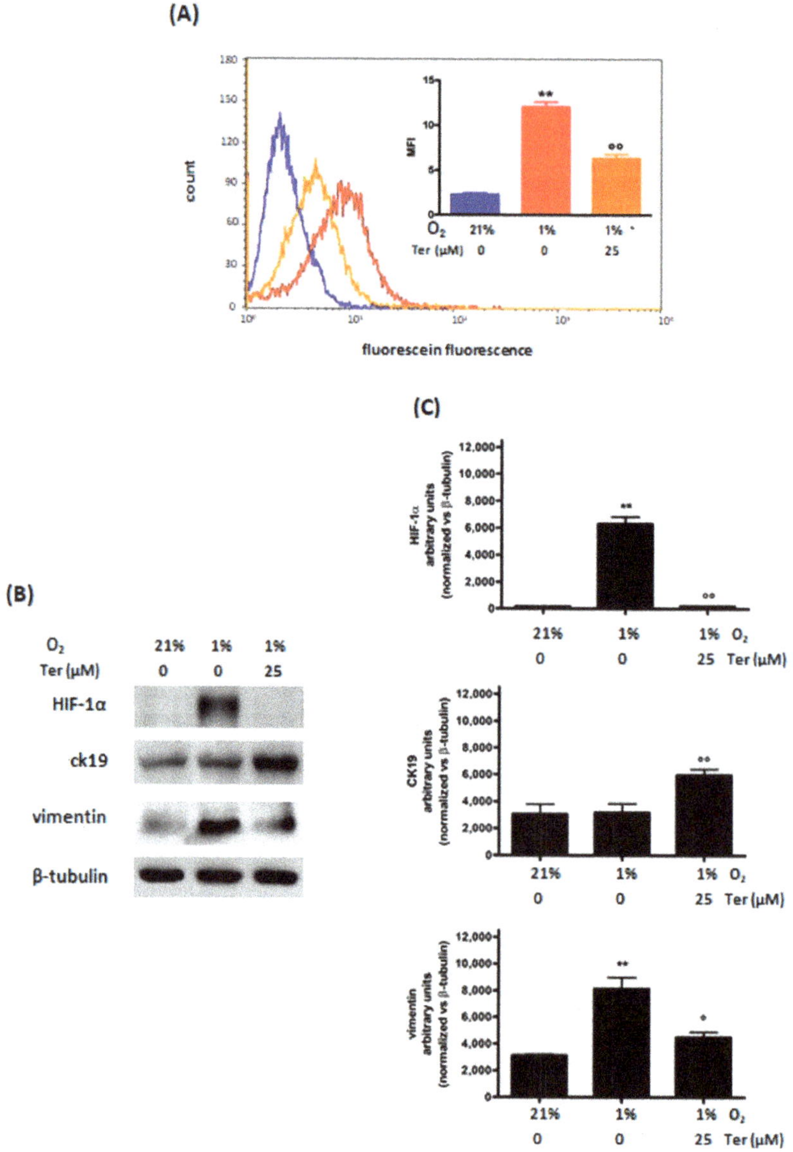

Figure 4. Effects of 48 h of treatment with Ter (25 μM) and incubation in normoxia (O_2 21%) or hypoxia (O_2 1%) in MDA−MB468 cells: ROS (**A**) and HIF−1α, cytokeratin 19, and vimentin (**B**) levels. The figure shows a representative flow cytometric and Western blot analysis out of 3 independent experiments with similar results. For flow cytometric analysis of ROS levels (**A**); blue: normoxic control; red: hypoxic control; orange: hypoxic Ter treated cells), fluorescence intensity was quantitated based on the Median Fluorescence Intensity (MFI), and results obtained in 3 independent experiments are reported in the graph. Densitometric analysis (**C**) was performed on all experiments (** $p < 0.01$ vs. normoxic control; °° $p < 0.01$ and ° $p < 0.05$ vs. hypoxic control).

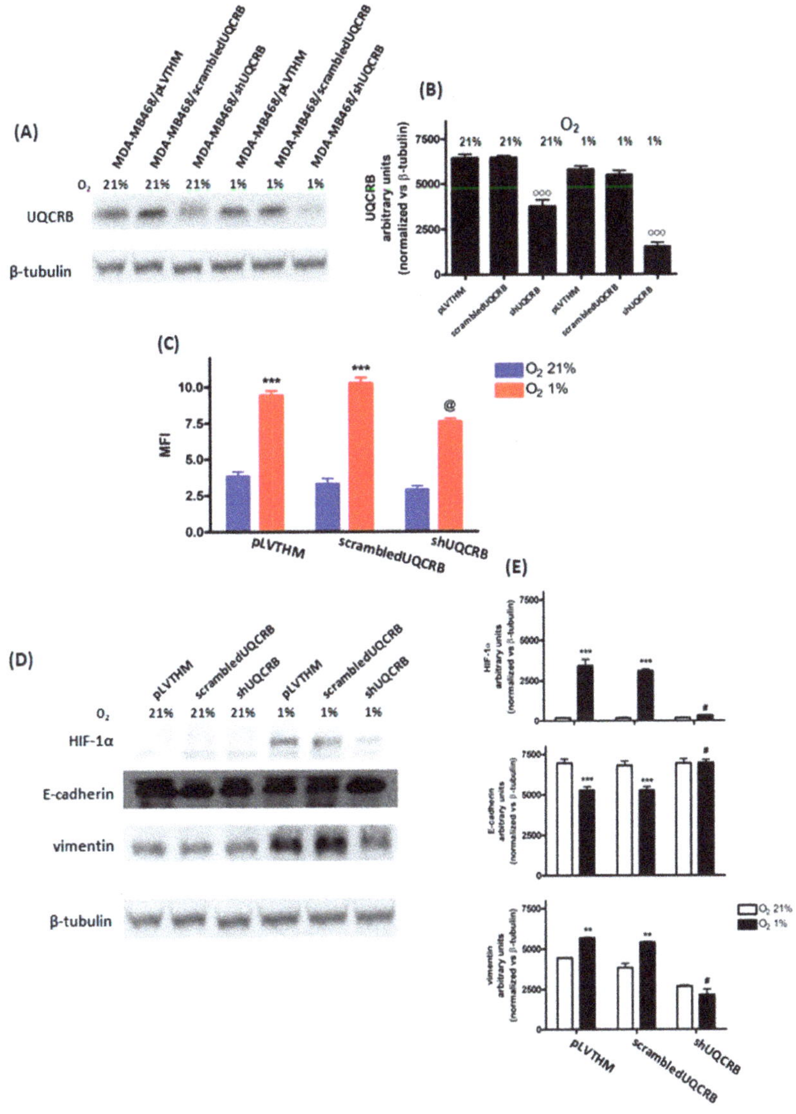

Figure 5. Effects of UQCRB silencing in MDA−MB468 cells, following 48 h incubation in normoxia (O_2 21%) or hypoxia (O_2 1%): UQCRB protein levels (**A**) and related densitometric analysis (**B**), ROS (**C**) and HIF-1α, E−cadherin and vimentin levels (**D**) and related densitometric analysis (**E**). The figure shows densitometric analysis of 3 independent western blot experiments. For flow cytometric analysis of ROS levels (**C**), fluorescence data are expressed as Median Fluorescence Intensity (MFI, blue: O_2 21%; red: O_2 1%). (°°° $p < 0.001$, *** $p < 0.001$ and ** $p < 0.01$ vs. normoxic control; @ $p < 0.01$ vs. hypoxic (O_2 1%) PLVTHMand scrambled; # $p < 0.01$ vs. hypoxic pLVTHMand scrambled).

UQCRB silencing abrogated HIF-1α stabilization and reversed both the decrease in E-cadherin and the increase in vimentin levels induced by hypoxia (Figure 5D,E). Snail levels were also decreased in hypoxic MDA-MB468/shUQCRB cells, compared to the two control cell lines (Figure S7).

The results from the Scratch Wound Healing and viability assays are reported in Figures 6 and 7, respectively, while representative images of the scratches are reported in Figure S8. Interestingly, UQCRB silencing results in a significant reduction of the migratory capability of MDA-MB468 cells mainly under hypoxic conditions, while in normoxia, only a modest decrease in migration was observed.

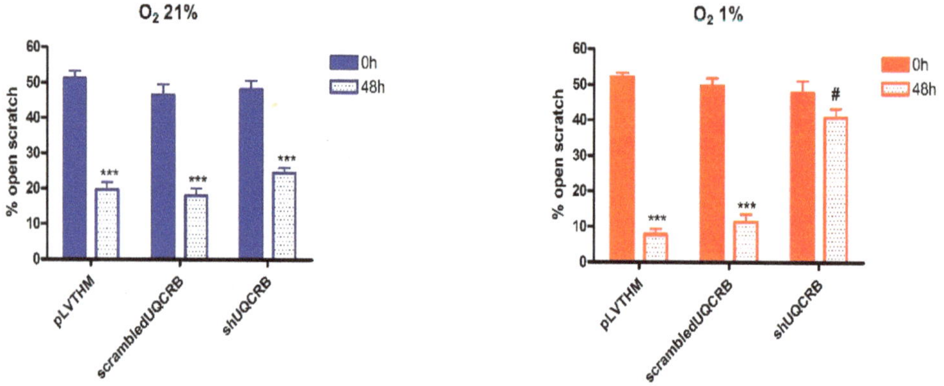

Figure 6. Percentage of open scratch wound in MDA-MB468/pLVTHM, MDA-MB468/scrambledUQCRB and MDA468/shUQCRB cells following 48 h incubation under normoxic (O_2 21%) and hypoxic conditions (O_2 1%) (mean S.E. of 3 independent experiments; two-ways ANOVA results: *** $p < 0.001$ vs. 0 h same condition; # $p < 0.01$ vs. MDA-MB468/pLVTHM and MDA-MB468/scrambledUQCRB 48 h).

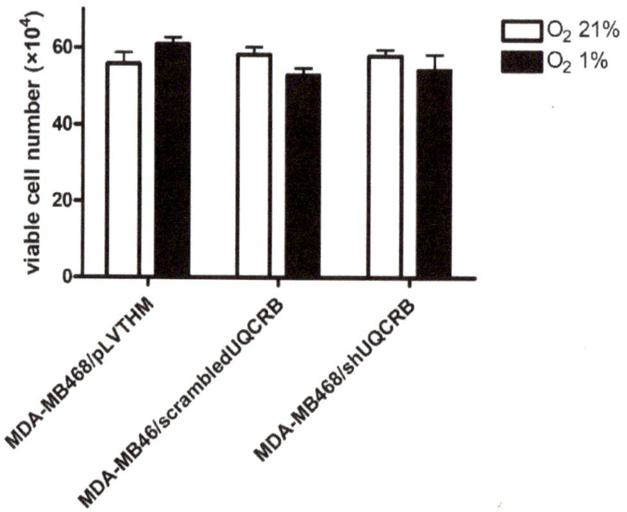

Figure 7. Effects of UQCRB silencing on MDA-MB468/pLVTHM, MDA-MB468/scrambledUQCRB, and MDA-MB468/shUQCRB cell viability in normoxia (O_2 21%) or hypoxia (O_2 1%) (mean ± S.E. of 3 independent experiments).

However, the viability of UQCRB-silenced cells was comparable to that of the two control cell lines, confirming that UQCRB inhibition does not significantly affect cell viability.

4. Discussion

The role of ROS in the activation of EMT, through HIF1α stabilization, is still unclear, despite the involvement of ROS and HIF-1α in this process which has been highlighted by a number of authors [40–42]. This evidence was also supported by data showing that antioxidants, such as SkQR1, NAC, Mito-CP, and Mito-TEMPO, are able to modulate EMT [41,43,44].

Recently, it has been demonstrated that hypoxia causes superoxide production by respiratory chain Complex III; the superoxide anion is then converted to H_2O_2, which inhibits PHD enzyme activity by direct attack and/or reduction of Fe (II) availability, causing HIF1α stabilization [27]. Thus, the hypoxic response, mediated by HIF-1α, modifies the expression of many target genes linked to EMT, including those encoding for the transcriptional modulators Snail, Slug, Twist, and Zeb [7–9].

Here, we have confirmed that the inhibition of hypoxia-induced mitochondrial ROS production through treatment with antioxidants leads to decreased HIF-1α stabilization and reverts the changes in some hypoxia-induced EMT markers in the TNBC cell line MDA-MB468. However, the aspecific effects of these antioxidants on both mitochondrial Complexes I and III also resulted in a cell viability reduction on cell viability. The novelty of the present work lies in the hypothesis that specific inhibition of hypoxia-induced mitochondrial ROS at the UQCRB site of the mitochondria Complex III, obtained both through Ter treatment or UQCRB silencing, results in the reversion of EMT, in addition to decreased HIF-1α stabilization and without significantly affecting cell viability.

To account for the discrepancies observed between this cell line and the other TNBC cell line characterized, namely MDA-MB231, it should be emphasized that TBNCs are far from homogeneous. While cell lines derived from this type of breast cancer are generally defined as basal-like, a first major distinction can be made based on gene clustering between the basal A group, displaying epithelial characteristics and often associated with *BRCA1* gene signatures, and the Basal B group, displaying mesenchymal and stem/progenitor-like characteristics. A more refined analysis of gene expression profiles has led to the classification of TNBCs into six different subtypes [45]. According to this classification, MDA-MB468 cells belong to the basal-like BL-1 subtype, displaying epithelial characteristics, whereas MDA-MB231 are assigned to the mesenchymal/stem-like subtype, which may well account for their more invasive behavior and for their pattern of expression of epithelial and mesenchymal markers even under normoxic conditions. Important differences have been detected in the genomic profiles of the two cell lines, including the presence of the activating $KRAS^{G13D}$ mutation in MDA-MB-231 (which per se might confer a more aggressive behavior) [46]. In addition, this latter cell line has been shown to constitutively express the urokinase receptor uPAR, and silencing this receptor has been shown to lead to a more epithelial phenotype; in contrast, MDAMB468 cells express low uPAR levels in normoxia, but under hypoxic conditions, they upregulate its expression in a HIF-1-dependent fashion, and uPAR overexpression has been shown to reversibly induce EMT in this cell line [47]. These features may help explain the differences observed between the two cell lines in the preliminary phase of the present study.

Results obtained with the mitochondrial ROS scavenger MitoTP indicated that in MDA-MB468 cells reduction of ROS levels lead to the reduction of HIF-1α cellular levels and consequently to the mesenchymal-epithelial switch (i.e., increase in E-cadherin and decrease in vimentin protein levels), confirming the interplay between ROS increase, HIF-1α stabilization, and EMT induction.

Mitochondria are essential to hypoxia-induced ROS increase [10]. Specifically, complexes I and III of the mitochondrial electron transport chain represent the main sites of ROS production; however, recently Chandel demonstrated, then it was confirmed by others, that Complex III is the site responsible for the hypoxic production of ROS [15,48]. Using the so-called Q cycle, Complex III transfers electrons from ubiquinol to cytochrome c and contributes to the generation of an electrochemical proton gradient resulting in the formation of a ubisemiquinone radical intermediate, which in some conditions is likely to give an electron to O_2 to form superoxide [16,37]. According to this model, mitochondrial ROS production is necessary and sufficient to induce HIF-1α stabilization, promoting HIF-1-regulated processes, including EMT. This finding is very important for cancer therapy since it identifies the mitochondrial respiratory chain Complex III as a potential target through which EMT, and therefore metastases formation and resistance to therapies, could be inhibited. However, not all Complex III inhibitors inhibit ROS production; as a matter of fact, inhibitors that act at the Qi site are known to increase superoxide anion levels [36,49].

In contrast, and in agreement with other authors [16,37], in our model, the Qo inhibitor Myxo has been shown to inhibit both hypoxia-induced ROS formation and hypoxia-induced HIF-1α stabilization, along with EMT. However, Myxo also inhibits mitochondrial respiration [27], which drastically reduces its potential for successful clinical application. As a matter of fact, the antiproliferative effect observed following Myxo treatment, both in normoxia and hypoxia, could be a confirmation of this last statement. In contrast, Terpestacin (Ter) was shown to bind the UQCRB subunit of complex III and to suppress hypoxia-induced mitochondrial ROS generation without inhibiting mitochondrial respiration and ATP generation [21]. This effect was also confirmed in MDA-MB468 cells.

The effects of UQCRB downregulation on ROS production and HIF-1α and EMT protein levels, evaluated on MDA-MB468 cells in which UQCRB was silenced, namely MDA-MB468/shUQCRB, support the hypothesis that the reduction of the hypoxia-induced ROS increase, through the inhibition of mitochondrial Complex III, can reverse EMT phenotype.

It was perhaps surprising that a comparatively modest, albeit significant, decrease in ROS generation observed in UQCRB-silenced cells should totally prevent HIF-1α stabilization and induce such drastic changes in E-cadherin and vimentin levels as compared to control/scrambled cells. Actually, both complex I and Complex III have been implicated as sources of ROS production during hypoxia [36]. UQCRB silencing selectively prevents ROS generation through Complex III; thus, our observations provide indirect evidence of the role played by Complex III-derived ROS hypoxia in HIF-α stabilization and EMT. Indeed, our results confirm and expand observations reported by several groups regarding the role played by mitochondrial ROS in HIF-1 activation and in the downstream hypoxic response [50–52]. Interestingly, a recent article on the same cell line used for this study partially contradicts the role played by HIF-1 in hypoxia-induced changes in some, but not all, EMT markers [42], a discrepancy that can possibly be explained by shorter exposures to hypoxia as compared to the present study and to studies by other groups [53–55].

5. Conclusions

In conclusion, our results are in agreement with recent data suggesting that the UQCRB subunit of Complex III in the mitochondrial respiratory chain plays a pivotal role in hypoxic signaling, and identifies UQCRB as a potential novel therapeutic target for the development of drugs able to counteract tumor progression due to the EMT, by blocking mitochondrial ROS production [21,23,54]. Interestingly, although the migratory capacity of MDA-MB468 cells was significantly decreased by silencing UQCRB, cell survival was not affected.

A number of authors have reported the effect of antioxidants, such as SkQR1, NAC, Mito-CP, and Mito-TEMPO [41,43,44] on the redox modulation of EMT; however, the involvement of the mitochondrial Complex III UQCRB site in the EMT reversion, due to the inhibition of the hypoxia-induced mitochondrial ROS production and the related decrease in HIF-1α stabilization, represents the novelty of this work and might provide a rationale for the development of drugs able to counteract tumor progression induced by the EMT by blocking mitochondrial ROS production at this site.

Finally, the cell line we have used for this study is a particularly interesting model, being representative of the triple negative adenocarcinoma, an aggressive and untreatable subset of breast cancer, that does not respond to endocrine therapy or other currently available targeted agents [56,57]. Novel therapies addressing this tumor subset are urgently needed, and targeting UQCRB might be an option worth investigating, at least for some forms of these tumor types.

Supplementary Materials: The following supporting information can be downloaded at: https://www.mdpi.com/article/10.3390/cimb44110359/s1, Figure S1—Slug and Snail protein levels in MDA-MB468 cell line, following 48 h hypoxia (O_2 1%); Figure S2—Effects of MitoTP (100 and 200 mM) treatment and 48 h incubation in normoxia (O_2 21%) or hypoxia (O_2 1%) on MDA-MB468 viability (mean ± S.E. of 3 independent experiments); Figure S3: N-cadherin, Slug and Snail protein levels in MDA-MB468 cells, following 48 h treatment with Myxo (1 mM) and incubation in hypoxia (O_2 1%); Figure S4—Effects of Myxo (1 mM) treatment and 48 h incubation in normoxia (O_2 21%) or hypoxia (O_2 1%) on MDA-MB468 viability (mean ± S.E. of 3 independent experiments); Figure S5—Effects of Ter (25 mM) treatment and 48 h incubation in normoxia (O_2 21%) or hypoxia (O_2 1%) on MDA-MB468 viability (mean ± S.E. of 3 independent experiments); Figure S6: Effects of UQCRB silencing on ROS levels in MDA-MB468 cells, following 48 h incubation in normoxia (O_2 21%) or hypoxia (O_2 1%). Blue line: MDA-MB468/pLVTHM O_2 21%; red line: MDA-MB468/scrambledUQCRB O_2 21%; purple line: MDA-MB468/shUQCRB O_2 21%; light blue line: MDA-MB468/pLVTHM O_2 1%; pink line: MDA-MB468/scrambledUQCRB O_2 1%; green line: MDA-MB468/shUQCRB O_2 1%; Figure S7: Snail protein levels in MDA-MB468/pLVTHM, MDA-MB468/scrambledUQCRB and MDA-MB468/shUQCRB cell lines, following 48 h incubation in normaxia or hypoxia (O_2 1%); Figure S8: representative images of the Scratch Wound Healing assay performed in MDA-MB468/pLVTHM, MDA-MB468/scrambledUQCRB and MDA-MB468/shUQCRB cell lines, following 48 h incubation in normoxia (O_2 21%) or hypoxia (O_2 1%).

Author Contributions: Conceptualization, E.M. (Elena Monti), A.M. and M.B.G.; Data curation, E.M. (Elena Monti) and M.B.G.; Funding acquisition, E.M. (Elena Monti), A.M. and M.B.G.; Investigation, E.M. (Emanuela Marras) and M.B.G.; Methodology, A.M., E.M. (Emanuela Marras) and M.B.G.; Supervision, E.M. (Elena Monti) and M.B.G.; Writing—original draft, E.M. (Elena Monti), A.M., E.M. (Emanuela Marras) and M.B.G.; Writing—review and editing, E.M. (Elena Monti), E.M. (Emanuela Marras) and M.B.G. All authors have read and agreed to the published version of the manuscript.

Funding: This research was funded by the University of Insubria, grant number FAR2018, and by a contribution from BioUp Sagl Lugano.

Institutional Review Board Statement: Not applicable.

Informed Consent Statement: Not applicable.

Data Availability Statement: Data are contained within the article and supplementary material and are available on request.

Conflicts of Interest: The authors declare no conflict of interest.

References

1. Lamouille, S.; Xu, J.; Derynck, R. Molecular mechanisms of epithelial-mesenchymal transition. *Nat. Rev. Mol. Cell Biol.* **2014**, *15*, 178–196. [CrossRef] [PubMed]
2. Giannoni, E.; Parri, M.; Chiarugi, P. EMT and oxidative stress: A bidirectional interplay affecting tumor malignancy. *Antioxid. Redox Signal* **2012**, *16*, 1248–1263. [CrossRef]
3. Derynck, R.; Weinberg, R.A. EMT and Cancer: More Than Meets the Eye. *Dev. Cell.* **2019**, *49*, 313–316. [CrossRef] [PubMed]
4. Tam, S.Y.; Wu, V.W.C.; Law, H.K.W. Hypoxia-Induced Epithelial-Mesenchymal Transition in Cancers: HIF-1α and Beyond. *Front. Oncol.* **2020**, *10*, 486. [CrossRef]
5. Semenza, G.L. HIF-1, upstream and downstream of cancer metabolism. *Curr. Opin. Genet. Dev.* **2010**, *20*, 51–56. [CrossRef] [PubMed]
6. Nagao, A.; Kobayashi, M.; Koyasu, S.; Chow, C.C.T.; Harada, H. HIF-1-dependent reprogramming of glucose metabolic pathway of cancer cells and its therapeutic significance. *Int. J. Mol. Sci.* **2019**, *20*, 238. [CrossRef] [PubMed]
7. Salnikov, A.V.; Liu, L.; Platen, M.; Gladkich, J.; Salnikova, O.; Ryschich, E.; Mattern, J.; Moldenhauer, G.; Werner, J.; Schemmer, P.; et al. Hypoxia Induces EMT in Low and Highly Aggressive Pancreatic Tumor Cells but Only Cells with Cancer Stem Cell Characteristics Acquire Pronounced Migratory Potential. *PLoS ONE* **2012**, *7*, e46391.
8. Liu, Z.-J.; Semenza, G.L.; Zhang, H.-F. Hypoxia-inducible factor 1 and breast cancer metastasis. *J. Zhejiang Univ. Sci. B* **2015**, *16*, 32–43. [CrossRef]
9. Semenza, G.L. Molecular mechanisms mediating metastasis of hypoxic breast cancer cells. *Trends Mol. Med.* **2012**, *18*, 534–543. [CrossRef]
10. Chandel, N.S.; Maltepe, E.; Goldwasser, E.; Mathieu, C.E.; Simon, M.C.; Schumacker, P.T. Mitochondrial reactive oxygen species trigger hypoxia-induced transcription. *Proc. Natl. Acad. Sci. USA* **1998**, *95*, 11715–11720. [CrossRef] [PubMed]

11. Bell, E.L.; Chandel, N.S. Mitochondrial oxygen sensing: Regulation of hypoxia-inducible factor by mitochondrial generated reactive oxygen species. *Essays Biochem.* **2007**, *43*, 17–27. [PubMed]
12. Brunelle, J.K.; Bell, E.L.; Quesada, N.M.; Vercauteren, K.; Tiranti, V.; Zeviani, M.; Scarpulla, R.C.; Chandel, N.S. Oxygen sensing requires mitochondrial ROS but not oxidative phosphorylation. *Cell Metab.* **2005**, *1*, 409–414. [CrossRef] [PubMed]
13. Fuhrmann, D.C.; Brüne, B. Mitochondrial composition and function under the control of hypoxia. *Redox Biol.* **2017**, *12*, 208–215. [CrossRef] [PubMed]
14. Chandel, N.S.; McClintock, D.S.; Feliciano, C.E.; Wood, T.M.; Melendez, J.A.; Rodriguez, A.M.; Schumacker, P.T. Reactive oxygen species generated at mitochondrial Complex III stabilize hypoxia-inducible factor-1α during hypoxia: A mechanism of O_2 sensing. *J. Biol. Chem.* **2000**, *275*, 25130–25138. [CrossRef]
15. Chandel, N.S. Mitochondrial complex III: An essential component of universal oxygen sensing machinery? *Respir. Physiol. Neurobiol.* **2010**, *174*, 175–181. [CrossRef] [PubMed]
16. Guzy, R.D.; Hoyos, B.; Robin, E.; Chen, H.; Liu, L.; Mansfield, K.D.; Simon, M.C.; Hammerling, U.; Schumacker, P.T. Mitochondrial complex III is required for hypoxia-induced ROS production and cellular oxygen sensing. *Cell Metab.* **2005**, *1*, 401–408. [CrossRef]
17. Cooley, J.W. Protein conformational changes involved in the cytochrome bc1 complex catalytic cycle. *Biochim. Biophys. Acta-Bioenerg.* **2013**, *1827*, 1340–1345. [CrossRef]
18. Cramer, W.A.; Hasan, S.S.; Yamashita, E. The Q cycle of cytochrome bc complexes: A structure perspective. *Biochim. Biophys. Acta-Bioenerg.* **2011**, *1807*, 788–802. [CrossRef]
19. Guzy, R.D.; Schumacker, P.T. Oxygen sensing by mitochondria at complex III: The paradox of increased reactive oxygen species during hypoxia. *Exp. Physiol.* **2006**, *91*, 807–819. [CrossRef]
20. Quinlan, C.L.; Gerencser, A.A.; Treberg, J.R.; Brand, M.D. The mechanism of superoxide production by the antimycin-inhibited mitochondrial Q-cycle. *J. Biol. Chem.* **2011**, *286*, 31361–31372. [CrossRef]
21. Jung, H.J.; Shim, J.S.; Lee, J.; Song, Y.M.; Park, K.C.; Choi, S.H.; Kim, N.D.; Yoon, J.H.; Mungai, P.T.; Schumacker, P.T.; et al. Terpestacin inhibits tumor angiogenesis by targeting UQCRB of mitochondrial complex III and suppressing hypoxia-induced reactive oxygen species production and cellular oxygen sensing. *J. Biol. Chem.* **2010**, *285*, 11584–11595.
22. Jung, H.J.; Kim, Y.; Chang, J.; Kang, S.W.; Kim, J.H.; Kwon, H.J. Mitochondrial UQCRB regulates VEGFR2 signaling in endothelial cells. *J. Mol. Med.* **2013**, *91*, 1117–1128. [CrossRef] [PubMed]
23. Jung, N.; Kwon, H.J.; Jung, H.J. Downregulation of mitochondrial UQCRB inhibits cancer stem cell-like properties in glioblastoma. *Int. J. Oncol.* **2018**, *52*, 241–251. [CrossRef] [PubMed]
24. Dai, X.; Cheng, H.; Bai, Z.; Li, J. Breast cancer cell line classification and Its relevance with breast tumor subtyping. *J. Cancer* **2017**, *8*, 3131–3141. [CrossRef] [PubMed]
25. Monti, E.; Marras, E.; Prini, P.; Gariboldi, M.B. Luteolin impairs hypoxia adaptation and progression in human breast and colon cancer cells. *Eur. J. Pharmacol.* **2020**, *881*, 173210. [CrossRef]
26. Ye, J.; Jiang, Z.; Chen, X.; Liu, M.; Li, J.; Liu, N. Electron transport chain inhibitors induce microglia activation through enhancing mitochondrial reactive oxygen species production. *Exp. Cell Res.* **2016**, *340*, 315–326. [CrossRef] [PubMed]
27. Young, T.A.; Cunningham, C.C.; Bailey, S.M. Reactive oxygen species production by the mitochondrial respiratory chain in isolated rat hepatocytes and liver mitochondria: Studies using myxothiazol. *Arch. Biochem. Biophys.* **2002**, *405*, 65–72. [CrossRef]
28. Luo, J.; Li, N.; Robinson, J.P.; Shi, R. The increase of reactive oxygen species and their inhibition in an isolated guinea pig spinal cord compression model. *Spinal Cord.* **2002**, *40*, 656–665. [CrossRef]
29. Zielonka, J.; Vasquez-Vivar, J.; Kalyanaraman, B. Detection of 2-hydroxyethidium in cellular systems: A unique marker product of superoxide and hydroethidine. *Nat. Protoc.* **2008**, *3*, 8–21. [CrossRef]
30. Osti, D.; Marras, E.; Ceriani, I.; Grassini, G.; Rubino, T.; Viganò, D.; Parolaro, D.; Perletti, G. Comparative analysis of molecular strategies attenuating positional effects in lentiviral vectors carrying multiple genes. *J. Virol. Methods* **2006**, *136*, 93–101. [CrossRef]
31. Neve, R.M.; Chin, K.; Fridlyand, J.; Yeh, J.; Baehner, F.L.; Fevr, T.; Clark, L.; Bayani, N.; Coppe, J.-P.; Tong, F.; et al. A collection of breast cancer cell lines for the study of functionally distinct cancer subtypes. *Cancer Cell* **2006**, *10*, 515–527. [CrossRef] [PubMed]
32. Koh, M.Y.; Spivak-Kroizman, T.R.; Powis, G. HIF-1 regulation: Not so easy come, easy go. *Trends Biochem. Sci.* **2008**, *33*, 526–534. [CrossRef]
33. Naranjo-Suarez, S.; Carlson, B.A.; Tsuji, P.A.; Yoo, M.-H.; Gladyshev, V.N.; Hatfield, D.L. HIF-independent regulation of thioredoxin reductase 1 contributes to the high levels of reactive oxygen species induced by hypoxia. *PLoS ONE* **2012**, *7*, e30470. [CrossRef]
34. Kallergi, G.; Papadaki, M.A.; Politaki, E.; Mavroudis, D.; Georgoulias, V.; Agelaki, S. Epithelial to mesenchymal transition markers expressed in circulating tumour cells of early and metastatic breast cancer patients. *Breast Cancer Res.* **2011**, *13*, R59. [CrossRef] [PubMed]
35. Medici, D.; Hay, E.D.; Olsen, B.R. Snail and slug promote epithelial-mesenchymal transition through β-catenin-T-cell factor-4-dependent expression of transforming growth factor-β3. *Mol. Biol. Cell.* **2008**, *19*, 4875–4887. [CrossRef] [PubMed]
36. Zhao, R.-Z.; Jiang, S.; Zhang, L.; Yu, Z.-B. Mitochondrial electron transport chain, ROS generation and uncoupling (Review). *Int. J. Mol. Med.* **2019**, *44*, 3–15. [CrossRef] [PubMed]
37. Lin, X.; David, C.A.; Donnelly, J.B.; Michaelides, M.; Chandel, N.S.; Huang, X.; Warrior, U.; Weinberg, F.; Tormos, K.V.; Fesik, S.W.; et al. A chemical genomics screen highlights the essential role of mitochondria in HIF-1 regulation. *Proc. Natl. Acad. Sci. USA* **2008**, *105*, 174–179. [CrossRef]

38. Le Gal, K.; Wiel, C.; Ibrahim, M.X.; Henricsson, M.; Sayin, V.I.; Bergo, M.O. Mitochondria-targeted antioxidants MitoQ and mitoTEMPO do not influence BRAF-driven malignant melanoma and KRAS-driven lung cancer progression in mice. *Antioxidants* **2021**, *10*, 163. [CrossRef]
39. Porporato, P.E.; Payen, V.L.; Pérez-Escuredo, J.; De Saedeleer, C.J.; Danhier, P.; Copetti, T.; Dhup, S.; Tardy, M.; Vazeille, T.; Bouzin, C.; et al. A mitochondrial switch promotes tumor metastasis. *Cell Rep.* **2014**, *8*, 754–766. [CrossRef]
40. Zhou, G.; Dada, L.A.; Wu, M.; Kelly, A.; Trejo, H.; Zhou, Q.; Varga, J.; Sznajder, J.I. Hypoxia-induced alveolar epithelial-mesenchymal transition requires mitochondrial ROS and hypoxia-inducible factor 1. *Am. J. Physiol.-Lung Cell Mol. Physiol.* **2009**, *297*, L1120–L1130. [CrossRef]
41. Jiang, J.; Wang, K.; Chen, Y.; Chen, H.; Nice, E.C.; Huang, C. Redox regulation in tumor cell epithelial–mesenchymal transition: Molecular basis and therapeutic strategy. *Signal Transduct. Target. Ther.* **2017**, *2*, 17036. [CrossRef] [PubMed]
42. Azimi, I.; Petersen, R.M.; Thompson, E.W.; Roberts-Thomson, S.J.; Monteith, G.R. Hypoxia-induced reactive oxygen species mediate N-cadherin and SERPINE1 expression, EGFR signalling and motility in MDA-MB-468 breast cancer cells. *Sci. Rep.* **2017**, *7*, 15140. [CrossRef]
43. Rakowski, M.; Porębski, S.; Grzelak, A. Silver nanoparticles modulate the epithelial-to-mesenchymal transition in estrogen-dependent breast cancer cells in vitro. *Int. J. Mol. Sci.* **2021**, *22*, 9203. [CrossRef] [PubMed]
44. Shagieva, G.; Domnina, L.; Makarevich, O.; Chernyak, B.; Skulachev, V.; Dugina, V. Depletion of mitochondrial reactive oxygen species downregulates epithelial-to-mesenchymal transition in cervical cancer cells. *Oncotarget* **2017**, *8*, 4901–4913. [CrossRef] [PubMed]
45. Lehmann, B.D.; Bauer, J.A.; Chen, X.; Sanders, M.E.; Chakravarthy, A.B.; Shyr, Y.; Pietenpol, J.A. Identification of human triple-negative breast cancer subtypes and preclinical models for selection of targeted therapies. *J. Clin. Invest.* **2011**, *121*, 2750–2767. [CrossRef]
46. Kim, R.-K.; Suh, Y.; Yoo, K.-C.; Cui, Y.-H.; Kim, H.; Kim, M.-J.; Kim, I.G.; Lee, S.-J. Activation of KRAS promotes the mesenchymal features of Basal-type breast cancer. *Exp. Mol. Med.* **2015**, *47*, e137. [CrossRef] [PubMed]
47. Jo, M.; Lester, R.D.; Montel, V.; Eastman, B.; Takimoto, S.; Gonias, S.L. Reversibility of epithelial-mesenchymal transition (EMT) induced in breast cancer cells by activation of urokinase receptor-dependent cell signaling. *J. Biol. Chem.* **2009**, *284*, 22825–22833. [CrossRef] [PubMed]
48. Bleier, L.; Dröse, S. Superoxide generation by complex III: From mechanistic rationales to functional consequences. *Biochim. Biophys. Acta-Bioenerg.* **2013**, *1827*, 1320–1331. [CrossRef] [PubMed]
49. Brand, M.D. The sites and topology of mitochondrial superoxide production. *Exp. Gerontol.* **2010**, *45*, 466–472. [CrossRef]
50. Patten, D.A.; Lafleur, V.N.; Robitaille, G.A.; Chan, D.A.; Giaccia, A.J.; Richard, D.E. Hypoxia-inducible factor-1 activation in nonhypoxic conditions: The essential role of mitochondrial-derived reactive oxygen species. *Mol. Biol. Cell.* **2010**, *21*, 3247–3257. [CrossRef]
51. Klimova, T.; Chandel, N.S. Mitochondrial complex III regulates hypoxic activation of HIF. *Cell Death Differ.* **2008**, *15*, 660–666. [CrossRef] [PubMed]
52. Bell, E.L.; Klimova, T.A.; Eisenbart, J.; Moraes, C.T.; Murphy, M.P.; Budinger, G.R.S.; Chandel, N.S. The Qo site of the mitochondrial complex III is required for the transduction of hypoxic signaling via reactive oxygen species production. *J. Cell. Biol.* **2007**, *177*, 1029–1036. [CrossRef] [PubMed]
53. Theys, J.; Jutten, B.; Habets, R.; Paesmans, K.; Groot, A.J.; Lambin, P.; Wouters, B.G.; Lammering, G.; Vooijs, M. E-Cadherin loss associated with EMT promotes radioresistance in human tumor cells. *Radiother. Oncol.* **2011**, *99*, 392–397. [CrossRef] [PubMed]
54. Zhu, L.; Zhao, Q. Hypoxia-inducible factor 1α participates in hypoxia-induced epithelial-mesenchymal transition via response gene to complement 32. *Exp. Ther. Med.* **2017**, *14*, 1825–1831. [CrossRef]
55. Tang, C.; Tianjie, L.I.U.; Wang, K.; Wang, X.; Shan, X.U.; Dalin, H.E.; Zeng, J. Transcriptional regulation of FoxM1 by HIF-1α mediates hypoxia-induced EMT in prostate cancer. *Oncol. Rep.* **2019**, *42*, 1307–1318. [CrossRef]
56. Lee, A.; Djamgoz, M.B.A. Triple negative breast cancer: Emerging therapeutic modalities and novel combination therapies. *Cancer Treat. Rev.* **2018**, *62*, 110–122. [CrossRef]
57. Yu, K.-D.; Zhu, R.; Zhan, M.; Rodriguez, A.A.; Yang, W.; Wong, S.; Makris, A.; Lehmann, B.D.; Chen, X.; Mayer, I.; et al. Identification of prognosis-relevant subgroups in patients with chemoresistant triple-negative breast cancer. *Clin. Cancer Res.* **2013**, *19*, 2723–2733. [CrossRef]

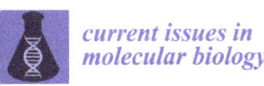

current issues in molecular biology

Article

Presence of Human Papillomavirus DNA in Malignant Neoplasia and Non-Malignant Breast Disease

Erika Maldonado-Rodríguez [1,2], Marisa Hernández-Barrales [2], Adrián Reyes-López [2], Susana Godina-González [1,2], Perla I. Gallegos-Flores [3], Edgar L. Esparza-Ibarra [3], Irma E. González-Curiel [1,2], Jesús Aguayo-Rojas [2], Adrián López-Saucedo [4], Gretel Mendoza-Almanza [5,*] and Jorge L. Ayala-Luján [1,2,*]

1 Master in Science and Chemical Technology, Autonomous University of Zacatecas, Zacatecas 98160, Mexico
2 Academic Unit of Chemical Sciences, Autonomous University of Zacatecas, Zacatecas 98160, Mexico
3 Academic Unit of Biological Sciences, Autonomous University of Zacatecas, Zacatecas 98068, Mexico
4 Health Sciences Area, Autonomous University of Zacatecas, Zacatecas 98160, Mexico
5 National Council of Science and Technology, Autonomous University of Zacatecas, Zacatecas 98000, Mexico
* Correspondence: grmendoza@uaz.edu.mx (G.M.-A.); jayala69@uaz.edu.mx (J.L.A.-L.)

Abstract: Breast cancer is the leading cause of cancer death among women worldwide. Multiple extrinsic and intrinsic factors are associated with this disease's development. Various research groups worldwide have reported the presence of human papillomavirus (HPV) DNA in samples of malignant breast tumors. Although its role in mammary carcinogenesis is not fully understood, it is known that the HPV genome, once inserted into host cells, has oncogenic capabilities. The present study aimed to detect the presence of HPV DNA in 116 breast tissue biopsies and classify them according to their histology. It was found that 50.9% of the breast biopsies analyzed were malignant neoplasms, of which 74.6% were histologically classified as infiltrating ductal carcinoma. In biopsies with non-malignant breast disease, fibroadenoma was the most common benign neoplasm (39.1%). Detection of HPV DNA was performed through nested PCR using the external primer MY09/11 and the internal primer GP5+/6+. A hybridization assay genotyped HPV. HPV DNA was identified in 20.3% (12/59) of malignant neoplasms and 35% non-malignant breast disease (16/46). It was also detected in 27.3% (3/11) of breast tissue biopsies without alteration. However, there are no statistically significant differences between these groups and the existence of HPV DNA ($p = 0.2521$). Its presence was more frequent in non-malignant alterations than in malignant neoplasias. The most frequent genotypes in the HPV-positive samples were low-risk (LR) HPV-42 followed by high-risk (HR) HPV-31.

Keywords: HPV DNA in breast; nested PCR; breast cancer

Citation: Maldonado-Rodríguez, E.; Hernández-Barrales, M.; Reyes-López, A.; Godina-González, S.; Gallegos-Flores, P.I.; Esparza-Ibarra, E.L.; González-Curiel, I.E.; Aguayo-Rojas, J.; López-Saucedo, A.; Mendoza-Almanza, G.; et al. Presence of Human Papillomavirus DNA in Malignant Neoplasia and Non-Malignant Breast Disease. *Curr. Issues Mol. Biol.* **2022**, *44*, 3648–3665. https://doi.org/10.3390/cimb44080250

Academic Editor: Dumitru A Iacobas

Received: 22 June 2022
Accepted: 3 August 2022
Published: 13 August 2022

1. Introduction

Breast cancer is the most common and fatal cancer among women in developed and developing countries. According to data from the World Health Organization, 2,261,419 new breast cancer cases were reported worldwide in 2020, as well as 684,996 deaths [1].

It is known that many factors are involved in the development of breast cancer, such as the environment, age, hormones, alcohol consumption, fat in the diet, a diet poor in fruits and vegetables, family history, obesity, tabaquism, alcoholism, number of offspring, breastfeeding, estrogen levels, estrogen receptors [2,3], among others.

Viruses are considered a controversial etiological risk factor for breast cancer. Viral DNA from human papillomaviruses (HPV), Epstein–Barr virus (EBV), human cytomegalovirus (HCMV), herpes simplex virus (HSV), and human herpesvirus type 8. (HHV-8) has been found in healthy and breast cancer samples [4,5]. However, these results show no pattern, even within the same country, and some are contradictory; moreover, there is no proof of viral breast carcinogenesis [6].

Since Bittner, in 1943, identified the mouse mammary tumor virus (MMTV) as the etiological agent of breast cancer in mice [7], several research groups around the world have been interested in finding a similar relationship between human breast cancer and a viral etiologic agent. In 1995, Wang et al. identified the *env* gene sequence, which codes for the MMTV envelope protein, in 38% of 314 breast neoplasms [8]. In subsequent years, the same research group has worked tirelessly to find a relationship between the onset of breast cancer and infection by the human mammary tumor virus (HMTV). Among other interesting data, they reported the expression of sequences of several proteins from the capsid and envelope of the HMTV virus in ten primary cultures of human breast cancer [9].

In 2017, Islam et al. reported a pattern in the presence of HPV in normal and benign tumors and a markedly increased presence in malignant breast tumors, indicating its pathological importance in breast cancer. HPV was also associated with poor hygienic conditions and patient malnutrition, together with ethnicity [10].

Integrating the HPV genome into the host genome may cause chromosomal instability and trigger carcinogenesis [6,11,12]. Identifying HPV DNA in breast cancer samples suggests the possible role of HPV as a mutagen that promotes breast oncogenesis. However, the prevalence of HPV in breast cancer samples reported by several research groups varies widely, ranging from 0% to 86%. It is often difficult to determine the presence of HPV due to the low viral load in samples or paraffin-embedded tissue, as well as the diversity of techniques employed such as hybridization in situ (HIS), Polymerase Chain Reaction (PCR), Nested PCR, quantitative real-time PCR (RT-qPCR), and Next Generation Sequencing (NGS), among others. Table 1 summarizes the results of HPV DNA found in breast cancer samples worldwide.

Table 1. HPV DNA found in breast cancer samples worldwide.

Country	Sample	Method	Control/+HPV	Cases/+HPV	Breast Pathology Predominant	VPH Predominant	Reference
UK	FST	PCR (L1, E7), SB	NS	80/0	IC	NS	[13]
USA	PET	PCR (E6), DB	15/0	28/0	PC	NS	[14]
India	PET	PCR (E6, URR), SB	NS	30/0	IDC	NS	[15]
Austria	PET	PCR (L1), DB	NS	20/0	PD	NS	[16]
Switzerland	PET	PCR (L1)	NS	81/0	IC	NS	[17]
France	PET	PCR (L1)	NS	50/0	IDC	NS	[18]
Tunisia	PET/FST	PCR (L1, E1, E6, E7), ISH	NS	123/0	IDC	NS	[19]
India	FST	PCR (L1, E6, E7)	NS	228/0	IDC	NS	[20]
Brazil	PET	PCR (L1)	NS	79/0	IDC	NS	[21]
China	FroST	PCR (L1)	77/0	77/0	IDC	NS	[22]
Spain	PET	PCR (L1), DEIA	2/0	76/0	IDC	NS	[23]
Greece	FroST	MA (E1)	NS	201/0	IDC	NS	[24]
Italy	PET	PCR (L1,E6), ISH	NS	40/12, 12/0	IC	16	[25]
Norway	PET	PCR (L1,E6)	NS	41/19	IC	16	[26]
Brazil	PET	PCR (E6)	41/0	101/25	IC	16	[27]
Australia	PET	PCR (L1)	NS	11/7	IC	16	[28]
USA	PET	PCR (L1), SEQ, ISH	NS	29/25	IC	11	[29]
Australia	FST	PCR (E6), SEQ	NS	50/24	IDC	18	[30]
Greece	FroST	PCR (L1,E6,E4) RFLP	NS	107/17	IDC	16	[31]
Turkey	FST	PCR (L1, E6,E7)	50/16	50/37	IDC	18	[32]
Syria	PET	PCR (E1), TMA	NS	113/69	IC	33	[33]
Japan	PET	PCR (E6)	11/0	124/26	IC	16	[34]
Mexico	PET	PCR, SEQ	40/0	67/3	IDC	16, 18, 31, 33, 6	[35]

Table 1. *Cont.*

Country	Sample	Method	Control/+HPV	Cases/+HPV	Breast Pathology Predominant	VPH Predominant	Reference
Mexico	PET	PCR (L1), SEQ	43/0	51/15	IDC	16	[36]
Australia	PET	PCR (L1), SEQ, ISPCR	17/3	26/8	IDC,	18	[37]
Mexico	PET	PCR (L1), RT-qPCR	NS	70/17	IDC	16	[38]
Chile	PET	PCR (L1), RT-qPCR	NS	46/4	IDC	16	[39]
China	FroTS	PCR (L1), DB, SEQ	46/0	62/4	IC	16	[40]
Australia	FroTS	PCR (L1), ISH, SEQ	NS	54/27	IDC	18	[41]
Iran	PET	PCR (L1), SEQ	41/1	58/1	IDC	16, 18	[42]
Mexico	PET	PCR (L1)	NS	20/8	MBC	16	[43]
Argentina	FST	PCR (L1)	NS	61/16	IDC	11	[44]
China	FST	HCA	37/6	224/48	IC	NS	[45]
Iraq	PET	ISH	24/320/0	129/60	IC	31	[46]
Italy	PET	INNO-LIPPA (L1)	40/0	40/6	IDC	16	[47]
Iran	PET	INNO-LIPPA (L1)	51/7	55/10	IC	16	[48]
China	DS	PCR, MS	50/0	100/2	IDC	18	[49]
Iran	PET	PCR (L1), SEQ	65/0	65/22	IDC	6	[50]
China	PET	PCR (E7), ISH	83/1	169/25	IDC	58	[51]
Australia	FroST	PCR (L1)	10/1	80/13	IDC	NS	[52]
China	PET	PCR (L1), SEQ	92/0	187/3	IDC	16	[53]
Venezuela	FST	INNO-LIPPA (L1)	NS	24/10	IDC	51	[54]
Australia	PET	PCR (L1), SEQ	18/3	28/13	IC	18	[55]
Corea	PET	PCR	NS	123/22	IDC	51	[56]
Pakistan	PET	PCR (L1)	NS	46/8	IDC	16	[57]
Iran	PET	PCR (L1)	NS	84/27	IDC	16	[58]
China	PET	PCR (E6, E7)	NS	76/23	IDC	18	[59]
Spain	PET	PCR (L1)	186/49	251/130	IC	16	[4]
Thailand	PET	PCR (L1)	350/10	350/15	IDC	16	[60]
India	FST	PCR (L1)	21/2	313/203	IDC	16	[10]
UK	FST	PCR (L1)	36/11	74/35	IC	16	[61]
China	FST	HCA	NS	81/14	IDC	NS	[62]
Pakistan	PET	PCR (L1)	NS	250/45	IDC	NS	[63]
Brazil	PET	PCR (L1)	95/15	103/51	NS	6/11	[64]
Iran	PET	PCR (L1), MA	NS	72/4	IDC	NS	[65]
Morocco	FroST	TS-MPG	12/1	76/19	IDC	11	[66]
Rwuanda	PET	PCR (L1)	NS	47/22	IDC	16	[67]
Denmark	PET	PCR (E6, E7), RH	100/3	93/1	IDC	16	[68]
Iran	PET	RT-qPCR (L1)	40/0	98/8	NS	16,18	[69]
Iran	FroST	PCR (L1, E7)	31/5	72/35	IDC	18	[70]
Italy	PET	PCR (L1), ISH, MS	NS	273/80	IC	16	[71]
USA	PET	PCR (L1), MA	27/8	18/8	IP	11	[72]
Egypt	FroST	RT-qPCR (E6)	15/0	20/4	IDC	16	[73]
Qatar	FST	TS-MPG	50/4	50/10	IDC	16, 35	[74]
Egypt	PET, FST	PCR (L1)	30/0	80/33	IDC	NS	[75]
Sudan	PET	PCR	NS	150/13	NS	16	[76]
Qatar	PET	PCR (E6,E7)	NS	74/48	IDC	52	[77]

Paraffin-embedded tissue: PET; Fresh samples tissue: FST; Frozen samples tissue: FroST; Polymerase Chain Reaction: PCR; In Situ Hybridization: ISH; Tissue Microarray: TMA; Hybrid Capture Assay: HCA; Type-Specific Polymerase Chain Reaction bead-based multiplex genotyping assay: TS-MPG; Microarray: MA; Quantitative Real-Time Polymerase Chain Reaction: RT-qPCR; Sequencing: SEQ; In situ PCR: ISPCR; Dot blot hybridization: DB; Reverse hybridization: RH; Diverse samples (blood, cancer tissue, axillary lymph nodes, normal tissue): DS; Mass spectrometry: MS; Intraductal papilloma: IP; Southern blot: SB; Papillary carcinoma: PC; Paget's disease: PD; DNA enzyme immunoassay: DEIA; Restriction fragment length polymorphism: RFLP; Upstream Regulatory Region: URR; Not Specified: NS; Invasive Carcinoma: IC; Invasive ductal carcinoma: IDC; Metaplasia breast carcinoma: MBC; Human Papillomavirus protein L1: L1.

Persistent infection with HR-HPV is considered one of the main causative biological factors in developing cervical cancer (CC). HR-HPV 16 and 18 are responsible for more than 65–75% of precancerous cervical lesions and CC. Furthermore, HPV is associated with carcinomas such as head and neck, anal, vulva, oral, vagina, and penile cancer [78].

Cervical cancer is the third most common type of malignant tumor and the fourth cause of cancer death among women worldwide. It is also one of the deadliest cancers among women in underdeveloped countries [1]. In Mexico, it is the second-highest cause of cancer death in women due mainly to poor clinical diagnosis in the early stages of the disease and the wide distribution of HR-HPV throughout the country. The first cause of cancer death among women in Mexico and worldwide is breast cancer. For this reason, finding that HPV is an etiological factor for breast cancer would have a high impact on public health programs in Mexico and countries with the highest rates of women mortality from cervical cancer and breast cancer.

However, the oncogenic role of HPV in the development of breast cancer has not yet been clarified, so this study aimed to determine the prevalence of high- and low-risk HPV in breast biopsies diagnosed with benign-alteration and malignant-alteration neoplasms from Mexican women.

2. Materials and Methods

2.1. Sample Collection and Classification

A total of 116 formalin-fixed, paraffin-embedded breast samples from 2009 to 2019 were used for the present study. The remaining tissues were donated to and collected by the Mexican Social Security Institute in Zacatecas. The diagnosis associated with each sample was confirmed by histopathological diagnosis using hematoxylin-eosin (HE) staining and classified according to the World Health Organization (WHO) classification system [1]. The study was approved by the Institutional Ethics Committee of the Autonomous University of Zacatecas and the Mexican Social Security Institute, Zacatecas, and carried out following the guidelines of the Helsinki Declaration.

2.2. Histological Diagnosis

Fresh biopsies were treated with formaldehyde immediately after surgical removal and processed for inclusion in paraffin. Tissue sections were cut and stained with hematoxylin-eosin (HE) for observation under an optical microscope. The pathology specialist performed the analysis and made the histological and clinical diagnoses.

2.3. DNA Extraction and Amplification

DNA purification was performed using a QIAmp® FFPE Tissue kit (QIAGEN, Hilden, Germany 56404). Ten 5 µm tissue sections of FFPE breast samples were cut, deparaffinized by incubation with xylene, and washed and rehydrated with ethanol. After complete deparaffinization, the samples were digested with proteinase K at 56 °C for one hour and inactivated at 90 °C. The amount and quality of the DNA were evaluated using a UV-VIS spectrophotometer Q500 (Quawell®) at 260–280 nm. The integrity of the extracted DNA and the absence of PCR inhibitors were assessed by polymerase chain reaction (PCR) amplification of the β-globin gene using 5 µM of primers KM29/PCO4 (Table 2) and 50 ng of DNA in a total reaction volume of 25 µL containing: 2.5 µL PCR Buffer (10×, 1.5 µL $MgCl_2$ (25 mM), 1 U Taq DNA polymerase (Thermo Fisher Scientific Waltham Massachussetts® EPO402), 0.5 µL of dNTP (10 mM), and water. The amplification of the β-globin gene was performed under the following conditions: initial activation of the enzyme at 95 °C for 2 min, followed by 40 cycles under the following conditions: 95 °C for 30 s, 55.4 °C for 30 s, and 72 °C for 30 s, with a final elongation step at 72 °C for 5 min. The amplicon was visualized in agarose gel (1.5%) stained with ethidium bromide. The images were digitally processed using the Electrophoresis Documentation and Analysis System 120 (Kodak Digital Science).

Table 2. Primers used to amplify ß globin fragment and L1 VPH fragment.

Primer	Sequence 5′-3′	Gene Fragment	Size (pb)
KM29	GGTTGGCCAATCTACTCCCAGG	β-globin	205
PCO4	CAACTTCATCCACGTTACCC	β-globin	205
MY09 *	CGTCCMARRGGAWACTGATC	L1 VPH	450
MY11 *	GCMCAGGGWCATAAYAATGG	L1 VPH	450
GP5+	TTTGTTACTGTGGTAGATACTAC	L1 VPH	140–150
GP6+	GAAAAATAAACTGTAAATCATATTC	L1 VPH	140–150

* M = A + C, W = A + T, Y = C + T, R = A + G.

2.4. Detection and Genotyping of HPV

The detection of HPV DNA was first carried out by screening all the samples by end-point PCR using the primers GP5+/6+, which generated a 150 pb fragment. Subsequently, a nested PCR was performed on the samples that tested negative for HPV to increase sensitivity.

Genomic DNA samples from the cervical cancer cell lines SiHa and Caski were used as positive controls for MY09/11 and GP5+/6+ amplification. A paraffin block without tissue and a PCR mix without DNA were negative controls. The primers used are reported in Table 2.

2.4.1. Nested PCR Conditions

For the nested PCR, MY09/11 primers were used to obtain the first amplicon of 450 bp. Subsequently, GP5+/GP6+ primers were used on the first amplicon. The first PCR reaction was carried out with 100 ng of DNA in a total reaction volume of 25 µL containing 2.5 µL of buffer (10×), 1.5 µL of MgCl2 (25 mM), 0.5 µL of each MY09/MY11 primer (10 µM) (Table 2), 0.5 µL of dNTPs (10 mM), 0.25 µL of Taq polymerase (5U/µL) (Thermo®, EPO402), and water. The amplification was performed under the following conditions: initial activation of the enzyme at 95 °C for 3 min, followed by 39 cycles under the following conditions: 95 °C for 30 s, 57 °C for 30 s, and 72 °C for 45 s with a final elongation step at 72 °C for 5 min. The second PCR reaction was performed with 5 µL of the first amplicon and the GP5+/6+ primers. The initial activation of the enzyme was performed at 95 °C for 3 min, followed by 39 cycles under the following conditions: 95 °C for 30 s, 48 °C for 30 s, and 72 °C for 30 s with an elongation step at 72 °C for 5 min. The amplicon products were visualized in agarose gel (2%) stained with ethidium bromide.

2.4.2. qPCR Conditions

Quantitative PCR was performed when the samples had a DNA concentration lower than 10 ng/ul. The first amplicon was amplified using primers MY09/11 followed by qPCR. The qPCR was carried out in a 7500 Fast Real-Time PCR System (Applied Biosystems Foster City, California™) in a total reaction volume of 25 µL containing 5 µL of the first amplicon, Platinum SYBR Green qPCR SuperMix-VGD (platinum Taq DNA polymerase, SYBR Green I dye, Tris-HCl, KCl, 6 mM MgCl2, 400 µM dNTPs, UDG), 0.5 µL of each the GP5+/6+ primers (10 µM), 0.1 ul of ROX Reference Dye Solution (25 µM), and water. The amplification was performed under the following conditions: 50 °C for 120 s, 95 °C for 120 s, followed by 40 cycles under the following conditions: 95 °C for 15 s, 48.4 °C for 30 s, 60 °C for 30 s. The data were analyzed using Applied Biosystems 7500 Software v2.0.6, Foster City, California.

2.4.3. HPV Genotyping

Samples positive for amplified HPV L1 gene DNA were subjected to genotyping using the LCD-Array HPV-Type 3.5 kit (Chipron GmbH, Berlin, Germany), which allows for the identification of 32 low-, intermediate-, and high-risk HPV types (6, 11, 16, 18, 31, 33, 35, 39, 42, 44, 45, 51, 52, 53, 54, 56, 58, 59, 61, 62, 66, 67, 68, 70, 72, 73, 81, 82, 83, 84, 91, and 91).

The amplicons generated previously from the biotinylated primers were used for hybridization on the chip: (1) MY09/MY11, which generated a fragment of approximately 450 bp. (2) "125" primers, an internal sequence of the 450 bp fragment (the kit's own). Both amplicons were combined before hybridization. The PCR was performed according to the manufacturer's protocol. The reaction mixture was prepared using 2.5 μL buffer (10×), 2 μL $MgCl_2$ (25 mM), 1 μL primer mix MY09/MY11 or 2 μL primer mix 125, 1 μL dNTPs (10 mM), 0.3 μL of Taq polymerase (5 U/μL) (Thermo®, EPO402), 100 ng of template, and water. Genomic DNA from the CaSki cell line was used as a positive control. The run was conducted as follows: 3 min at 95 °C, followed by 41 cycles of 1 min at 94 °C in denaturation, 1.5 min at 45 °C in alignment, and 1.5 min at 72 °C in extension, and a final extension of 3 min at 72 °C.

Hybridization was carried out according to the manufacturer's protocol. Biotin-labeled PCR products were hybridized with HPV subtype-specific capture probes immobilized on the surface of the LCD chip. After washing, each field was incubated with a secondary solution (enzyme conjugate). The PCR fragments were then hybridized with capture probes, and the place where they joined was revealed with an enzyme substrate that generated a blue precipitate. Data reading was performed using the LCD SlideReader V9 software.

2.5. Statistical Analysis

The breast samples were grouped according to the histological diagnosis. The group measured the presence/absence distribution of HPV DNA by simple counting. Chi-squared tests and Fisher's exact test were used to compare the presence/absence of HPV DNA between histological diagnosis, sex of patients, tumor size (TMN), and clinical stage. SBR scales were compared using Mann–Whitney tests. The age of patients and tumor size (in cm) between HPV-positive and -negative samples were compared using t-student and Mann–Whitney tests, respectively. All statistical tests were performed in GraphPad Prism version 6. Differences were considered significant when the p-value was less than 0.05.

3. Results

3.1. Histopathological Diagnosis

The histopathological diagnosis results from 116 breast samples were classified as shown in Figure 1 and Table 3.

Table 3. Histopathological classification according to World Health Organization.

Classification	n	HPV+	HPV−	% HPV+
Normal mammary tissue	**11**	**3**	**8**	**27.3**
Normal breast tissue	10	2	8	20
Normal mammary lymph node	1	1	0	100
Malignant neoplasm	**59**	**12**	**47**	**20.3**
Infiltrating ductal carcinoma	44	8	36	18.2
Infiltrating lobular carcinoma	8	2	6	25
Mucinous carcinom	1	1	0	100
Ductal carcinoma in situ	5	0	5	0
Metaplastic carcinoma	1	1	0	100
Non-cancerous breast disease	**46**	**16**	**30**	**34.8**
Phyllodes tumors	1	0	1	0
Fibroadenoma	18	7	11	38.9
Adenomyoepithelioma	1	0	1	0
Intraductal papilloma	1	0	1	0
Hyperplasia	4	0	4	0
Mastitis	6	3	3	50
Fibrocystic mastopathy	15	6	9	40
Total	**116**	**31**	**85**	**26.7%**

% HPV+ per classification; number of samples: n.

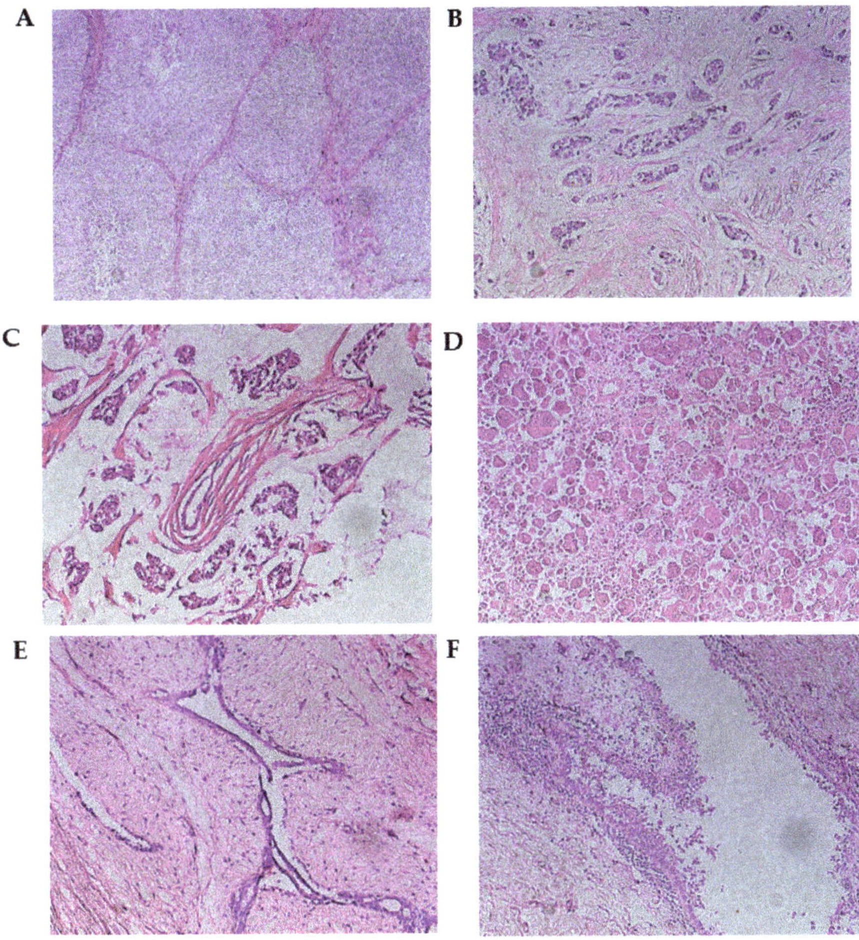

Figure 1. Malignant breast neoplasms. (**A**) Infiltrating ductal carcinoma (solid pattern) separated by connective tissue septa. (**B**) Infiltrating lobular carcinoma. This is characterized by the invasion of the stroma in the form of fine cell cords, called Indian row cords. (**C**) Mucinous carcinoma. Tumor cells are seen within lakes of mucin. (**D**) Metaplastic carcinoma. (**E**) Fibroadenoma. The proliferation of cells can be observed, creating well-defined borders concerning the surrounding normal tissue. (**F**) Mastopathy with hyperplasia. A duct with apocrine metaplasia and foci of hyperplasia can be observed. Hematoxylin and eosin staining. Optical microscopy magnification 20×.

Of all the samples analyzed, 50.9% (59/116) were malignant neoplasms, 4.3% (5/116) were in situ neoplasms, 0.9% (1/116) were borderline neoplasms, and 17.2% (20/116) were benign neoplasms. Benign or non-neoplastic alterations of the mammary gland were also diagnosed, 21.6% (25/116) of the analyzed tissue samples. Breast tissue samples and axillary lymph nodes without alterations were also found, 9.5% (10/116 and 1/116, respectively) of the total.

The distribution of the types of malignant neoplasms diagnosed was as follows: 74.6% (44/59) infiltrating ductal carcinoma, 13.6% (8/59) infiltrating lobular carcinoma, 1.7% (1/59) mucinous carcinoma, and 1.7% (1/59) metaplastic carcinoma. The distribution of benign neoplasms was as follows: 90% (18/20) fibroadenoma, 5% (1/20) adenomyoepithelioma, and 5% (1/20) intraductal papilloma. Regarding non-neoplastic alterations, 60% (15/25) of the samples were diagnosed with cystic fibrous mastopathy, 24% (6/25) corresponded to mastitis, and 16% (4/25) with hyperplasia.

3.2. HPV Presence in Breast Samples

HPV DNA was identified in 20.3% (12/59) of malignant neoplasms and 35% of benign neoplasms (16/46). It was also detected in 27.3% (3/11) of breast tissue biopsies without alteration. It was not detected in in situ neoplasms or borderline neoplasms. Regarding malignant neoplasms, HPV was detected in 18.2% (8/44) of the diagnosed biopsies as infiltrating ductal carcinoma, in 25% (2/8) of the biopsies of infiltrating lobular carcinoma, and was in the only sample diagnosed as mucinous and metaplastic carcinoma. Among benign neoplasms, HPV DNA was only identified in 38.9% (7/18) of the fibroadenoma samples. Regarding benign alterations, HPV was detected in 40% (6/15) of the biopsies with cystic fibrous mastopathy and 50% (3/6) of the biopsies of mastitis (Table 3). No relationship was found between the characteristics of the tumor and the presence of HPV in the sample (Table 4).

Table 4. Relationship between HPV and clinicopathological parameters of breast biopsies.

		VPH		*p* Value
	n (%)	**Positive n (%)**	**Negative n (%)**	
Number of samples	**116 (100)**	**31 (26.7)**	**85 (73.3)**	
Sex				
Male	2 (100)	1 (50)	1 (50)	0.4648 [a]
Female	114 (100)	30 (26.3)	84 (73.7)	
Age (years)	48.9 ± 13.1	46.9 ± 14.4	49.8 ± 12.6	
CI 95%	(45.8–52.1)	(40.4–53.5)	(46.2–53.4)	0.4044 [c]
Range	(17–76)	(17–74)	(18–76)	
Malignant neoplasm	59 (100)	12 (20.3)	47 (79.7)	
Non-cancerous breast disease	46 (100)	16 (34.8)	30 (65.2)	0.2521 [b]
Normal breast	11 (100)	3 (27.3)	8 (72.7)	
Tumor size (cm)	3.5 (16.0–1.0)	3.5 (8.0–1.2)	3.7 (16.0–1.0)	0.6788 [d]
CI 95%	(3.0–4.0)	(2.0–7.8)	(3.0–4.0)	
Tumor size (TNM)				
T1 (≤2 cm)	11 (100)	3 (27.3)	(72.7)	
T2 (>2 cm–5 cm)	28 (100)	6 (21.4)	22 (78.6)	0.4422 [b]
T3 (>5 cm)	6 (100)	2 (33.3)	4 (66.7)	
Not available	9	1	8	
Clinique stage				
EII	39 (100)	9 (23.1)	30 (76.9)	0.6242 [a]
EIII	6 (100)	2 (33.3)	4 (66.7)	
Not available	9	1	8	
Scale SBR	8 (9–4)	8 (9–4)	8 (9–6)	0.2470 [d]
CI 95%	(8–9)	6–9	8–9	
SBR				
3–5: Stage I (well differentiated)	2 (100)	2 (100)	0	U
6–7: Stage II (moderately differentiated)	11 (100)	2 (18.2)	9 (81.8)	U
8–9: Stage III (poorly differentiated)	41 (100)	8 (19.5)	33 (80.5)	U

The Scarff–Bloom–Richardson grade: SBR; Primary Tumor, Regional Lymph Nodes, Distant Metastasis: TNM; number of samples: n; centimeter: cm; Confidence Interval: CI; Probability value: *p*; Not significant *p*-value > 0.05; Significant *p*-value < 0.05; *p* = probability value *p* < 0.05; [a] Fisher test; [b] Pearson test; [c] Unpaired Student's *t* test; [d] Mann–Whitney U test; undeterminated: U.

The most frequent HPV genotype in the samples was LR-HPV 42, which was identified in 19% (6/31) of the analyzed samples, followed by HR-HPV 31 (4/31, 13%) and HR-HPV 59 (3/31, 10%). LR-HPV 44 and HR-HPV 58 were identified in only 6% (2/31) of samples, and HR-HPV 51 in 3% (1/31). Ten percent (3/31) of all HPV-positive samples had mono-infection, while 16% (5/31) had co-infections. Two samples were positive for two virus genotypes, HPV 58/51 (1/31) and HPV 31/42 (1/31) (Table 5). Three samples were positive for more than two genotypes; the most frequent combinations were HPV 31/59/42 (2/31), followed by HPV 42/31/59/44/58 (1/31). The most frequent co-infection was HPV 42/31 (4/31), regardless of the number of genotypes detected per sample.

Table 5. Genotyping of VHP in FFPE breast samples.

	Histological Type	HPV Genotypes Detected
Single genotype n = 3 (10%)	Mucinous carcinoma	44
	Fibroadenoma	42
	Mastitis	42
Multiple genotype n = 5 (16%)	Fibroadenoma	58 + 51
	Fibroadenoma	31 + 42 + 59
	Cystic fibrous mastopathy	31 + 59 + 42
	Cystic fibrous mastopathy	31 + 42
	Mastitis	42 + 31 + 59 + 44 + 58

number of samples: n.

4. Discussion

Breast and cervical cancer are the leading cause of death for women worldwide, mainly in developing countries [1]. HPV is estimated to be associated with more than 5% of all types of carcinomas in humans. High-risk HPV infection has been recognized as an essential factor in developing cervical cancer. It has also been associated with 99.7% of cases of cervical cancer, 50% of head and neck squamous cell carcinomas, and 25% of oropharyngeal cancer [78]. Integration of HPV DNA into the host cell genome is critical in HPV-mediated carcinogenesis, leading to abnormal cell proliferation and malignant progression [10].

In breast cancer, HPV has been proposed in several studies as a probable causative agent of breast cancer carcinogenesis [4,6].

A controversial fact is that HPV DNA has been reported in healthy breast samples. It would be very interesting to make a follow-up study on samples donated by women with healthy breast tissue but positive for HPV DNA to observe if, over the years, they developed some mammary carcinoma, which would support the hypothesis that HPV is an oncogenic factor in breast cancer (Table 1).

The role of HPV in breast cancer carcinogenesis remains controversial due to inconsistent data on the presence of HPV DNA in tumor samples from patients with breast cancer and a lack of clarity regarding the route of HPV transmission from one organ to the other.

The variability of the reported results within the same country could be explained by the quantity and quality of the samples analyzed, considering that breast cancer samples have a lower viral load, making HPV challenging to detect. Other factors that may introduce noise in the study of this subject include the preprocessing of the examined samples, the HPV DNA detection method, and the distribution of HPV among women in each country.

Our results show the presence of HPV DNA in 26.7% (31/116) of the samples, which is in accordance with the findings of other authors in different Latin American countries, whose detection rate ranges from 0 to 49% and the average frequency is 25% [21,27,39,44,54,64] (Table 1). There is a wide range of distribution of genotypes in breast tissue depending on the geographical region. Previous studies in Mexico identified HPV-16, 18, and 33 [35,36,38,43]. Regarding Table 1, it was determined that the five most common genotypes in breast tissue in decreasing order of prevalence are HPV-16, 33, 11, 18, and 6. Similar to the present work, studies conducted in Venezuela and Brazil identified high-risk genotypes HPV-31 and 51 (54,64).

Classified according to their oncogenic characteristics, the prevalence of high-risk HPV types was higher than those with low risk. However, none of these genotypes were identified in this study, and the prevalence of low-risk genotypes was higher than high-risk genotypes. Table 6.

Table 6. Comparative table of HPV reported in breast tissue samples in Latin America.

Country	Sample	Method	Cases/+HPV (%)	HPV Genotype						Reference
Venezuela	FST	PCR (L1), RH	24/10 (41.7)	51	18	33	6	11		[54]
Brazil	PET	PCR (L1), TS-MPG, SEQ	103/51 (49.5)	6	11	18	31	33	52	[64]
Brazil	PET	PCR (E6)	101/25 (24.8)	16	18					[27]
Brazil	PET	PCR (L1)	79/0 (0)							[21]
Argentina	FST	PCR (L1), RT-qPCR, SEQ	61/16 (26.2)	11	16					[44]
Chile	PET	PCR (L1), RT-qPCR	46/4 (8.7)	16						[39]
Mexico	PET	PCR (E1), SEQ	67/3 (4.5)	16	18	33				[35]
Mexico	PET	PCR (L1), SEQ	51/15 (29.4)	16	18					[36]
Mexico	PET	PCR (L1), RT-qPCR	70/17 (24.3)	16	33					[38]
Mexico	PET	PCR (L1), RT-qPCR	20/8 (40)	16	18					[43]
Mexico	PET	PCR (L1), MA	116/31 (26.7)	42	31	59	58	44	51	Present study

Paraffin-embedded tissue: PET; Fresh samples tissue: FST; Polymerase Chain Reaction: PCR; Type-Specific Polymerase Chain Reaction bead-based multiplex genotyping assay: TS-MPG; Microarray: MA; Quantitative Real-Time Polymerase Chain Reaction: RT-qPCR; Sequencing: SEQ; Reverse hybridization: RH; Human Papillomavirus protein L1: L1.

Methodological diversity may partly explain the differences in HPV positivity between studies. However, more importantly, it has been suggested that the viral load of HPV in breast cancer is low [79]. Once cell transformation occurs, viral replication stops, and integration of the viral genome into the host occurs [80]. Under these circumstances, the number of HPV copies decreases sharply. It has been shown that after genome integration, HPV replication decreases; therefore, the choice of detection method and its sensitivity are essential factors to consider since they influence the HPV detection rate [81].

The low prevalence of HPV reported by some studies results from low sensitivity. Therefore, the present study used two variants of the PCR technique to increase the sensitivity and reduce the risk of false negatives. HPV-specific amplicons were detected in 13.8% (16/116) of samples when analyzed by one-step PCR, while the real-time PCR approach increased the positivity rate to 26.7% (31/116).

The differences in HPV prevalence between studies can also be explained by false-positive results, in which contamination is a crucial factor. The present study followed a strict quality control procedure, and the results showed no signs of cross-contamination.

The use of broad-spectrum primers versus specific primers is somewhat controversial since broad-range primers target the HPV L1 gene sequence that could be lost during the integration of the virus into the host genome [27].

It has been suggested that HPV virions present in paraffin-embedded tissue samples may be destroyed during fixation and sample processing. Therefore, HPV may be difficult to detect in tissues preserved for long periods of storage [82]. Some authors suggest fresh tissues may be associated with a higher HPV detection rate compared to samples embedded in paraffin. However, some studies indicate that the low viral load is not a result of tissue samples' fixation and paraffin inclusion since higher viral loads have been found in formalin-fixed, paraffin-embedded samples than in fresh-frozen cervical cancer tissue samples [83]. Several studies confirm that the type of high-risk HPV and the stage at which the cervical intraepithelial lesion is diagnosed could be triggering factors in the development of breast cancer. According to Table 1, HPV16 is the most common genotype detected in both benign and breast cancer tumors. Among the different types of carcinomas, invasive ductal carcinoma is the breast carcinoma in which HPV DNA is most commonly found.

In 1999, Henning et al. reported that 46% of women with a history of HPV-16-positive high-grade cervical intraepithelial neoplasia (CIN III) lesions were correlated with both ductal and lobular breast carcinomas [26]. Widschwendter et al. and Damin et al. found that the presence of HPV-16 DNA in breast cancer is more frequent in women with a history of cervical cancer [27,28].

Yasmeen et al. made an important observation about breast cancer behavior and HPV. They reported that HPV16 is frequently present in invasive and metastatic breast cancer and less frequently in in situ breast cancer [84].

In a retrospective study, Atique et al. (2017) reported that the incidence of breast cancer was higher among 800,000 HPV-infected patients than among the non-HPV-infected population [85].

The mechanism by which HPV can infect mammary gland cells is still unknown. However, two main hypotheses have been proposed; they are summarized in Figure 2. The first one explains that HPV arrives in mammary glands via the lymphatic or blood system through HPV-carrying mononuclear cells present in women with cervical intraepithelial lesions [86]. Other authors conclude that because the HPV life cycle occurs in the epithelial layers, HPV viremia is impossible [43]. The second hypothesis suggests that the mammary gland can be infected with HPV through the skin of the nipple, as demonstrated in the work of Villiers et al. [29], who proposed a retrograde ductal pattern of viral propagation. The exposure of the mammary ducts to the external environment increases the risk of HPV infection since the mammary ducts are open ducts and could serve as an entry point for viral infection. Furthermore, most mammary neoplasms originate from the epithelium of these structures [81]. Sexual transmission is the generally accepted transmission route, although it does not seem to be the only one. Some studies suggest that transmission can occur through hand-mediated contact between the female perineum and the mammary gland, which could occur during sexual activity, or through contact of bodily fluids with nipple fissures, which could serve as an entry point for HPV [4].

Once they have managed to approach the mammary gland cells, the next question is how do HPV viruses penetrate the cells? One hypothesis explains that oncogenic HPV types of the alpha genus use a complex network of proteins for endocytosis and cellular transport, the latter organized by a specific subset of tetraspanins, annexins, and associated proteins such as integrins and EGFRs [87]. Integrins are the extracellular matrix's central receptors and participate in cell–cell interactions [88]. The α6 integrin has been proposed as the main receptor for HPV-16 in cervical cells [43,89]. In breast tissue, α6 integrins are essential molecules that regulate the growth and differentiation of epithelial cells. Their ability to promote cell anchoring, proliferation, survival, migration, and the activation of extracellular matrix-degrading enzymes suggests that they play an essential role in normal mammary morphogenesis and indicates their potential as HPV receptors and tumor progression promoters [43,88]. Another hypothesis is based on the activity of the extracellular vesicles, including exosomes (Exos), microvesicles (MV), and apoptotic bodies (AB), that are released into biofluids by virtually all live cells (Figure 3).

In 2019, de Carolis et al. detected the same HPV genotype in the same patient's extracellular vesicles from serum, breast, and cervical tissue. Therefore, the authors suggested that HPV DNA was associated with mammary malignancies and was transferred to the stromal cells of the gland by extracellular vesicles [71].

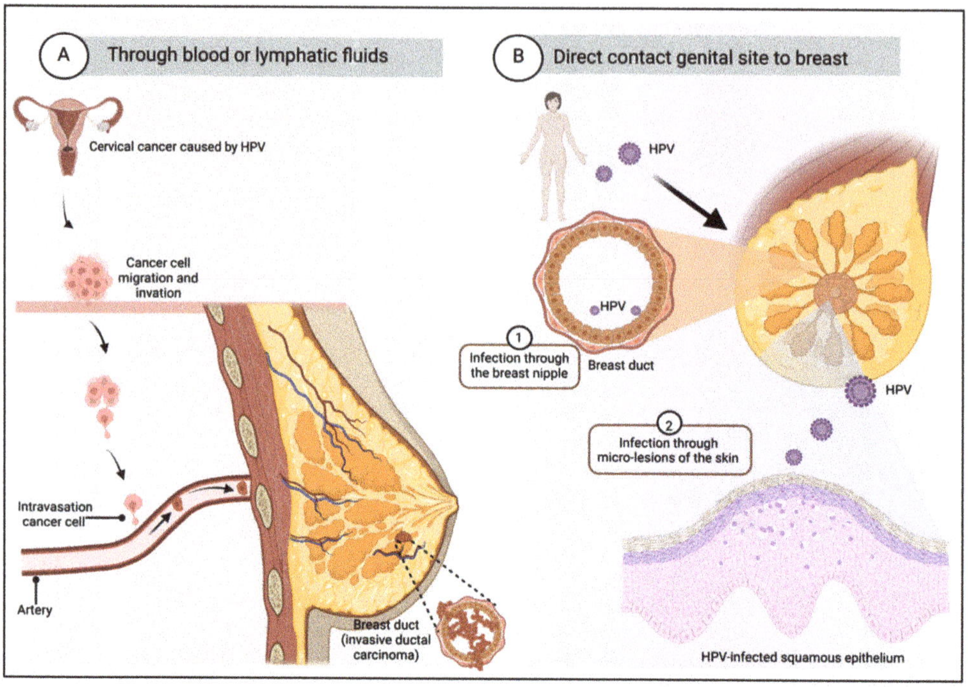

Figure 2. Transmission route of HPV to breast tissue. There are mainly two possible mechanisms: (**A**) Through the blood or lymphatic fluids from the primary site of infection. It is suggested that malignant transformation results from transfection of cells from a primary tumor by plasma flow or that HPV virions can be transported from the site of initial infection to other organs. (**B**) Through the skin of the nipple, by direct contact between the genitals and the breast. The mammary ducts are open ducts and could represent an entry point for virus infection. Transmission can occur through hand contact with the genitals and the mammary gland, which could happen during sexual activity, or through contact of bodily fluids with the fissures of the nipple, which can serve as an entry point for HPV. Created by Biorender.

In an in vitro study, the MCF-10A cell line was transfected with HPV-18 to determine whether HPV may be the starting point of carcinogenesis in breast cancer through APOBEC3B (A3B) overexpression. Their results demonstrated that HPV infection induces upregulation of A3B mRNA and that infected cells exhibited a more malignant phenotype than parental cells since A3B overexpression caused γH2AX foci formation and DNA breakage. The expression of these malignant phenotypes was restricted by shRNA to HPV E6, E7, and A3B. These results suggest an active involvement of HPV in the early stage of breast cancer carcinogenesis through the induction of A3B [90]. A3B expression levels have been reported to be low in most healthy tissues; however, Vieira et al. [91] demonstrated that the E6 protein induces upregulation of A3B RNA in high-risk HPV. It was later observed that in samples from patients with head and neck cancer, there was an overexpression of A3B in HPV-positive tumors.

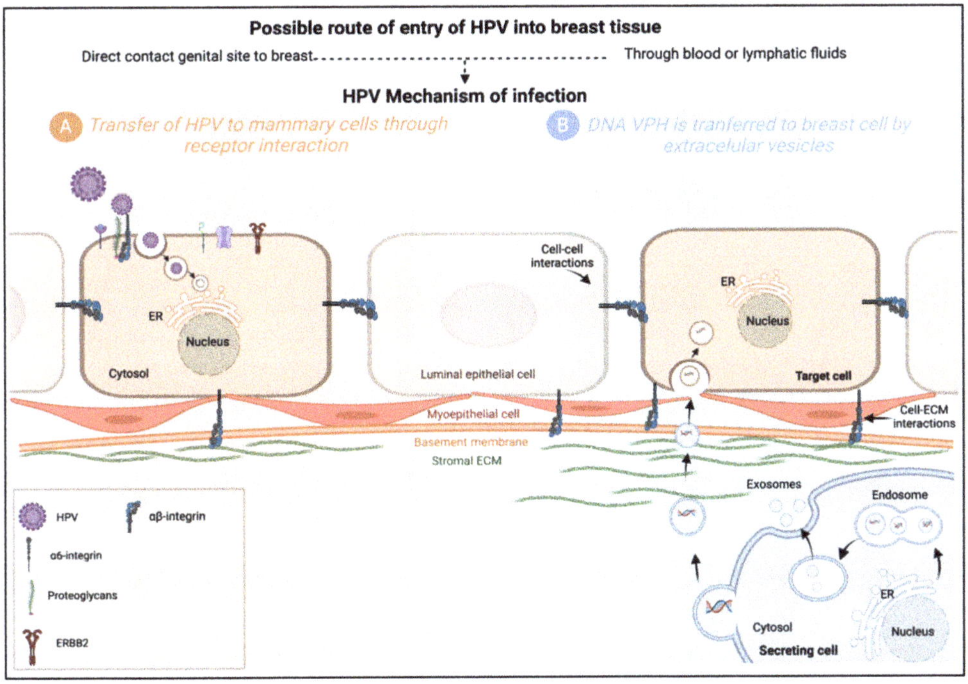

Figure 3. Possible mechanisms of HPV infection in the mammary gland. (**A**) Transfer of HPV to mammary cells through receptor interaction. The HPV-16 capsid interacts with the entry receptor complex composed of growth factor receptors, integrins, and proteoglycans, among others. After HPV binds to this complex, an endocytic process begins. Internalized viruses reside in vesicles directed to acidified multivesicular bodies for capsid disassembly. Viral genomes are transported to the TGN, ER, or core. (**B**) DNA HPV is transferred to breast cells by extracellular vesicles. Transfer of HPV DNA to cells lacking the HPV receptor could be carried out by extracellular vesicles (EVs), microvesicles (EVs), exosomes (Exos), or apoptotic bodies (ABs), which serve as vehicles for cell communication. Cell-to-cell, from a primary site of infection through the transfer of bioactive molecules (proteins, lipids, and nucleic acids). Extracellular vesicles produced from a secretory cell may be internalized by fusion, endocytosis, or phagocytosis, or interact with target cell membrane proteins. Created by Biorender.

5. Conclusions

The presence of HPV in breast tissue is an important finding, but it is not a sufficient condition to establish an etiological role for this virus in developing breast cancer or any other breast pathology. However, the results suggest a possible role of HPV in breast pathologies as a co-participant in molecular pathogenesis processes that differ from other HPV-associated neoplasms.

According to the results reported in this work, it was possible to detect HPV in breast tissue with malignant neoplasms but also normal tissue. The possible mechanisms by which HPV is present in the breast tissue that would respond have been proposed for its presence in healthy tissue. However, still, there is a pending question. Is it possible that persistent HPV infections in the mammary gland could achieve carcinogenic processes as in the case of cervical cancer?

Since 1992, HPV infection has been proposed as a possible risk factor for the development of breast cancer. Several authors have suggested that the increased incidence of HPV infection may be linked to environmental factors. These observations support the hypothesis of a possible infectious etiology in the development of sporadic breast cancer based on studies that report the presence of sequences of different types of high-risk HPV (oncogenic) in breast carcinoma tissues. However, to date, the results have been controversial and inconclusive. Further studies are required to demonstrate an association between HPV and breast cancer.

Author Contributions: Conceptualization, J.L.A.-L. and G.M.-A.; methodology, E.M.-R. and A.L.-S.; software, E.M.-R. and A.R.-L.; formal analysis, A.L.-S., A.R.-L., J.A.-R., I.E.G.-C., S.G.-G. and P.I.G.-F.; investigation, E.M.-R., E.L.E.-I., J.A.-R., S.G.-G. and M.H.-B.; writing—review and editing, E.M.-R., G.M.-A. and J.L.A.-L.; supervision, M.H.-B. and J.L.A.-L. All authors have read and agreed to the published version of the manuscript.

Funding: This research received no external funding.

Institutional Review Board Statement: The study was conducted in accordance with the Declaration of Helsinki, and approved by the Bioethics Committee of the Health Sciences Area belonging to the Autonomous University of Zacatecas (Of. N. ACS/UAZ/161/2019). The protocol was designed in accordance with the guidelines of the Official Mexican Standard NOM-012-SSA3-2012, which guarantees respect for all human beings and protects their health, their rights and the confidentiality of personal information.

Informed Consent Statement: Not applicable.

Data Availability Statement: Not applicable.

Acknowledgments: We acknowledge the Department of Pathological Anatomy of the IMSS Zacatecas for the donation of tissue for the experimental analysis.

Conflicts of Interest: The authors declare no conflict of interest.

References

1. World Health Organization. Global Cancer Observatory. Available online: https://gco.iarc.fr (accessed on 5 June 2022).
2. Mendoza-Almanza, G.; Burciaga-Hernández, L.; Maldonado, V.; Melendez-Zajgla, J.; Olmos, J. Role of platelets and breast cancer stem cells in metastasis. *World J. Stem Cells* **2020**, *12*, 1237–1254. [CrossRef] [PubMed]
3. Grabinski, V.F.; Brawley, O.W. Disparities in Breast Cancer. *Obstet. Gynecol. Clin. N. Am.* **2022**, *49*, 149–165. [CrossRef] [PubMed]
4. Delgado-Garcia, S.; Martinez-Escoriza, J.C.; Alba, A.; Martin-Bayon, T.A.; Ballester-Galiana, H.; Peiro, G.; Caballero, P.; Ponce Lorenzo, J. Presence of human papillomavirus DNA in breast cancer: A Spanish case-control study. *BMC Cancer* **2017**, *17*, 320. [CrossRef] [PubMed]
5. Afzal, S.; Fiaz, K.; Noor, A.; Sindhu, A.S.; Hanif, A.; Bibi, A.; Asad, M.; Nawaz, S.; Zafar, S.; Ayub, S.; et al. Interrelated Oncogenic Viruses and Breast Cancer. *Front. Mol. Biosci.* **2022**, *9*, 781111. [CrossRef]
6. Blanco, R.; Carrillo-Beltrán, D.; Muñoz, J.P.; Corvalán, A.H.; Calaf, G.M.; Aguayo, F. Human Papillomavirus in Breast Carcinogenesis: A Passenger, a Cofactor, or a Causal Agent? *Biology* **2021**, *10*, 804. [CrossRef]
7. Bittner, J.J. Possible relationship of the estrogenic hormones, genetic susceptibility, and milk influence in the production of mammary cancer in mice. *Cancer Res.* **1942**, *2*, 710–721.
8. Wang, Y.; Holland, J.F.; Bleiweiss, I.J.; Melana, S.; Liu, X.; Pelisson, I.; Cantarella, A.; Stellrecht, K.; Mani, S.; Pogo, B.G. Detection of mammary tumor virus env gene-like sequences in human breast cancer. *Cancer Res.* **1995**, *55*, 5173–5179.
9. Melana, S.M.; Nepomnaschy, I.; Hasa, J.; Djougarian, A.; Holland, J.F.; Pogo, B.G. Detection of human mammary tumor virus proteins in human breast cancer cells. *J. Virol. Methods* **2010**, *163*, 157–161. [CrossRef]
10. Islam, S.; Dasgupta, H.; Roychowdhury, A.; Bhattacharya, R.; Mukherjee, N.; Roy, A.; Mandal, G.K.; Alam, N.; Biswas, J.; Mandal, S.; et al. Study of association and molecular analysis of human papillomavirus in breast cancer of Indian patients: Clinical and prognostic implication. *PLoS ONE* **2017**, *12*, e0172760. [CrossRef]
11. Karbalaie Niya, M.H.; Keyvani, H.; Safarnezhad Tameshkel, F.; Salehi-Vaziri, M.; Teaghinezhad, S.S.; Bokharaei Salim, F.; Monavari, S.H.R.; Javanmard, D. Human Papillomavirus Type 16 Integration Analysis by Real-time PCR Assay in Associated Cancers. *Transl. Oncol.* **2018**, *11*, 593–598. [CrossRef]
12. Hsu, C.R.; Lu, T.M.; Chin, L.W.; Yang, C.C. Possible DNA viral factors of human breast cancer. *Cancers* **2010**, *2*, 498–512. [CrossRef] [PubMed]
13. Wrede, D.; Luqmani, Y.A.; Coombes, R.C.; Vousden, K.H. Absence of HPV 16 and 18 DNA in breast cancer. *Br. J. Cancer* **1992**, *65*, 891–894. [CrossRef] [PubMed]
14. Bratthauer, G.L.; Tavassoli, F.A.; O'Leary, T.J. Etiology of breast carcinoma: No apparent role for papillomavirus types 6/11/16/18. *Pathol. Res. Pract.* **1992**, *188*, 384–386. [CrossRef]

15. Gopalkrishna, V.; Singh, U.R.; Sodhani, P.; Sharma, J.K.; Hedau, S.T.; Mandal, A.K.; Das, B.C. Absence of human papillomavirus DNA in breast cancer as revealed by polymerase chain reaction. *Breast Cancer Res Treat.* **1996**, *39*, 197–202. [CrossRef] [PubMed]
16. Czerwenka, K.; Heuss, F.; Hosmann, J.W.; Manavi, M.; Lu, Y.; Jelincic, D.; Kubista, E. Human papilloma virus DNA: A factor in the pathogenesis of mammary Paget's disease? *Breast Cancer Res. Treat.* **1996**, *41*, 51–57. [CrossRef]
17. Lindel, K.; Forster, A.; Altermatt, H.J.; Greiner, R.; Gruber, G. Breast cancer and human papillomavirus (HPV) infection: No evidence of a viral etiology in a group of Swiss women. *Breast* **2007**, *16*, 172–177. [CrossRef]
18. De Cremoux, P.; Thioux, M.; Lebigot, I.; Sigal-Zafrani, B.; Salmon, R.; Sastre-Garau, X. No evidence of human papillomavirus DNA sequences in invasive breast carcinoma. *Breast Cancer Res. Treat.* **2008**, *109*, 55–58. [CrossRef]
19. Hachana, M.; Ziadi, S.; Amara, K.; Toumi, I.; Korbi, S.; Trimeche, M. No evidence of human papillomavirus DNA in breast carcinoma in Tunisian patients. *Breast* **2010**, *19*, 541–544. [CrossRef]
20. Hedau, S.; Kumar, U.; Hussain, S.; Shukla, S.; Pande, S.; Jain, N.; Tyagi, A.; Deshpande, T.; Bhat, D.; Mir, M.M.; et al. Breast cancer and human papillomavirus infection: No evidence of HPV etiology of breast cancer in Indian women. *BMC Cancer* **2011**, *11*, 27. [CrossRef]
21. Silva, R.G., Jr.; da Silva, B.B. No evidence for an association of human papillomavirus and breast carcinoma. *Breast Cancer Res. Treat.* **2011**, *125*, 261–264. [CrossRef]
22. Zhou, Y.; Li, J.; Ji, Y.; Ren, M.; Pang, B.; Chu, M.; Wei, L. Inconclusive role of human papillomavirus infection in breast cancer. *Infect. Agent. Cancer* **2015**, *10*, 36. [CrossRef] [PubMed]
23. Vernet-Tomas, M.; Mena, M.; Alemany, L.; Bravo, I.; De Sanjose, S.; Nicolau, P.; Bergueiro, A.; Corominas, J.M.; Serrano, S.; Carreras, R.; et al. Human papillomavirus and breast cancer: No evidence of association in a Spanish set of cases. *Anticancer Res.* **2015**, *35*, 851–856. [PubMed]
24. Kouloura, A.; Nicolaidou, E.; Misitzis, I.; Panotopoulou, E.; Kassiani, T.; Smyrniotis, V.; Corso, G.; Veronesi, P.; Arkadopoulos, N. HPV infection and breast cancer. Results of a microarray approach. *Breast* **2018**, *40*, 165–169. [CrossRef] [PubMed]
25. Di Lonardo, A.; Venuti, A.; Marcante, M.L. Human papillomavirus in breast cancer. *Breast Cancer Res. Treat.* **1992**, *21*, 95–100. [CrossRef]
26. Hennig, E.M.; Suo, Z.; Thoresen, S.; Holm, R.; Kvinnsland, S.; Nesland, J.M. Human papillomavirus 16 in breast cancer of women treated for high grade cervical intraepithelial neoplasia (CIN III). *Breast Cancer Res Treat.* **1999**, *53*, 121–135. [CrossRef]
27. Damin, A.P.; Karam, R.; Zettler, C.G.; Caleffi, M.; Alexandre, C.O. Evidence for an association of human papillomavirus and breast carcinomas. *Breast Cancer Res. Treat.* **2004**, *84*, 131–137. [CrossRef]
28. Widschwendter, A.; Brunhuber, T.; Wiedemair, A.; Mueller-Holzner, E.; Marth, C. Detection of human papillomavirus DNA in breast cancer of patients with cervical cancer history. *J. Clin. Virol.* **2004**, *31*, 292–297. [CrossRef]
29. de Villiers, E.M.; Sandstrom, R.E.; zur Hausen, H.; Buck, C.E. Presence of papillomavirus sequences in condylomatous lesions of the mamillae and in invasive carcinoma of the breast. *Breast Cancer Res.* **2005**, *7*, R1–R11. [CrossRef]
30. Kan, C.Y.; Iacopetta, B.J.; Lawson, J.S.; Whitaker, N.J. Identification of human papillomavirus DNA gene sequences in human breast cancer. *Br. J. Cancer* **2005**, *93*, 946–948. [CrossRef]
31. Kroupis, C.; Markou, A.; Vourlidis, N.; Dionyssiou-Asteriou, A.; Lianidou, E.S. Presence of high-risk human papillomavirus sequences in breast cancer tissues and association with histopathological characteristics. *Clin. Biochem.* **2006**, *39*, 727–731. [CrossRef]
32. Gumus, M.; Yumuk, P.F.; Salepci, T.; Aliustaoglu, M.; Dane, F.; Ekenel, M.; Basaran, G.; Kaya, H.; Barisik, N.; Turhal, N.S. HPV DNA frequency and subset analysis in human breast cancer patients'normal and tumoral tissue samples. *J. Exp. Clin. Cancer Res. CR* **2006**, *25*, 515–521. [PubMed]
33. Akil, N.; Yasmeen, A.; Kassab, A.; Ghabreau, L.; Darnel, A.D.; Al Moustafa, A.E. High-risk human papillomavirus infections in breast cancer in Syrian women and their association with Id-1 expression: A tissue microarray study. *Br. J. Cancer* **2008**, *99*, 404–407. [CrossRef] [PubMed]
34. Khan, N.A.; Castillo, A.; Koriyama, C.; Kijima, Y.; Umekita, Y.; Ohi, Y.; Higashi, M.; Sagara, Y.; Yoshinaka, H.; Tsuji, T.; et al. Human papillomavirus detected in female breast carcinomas in Japan. *Br. J. Cancer* **2008**, *99*, 408–414. [CrossRef] [PubMed]
35. Mendizabal-Ruiz, A.P.; Morales, J.A.; Ramirez-Jirano, L.J.; Padilla-Rosas, M.; Moran-Moguel, M.C.; Montoya-Fuentes, H. Low frequency of human papillomavirus DNA in breast cancer tissue. *Breast Cancer Res. Treat.* **2009**, *114*, 189–194. [CrossRef]
36. De Leon, D.C.; Montiel, D.P.; Nemcova, J.; Mykyskova, I.; Turcios, E.; Villavicencio, V.; Cetina, L.; Coronel, A.; Hes, O. Human papillomavirus (HPV) in breast tumors: Prevalence in a group of Mexican patients. *BMC Cancer* **2009**, *9*, 26.
37. Heng, B.; Glenn, W.K.; Ye, Y.; Tran, B.; Delprado, W.; Lutze-Mann, L.; Whitaker, N.J.; Lawson, J.S. Human papilloma virus is associated with breast cancer. *Br. J. Cancer* **2009**, *101*, 1345–1350. [CrossRef]
38. Herrera-Goepfert, R.; Khan, N.A.; Koriyama, C.; Akiba, S.; Perez-Sanchez, V.M. High-risk human papillomavirus in mammary gland carcinomas and non-neoplastic tissues of Mexican women: No evidence supporting a cause and effect relationship. *Breast* **2011**, *20*, 184–189. [CrossRef]
39. Aguayo, F.; Khan, N.; Koriyama, C.; Gonzalez, C.; Ampuero, S.; Padilla, O.; Solis, L.; Eizuru, Y.; Corvalan, A.; Akiba, S. Human papillomavirus and Epstein-Barr virus infections in breast cancer from chile. *Infect. Agent. Cancer* **2011**, *6*, 7. [CrossRef]
40. Mou, X.; Chen, L.; Liu, F.; Shen, Y.; Wang, H.; Li, Y.; Yuan, L.; Lin, J.; Teng, L.; Xiang, C. Low prevalence of human papillomavirus (HPV) in Chinese patients with breast cancer. *J. Int. Med. Res.* **2011**, *39*, 1636–1644. [CrossRef]

41. Antonsson, A.; Spurr, T.P.; Chen, A.C.; Francis, G.D.; McMillan, N.A.; Saunders, N.A.; Law, M.; Bennett, I.C. High prevalence of human papillomaviruses in fresh frozen breast cancer samples. *J. Med. Virol.* **2011**, *83*, 2157–2163. [CrossRef]
42. Sigaroodi, A.; Nadji, S.A.; Naghshvar, F.; Nategh, R.; Emami, H.; Velayati, A.A. Human papillomavirus is associated with breast cancer in the north part of Iran. *Sci. World J.* **2012**, *2012*, 837191. [CrossRef] [PubMed]
43. Herrera-Goepfert, R.; Vela-Chavez, T.; Carrillo-Garcia, A.; Lizano-Soberon, M.; Amador-Molina, A.; Onate-Ocana, L.F.; Hallmann, R.S. High-risk human papillomavirus (HPV) DNA sequences in metaplastic breast carcinomas of Mexican women. *BMC Cancer* **2013**, *13*, 445. [CrossRef] [PubMed]
44. Pereira Suarez, A.L.; Lorenzetti, M.A.; Gonzalez Lucano, R.; Cohen, M.; Gass, H.; Martinez Vazquez, P.; Gonzalez, P.; Preciado, M.V.; Chabay, P. Presence of human papilloma virus in a series of breast carcinoma from Argentina. *PLoS ONE* **2013**, *8*, e61613.
45. Liang, W.; Wang, J.; Wang, C.; Lv, Y.; Gao, H.; Zhang, K.; Liu, H.; Feng, J.; Wang, L.; Ma, R. Detection of high-risk human papillomaviruses in fresh breast cancer samples using the hybrid capture 2 assay. *J. Med. Virol.* **2013**, *85*, 2087–2092. [CrossRef] [PubMed]
46. Ali, S.H.; Al-Alwan, N.A.; Al-Alwany, S.H. Detection and genotyping of human papillomavirus in breast cancer tissues from Iraqi patients. *East Mediterr. Health J.* **2014**, *20*, 372–377. [CrossRef]
47. Piana, A.F.; Sotgiu, G.; Muroni, M.R.; Cossu-Rocca, P.; Castiglia, P.; De Miglio, M.R. HPV infection and triple-negative breast cancers: An Italian case-control study. *Virol. J.* **2014**, *11*, 190. [CrossRef]
48. Manzouri, L.; Salehi, R.; Shariatpanahi, S.; Rezaie, P. Prevalence of human papilloma virus among women with breast cancer since 2005-2009 in Isfahan. *Adv. Biomed. Res.* **2014**, *3*, 75.
49. Peng, J.; Wang, T.; Zhu, H.; Guo, J.; Li, K.; Yao, Q.; Lv, Y.; Zhang, J.; He, C.; Chen, J.; et al. Multiplex PCR/mass spectrometry screening of biological carcinogenic agents in human mammary tumors. *J. Clin. Virol.* **2014**, *61*, 255–259. [CrossRef]
50. Ahangar-Oskouee, M.; Shahmahmoodi, S.; Jalilvand, S.; Mahmoodi, M.; Ziaee, A.A.; Esmaeili, H.A.; Keshtvarz, M.; Pishraft-Sabet, L.; Yousefi, M.; Mollaei-Kandelous, Y.; et al. No detection of 'high-risk' human papillomaviruses in a group of Iranian women with breast cancer. *Asian Pac. J. Cancer Prev.* **2014**, *15*, 4061–4065. [CrossRef]
51. Fu, L.; Wang, D.; Shah, W.; Wang, Y.; Zhang, G.; He, J. Association of human papillomavirus type 58 with breast cancer in Shaanxi province of China. *J. Med. Virol.* **2015**, *87*, 1034–1040. [CrossRef]
52. Gannon, O.M.; Antonsson, A.; Milevskiy, M.; Brown, M.A.; Saunders, N.A.; Bennett, I.C. No association between HPV positive breast cancer and expression of human papilloma viral transcripts. *Sci. Rep.* **2015**, *5*, 18081. [CrossRef] [PubMed]
53. Li, J.; Ding, J.; Zhai, K. Detection of Human Papillomavirus DNA in Patients with Breast Tumor in China. *PLoS ONE* **2015**, *10*, e0136050. [CrossRef] [PubMed]
54. Fernandes, A.; Bianchi, G.; Feltri, A.P.; Perez, M.; Correnti, M. Presence of human papillomavirus in breast cancer and its association with prognostic factors. *Ecancermedicalscience* **2015**, *9*, 548. [CrossRef]
55. Lawson, J.S.; Glenn, W.K.; Salyakina, D.; Clay, R.; Delprado, W.; Cheerala, B.; Tran, D.D.; Ngan, C.C.; Miyauchi, S.; Karim, M.; et al. Human Papilloma Virus Identification in Breast Cancer Patients with Previous Cervical Neoplasia. *Front. Oncol.* **2015**, *5*, 298. [CrossRef] [PubMed]
56. Choi, J.; Kim, C.; Lee, H.S.; Choi, Y.J.; Kim, H.Y.; Lee, J.; Chang, H.; Kim, A. Detection of Human Papillomavirus in Korean Breast Cancer Patients by Real-Time Polymerase Chain Reaction and Meta-Analysis of Human Papillomavirus and Breast Cancer. *J. Pathol. Transl. Med.* **2016**, *50*, 442–450. [CrossRef]
57. Ilahi, N.E.; Anwar, S.; Noreen, M.; Hashmi, S.N.; Murad, S. Detection of human papillomavirus-16 DNA in archived clinical samples of breast and lung cancer patients from North Pakistan. *J. Cancer Res. Clin. Oncol.* **2016**, *142*, 2497–2502. [CrossRef]
58. Mohtasebi, P.; Rassi, H.; Maleki, F.; Hajimohammadi, S.; Bagheri, Z.; Fakhar Miandoab, M.; Naserbakht, M. Detection of Human Papillomavirus Genotypes and Major BRCA Mutations in Familial Breast Cancer. *Monoclon. Antib. Immunodiagn. Immunother.* **2016**, *35*, 135–140. [CrossRef]
59. Yan, C.; Teng, Z.P.; Chen, Y.X.; Shen, D.H.; Li, J.T.; Zeng, Y. Viral Etiology Relationship between Human Papillomavirus and Human Breast Cancer and Target of Gene Therapy. *Biomed. Environ. Sci.* **2016**, *29*, 331–339.
60. Ngamkham, J.; Karalak, A.; Chaiwerawattana, A.; Sornprom, A.; Thanasutthichai, S.; Sukarayodhin, S.; Mus-u-Dee, M.; Boonmark, K.; Phansri, T.; Laochan, N. Prevalence of Human Papillomavirus Infection in Breast Cancer Cells from Thai Women. *Asian Pac. J. Cancer Prev.* **2017**, *18*, 1839–1845.
61. Salman, N.A.; Davies, G.; Majidy, F.; Shakir, F.; Akinrinade, H.; Perumal, D.; Ashrafi, G.H. Association of High Risk Human Papillomavirus and Breast cancer: A UK based Study. *Sci. Rep.* **2017**, *7*, 43591. [CrossRef]
62. Wang, Y.W.; Zhang, K.; Zhao, S.; Lv, Y.; Zhu, J.; Liu, H.; Feng, J.; Liang, W.; Ma, R.; Wang, J. HPV Status and Its Correlation with BCL2, p21, p53, Rb, and Survivin Expression in Breast Cancer in a Chinese Population. *Biomed. Res. Int.* **2017**, *2017*, 6315392. [CrossRef] [PubMed]
63. Naushad, W.; Surriya, O.; Sadia, H. Prevalence of EBV, HPV and MMTV in Pakistani breast cancer patients: A possible etiological role of viruses in breast cancer. *Infect. Genet. Evol.* **2017**, *54*, 230–237. [CrossRef] [PubMed]
64. Cavalcante, J.R.; Pinheiro, L.G.P.; Almeida, P.R.C.; Ferreira, M.V.P.; Cruz, G.A.; Campelo, T.A.; Silva, C.S.; Lima, L.; Oliveira, B.M.K.; Lima, L.M.; et al. Association of breast cancer with human papillomavirus (HPV) infection in Northeast Brazil: Molecular evidence. *Clinics* **2018**, *73*, e465. [CrossRef]

65. Ghaffari, H.; Nafissi, N.; Hashemi-Bahremani, M.; Alebouyeh, M.R.; Tavakoli, A.; Javanmard, D.; Bokharaei-Salim, F.; Mortazavi, H.S.; Monavari, S.H. Molecular prevalence of human papillomavirus infection among Iranian women with breast cancer. *Breast Dis.* **2018**, *37*, 207–213. [CrossRef] [PubMed]
66. ElAmrani, A.; Gheit, T.; Benhessou, M.; McKay-Chopin, S.; Attaleb, M.; Sahraoui, S.; El Mzibri, M.; Corbex, M.; Tommasino, M.; Khyatti, M. Prevalence of mucosal and cutaneous human papillomavirus in Moroccan breast cancer. *Papillomavirus Res.* **2018**, *5*, 150–155. [CrossRef] [PubMed]
67. Habyarimana, T.; Attaleb, M.; Mazarati, J.B.; Bakri, Y.; El Mzibri, M. Detection of human papillomavirus DNA in tumors from Rwandese breast cancer patients. *Breast Cancer* **2018**, *25*, 127–133. [CrossRef]
68. Bonlokke, S.; Blaakaer, J.; Steiniche, T.; Hogdall, E.; Jensen, S.G.; Hammer, A.; Balslev, E.; Strube, M.L.; Knakkergaard, H.; Lenz, S. Evidence of No Association Between Human Papillomavirus and Breast Cancer. *Front. Oncol.* **2018**, *8*, 209. [CrossRef]
69. Malekpour Afshar, R.; Balar, N.; Mollaei, H.R.; Arabzadeh, S.A.; Iranpour, M. Low Prevalence of Human Papilloma Virus in Patients with Breast Cancer, Kerman; Iran. *Asian Pac. J. Cancer Prev.* **2018**, *19*, 3039–3044. [CrossRef]
70. Khodabandehlou, N.; Mostafaei, S.; Etemadi, A.; Ghasemi, A.; Payandeh, M.; Hadifar, S.; Norooznezhad, A.H.; Kazemnejad, A.; Moghoofei, M. Human papilloma virus and breast cancer: The role of inflammation and viral expressed proteins. *BMC Cancer* **2019**, *19*, 61. [CrossRef]
71. De Carolis, S.; Storci, G.; Ceccarelli, C.; Savini, C.; Gallucci, L.; Sansone, P.; Santini, D.; Seracchioli, R.; Taffurelli, M.; Fabbri, F.; et al. HPV DNA Associates With Breast Cancer Malignancy and It Is Transferred to Breast Cancer Stromal Cells by Extracellular Vesicles. *Front. Oncol.* **2019**, *9*, 860. [CrossRef]
72. Balci, F.L.; Uras, C.; Feldman, S.M. Is human papillomavirus associated with breast cancer or papilloma presenting with pathologic nipple discharge? *Cancer Treat. Res. Commun.* **2019**, *19*, 100122. [CrossRef] [PubMed]
73. Tawfeik, A.M.; Mora, A.; Osman, A.; Moneer, M.M.; El-Sheikh, N.; Elrefaei, M. Frequency of CD4+ regulatory T cells, CD8+ T cells, and human papilloma virus infection in Egyptian Women with breast cancer. *Int. J. Immunopathol. Pharmacol.* **2020**, *34*, 2058738420966822. [CrossRef] [PubMed]
74. Sher, G.; Salman, N.A.; Kulinski, M.; Fadel, R.A.; Gupta, V.K.; Anand, A.; Gehani, S.; Abayazeed, S.; Al-Yahri, O.; Shahid, F. Prevalence and type distribution of high-risk Human Papillomavirus (HPV) in breast cancer: A Qatar based study. *Cancers* **2020**, *12*, 1528. [CrossRef]
75. Metwally, S.A.; Abo-Shadi, M.A.; Abdel Fattah, N.F.; Barakat, A.B.; Rabee, O.A.; Osman, A.M.; Helal, A.M.; Hashem, T.; Moneer, M.M.; Chehadeh, W.; et al. Presence of HPV, EBV and HMTV Viruses Among Egyptian Breast Cancer Women: Molecular Detection and Clinical Relevance. *Infect. Drug Resist.* **2021**, *14*, 2327–2339. [CrossRef]
76. Elagali, A.M.; Suliman, A.A.; Altayeb, M.; Dannoun, A.I.; Parine, N.R.; Sakr, H.I.; Suliman, H.S.; Motawee, M.E. Human papillomavirus, gene mutation and estrogen and progesterone receptors in breast cancer: A cross-sectional study. *Pan. Afr. Med. J.* **2021**, *38*, 43. [CrossRef] [PubMed]
77. Gupta, I.; Jabeen, A.; Al-Sarraf, R.; Farghaly, H.; Vranic, S.; Sultan, A.A.; Al Moustafa, A.E.; Al-Thawadi, H. The co-presence of high-risk human papillomaviruses and Epstein-Barr virus is linked with tumor grade and stage in Qatari women with breast cancer. *Hum. Vaccin. Immunother.* **2021**, *17*, 982–989. [CrossRef] [PubMed]
78. Mendoza-Almanza, G.; Ortíz-Sánchez, E.; Rocha-Zavaleta, L.; Rivas-Santiago, C.; Esparza-Ibarra, E.; Olmos, J. Cervical cancer stem cells and other leading factors associated with cervical cancer development. *Oncol. Lett.* **2019**, *18*, 3423–3432. [CrossRef] [PubMed]
79. Glenn, W.K.; Heng, B.; Delprado, W.; Iacopetta, B.; Whitaker, N.J.; Lawson, J.S. Epstein-Barr virus, human papillomavirus and mouse mammary tumour virus as multiple viruses in breast cancer. *PLoS ONE* **2012**, *7*, e48788. [CrossRef]
80. Zur Hausen, H. Papillomaviruses in the causation of human cancers—A brief historical account. *Virology* **2009**, *384*, 260–265. [CrossRef]
81. Wang, T.; Chang, P.; Wang, L.; Yao, Q.; Guo, W.; Chen, J.; Yan, T.; Cao, C. The role of human papillomavirus infection in breast cancer. *Med. Oncol.* **2012**, *29*, 48–55. [CrossRef]
82. Malhone, C.; Longatto-Filho, A.; Filassi, J.R. Is Human Papilloma Virus Associated with Breast Cancer? A Review of the Molecular Evidence. *Acta Cytol.* **2018**, *62*, 166–177. [CrossRef]
83. Vaccarella, S.; Franceschi, S.; Engholm, G.; Lönnberg, S.; Khan, S.; Bray, F. 50 years of screening in the Nordic countries: Quantifying the effects on cervical cancer incidence. *Br. J. Cancer* **2014**, *111*, 965–969. [CrossRef] [PubMed]
84. Yasmeen, A.; Bismar, T.A.; Kandouz, M.; Foulkes, W.D.; Desprez, P.Y.; Al Moustafa, A.E. E6/E7 of HPV type 16 promotes cell invasion and metastasis of human breast cancer cells. *Cell Cycle* **2007**, *6*, 2038–2042. [CrossRef] [PubMed]
85. Atique, S.; Hsieh, C.H.; Hsiao, R.T.; Iqbal, U.; Nguyen, P.A.A.; Islam, M.M.; Li, Y.J.; Hsu, C.Y.; Chuang, T.W.; Syed-Abdul, S. Viral warts (Human Papilloma Virus) as a potential risk for breast cancer among younger females. *Comput. Methods Programs Biomed.* **2017**, *144*, 203–207. [CrossRef] [PubMed]
86. Pao, C.C.; Hor, J.J.; Yang, F.P.; Lin, C.Y.; Tseng, C.J. Detection of human papillomavirus mRNA and cervical cancer cells in peripheral blood of cervical cancer patients with metastasis. *J. Clin. Oncol.* **1997**, *15*, 1008–1012. [CrossRef]
87. Mikulicic, S.; Florin, L. The endocytic trafficking pathway of oncogenic papillomaviruses. *Papillomavirus Res.* **2019**, *7*, 135–137. [CrossRef] [PubMed]
88. Taddei, I.; Faraldo, M.M.; Teulière, J.; Deugnier, M.-A.; Thiery, J.P.; Glukhova, M.A. Integrins in mammary gland development and differentiation of mammary epithelium. *J. Mammary Gland. Biol. Neoplasia* **2003**, *8*, 383–394. [CrossRef]

89. Yoon, C.-S.; Kim, K.-D.; Park, S.-N.; Cheong, S.-W. α6 integrin is the main receptor of human papillomavirus type 16 VLP. *Biochem. Biophys. Res. Commun.* **2001**, *283*, 668–673. [CrossRef]
90. Ohba, K.; Ichiyama, K.; Yajima, M.; Gemma, N.; Nikaido, M.; Wu, Q.; Chong, P.; Mori, S.; Yamamoto, R.; Wong, J.E.; et al. In vivo and in vitro studies suggest a possible involvement of HPV infection in the early stage of breast carcinogenesis via APOBEC3B induction. *PLoS ONE* **2014**, *9*, e97787.
91. Vieira, V.C.; Leonard, B.; White, E.A.; Starrett, G.J.; Temiz, N.A.; Lorenz, L.D.; Lee, D.; Soares, M.A.; Lambert, P.F.; Howley, P.M.; et al. Human papillomavirus E6 triggers upregulation of the antiviral and cancer genomic DNA deaminase APOBEC3B. *mBio* **2014**, *5*, e02234-14. [CrossRef]

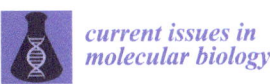

current issues in molecular biology

MDPI

Article

«Salivaomics» of Different Molecular Biological Subtypes of Breast Cancer

Lyudmila V. Bel'skaya * and Elena A. Sarf

Biochemistry Research Laboratory, Omsk State Pedagogical University, 644043 Omsk, Russia; nemcha@mail.ru
* Correspondence: belskaya@omgpu.ru

Abstract: The aim of the study was to determine the metabolic characteristics of saliva depending on the molecular biological subtype of breast cancer, as well as depending on the expression levels of HER2, estrogen receptors (ER), and progesterone receptors (PR). The study included 487 patients with morphologically verified breast cancer and 298 volunteers without breast pathologies. Saliva samples were obtained from all patients strictly before the start of treatment and the values of 42 biochemical indicators were determined. It has been established that the saliva of healthy volunteers and patients with various molecular biological subtypes of breast cancer differs in 12 biochemical indicators: concentrations of protein, urea, nitric oxide, malondialdehyde, total amino acid content, and activity of lactate dehydrogenase, alkaline phosphatase, gamma-glutamyltransferase, catalase, amylase, superoxide dismutase, and peroxidases. The saliva composition of patients with basal-like breast cancer differs from other subtypes in terms of the maximum number of indicators. Changes in biochemical indicators indicated an increase in the processes of lipid peroxidation and endogenous intoxication and a weakening of antioxidant protection, which correlates with the severity of the disease and the least favorable prognosis for this subtype of breast cancer. An analysis was made of the individual contribution of the expression level of HER2, estrogen, and progesterone receptors to changes in the biochemical composition of saliva. The HER2 (−)/HER2 (+) group, which should be considered as a single group, as well as ER-positive breast cancer, differ statistically significantly from the control group. For ER/PR-positive breast cancer, a more favorable ratio of saliva biochemical indicators was also noted compared to ER/PR-negative breast cancer.

Keywords: salivaomics; breast cancer; biomarkers; saliva; molecular biological subtype; HER2 status; estrogen receptors; progesterone receptors

Citation: Bel'skaya, L.V.; Sarf, E.A. «Salivaomics» of Different Molecular Biological Subtypes of Breast Cancer. *Curr. Issues Mol. Biol.* **2022**, *44*, 3053–3074. https://doi.org/10.3390/cimb44070211

Academic Editor: Dumitru A. Iacobas

Received: 11 June 2022
Accepted: 4 July 2022
Published: 5 July 2022

1. Introduction

Breast cancer is the most common female cancer worldwide [1–4]. Despite the improvement in early diagnosis and the active use of adjuvant drug treatment, only 59% of patients in Russia survive the 5-year follow-up period [5], and mortality from breast cancer in Russia does not decrease due to late detection of the disease [6,7]. The proportion of early breast cancer (cancer in situ and stage I) is critically small: the proportion of non-invasive breast cancer was less than 1%, and stage I breast cancer was only 18.3%, which focuses attention on the existing problem of early diagnosis of the disease [8]. Nevertheless, the current level of knowledge about the molecular mechanisms of the onset and development of breast cancer, its sensitivity or resistance to various drugs, allows the transition from averaged standard therapy regimens to the so-called "personalized medicine" [9,10], i.e., the appointment treatment in accordance with the individual characteristics of the patient and the biological characteristics of the tumor. Breast cancer is a heterogeneous disease [11]. This heterogeneity, which has been characterized at the histological level for decades, is now being assessed at the molecular genetic level, so that each type of tumor is an independent disease. The high heterogeneity of breast cancer makes its molecular characterization fundamentally important, based not only on the determination of gene mutations and gene expression profile, but also on biological markers [12]. Examples of such markers include: expression of estrogen receptors (ER) and progesterone receptors (PR), expression of the proliferation marker Ki-67 in the active phase of the cell cycle (G1, S, G2, and mitosis) and its absence in resting cells (G0), and the expression of the type 2 human epidermal growth factor receptor (HER2) are also isolated [13–15]. Determination of these characteristics is possible only after surgical treatment or tumor biopsy, and the results can be significantly distorted after preoperative chemotherapy and radiotherapy. For some patients, data on the molecular characteristics of the tumor cannot be obtained for a number of reasons. In this regard, it is necessary to search for alternative non-invasive markers that can characterize individual types of tumors and act as diagnostic and prognostic signs [16].

Recently, evidence has been accumulating demonstrating the diagnostic and prognostic value of saliva as a promising alternative to liquid biopsy [17–23]. Saliva is a complex body fluid that contains a wide range of proteins, as well as DNA, mRNA, microRNA (miRNA/miR), metabolites, and microbiota [24]. As a diagnostic approach, saliva has many biochemical advantages over blood and tissues, such as non-invasiveness, ease of storage, cost-effectiveness of collection, and dynamic availability for monitoring with less discomfort for the patient [25]. Continuous progress in saliva research has allowed the scientific community to coin the term "salivaomics" [26,27]. Changes in the genome, microbiome, epigenome, transcriptome, proteome, and metabolome of saliva can be used for diagnosis, assessment of individual risk, prognosis, and disease monitoring [27].

The literature provides data on the study of the composition of saliva in breast cancer [28–44]. Saliva biomarkers have been shown to achieve a sensitivity of 73% (72–74) and a specificity of 74% (72–76) in the diagnosis of breast cancer [45]. However, only one study showed the relationship between the saliva metabolome and the molecular biological subtype of breast cancer [46]. Previously, we have shown that there are changes in the metabolic profile of saliva in breast cancer [47]. It has been shown that concentration of total protein, urea, uric acid (UA), the total content of α-amino acids and lipid peroxidation products, and the activity of metabolic and antioxidant enzymes (in particular catalase) of saliva changed significantly in breast cancer. This study is one of the largest to date and includes patients with early stages of breast cancer (226/487). The metabolic features of the composition of the saliva of patients depending on the prevalence of the process and the histological type of breast cancer are considered [47]. In this work, we analyze the changes in 42 biochemical indicators of saliva depending on the molecular biological subtype of breast cancer, as well as depending on the expression levels of HER2, estrogen receptors (ER), and progesterone receptors (PR).

2. Materials and Methods

2.1. Study Design and Group Description

The study included 487 patients of the Clinical Oncological Dispensary in Omsk. The sample size of this study was the number we could recruit within the study periods (January 2015–May 2017). All patients had histologically diagnosed with breast cancer. None had received any prior treatment, including hormone therapy, chemotherapy, molecularly targeted therapy, radiotherapy, surgery, etc. The inclusion criteria that were considered include: the age of patients 30–70 years, the absence of any treatment at the time of the study, the absence of signs of active infection (including purulent processes), and good oral hygiene. The volunteers included in the study did not reveal any clinically significant concomitant diseases other than cancer pathology (in particular, diabetes mellitus, cardiovascular pathologies, etc.) that could affect the results of the study. Exclusion criteria included lack of histological verification of the diagnosis. The control group consisted of 298 healthy patients, in whom no breast pathology was detected during routine clinical examination. A detailed description of the study group is given in Table 1.

Table 1. The structure of the study group.

Feature	Breast Cancer, *n* = 487	Control Group, *n* = 298
Age, years	54.5 [47.0; 56.0]	49.3 [43.8; 56.1]
Histological type		
Ductal	227 (46.6%)	-
Lobular	86 (17.7%)	-
Mixed (Ductal + Lobular)	12 (2.5%)	-
Rare forms	58 (11.9%)	-
Unknown	104 (21.3%)	-
Clinical Stage		
Stage I	119 (24.4%)	-
Stage IIa	123 (25.3%)	-
Stage IIb	88 (18.1%)	-
Stage IIIa	55 (11.3%)	-
Stage IIIb	47 (9.6%)	-
Stage IV	55 (11.3%)	-
Subtype		
Luminal A-like	64 (13.1%)	-
Luminal B-like (HER2+)	230 (47.4%)	-
Luminal B-like (HER2−)	63 (12.9%)	-
Non-Luminal (HER2+)	38 (7.8%)	-
Basal-like (Triple-negative)	28 (5.7%)	-
Unknown	64 (13.1%)	-
HER2-status		
HER2-negative HER2 (−)	156 (36.1%)	-
HER2-positive	276 (63.9%)	-
HER2 (+)	124 (44.9%)	-
HER2 (++)	83 (30.1%)	-
HER2 (+++)	69 (25.0%)	-
ER-status		
ER-negative ER (−)	77 (17.7%)	-
ER-positive	359 (82.3%)	-
ER (+)	60 (16.7%)	-
ER (++)	77 (21.4%)	-
ER (+++)	222 (61.9%)	-
PR-status		
PR-negative PR (−)	125 (28.7%)	-
PR-positive	310 (71.3%)	-
PR (+)	64 (20.6%)	-
PR (++)	79 (25.5%)	-
PR (+++)	167 (53.9%)	-

2.2. *Determination of the Expression of the Receptors for Estrogen, Progesterone and HER2*

The Allred Scoring Guideline was used to assess the expression level of estrogen receptors (ER) and progesterone (PR) (Table 1) [48]. The calculated integrative indicator allows us to define the case under study in one of four main groups: a group with an expression level of 0 points (complete absence of stained nuclei, indicated by "−"), a group with a weak color level (index from 2 to 4 points, indicated by "+"), a group with an average level of expression (index from 5 to 6 points, indicated by "++"), and a group with a high level of expression (index is from 7 to 8 points, indicated by "+++"). When determining one of the four categories of the receptors for estrogen, progesterone, and HER2 expression levels (−, +, ++, +++), the recommendations of ASCO/CAP were followed [49]. Determination of HER2 expression was carried out with immunohistochemical method, with an indeterminate result (++) used to confirm the HER2 status. Following this, a study was carried out with in situ hybridization (FISH). HER2-status assessed as "−" and "+" was considered negative, assessed as "+++" was considered positive, and assessed as "++" was assigned to an undefined level. Additionally, breast cancer sub-classification differentiates these tumors into five groups: basal-like (BL, Triple-negative), luminal A-like, luminal B-like (HER2-negative), luminal B-like (HER2-positive), and non-luminal (Table 1). The determination of the molecular biological subtype was carried out as standard with a combination of the status of HER2, estrogen, and progesterone receptors and the level of Ki67. [7].

2.3. *Saliva Collection and Analysis*

Saliva (5 mL) was collected from all participants prior to treatment. Collection of saliva samples was carried out on an empty stomach after rinsing the mouth with water in the interval of 8–10 am by spitting into sterile polypropylene tubes; the salivation rate (mL/min) was calculated. We did not find significant differences in the salivary flow rate in the studied groups, so they were not shown in the tables below. Saliva samples were centrifuged (10,000× *g* for 10 min) (CLb-16, Moscow, Russia), and the supernatant fraction was used for subsequent analysis. Biochemical analysis was immediately performed without storage and freezing using the StatFax 3300 semi-automatic biochemical analyzer (Awareness Technology, Palm City, FL, USA) [50]. A full cycle of studies was performed within 3–4 h from the moment of collection. Protease inhibitors were not used.

The pH, mineral composition (calcium, phosphorus, sodium, potassium, magnesium, chlorides), the content of urea, total protein, albumin, uric acid, α-amino acids, imidazole compounds, seromucoids, nitric oxide—NO, lactic, pyruvic, and sialic acids, as well as the activity of enzymes (aminotransferases—ALT, AST; alkaline phosphatase—ALP; lactate dehydrogenase—LDH; gamma-glutamyl transpeptidase—GGT; α-amylase), were determined in all samples. The content of substrates for lipid peroxidation processes (diene conjugates—DC, triene conjugates—TC, Schiff bases—SB, malondialdehyde—MDA) and indicators of endogenous intoxication (MM—middle molecules) were determined. We determined the MM at wavelengths of 254 and 280 nm; they are designated MM 254 and MM 280, respectively [51]. Additionally, we assessed the activity of antioxidant enzymes (catalase—CAT, superoxide dismutase—SOD, antioxidant activity, peroxidase). The potential value of calculating a number of ratios has been previously shown, for example, Na/K, Ca/P, AST/ALT, SOD/Catalase, SOD/Peroxidase, SB/(DC+TC), SB/TC, and MM 280/254. In addition to the direct evaluation of 34 biochemical salivary indicators, we additionally evaluated the values of 8 ratios, so the total number of indicators was 42.

2.4. *Statistical Analysis*

Statistical analysis was performed using Statistica 13.3 EN software (StatSoft, Tulsa, OK, USA); R version 3.6.3; RStudio Version 1.2.5033; FactoMineR version 2.3. (RStudio, version 3.2.3, Boston, MA, USA) with a nonparametric method using the Mann–Whitney U-test and the Kruskal–Wallis H-test. The description of the sample was made by calculating the median (Me) and the interquartile range as the 25th and 75th percentiles [LQ; UQ]. Differences were considered statistically significant at $p > 0.05$.

A principal component analysis (PCA) was performed using the PCA program in R. The choice of variables for the PCA method was carried out according to the results of comparison of biochemical indicators in the studied groups. When comparing two groups, we used the Mann–Whitney test; when comparing three groups or more, we used the Kruskal–Wallis test. Next, we selected indicators for which the differences between all groups are significant at the $p < 0.10$ level. PCA results were presented in the form of factor planes and corresponding correlation circles. In each case, the figures show only the first two principal components (PC1 and PC2). The color of the arrows on the correlation circle changed from blue (weak correlation) to red (strong correlation) as shown on the color bar. The orientation of the arrows characterized positive and negative correlations (for the first principal component, we analyzed the location of the arrows relative to the vertical axis; for the second principal component, relative to the horizontal axis). The significance of the correlation was determined by the correlation coefficient (r): strong-r = ± 0.700 to ± 1.00, medium-r = ± 0.300 to ± 0.699, weak-r = 0.00 to ± 0.299.

3. Results

3.1. Changes in the Biochemical Composition of the Saliva of Patients with Breast Cancer, Depending on Its Molecular Biological Subtype

At the first stage of the study, it was shown that the biochemical composition of saliva in different molecular biological subtypes of breast cancer had differences (Supplementary Tables S1 and S2). The values of biochemical indicators of saliva in the control group, as well as in various molecular biological subtypes of breast cancer, are given in Supplementary Tables S1 and S2. Table 2 below shows the deviation values of the average content of each indicator of saliva from the corresponding values for the control group. We have identified 12 biochemical indicators for which the differences between the groups are statistically significant (Table 2). Selected biochemical indicators were used to compare groups by PCA analysis (Figure 1).

Table 2. Changes in biochemical indicators of saliva compared with the control group, %.

No.	Indicators	Lum A	Lum B (+)	Lum B (−)	BL	Non-Lum	Kruskal–Wallis Test (H, *p*)
1	pH	0.6	−0.1	−0.2	1.1	−1.3	4.125; 0.5316
2	Calcium, mmol/L	8.7	−4.2	−1.2	−11.5	−9.7	6.424; 0.2671
3	Phosphorus, mmol/L	0.7	7.5	8.7	9.7	−5.4	7.439; 0.1900
4	Ca/P-ratio, c.u.	0.5	−8.4	−12.4	−20.2	−6.0	8.226; 0.1442
5	Sodium, mmol/L	−21.4	−12.5	7.9	5.0	−15.2	6.519; 0.2589
6	Potassium, mmol/L	−1.3	2.1	21.6	23.0	5.3	7.879; 0.1631
7	Na/K-ratio, c.u.	−15.5	−14.0	−13.4	−17.0	3.3	5.517; 0.3560
8	Chlorides, mmol/L	−3.9	−2.0	6.6	12.8	−2.1	7.662; 0.1759
9	Magnesium, mmol/L	−6.5	−0.7	0.1	−7.4	−1.4	1.992; 0.8503
10	NO, μmol/L	22.4	21.6	40.7	19.0	49.3	16.02; 0.0068 *
11	Protein, mg/mL	−24.5	−21.4	−12.8	−5.7	−10.7	70.13; 0.0000 *
12	Urea, mmol/L	33.0	42.6	46.5	41.0	36.0	66.45; 0.0000 *
13	Uric acid, μmol/L	−28.8	−34.0	−21.9	2.6	−9.2	7.819; 0.1665
14	Lactic acid, mmol/L	10.4	−3.0	−6.4	−1.5	−4.9	4.204; 0.5205
15	Pyruvic acid, μmol/L	−1.8	1.8	10.7	8.9	5.4	4.440; 0.4879
16	Albumin, mg/mL	14.7	8.8	40.1	0.5	29.5	6.786; 0.2371
17	α-Aminoacids, mmol/L	3.8	4.8	9.1	6.5	5.2	29.84; 0.0000 *
18	Imidazole compounds, mmol/L	−7.9	−3.9	13.2	3.9	−2.6	6.158; 0.2912
19	Sialic acids, mmol/L	0.0	13.8	13.8	37.9	−5.2	7.038; 0.2179
20	Seromucoids, c.u.	8.8	6.6	8.2	8.8	2.7	6.123; 0.2944
21	ALT, U/L	2.0	10.0	6.0	10.0	16.0	0.9695; 0.9650
22	AST, U/L	16.4	10.4	16.4	26.9	10.4	9.703; 0.0841
23	AST/ALT-ratio, c.u.	16.7	1.4	19.0	26.0	11.8	8.854; 0.1150
24	LDH, U/L	43.1	31.8	4.6	21.3	65.3	18.78; 0.0021 *
25	ALP, U/L	13.8	6.9	44.8	34.7	24.1	21.68; 0.0006 *
26	GGT, U/L	8.7	12.3	17.3	13.2	2.4	40.03; 0.0000 *

Table 2. *Cont.*

No.	Indicators	Lum A	Lum B (+)	Lum B (−)	BL	Non-Lum	Kruskal–Wallis Test (H, *p*)
27	Catalase, nkat/mL	−14.9	−22.5	7.7	-18.2	−10.8	15.24; 0.0094 *
28	Superoxide dismutase, c.u.	2.3	31.8	45.5	27.3	15.9	10.79; 0.0557
29	α-Amylase, U/L	60.7	65.1	141.0	60.5	20.6	17.41; 0.0038 *
30	Antioxidant activity, mmol/L	−14.1	−0.8	4.9	2.4	−11.0	2.688; 0.7479
31	Peroxidase, c.u.	−9.6	9.6	16.4	20.5	86.3	7.194; 0.2066
32	SOD/Catalase-ratio, c.u.	3.2	53.6	53.3	80.3	36.9	12.12; 0.0332 *
33	SOD/Peroxidase-ratio, c.u.	−9.6	12.4	88.9	51.8	−35.0	13.97; 0.0158 *
34	Diene conjugates, c.u.	−3.1	−0.3	−2.8	1.7	0.7	8.617; 0.1254
35	Triene conjugates, c.u.	6.6	−0.7	−0.9	1.1	2.0	3.988; 0.5511
36	Schiff bases, c.u.	10.2	0.6	−3.2	−3.3	3.1	7.175; 0.2079
37	MDA, µmol/L	1.3	6.4	9.6	35.9	4.5	22.95; 0.0003 *
38	SB/(DC+TC)-ratio, c.u.	5.9	−0.4	−2.2	−0.7	−0.2	9.043; 0.1074
39	SB/TC-ratio, c.u.	7.2	1.6	−1.1	3.0	7.2	10.05; 0.0739
40	MM 254, c.u.	−18.8	−5.4	5.4	24.1	1.7	7.962; 0.1583
41	MM 280, c.u.	−16.7	−8.0	−10.3	22.0	−5.9	6.170; 0.2901
42	MM 280/254	2.8	0.9	3.1	4.5	1.0	4.729; 0.4498

Note. *—differences with the control group are statistically significant at $p < 0.05$.

It was shown by PCA analysis that there was no complete separation of all the studied groups (Figure 1). The first principal component (PC1) separated the control group (to the left of the vertical axis) and all groups of patients with breast cancer (to the right of the vertical axis) (Figure 1A). The maximum contribution to the separation was made by protein ($r = 0.6587$), the total content of α-amino acids ($r = 0.5804$) and urea ($r = 0.5088$), as well as the activities of catalase ($r = 0.5933$), GGT ($r = 0.5756$), ALP ($r = 0.5732$), LDH ($r = 0.5451$), and α-amylase ($r = 0.4008$). The separation by the first principal component was statistically significant ($p = 0.0052$). The division of groups relative to the horizontal axis was due to the contribution of lipid peroxidation indicators, high correlation coefficients were determined for the SB/TC-ratio and SB/(DC+TC)-ratio and amounted to 0.8241 and 0.7935, respectively (Figure 1B). At the same time, the groups of luminal A and non-luminal breast cancer, as well as both subgroups of luminal B breast cancer, turned out to be close to each other (Figure 1A). If we compare only breast cancer patients with each other, then the trend persisted (Figure 1C). Subgroups of luminal A and non-luminal breast cancer were distinguished on the diagram by a single field (Figure 1C). The vertical axis made it possible to distinguish groups of luminal B (−) and basal-like breast cancer (to the right of the axis) from the rest (Figure 1D). The contribution to the separation was made by the same parameters as when taking into account the control group (Figure 1D); however, in this case, the separation was not statistically significant. The horizontal axis also divided luminal A and B breast cancers (Figure 1C). The division was characterized by lipid peroxidation indices and was statistically significant ($p = 0.0083$).

The values of biochemical indicators, which significantly change in the studied groups, are shown in Figure 2.

It was shown that the total protein content decreased in all groups, however, it was statistically significant only for the luminal subtypes. Against the background of a decrease in protein content, the content of α-amino acids and urea increased. The activity of enzymes changed ambiguously, so the subgroups of luminal A and B (+) breast cancer are similar in the nature of changes in the activity of enzymes (Figure 2). These subgroups were characterized by a slight increase in the activity of ALP, an increase in the activity of LDH and GGT, as well as a sharp decrease in the activity of catalase. For luminal B (−) breast cancer, ALP and GGT activities reached maximum values, while LDH and catalase activities remained practically unchanged. For all luminal subtypes of breast cancer, a statistically significant increase in α-amylase activity was shown. For all subgroups except for non-luminal breast cancer, an increase in the SOD/Catalase-ratio was shown, while for the non-luminal subgroup, a decrease in the SOD/Peroxidase-ratio was statistically significant (Figure 2). For basal-like cancer, the highest content of toxic products of lipid peroxidation, in particular MDA, was noted against the background of minimal catalase activity.

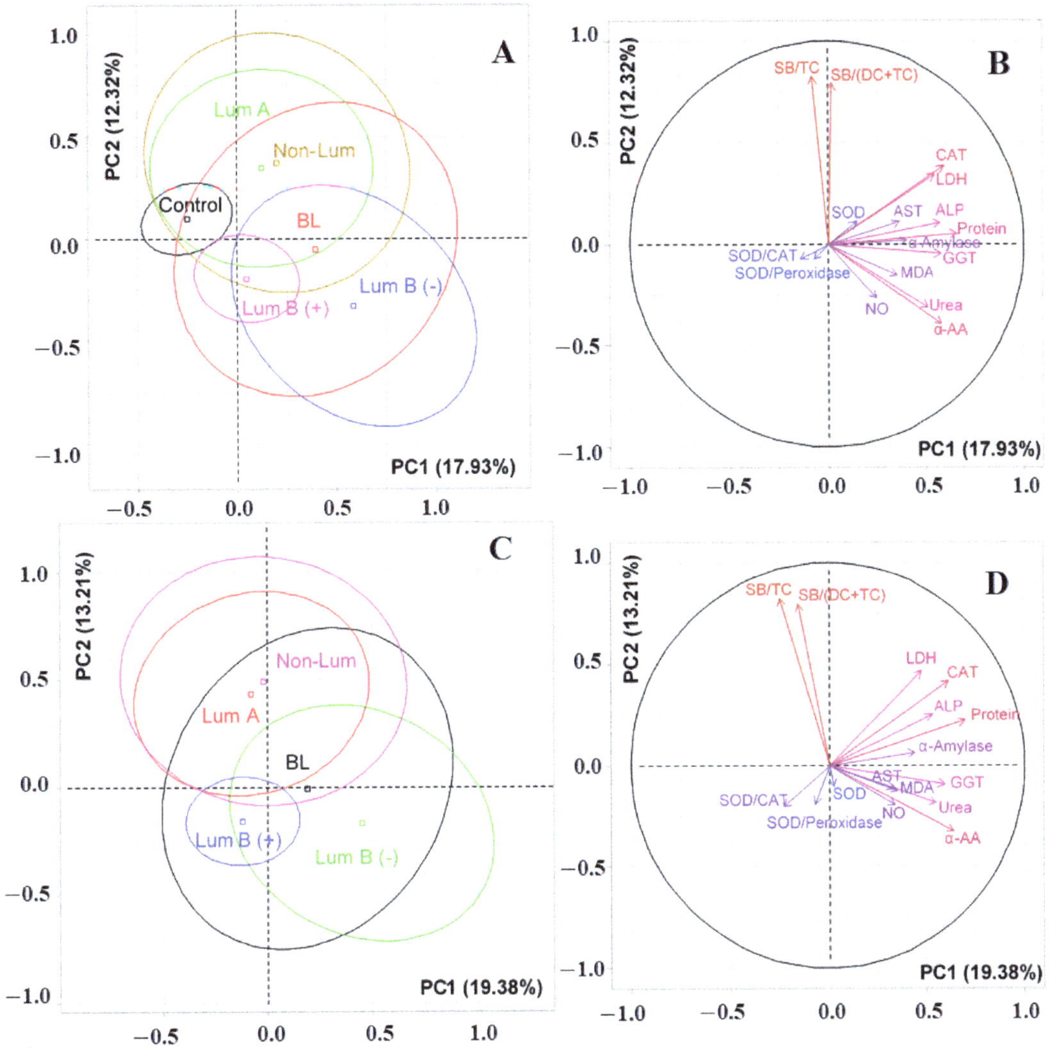

Figure 1. Individuals factor map (PCA) with control group (**A**) and without control group (**C**); variables factor map with control group (**B**) and without control group (**D**). LDH—lactate dehydrogenase, CAT—catalase, ALP—alkali phosphatase, AST—aspartate aminotransferase, MDA—malondialdehyde, GGT—gamma glutamyltransferase, α-AA—α-Amino acids, SB—Schiff Bases, TC—triene conjugates, DC—diene conjugates, SOD—superoxide dismutase.

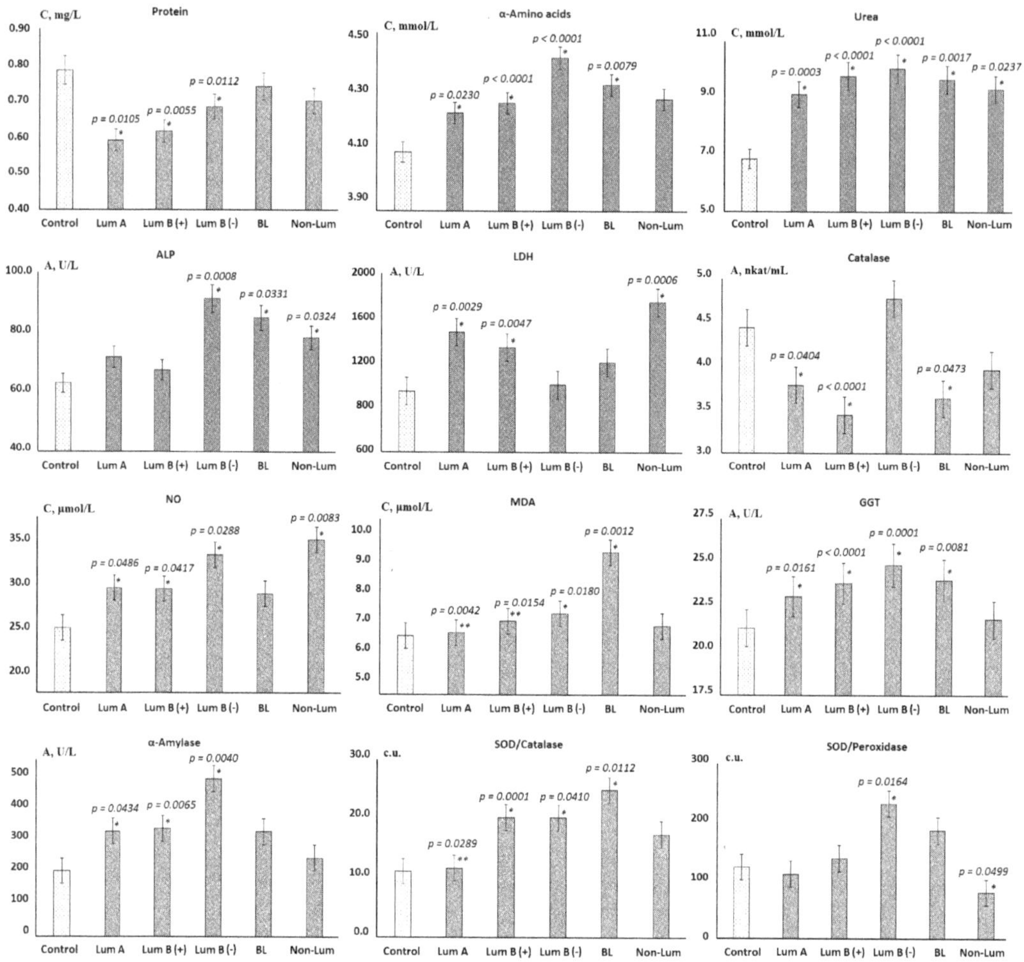

Figure 2. Biochemical composition of saliva depending on the molecular biological subtype of breast cancer. Differences between groups were calculated using the Wilcoxon matched pairs test with the Bonferoni correction at $p < 0.05$; *—differences with the control group are statistically significant, **—differences with BL are statistically significant. C—concentration, A—activity.

3.2. Changes in the Salivary Biochemical Composition of Breast Cancer Patients Depending on the HER2 Status

At the next stage, we tried to figure out which parameter determined the differences between the identified molecular biological subtypes of breast cancer. Table 3 shows the values of the Kruskal–Wallis criterion when separating groups according to the level of expression of HER2, estrogen, and progesterone receptors. We identified biochemical indicators whose differences between subgroups were significant at the level of 0.05 and 0.10 (Table 3). These parameters were subsequently used to compare groups with PCA analysis. A complete list of salivary biochemical indicator values for each of the subgroups is given in Supplementary Tables S3–S8.

Figure 3A shows that significant differences were observed only between the control group and HER2-negative breast cancer ($p = 0.0178$). The separation was due to the contribution of albumin ($r = 0.7412$), total protein ($r = 0.6717$), catalase ($r = 0.6123$), GGT ($r = 0.6058$), ALP ($r = 0.5767$), LDH ($r = 0.5409$), α-amino acids ($r = 0.5314$), and urea ($r = 0.4334$) (Figure 3B). The horizontal axis separated HER2-positive and HER2-negative breast cancer; however, the differences between the groups were not statistically significant (Figure 3A). In this case, positive correlations were noted for AST/ALT-ratio ($r = 0.7608$) and AST ($r = 0.7048$), while negative correlations were noted for uric acid ($r = -0.3065$) and α-amylase ($r = -0.3076$) (Figure 3B). If we consider the division without a control group (Figure 3C), then the vertical axis allowed the formation of two subgroups: HER2 (−) and HER2 (+), as well as HER2 (++) and HER2 (+++) ($p = 0.0189$). The horizontal axis divided the HER2 (++) and HER2 (+++) groups. In this case, α-amylase ($r = 0.4028$) was added to the list of parameters by which separation occurs for PC1, as was catalase ($r = -0.3205$) for PC2. Meanwhile, the effect of uric acid increased ($r = -0.4758$) (Figure 3D).

Figure 4 shows the relative change in each of the 42 biochemical indicators in the HER2-positive and HER2-negative breast cancer groups (Supplementary Table S3). We found that most of the indicators change in the same direction. The exception was calcium, chlorides, diene conjugates, and MM 254 nm (Figure 4). Statistically significant differences between HER2-positive and HER2-negative breast cancer were observed in AST/ALT-ratio and activity of ALP and α-amylase, as well as the SOD/Peroxidase-ratio. In general, deviations from the control group were more pronounced for HER2-positive breast cancer (Figure 4). The only indicator that statistically significantly differed between subgroups with different HER2 expression was albumin (Supplementary Table S4).

Table 3. Values of the Kruskal–Wallis test when comparing subgroups by the level of expression of HER2, estrogen receptors and progesterone, taking into account the control group.

No.	Indicators	Kruskal–Wallis Test (H, *p*) + Control Group		
		HER2	**ER**	**PR**
1	pH	2.858; 0.5818	3.467; 0.4829	0.7895; 0.9399
2	Calcium, mmol/L	5.365; 0.2518	8.055; 0.0896 **	7.971; 0.0926 **
3	Phosphorus, mmol/L	6.013; 0.1982	9.168; 0.0570 **	3.688; 0.4498
4	Ca/P-ratio, c.u.	6.397; 0.1714	5.822; 0.2128	6.498; 0.1649
5	Sodium, mmol/L	3.038; 0.5515	4.154; 0.3856	3.286; 0.5112
6	Potassium, mmol/L	5.475; 0.2419	1.370; 0.8494	1.119; 0.8913
7	Na/K-ratio, c.u.	5.651; 0.2268	3.911; 0.4182	6.548; 0.1618
8	Chlorides, mmol/L	4.632; 0.3272	5.517; 0.2382	5.454; 0.2483
9	Magnesium, mmol/L	0.6162; 0.9612	1.312; 0.8593	0.6850; 0.9532

Table 3. *Cont.*

No.	Indicators	Kruskal–Wallis Test (H, p) + Control Group		
		HER2	**ER**	**PR**
10	NO, µmol/L	16.04; 0.0030 *	17.23; 0.0017 *	20.47; 0.0004 *
11	Protein, mg/mL	75.09; 0.0000 *	89.55; 0.0000 *	79.20; 0.0000 *
12	Urea, mmol/L	65.41; 0.0000 *	66.16; 0.0000 *	65.78; 0.0000 *
13	Uric acid, µmol/L	8.990; 0.0614 **	11.65; 0.0202 *	9.063; 0.0596 **
14	Lactic acid, mmol/L	4.819; 0.3064	2.271; 0.6860	4.215; 0.3776
15	Pyruvic acid, µmol/L	5.060; 0.2812	2.375; 0.6672	4.344; 0.3614
16	Albumin, mg/mL	8.648; 0.0705 **	7.980; 0.0923 **	12.68; 0.0130 *
17	α-Aminoacids, mmol/L	24.58; 0.0001 *	25.68; 0.0000 *	26.72; 0.0000 *
18	Imidazole compounds, mmol/L	4.478; 0.3452	4.757; 0.3132	4.123; 0.3896
19	Sialic acids, mmol/L	0.7944; 0.9392	2.169; 0.7047	1.640; 0.8017
20	Seromucoids, c.u.	6.399; 0.1713	7.597; 0.1075	10.70; 0.0302 *
21	ALT, U/L	3.192; 0.5263	0.7748; 0.9418	3.865; 0.4245
22	AST, U/L	9.008; 0.0609 **	3.668; 0.4528	5.101; 0.2771
23	AST/ALT-ratio, c.u.	9.652; 0.0467 *	3.293; 0.5101	9.622; 0.0473 *
24	LDH, U/L	13.83; 0.0078 *	14.50; 0.0059 *	12.35; 0.0149 *
25	ALP, U/L	14.46; 0.0060 *	26.72; 0.0000 *	18.83; 0.0008 *
26	GGT, U/L	39.01; 0.0000 *	36.20; 0.0000 *	42.78; 0.0000 *
27	α-Amylase, U/L	17.76; 0.0014 *	16.05; 0.0030 *	19.30; 0.0007 *
28	Catalase, nkat/mL	14.51; 0.0058 *	16.04; 0.0030 *	19.41; 0.0007 *
29	Superoxide dismutase, c.u.	6.113; 0.1908	5.349; 0.2533	6.311; 0.1771
30	Antioxidant activity, mmol/L	1.095; 0.8950	1.206; 0.8772	2.973; 0.5623
31	Peroxidase, c.u.	3.604; 0.4622	5.628; 0.2287	6.264; 0.1803
32	SOD/Catalase-ratio, c.u.	7.923; 0.0944 **	5.605; 0.2307	10.07; 0.0393 *
33	SOD/Peroxidase-ratio, c.u.	5.470; 0.2424	6.314; 0.1769	11.47; 0.0217 *
34	Diene conjugates, c.u.	7.459; 0.1135	9.019; 0.0606 **	2.224; 0.6946
35	Triene conjugates, c.u.	2.874; 0.5791	5.884; 0.2080	2.649; 0.6181
36	Schiff bases, c.u.	2.215; 0.6962	3.841; 0.4280	4.934; 0.2942
37	MDA, µmol/L	16.40; 0.0025 *	19.98; 0.0005 *	19.17; 0.0007 *
38	SB/(DC+TC)-ratio, c.u.	2.115; 0.7146	8.384; 0.0785 **	4.159; 0.3849
39	SB/TC-ratio, c.u.	2.369; 0.6682	5.409; 0.2478	3.029; 0.5529
40	MM 254, c.u.	5.346; 0.2536	8.585; 0.0724 **	6.688; 0.1533
41	MM 280, c.u.	4.588; 0.3323	6.409; 0.1706	5.865; 0.2095
42	MM 280/254	4.545; 0.3372	5.853; 0.2104	4.260; 0.3720

Note. *—differences are statistically significant at $p < 0.05$; **—differences are statistically significant at $p < 0.10$.

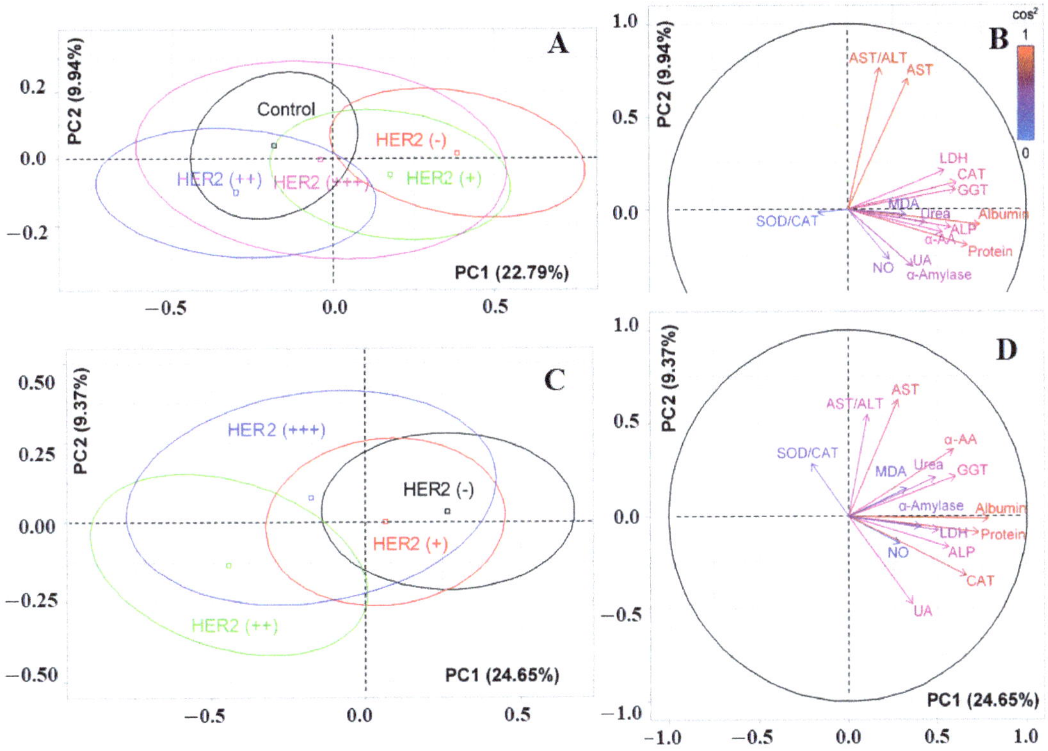

Figure 3. Individual factor map (PCA) with control group (**A**) and without control group (**C**); variables factor map with control group (**B**) and without control group (**D**). UA—uric acid, LDH—lactate dehydrogenase, CAT—catalase, ALP—alkali phosphatase, ALT—alanine aminotransferase, AST—aspartate aminotransferase, MDA—malondialdehyde, GGT—gamma glutamyltransferase, α-AA—α-Amino acids, SOD—superoxide dismutase.

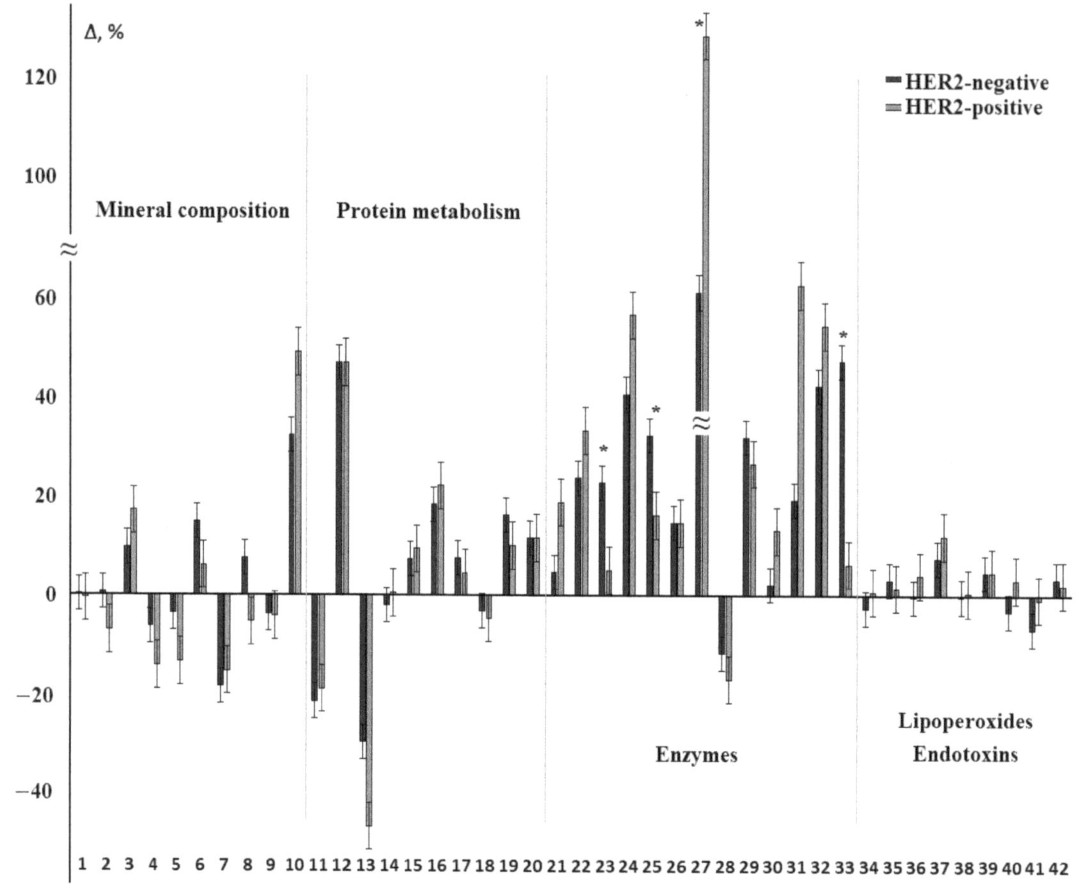

Figure 4. Relative change in biochemical indicators of saliva depending on the status of HER2 receptors. The interval of variation is given in comparison with the control group. The numbers of biochemical indicators correspond to the serial number in Tables 2 and 3. *—differences between groups with HER2-positive and HER2-negative status are statistically significant, $p < 0.05$.

3.3. *Changes in the Salivary Biochemical Composition of Breast Cancer Patients Depending on the ER Status*

According to PCA, the first principal component made it possible to distinguish the control group (to the left of the vertical axis), while the horizontal axis made it possible to distinguish ER-positive patients with breast cancer (Figure 5A). For PC1, albumin ($r = 0.7436$), protein ($r = 0.6780$), phosphorus ($r = 0.6458$), MM 254 ($r = 0.6380$), GGT ($r = 0.5587$), α-amino acids ($r = 0.5548$), catalase ($r = 0.5503$), ALP ($r = 0.5385$), LDH ($r = 0.5070$), urea ($r = 0.4554$), and calcium ($r = 0.4274$) contributed to the separation (Figure 5B). Separation by PC2 was determined by uric acid ($r = 0.5374$), diene conjugates ($r = 0.5273$), catalase ($r = 0.4276$), α-amino acids ($r = -0.4394$), and urea ($r = -0.5883$). When taking into account the degree of expression of estrogen receptors, it was shown that significant differences with the control group remained for all ER-positive subgroups, but the ER (+++) subgroup stood out separately ($p = 0.0192$, Figure 5C).

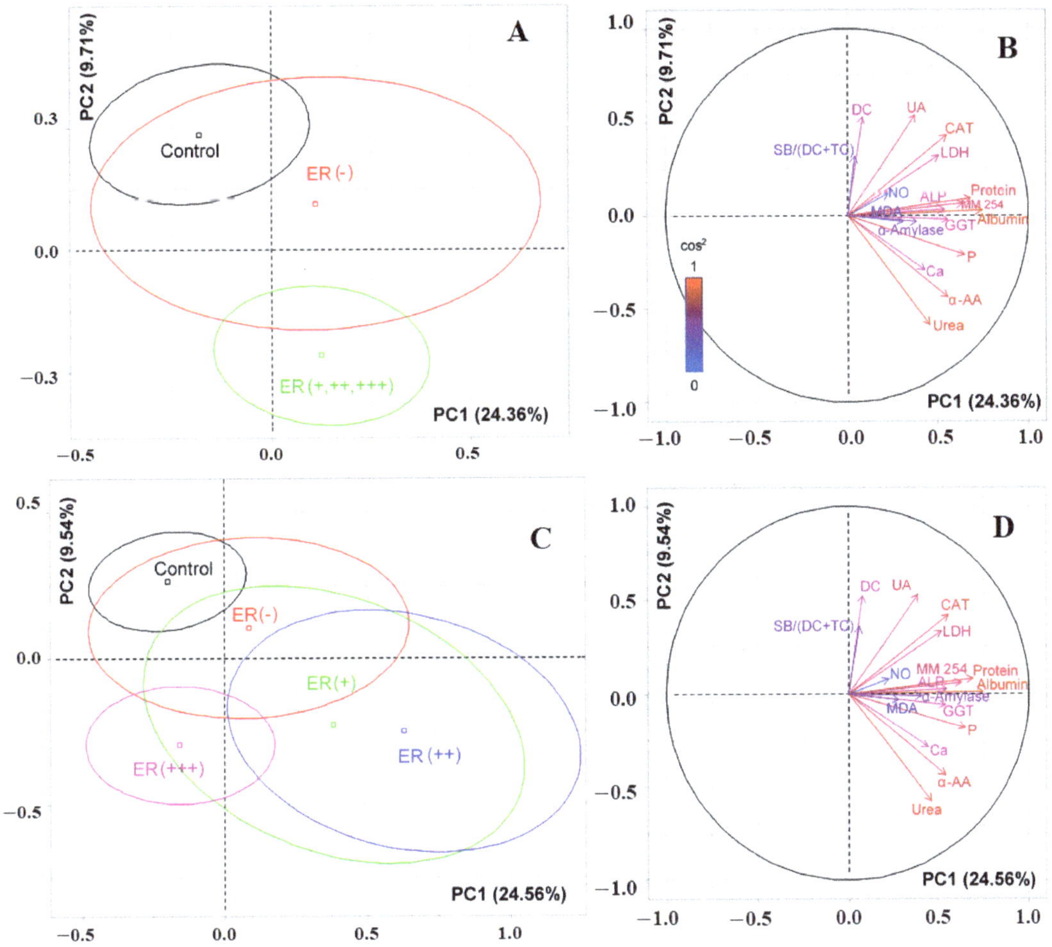

Figure 5. Individuals factor map (PCA) with control group (**A**) and without control group (**C**); variables factor map with control group (**B**) and without control group (**D**). UA—uric acid, LDH—lactate dehydrogenase, CAT—catalase, ALP—alkali phosphatase, MDA—malondialdehyde, SB—Schiff Bases, TC—triene conjugates, GGT—gamma glutamyltransferase, α-AA—α-Amino acids, P—phosphorus, MM—middle molecules.

For ER-positive and ER-negative breast cancer, the SOD/Peroxidase ratio and the content of diene conjugates and MM 254 and 280 changed in different directions (Figure 6, Supplementary Table S5). Also statistically significant was an increase in the activity of ALP and salivary peroxidase, the level of diene conjugates, and MDA for ER-negative breast cancer. At the same time, α-amylase activity and the SOD/Peroxidase-ratio were higher for the subgroup of ER-positive breast cancer (Figure 6). When comparing subgroups with different expression of estrogen receptors, it was shown that the ER (+) and ER (+++) groups differed in the content of total protein (−31.3%, $p = 0.0002$), triene conjugates (−5.7%, $p = 0.0192$), and SB/(DC+TC)-ratio (−3.5%, $p = 0.0161$). No differences were found between the ER (+) and ER (++) groups (Supplementary Table S6).

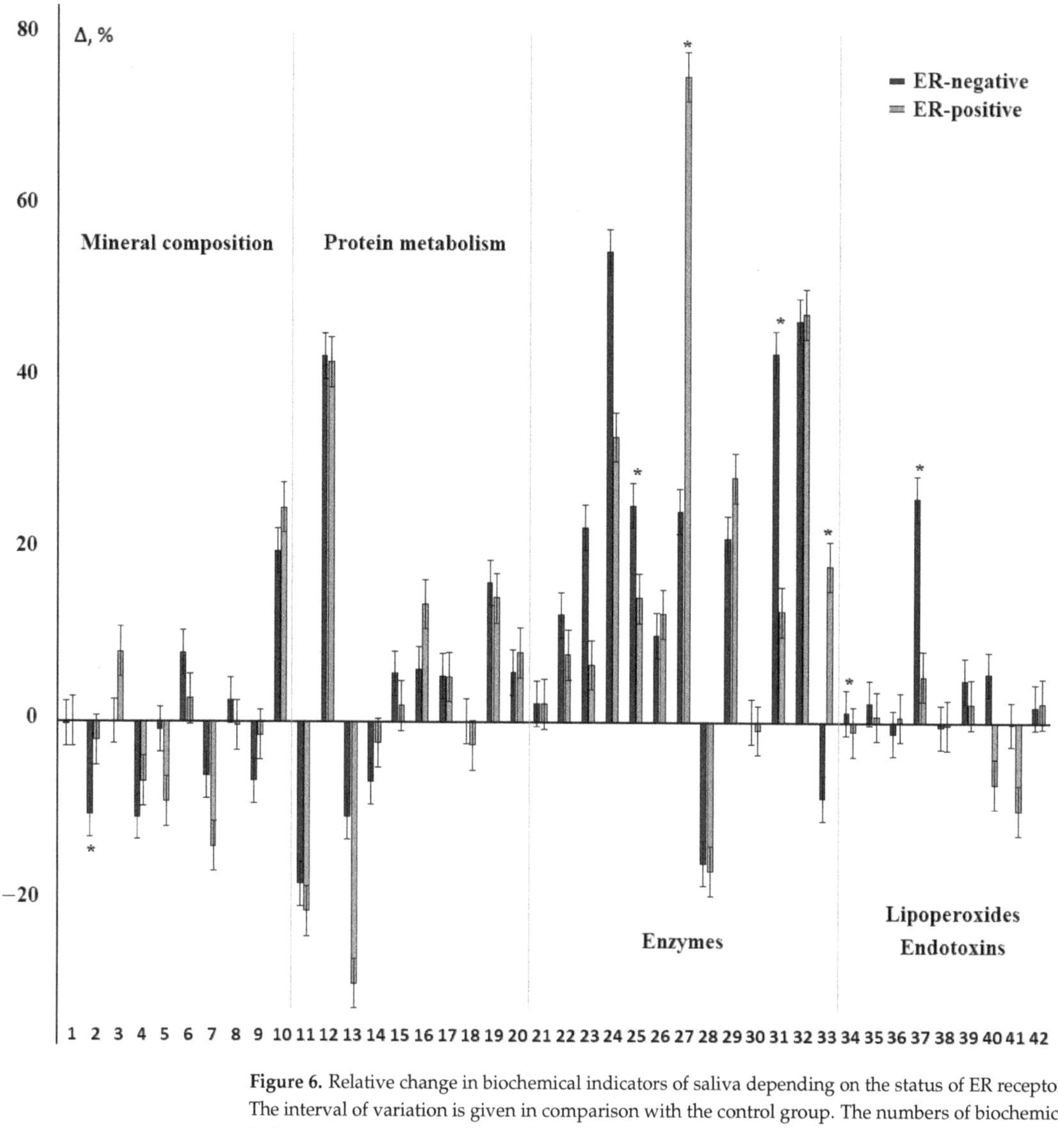

Figure 6. Relative change in biochemical indicators of saliva depending on the status of ER receptors. The interval of variation is given in comparison with the control group. The numbers of biochemical indicators correspond to the serial number in Tables 2 and 3. *—differences between groups with ER-positive and ER-negative status are statistically significant, $p < 0.05$.

3.4. Changes in the Salivary Biochemical Composition of Breast Cancer Patients Depending on the PR Status

When taking into account the expression of progesterone receptors, it was shown that PR-positive and PR-negative subgroups practically did not differ from each other, but significantly differed from the control group ($p < 0.0001$, Figure 7A). Albumin ($r = 0.7587$), total protein ($r = 0.6834$), seromucoids ($r = 0.6816$), catalase ($r = 0.6092$), GGT ($r = 0.5642$), ALP ($r = 0.5552$), LDH ($r = 0.5456$), α-amino acids ($r = 0.5192$), urea ($r = 0.4215$), and α-amylase ($r = 0.4009$) made the main contributions to the separation by PC1. Separation by PC2 was due to the contribution of urea ($r = 0.5290$), SOD/Catalase-ratio ($r = 0.4955$), α-amino acids ($r = 0.4829$), and catalase ($r = -0.4482$) (Figure 7B). When taking into account the degree of expression of progesterone receptors, it was shown that the vertical axis separated the PR (−), PR (+), and PR (++) groups from the PR (+++) group and the control group ($p = 0.0048$, Figure 7C). The horizontal axis separated patients with breast cancer from controls ($p<0.0001$). The contribution of biochemical indicators to the division of subgroups practically did not change (Figure 7B vs. Figure 7D).

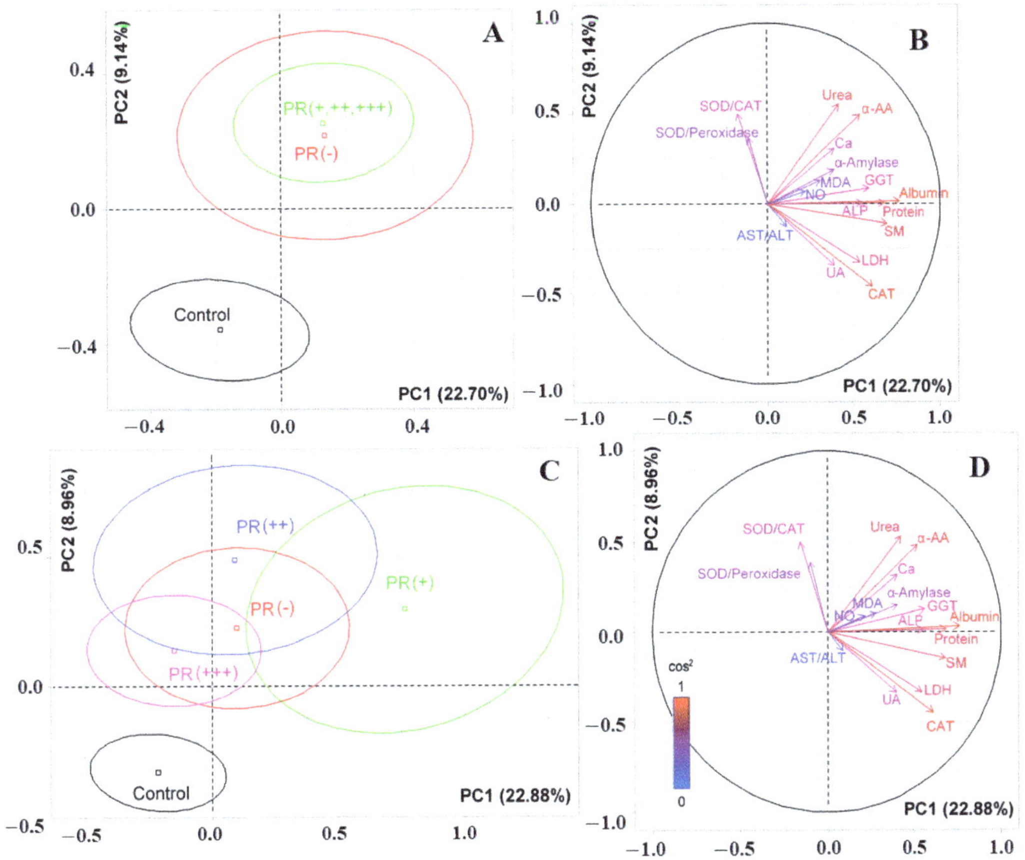

Figure 7. Individuals factor map (PCA) with control group (**A**) and without control group (**C**); variables factor map with control group (**B**) and without control group (**D**). UA—uric acid, LDH—lactate dehydrogenase, CAT—catalase, ALP—alkali phosphatase, ALT—alanine aminotransferase, AST—aspartate aminotransferase, MDA—malondialdehyde, GGT—gamma glutamyltransferase, α-AA—α-Amino acids, SOD—superoxide dismutase, SM—seromucoids.

Differences between PR-positive and PR-negative breast cancer were statistically significant for calcium, uric acid, and α-amylase (Figure 8, Supplementary Table S7). The content of sodium, chlorides, AOA, Schiff bases, and MM 254 and 280 nm changed in different directions compared to the control group (Figure 8).

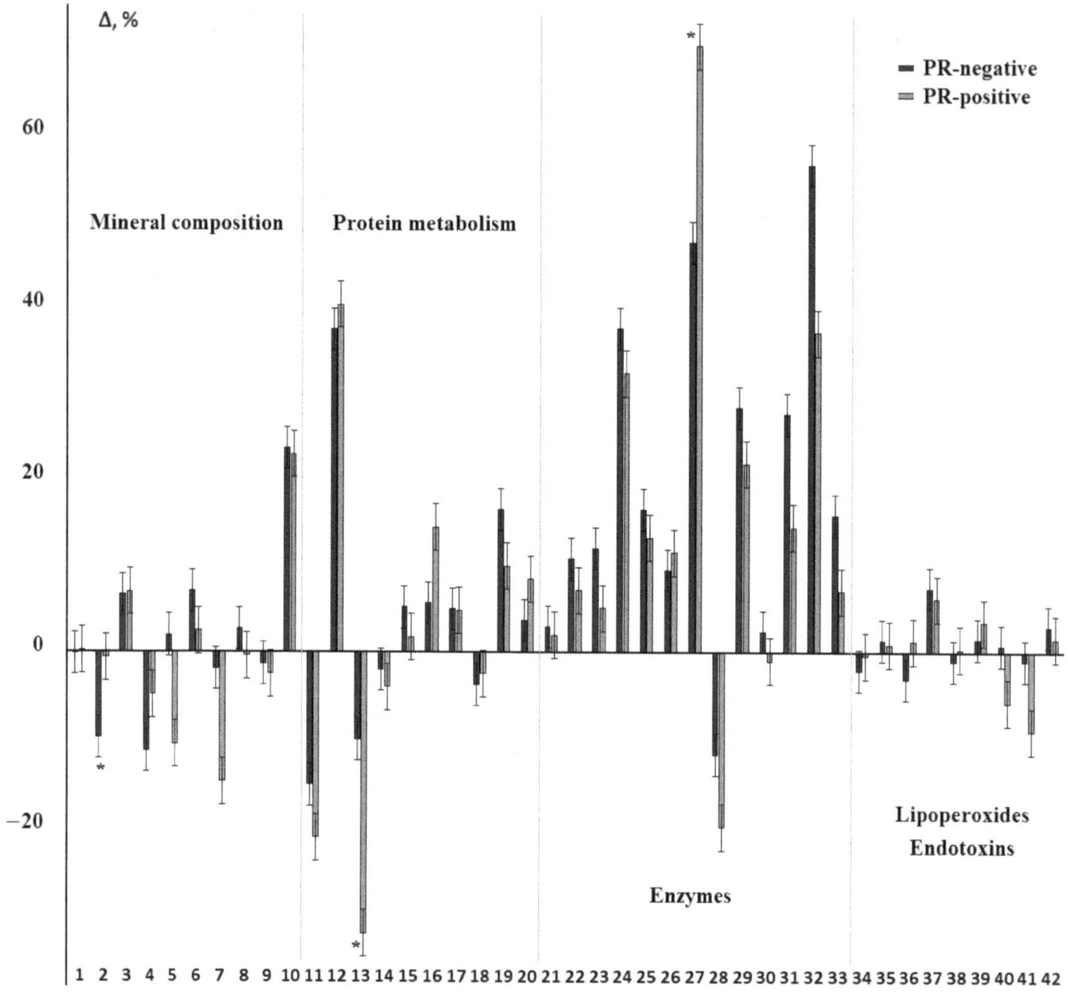

Figure 8. Relative change in biochemical indicators of saliva depending on the status of PR receptors. The interval of variation is given in comparison with the control group. The numbers of biochemical indicators correspond to the serial number in Tables 2 and 3. *—differences between groups with PR-positive and PR-negative status are statistically significant, $p < 0.05$.

Differences between PR (+) and PR (++) were significant in terms of ALP activity (−17.9%, p = 0.0353), catalase (−29.1%, p = 0.0125), and seromucoids level (−18.6%, p = 0.0196) (Supplementary Table S8). The same parameters determined the difference between the PR (+) and PR (+++) groups; however, the content of protein, albumin, AST/ALT-ratio, GGT, and α-amylase were also added. All of the listed indicators showed a decrease in values for the PR (+) and PR (+++) groups.

Simultaneous consideration of the positive and negative status of estrogen and progesterone receptors gave similar results with division by molecular biological subtypes of breast cancer (Supplementary Table S9). Thus, the ER/PR-negative subgroup united the subgroups of basal-like and non-luminal breast cancer, while the ER/PR-positive subgroup united the luminal subtypes of breast cancer (Supplementary Tables S1 and S2). Differences between ER/PR-negative and ER/PR-positive breast cancer were identified in the content of calcium (−10.0 and 0.0%), sodium (−2.2 and −11.4%), and MDA (+25, 6 and +6.4%), as well as the activity of ALP (+31.0% and +13.8%) and peroxidase (+57.5 and +17.8% for ER/PR-negative and ER/PR-positive breast cancer respectively. Changes are shown compared to the control group).

4. Discussion

Most studies on the analysis of saliva in breast cancer were aimed at identifying biomarkers that can differentiate patients with breast cancer from healthy controls [28–44]. Previously, we identified 11 metabolites that allow us to do this with a sensitivity of up to 91% [47]. These indicators included urea, total protein, total content of α-amino acids, MDA, NO, Na/K-ratio, SB/TC-ratio, as well as ALP and GGT activity. When taking into account the molecular biological subtype of breast cancer, Na/K-ratio and SB/TC-ratio do not contribute to the division of subgroups; however, α-amylase, LDH, catalase, SOD/Catalase-ratio, and SOD/Peroxidase-ratio become significant (Table 3). Within the breast cancer group, we showed the maximum differences for basal-like cancer in terms of increased levels of endogenous toxins (MM 254 and 280) and lipid peroxidation products (DC, MDA), as well as SOD/Catalase. Complex metabolic disorders and nonspecific clinical manifestations that accompany the development of malignant neoplasms are characterized as endogenous intoxication syndrome [52–54]. An increase in the ratio of MM 280/254 nm is indirect evidence of the excessive generation of active oxygen metabolites, superoxide radicals, and hydrogen peroxide [55]. Hydroxyl radicals are capable of damaging the phosphoglyceride membrane structures of cell membranes and their organoids. The object of exposure to active oxygen metabolites is arachidonic acid containing four double bonds separated by CH_2 groups. When exposed to hydroxyl radicals, double bonds become conjugated and diene conjugates are formed, which later turn into lipid hydroperoxides [51,56]. This situation reflects the fact that the accumulation of endogenous toxins and lipid peroxides occurs at a faster rate than their inactivation by the antioxidant defense system. It is significant that such a picture was observed for the BC subtype, which has the least favorable prognosis [57]. Differences in the largest number of indicators were found for basal-like and luminal A subtypes of breast cancer. For other subtypes of breast cancer, significant differences were found only in comparison with the control group.

Based on data reported in a study [46], the levels of five metabolites differed significantly between the luminal A-like and B-like subtypes (cadaverine, 5-aminovalerate, gamma-butyrobetaine, 2-hydroxy-4-methylpentanoate, alanine), while N-acetylneuraminate was only significantly differentiating between the luminal A-like and triple negative subtypes. For other metabolites, no differences were found between breast cancer subtypes [46]. The only indirect intersection in the list of determined parameters refers to alanine, since this amino acid is included in the indicator of the total content of α-amino acids determined by us. Nevertheless, both studies confirm that there are metabolic features of saliva depending on the molecular biological subtype of breast cancer, which shows the promise of research in this direction. The need for research is confirmed by the fact that it is not always possible to determine the molecular biological subtype of a tumor. Thus, in our sample,

13.1% of patients with breast cancer lack the results of immunohistochemistry of the tumor, which imposes certain restrictions on the choice of treatment tactics and determining the prognosis of the disease.

We tried to analyze the influence of each factor that determines the assignment to one or another subtype of breast cancer separately. The results obtained have not been described previously in the literature. Thus, it is considered that patients whose samples were assessed as HER2 (+++) have a positive HER2 status, and HER2 (−)/(+) have a negative status. HER2 (++) samples are considered indeterminate and should be retested by in situ hybridization. According to our data, samples with HER2 (−) and HER2 (+) status had no differences (Figure 3A,C) and were singled out on the factor diagram by one field, which once again confirms the legitimacy of considering these subgroups as one HER2-negative groups. Differences with the control group in this case were expressed as much as possible. The most important biochemical indicators that determine the division into HER2-positive and HER2-negative subgroups were the metabolic enzymes ALT and AST, as well as uric acid and the SOD/catalase-ratio.

Changes in the biochemical composition of saliva at different levels of estrogen and progesterone receptors were more pronounced. For ER-positive breast cancer, the salivary composition was significantly different from the control, but the differences between ER-positive and ER-negative BC were noticeable. Significant differences were found in terms of lipid peroxidation (DC, MDA), peroxidase activity, SOD/Peroxidase-ratio, ALP, and LDH. It is known that salivary peroxidase plays a dual role: it is responsible for the breakdown of cytotoxic hydrogen peroxide and has bactericidal activity against the oral microbiota [58]. Salivary peroxidase is the only antioxidant synthesized exclusively in the salivary glands [59]. Thus, salivary peroxidase activity reflects the effectiveness of the salivary glands in preventing oxidative stress. An increase in peroxidase activity in the saliva of patients with breast cancer indicates an increase in the enzymatic antioxidant defense that protects the salivary glands and the entire oral cavity from oxidative damage. An increase in peroxidase activity occurred against the background of an increase in the level of lipid peroxidation products and was characteristic of ER-negative breast cancer, which has the least favorable prognosis. It is known that ER-regulated overexpression of the HER2 protein is combined with increased activity in the tumor of the muscle isoform of LDH, one of the key enzymes of the glycolytic pathway of glucose oxidation, while LDH activity was higher in the blood of patients with ER-negative tumors [60]. Previously, we showed the existence of a correlation between the activity of LDH in saliva and blood plasma [61]. For PR receptors, no differences between PR-positive and PR-negative breast cancer in saliva were found. When considering combinations of the level of ER and PR receptors, it was shown that for ER/PR-positive breast cancer in saliva, the activity of metabolic enzymes (ALP, LDH) was statistically lower, the level of lipid peroxides (DC, MDA) was lower, as was the content of uric acid, catalase, and peroxidases. This indicates a less pronounced intensity of lipid peroxidation processes and a balanced work of the antioxidant defense system (both its enzymatic and non-enzymatic links). Since only about 10% of all breast cancers have the status of ER (+)/PR (−) and about 5% of ER (−)/PR (+), we did not consider such combinations due to the small number of patients in each [62]. There are data in the literature on the direct determination of HER2 in saliva [63–65]. Current research suggests that soluble fragments of the HER2 oncogene may be released from the cell surface and found in patients with breast carcinoma. The salivary HER2 protein assay has been shown to be reliable and may have potential applications in the initial detection and subsequent screening of recurrent breast cancer [63]. HER2 has been found in the saliva of women with benign breast lesions and women diagnosed with breast cancer. HER2 levels in cancer patients were significantly higher than those in saliva of healthy controls and patients with benign tumors [64,65]. Nevertheless, there are a number of unanswered questions, including how the level of HER2 in saliva correlates with its content in the tumor tissue and whether the level of HER2 allows the tumor to be assigned to a specific molecular biological subtype, which determines the choice of treatment tactics.

It should be noted that the correlation between salivary hormonal status and tumor receptor status has also not been proven; therefore, tumor biopsy is still a necessary step in diagnosis and treatment. However, for treatment progression and recurrence monitoring, the choice of indirect salivary indicators associated with a particular tumor type may be important.

The limitations of the study were related to the fact that we initially chose metabolites that can be determined using a biochemical analyzer. At this stage, we have shown potential directions for research; in particular, the important role of the total content of α-amino acids shows the need to determine the amino acid profile. In continuation of the research, we plan to analyze the ALP and LDH isoenzymes. The limitations included the fact that we did not conduct a parallel determination of HER2 in saliva and did not assess the hormonal status of saliva. In this work, we did not determine the prognostic significance of the selected saliva biochemical indicators and did not evaluate their change during treatment.

5. Conclusions

Our study showed that various molecular biological subtypes of breast cancer are characterized by changes in the metabolic profile of saliva. It was shown that the composition of the saliva of patients with basal-like breast cancer differed from the control group as much as possible. The biochemical composition of saliva varies more depending on the HER2 status and the status of estrogen receptors and, to a lesser extent, on the status of progesterone receptors. It was found that the HER2 (−)/HER2 (+) group, which should be considered as a single group, as well as ER-positive breast cancer, differed statistically significantly from the control group. For ER/PR-positive breast cancer, a more favorable ratio of biochemical indicators of saliva was also noted. Thus, the composition of saliva reacts very subtly to changes in the human body, including the ability to assess metabolic changes in different molecular biological subtypes of breast cancer. All this emphasizes the prospects for continuing research in this direction.

Supplementary Materials: The following supporting information can be downloaded at: https://www.mdpi.com/article/10.3390/cimb44070211/s1, Table S1: Biochemical composition of saliva for Luminal B-like breast cancer compared with the control group. Table S2: Biochemical composition of saliva for Luminal A-like, Non-Luminal and Basal-like breast cancer compared with the control group. Table S3: Biochemical composition of saliva in HER2-positive and HER2-negative breast cancer. Table S4: Biochemical composition of saliva in HER2-positive (+, ++, +++) breast cancer. Table S5: Biochemical composition of saliva in ER-positive and ER-negative breast cancer. Table S6: Biochemical composition of saliva in ER-positive (+, ++, +++) breast cancer. Table S7: Biochemical composition of saliva in PR-positive and PR-negative breast cancer. Table S8: Biochemical composition of saliva in PR-positive (+, ++, +++) breast cancer. Table S9: Biochemical composition of saliva in ER/PR-positive and ER/PR-negative breast cancer.

Author Contributions: Conceptualization, L.V.B. and E.A.S.; methodology, L.V.B.; validation, L.V.B.; formal analysis, L.V.B.; investigation, E.A.S.; resources, L.V.B.; data curation, E.A.S.; writing—original draft preparation, E.A.S.; writing—review and editing, L.V.B.; visualization, E.A.S.; supervision, L.V.B.; project administration, L.V.B. All authors have read and agreed to the published version of the manuscript.

Funding: This research received no external funding.

Institutional Review Board Statement: The study was conducted in accordance with the Declaration of Helsinki, and approved by the Ethics Committee of the Omsk Regional Clinical Hospital "Clinical Oncology Center" (21 July 2016, protocol No. 15).

Informed Consent Statement: Informed consent was obtained from all subjects involved in the study.

Data Availability Statement: The data presented in this study are available on request from the corresponding author. The data are not publicly available because they are required for the preparation of a Ph.D. Thesis.

Conflicts of Interest: The authors declare no conflict of interest.

References

1. Sung, H.; Ferlay, J.; Siegel, R.L.; Laversanne, M.; Soerjomataram, I.; Jemal, A.; Bray, F. Global cancer statistics 2020: GLOBOCAN estimates of incidence and mortality worldwide for 36 cancers in 185 countries. *CA Cancer J. Clin.* **2021**, *71*, 209–249. [CrossRef] [PubMed]
2. Global Burden of Disease Cancer Collaboration. Global, Regional, and National Cancer Incidence, Mortality, Years of Life Lost, Years Lived with Disability and Disability-Adjusted Life-Years for 29 Cancer Groups, 1990 to 2017. A Systematic Analysis for the Global Burden of Disease Study. *JAMA Oncol.* **2019**, *5*, 1749–1768. [CrossRef] [PubMed]
3. Veronesi, U.; Boyle, P. Breast Cancer. In *International Encyclopedia of Public Health*, 2nd ed.; Reference Module in Biomedical Sciences; Academic Press: Cambridge, MA, USA, 2017; pp. 272–280.
4. Wark, P.A.; Peto, J. *Cancer Epidemiology*, 2nd ed.; Reference Module in Biomedical Sciences International Encyclopedia of Public Health; Academic Press: Oxford, UK, 2017; pp. 339–346.
5. Grabinski, V.F.; Brawley, O.W. Disparities in Breast Cancer. *Obstet. Gynecol. Clin. N. Am.* **2022**, *49*, 149–165. [CrossRef]
6. Sullivan, C.L.; Butler, R.; Evans, J. Impact of a Breast Cancer Screening Algorithm on Early Detection. *J. Nurse Pract.* **2021**, *17*, 1133–1136. [CrossRef]
7. Poddubnaya, I.V.; Kolyadina, I.V.; Kalashnikov, N.D.; Borisov, D.A.; Makarova, M.V. Population "portrait" of breast cancer in Russia: Analysis of Russian registry data. *Sovrem. Onkol.* **2015**, *17*, 25–29.
8. Loi, S. The ESMO clinical practise guidelines for early breast cancer: Diagnosis, treatment and follow-up: On the winding road to personalized medicine. *Ann. Oncol.* **2019**, *30*, 1183–1184. [CrossRef] [PubMed]
9. Malla, R.R.; Kiran, P. Tumor microenvironment pathways: Cross regulation in breast cancer metastasis. *Genes Dis.* **2022**, *9*, 310–324. [CrossRef]
10. Kushlinskiy, N.Y.; Krasil'nikov, M.A. *Biological Markers of Tumors: Fundamental and Clinical Studies*; M: Izd-vo RAMN: Moscow, Russia, 2017; p. 632.
11. Polyak, K. Heterogeneity in breast cancer. *J. Clin. Investig.* **2011**, *121*, 3786–3788. [CrossRef] [PubMed]
12. Nuyten, D.S.A.; Hastie, T.; Chi, J.-T.A.; Chang, H.Y.; van de Vijver, M.J. Combining biological gene expression signatures in predicting outcome in breast cancer: An alternative to supervised classification. *Eur. J. Cancer* **2008**, *44*, 2319–2329. [CrossRef]
13. Barzaman, K.; Karami, J.; Zarei, Z.; Hosseinzadeh, A.; Kazemi, M.H.; Moradi-Kalbolandi, S.; Safari, E.; Farahmand, L. Breast cancer: Biology, biomarkers, and treatments. *Int. Immunopharmacol.* **2020**, *84*, 106535. [CrossRef]
14. Luond, F.; Tiede, S.; Christofori, G. Breast cancer as an example of tumour heterogeneity and tumour cell plasticity during malignant progression. *Br. J. Cancer* **2021**, *125*, 164–175. [CrossRef] [PubMed]
15. Bertos, N.R.; Park, M. Breast cancer-one term, many entities? *J. Clin. Investig.* **2011**, *121*, 3789–3796. [CrossRef] [PubMed]
16. Kolyadina, I.V.; Poddubnaya, I.V.; Frank, G.A.; Komov, D.V.; Ozhereliev, A.S.; Karseladze, A.I.; Ermilova, V.D.; Vishnevskaya, Y.V.; Makarenko, N.P.; Kerimov, R.A.; et al. The Prognostic Significance of Receptor Status in Patients with Early Breast Cancer. *Mod. Technol. Med.* **2012**, *4*, 48–53.
17. Rapado-González, Ó.; Majem, B.; Muinelo-Romay, L.; López-López, R.; Suarez-Cunqueiro, M.M. Cancer Salivary Biomarkers for Tumours Distant to the Oral Cavity. *Int. J. Mol. Sci.* **2016**, *17*, 1531. [CrossRef]
18. Bel'skaya, L.V. Possibilities of using saliva for the diagnosis of cancer. *Klin. Lab. Diagn. (Russ. Clinical Lab. Diagn.)* **2019**, *64*, 333–336. [CrossRef]
19. Kaczor-Urbanowicz, K.E.; Wei, F.; Rao, S.L.; Kim, J.; Shin, H.; Cheng, J.; Tu, M.; Wong, D.T.W.; Kim, Y. Clinical validity of saliva and novel technology for cancer detection. *BBA-Rev. Cancer* **2019**, *1872*, 49–59. [CrossRef]
20. Roblegg, E.; Coughran, A.; Sirjani, D. Saliva: An all-rounder of our body. *Eur. J. Pharm. Biopharm.* **2019**, *142*, 133–141. [CrossRef]
21. Tong, P.; Yuan, C.; Sun, X.; Yue, Q.; Wang, X.; Zheng, S. Identification of salivary peptidomic biomarkers in chronic kidney disease patients undergoing hemodialysis. *Clin. Chim. Acta* **2019**, *489*, 154–161. [CrossRef]
22. Khurshid, Z.; Warsi, I.; Moin, S.F.; Slowey, P.D.; Latif, M.; Zohaib, S.; Zafar, M.S. Biochemical analysis of oral fluids for disease detection. *Adv. Clin. Chem.* **2021**, *100*, 205–253.
23. Navazesh, M.; Dincer, S. *Salivary Bioscience and Cancer*; Salivary Bioscience; Springer Nature: Cham, Switzerland, 2020; pp. 449–467.
24. Bonne, N.J.; Wong, D.T. Salivary biomarker development using genomic, proteomic and metabolomic approaches. *Genome Med.* **2012**, *4*, 82. [CrossRef]
25. Yoshizawa, J.M.; Schafer, C.A.; Schafer, J.J.; Farrell, J.J.; Paster, B.J.; Wong, D.T. Salivary biomarkers: Toward future clinical and diagnostic utilities. *Clin. Microbiol. Rev.* **2013**, *26*, 781–791. [CrossRef] [PubMed]
26. Wong, D.T.W. Salivaomics. *J. Am. Dent. Assoc.* **2012**, *143*, 19S–24S. [CrossRef] [PubMed]
27. Cheng, J.; Nonaka, T.; Ye, Q.; Wei, F.; Wong, D.T.W. *Salivaomics, Saliva-Exosomics, and Saliva Liquid Biopsy*; Salivary Bioscience; Springer Nature: Cham, Switzerland, 2020; pp. 157–175.
28. Mockus, M.; Prebil, L.; Ereman, R.; Dollbaum, C.; Powell, M.; Yau, C.; Benz, C.C. First Pregnancy Characteristics, Postmenopausal Breast Density, and Salivary Sex Hormone Levels in a Population at High Risk for Breast Cancer. *BBA Clin.* **2015**, *3*, 189–195. [CrossRef] [PubMed]
29. Oakman, C.; Tenori, L.; Biganzoli, L.; Santarpia, L.; Cappadona, S.; Luchinat, C.; Di Leo, A. Uncovering the metabolomic fingerprint of breast cancer. *Int. J. Biochem. Cell Biol.* **2011**, *43*, 1010–1020. [CrossRef] [PubMed]

30. Porto-Mascarenhas, E.C.; Assad, D.X.; Chardin, H.; Gozal, D.; De Luca Canto, G.; Acevedo, A.C.; Guerra, E.N. Salivary biomarkers in the diagnosis of breast cancer: A review. *Crit. Rev. Oncol./Hematol.* **2017**, *110*, 62–73. [CrossRef]
31. Zhong, L.; Cheng, F.; Lu, X.; Duan, Y.; Wang, X. Untargeted saliva metabonomics study of breast cancer based on ultraperformance liquid chromatography coupled to mass spectrometry with HILIC and RPLC separations. *Talanta* **2016**, *158*, 351–360. [CrossRef]
32. Takayama, T.; Tsutsui, H.; Shimizu, I.; Toyama, T.; Yoshimoto, N.; Endo, Y.; Inoue, K.; Todoroki, K.; Min, J.Z.; Mizuno, H.; et al. Diagnostic approach to breast cancer patients based on target metabolomics in saliva by liquid chromatography with tandem mass spectrometry. *Clin. Chim. Acta* **2016**, *452*, 18–26. [CrossRef]
33. Jinno, H.; Murata, T.; Sunamura, M.; Sugimoto, M.; Kitagawa, Y. Breast cancer-specific signatures in saliva metabolites for early diagnosis. *Poster Abstr. I/Breast* **2015**, *24* (Suppl. S1), S26–S86. [CrossRef]
34. Cheng, F.; Wang, Z.; Huang, Y.; Duan, Y.; Wang, X. Investigation of salivary free amino acid profile for early diagnosis of breast cancer with ultra-performance liquid chromatography-mass spectrometry. *Clin. Chim. Acta* **2015**, *447*, 23–31. [CrossRef]
35. Liu, X.; Yu, H.; Qiao, Y.; Yang, J.; Shu, J.; Zhang, J.; Zhang, Z.; He, J.; Li, Z. Salivary Glycopatterns as Potential Biomarkers for Screening of Early-Stage Breast Cancer. *eBioMedicine* **2018**, *28*, 70–79. [CrossRef]
36. Pereira, J.A.M.; Taware, R.; Porto-Figueira, P.; Rapole, S.; Câmara, J.S. Chapter 29—The salivary volatome in breast cancer. In *Precision Medicine for Investigators, Practitioners and Providers*; Faintuch, J., Faintuch, S., Eds.; Academic Press: Cambridge, MA, USA, 2020; pp. 301–307.
37. López-Jornet, P.; Aznar, C.; Ceron, J.J.; Tvarijonaviciute, A. Salivary biomarkers in breast cancer: A cross-sectional study. *Support Care Cancer* **2021**, *29*, 889–896. [CrossRef] [PubMed]
38. Assad, D.X.; Mascarenhas, E.C.P.; de Lima, C.L.; de Toledo, I.P.; Chardin, H.; Combes, A.; Acevedo, A.C.; Guerra, E.N.S. Salivary metabolites to detect patients with cancer: A systematic review. *Int. J. Clin. Oncol.* **2020**, *25*, 1016–1036. [CrossRef] [PubMed]
39. Yang, J.; Liu, X.; Shu, J.; Hou, Y.; Chen, M.; Yu, H.; Ma, T.; Du, H.; Zhang, J.; Qiao, Y.; et al. Abnormal Galactosylated–Glycans recognized by Bandeiraea Simplicifolia Lectin I in saliva of patients with breast Cancer. *Glycoconj. J.* **2020**, *37*, 373–394. [CrossRef]
40. Sugimoto, M.; Wong, D.T.; Hirayama, A.; Soga, T.; Tomita, M. Capillaryelectrophoresis mass spectrometry-based saliva metabolomics identified oral, breast and pancreatic cancer-specific profiles. *Metabolomics* **2010**, *6*, 78–95. [CrossRef] [PubMed]
41. Bigler, L.R.; Streckfus, C.F.; Copeland, L.; Burns, R.; Dai, X.; Kuhn, M.; Martin, P.; Bigler, S.A. The potential use of saliva to detect recurrence of disease in women with breast carcinoma. *J. Oral Pathol. Med.* **2002**, *31*, 421–431. [CrossRef]
42. Streckfus, C.F. Salivary Biomarkers to Assess Breast Cancer Diagnosis and Progression: Are We There Yet? In *Saliva and Salivary Diagnostics*; IntechOpen: London, UK, 2019. [CrossRef]
43. Streckfus, C.; Bigler, L. A Catalogue of Altered Salivary Proteins Secondary to Invasive Ductal Carcinoma: A Novel In Vivo Paradigm to Assess Breast Cancer Progression. *Sci. Rep.* **2016**, *6*, 30800. [CrossRef]
44. Ragusa, A.; Romano, P.; Lenucci, M.S.; Civino, E.; Vergara, D.; Pitotti, E.; Neglia, C.; Distante, A.; Romano, G.D.; Di Renzo, N.; et al. Differential Glycosylation Levels in Saliva from Patients with Lung or Breast Cancer: A Preliminary Assessment for Early Diagnostic Purposes. *Metabolites* **2021**, *11*, 566. [CrossRef]
45. Rapado-González, Ó.; Martínez-Reglero, C.; Salgado-Barreira, Á.; Takkouche, B.; López-López, R.; Suárez-Cunqueiro, M.M.; Muinelo-Romay, L. Salivary biomarkers for cancer diagnosis: A meta-analysis. *Ann. Med.* **2020**, *52*, 131–144. [CrossRef]
46. Murata, T.; Yanagisawa, T.; Kurihara, T.; Kaneko, M.; Ota, S.; Enomoto, A.; Tomita, M.; Sugimoto, M.; Sunamura, M.; Hayashida, T.; et al. Salivary metabolomics with alternative decision tree-based machine learning methods for breast cancer discrimination. *Breast Cancer Res. Treat.* **2019**, *177*, 591–601. [CrossRef]
47. Bel'skaya, L.V.; Sarf, E.A.; Solomatin, D.V.; Kosenok, V.K. Metabolic Features of Saliva in Breast Cancer Patients. *Metabolites* **2022**, *12*, 166. [CrossRef]
48. Ilić, I.R.; Stojanović, N.M.; Radulović, N.S.; Živković, V.V.; Randjelović, P.J.; Petrović, A.S.; Božić, M.; Ilić, R.S. The Quantitative ER Immunohistochemical Analysis in Breast Cancer: Detecting the 3 + 0, 4 + 0, and 5 + 0 Allred Score Cases. *Medicina* **2019**, *55*, 461. [CrossRef] [PubMed]
49. Wolff, A.C.; Hammond, M.E.H.; Allison, K.H.; Harvey, B.E.; Mangu, P.B.; Bartlett, J.M.; Dowsett, M.; Bilous, M.; Ellis, I.O.; Fitzgibbons, P.; et al. Human Epidermal Growth Factor Receptor 2 Testing in Breast Cancer: American Society of Clinical Oncology/College of American Pathologists Clinical Practice Guideline Focused Update. *J. Clin. Oncol.* **2018**, *36*, 2105–2122. [CrossRef] [PubMed]
50. Bel'skaya, L.V.; Kosenok, V.K.; Sarf, E.A. Chronophysiological features of the normal mineral composition of human saliva. *Arch. Oral Biol.* **2017**, *82*, 286–292. [CrossRef] [PubMed]
51. Bel'skaya, L.V.; Kosenok, V.K.; Massard, G. Endogenous Intoxication and Saliva Lipid Peroxidation in Patients with Lung Cancer. *Diagnostics* **2016**, *6*, 39. [CrossRef] [PubMed]
52. Smolyakova, R.M.; Prokhorova, V.I.; Zharkov, V.V.; Lappo, S.V. Assessment of the binding capacity and transport function of serum albumin in patients with lung cancer. *Nov. Khirurgii* **2005**, *13*, 78–84.
53. Chesnokova, N.P.; Barsukov, V.Y.; Ponukalina, E.V.; Agabekov, A.I. Regularities of changes in free radical destabilization processes of biological membranes in cases of colon ascendens adenocarcinoma and the role of such regularities in neoplastic proliferation development. *Fundam. Res.* **2015**, *1*, 164–168.
54. Pankova, O.V.; Perelmuter, V.M.; Savenkova, O.V. Characteristics of proliferation marker expression and apoptosis regulation depending on the character of disregenerator changes in bronchial epithelium of patients with squamous cell lung cancer. *Sib. Oncol. J.* **2010**, *41*, 36–41.

55. Sato, E.F.; Choudhury, T.; Nishikawa, T.; Inoue, M. Dynamic aspect of reactive oxygen and nitric oxide in oral cavity. *J. Clin. Biochem. Nutr.* **2008**, *42*, 8–13. [CrossRef]
56. Bel'skaya, L.V.; Sarf, E.A.; Kosenok, V.K.; Gundyrev, I.A. Biochemical Markers of Saliva in Lung Cancer: Diagnostic and Prognostic Perspectives. *Diagnostics* **2020**, *10*, 186. [CrossRef]
57. Smirmova, O.V.; Borisov, V.I.; Guens, G.P. Immediate and long-term outcomes of drug treatment in patients with metastatic triple negative breast cancer. *Malig. Tumours* **2018**, *8*, 68–77. [CrossRef]
58. Knaś, M.; Maciejczyk, M.; Waszkiel, D.; Zalewska, A. Oxidative stress and salivary antioxidants. *Dent. Med. Probl.* **2013**, *50*, 461–466.
59. Bel'skaya, L.V.; Sarf, E.A.; Kosenok, V.K. Indicators of L-arginine metabolism in saliva: A focus on breast cancer. *J. Oral Biosci.* **2021**, *63*, 52–57. [CrossRef] [PubMed]
60. Zhao, Y.H.; Zhou, M.; Liu, H.; Ding, Y.; Khong, H.T.; Yu, D.; Fodstad, O.; Tan, M. Upregulation of lactate dehydrogenase A by ErbB2 through heat shock factor 1 promotes breast cancer cell glycolysis and growth. *Oncogene* **2009**, *28*, 3689–3701. [CrossRef] [PubMed]
61. Bel'skaya, L.V.; Sarf, E.A.; Kosenok, V.K. Age and gender characteristics of the biochemical composition of saliva: Correlations with the composition of blood plasma. *J. Oral Biol. Craniofacial Res.* **2020**, *10*, 59–65. [CrossRef] [PubMed]
62. Almasri, N.M.; Al Hamad, M. Immunohistochemical evaluation of human epidermal growth factor receptor 2 and estrogen and progesterone receptors in breast carcinoma. *Breast Cancer Res.* **2005**, *7*, 598–604. [CrossRef]
63. Streckfus, C.; Bigler, L.; Dellinger, T.; Dai, X.; Cox, W.J.; McArthur, A.; Kingman, A.; Thigpen, J.T. Reliability assessment of soluble c-erb B-2 concentrations in the saliva of healthy women and men. *Oral Surg. Oral Med. Oral Pathol. Oral Radiol. Endodontol.* **2001**, *91*, 174–179. [CrossRef]
64. Streckfus, C.; Bigler, L.; Tucci, M.; Thigpen, J.T. A preliminary study of CA15-3, c-erbB-2, epidermal growth factor receptor, cathepsin-D, and p53 in saliva among women with breast carcinoma. *Cancer Investig.* **2000**, *18*, 101–109. [CrossRef]
65. Streckfus, C.; Bigler, L. The Use of Soluble, Salivary c-erb B-2 for the Detection and Post-operative Follow-up of Breast Cancer in Women: The Results of a Five-year Translational Research Study. *Adv. Dent. Res.* **2005**, *18*, 17–24. [CrossRef]

current issues in molecular biology

MDPI

Article

Parkin, as a Regulator, Participates in Arsenic Trioxide-Triggered Mitophagy in HeLa Cells

Zhewen Zhang †, Juan Yi †, Bei Xie, Jing Chen, Xueyan Zhang, Li Wang, Jingyu Wang, Jinxia Hou and Hulai Wei *

School of Basic Medical Sciences, Lanzhou University, Lanzhou 730000, China; zhangzhw@lzu.edu.cn (Z.Z.); yij@lzu.edu.cn (J.Y.); xieb@lzu.edu.cn (B.X.); chenjing@lzu.edu.cn (J.C.); zhangxy@lzu.edu.cn (X.Z.); wangli15@lzu.edu.cn (L.W.); wangjingyu@lzu.edu.com (J.W.); houjx17@lzu.edu.com (J.H.)

* Correspondence: weihulai@lzu.edu.cn; Tel.: +86-13893158796

† These authors contribute equally to this work.

Abstract: Parkin is a well-established synergistic mediator of mitophagy in dysfunctional mitochondria. Mitochondria are the main target of arsenic trioxide (ATO) cytotoxicity, and the effect of mitophagy on ATO action remains unclear. In this study, we used stable Parkin-expressing (YFP-Parkin) and Parkin loss-of-function mutant (Parkin C431S) HeLa cell models to ascertain whether Parkin-mediated mitophagy participates in ATO-induced apoptosis/cell death. Our data showed that the overexpression of Parkin significantly sensitized HeLa cells to ATO-initiated proliferation inhibition and apoptosis; however, the mutation of Parkin C431S significantly weakened this Parkin-mediated responsiveness. Our further investigation found that ATO significantly downregulated two fusion proteins (Mfn1/2) and upregulated fission-related protein (Drp1). Autophagy was also activated as evidenced by the formation of autophagic vacuoles and mitophagosomes, increased expression of PINK1, and recruitment of Parkin to impaired mitochondria followed by their degradation, accompanied by the increased transformation of LC3-I to LC3-II, increased expression of Beclin1 and decreased expression of P62 in YFP-Parkin HeLa cells. Enhanced mitochondrial fragmentation and autophagy indicated that mitophagy was activated. Furthermore, during the process of mitophagy, the overproduction of ROS implied that ROS might represent a key factor that initiates mitophagy following Parkin recruitment to mitochondria. In conclusion, our findings indicate that Parkin is critically involved in ATO-triggered mitophagy and functions as a potential antiproliferative target in cancer cells.

Keywords: arsenic trioxide; Parkin; HeLa; apoptosis; autophagy

Citation: Zhang, Z.; Yi, J.; Xie, B.; Chen, J.; Zhang, X.; Wang, L.; Wang, J.; Hou, J.; Wei, H. Parkin, as a Regulator, Participates in Arsenic Trioxide-Triggered Mitophagy in HeLa Cells. *Curr. Issues Mol. Biol.* **2022**, *44*, 2759–2771. https://doi.org/10.3390/cimb44060189

Academic Editors: Dumitru A. Iacobas and Anna Kawiak

Received: 26 February 2022
Accepted: 16 June 2022
Published: 20 June 2022

1. Introduction

Parkin is an E3 ubiquitin ligase involved in the elimination of damaged mitochondria [1–3]. Parkin is normally localized to the cytosol and translocated to depolarized mitochondria in a PTEN-induced kinase (PINK1)-dependent manner [4–6]. Damaged mitochondria accumulate PINK1 on their outer membrane, and after phosphorylation, recruit and activate Parkin to induce mitophagy [7,8]. Mitophagy is a specific autophagic process for the removal of damaged or depolarized mitochondria through the selective targeting of such mitochondria to the autophagic pathway [9]. It remains controversial whether mitophagy is pro-survival or pro-death during cancer therapies. Carroll et al. showed that wild-type Parkin was greatly sensitized toward apoptosis induced by mitochondrial depolarization but not by proapoptotic stimuli that failed to activate Parkin [10,11]. Johnson et al. provided evidence of a specific ubiquitin E3 ligase that might inactive Bax to promote cell survival [12–14]. Furthermore, the mechanism by which Parkin influences cell death remains to be elucidated.

Arsenic trioxide (ATO) is an effective agent for treating acute promyelocytic leukemia. Its anticancer activity has been supported by numerous studies [15–17]. Our previous studies demonstrated that ATO leads to cancer cell death through different molecular mechanisms, including apoptosis, cell cycle arrest, autophagy, and excessive reactive oxygen species (ROS) [18–20]. Moreover, other studies found sufficient evidence for ATO-induced glutathione (GSH) level changes, DNA methyltransferase (DNMT) inhibition, nuclear factor kappa B (NF-κB) signaling pathway alterations, and so on [21]. Mechanistic studies of ATO in cancer cells are challenging, probably because of the complexity of the processes mediating its anticancer effects.

Mitochondria are one of the primary targets of ATO cytotoxicity [16,17]. Niu et al. indicated that ATO inhibited the proliferation of HepG2 cells by initiating mitophagy [22]. Watanabe et al. demonstrated that ATO induces the mitochondrial translocation of Parkin and ubiquitination of the voltage-dependent anion channel 1 (VDAC1) in HL-1 cardiomyocytes [23]. It has been shown that both mitophagy and the essential role of Parkin in mitochondrial quality control are critically involved in ATO-induced toxicity. Based on these reports, the present study was conducted to examine the role of Parkin in ATO-induced HeLa cell injury by observing its effects on mitophagy, mitochondrial function and apoptosis.

2. Materials and Methods

2.1. Cell Culture

Hela cells were obtained from the Experimental Center of School of Basic Medical, Lanzhou University. YFP-Parkin HeLa cells were retrieved from Hanming Shen's laboratory, and were originally provided as a gift from Dr. Richard Youle. Cells were cultured in Dulbecco's Modified Eagle Medium (DMEM) with 10% fetal calf serum and maintained in culture flasks at 37 °C in a humidified chamber containing 5% CO_2.

2.2. Plasmid Constructs

Parkin mutant (C431S) complementary DNAs (cDNAs) were generated by PCR using pEGFP-Parkin C431S (Addgene, 45877) as templates and inserted into pLVX-AcGFP1-N1 Vector (TaKaRa, 632154) with a C-terminal AcGFP1 protein (AcGFP1-Parkin C431S).

2.3. Establishment of the AcGFP1-Parkin Mutant (C431S) HeLa Stable Cell Lines

The three plasmids, AcGFP1-Parkin C431S, pMD2.G (Addgene, 12259) and psPAX2 (Addgene, 12260), were co-transfected into the virus packaging cell line 293T using Lipo2000. After 48 h and 72 h, the supernatant was harvested and concentrated. Then, HeLa cells were infected with the lentiviral supernatant expressing AcGFP1-Parkin C431S. After 48 h of infection, selection reagent puromycin (4 μg/mL) was added to the cells in fresh medium. After 3 days, the cells were reseeded at a low density with selection reagent (puromycin, 1 μg/mL) and incubated for 10 days. Lastly, single clones were expanded in a new dish with fresh medium.

2.4. Reagents and Antibodies

ATO (GB673-77) was purchased from Beijing Chemical (Beijing, China). DMEM and fetal bovine serum (FBS) were purchased from Hyclone (South Logan, UT, USA). The 3-(4,5-dimethylthiazol-2-yl)-2,5-diphenyl-2,5-tetrazolium bromide (MTT, M8180), dimethyl sulfoxide (DMSO, D8731), carbonyl cyanide m-chlorophenylhydrazone (CCCP, C6700) and Propidium Iodide (PI, C0080) were purchased from Solabio. Acetylcysteine (NAC, S1623) was purchased from Selleck. Annexin V-PE/7-AAD Apoptosis Detection Kit (40310ES20), MitoTracher®Red CMXRos (40741ES50) and 4′,6-diamidino-2-phenylindole (DAPI, 40728ES03) were from Yeasen. Fluorometric intracellular ROS Kit (MAK145) was from Sigma. Thr primary antibodies we used were Bcl-2 (Cell Signaling Technology,15071, 1:500 dilution), Bax (Cell Signaling Technology, 2772, 1:1000 dilution), GAPDH (Immunoway, YM3215, 1:5000 dilution), COX4 (GeneTex, GTX114330, 1:1000 dilution), Caspase-3 (Cell Signaling Technology, 9661, 1:500 dilution), Caspase-9 (Cell Signaling Technology, 9505, 1:500 dilution), Beclin 1 (Cell Signaling Technology, 3738, 1:1000 dilution),

P62 (GeneTex, GTX00955, 1:1000 dilution), LC3 (GeneTex, GTX100240, 1:1000 dilution), Parkin (Sigma, P5748, 1:1000 dilution), PINK1 (GeneTex, GTX107851, 1:1000 dilution), Drp1 (CST, CST5391, 1:1000 dilution), Mfn1 (CST, CST147395, 1:1000 dilution) and Mfn2 (CST, CST11925, 1:1000 dilution). Secondary antibodies were all from Immunoway and included anti-mouse-HRP (RS0001, 1:5000 dilution) and anti-rabbit-HRP (RS0002, 1:5000 dilution).

2.5. *Cytotoxicity Assay*

The MTT assay was carried out to evaluate the cytotoxicity of ATO. Briefly, cells (5×10^3) were seeded in a 96-well plate and allowed to adhere overnight. Then, the cells were treated with different concentrations of ATO for 24 h and 48 h. A total of 10 μL of the MTT reagent (5 mg/mL) was added to each well. After an additional incubation for 4 h, 100 μL 10% SDS was added to dissolve formazan crystals. Absorbance values of 570 nm were recorded with a microplate reader (Bio-Tek, Winooski, VT, USA). All assays were repeated three times and data were presented as mean ± SD.

2.6. *Colony Formation Assays*

Cells were plated in 12-well plates (1×10^3 cells per well) and treated with the indicated concentrations of ATO for 14 days. Then, cells were washed in PBS, fixed with 10% formaldehyde for 15 min and stained withed 1% crystal for 5 min before counting the number of colonies.

2.7. *Analysis of Cell Cycle and Apoptosis*

The distribution of the cell cycle was explored by staining DNA with Propidium Iodide (PI). After treatment, 1×10^6 cells were collected and fixed with 70% ethanol overnight. Cells were again washed with PBS and incubated at 37 °C for 30 min with RNase A (0.1 mg/mL) and PI (5 mg/mL). Then, cells' suspension was analyzed using flow cytometry (Beckman Coulter, Miami, FL, USA). The Annexin V-PE/7-AAD Apoptosis Detection Kit was used to determine whether ATO induced apoptosis in Hela cells. Similarly, after treatment, 1×10^6 cells were collected and suspended in binding buffer and stained with Annexin V/PE (5 μL) and 7-AAD (10 μL) at room temperature in the dark for 15 min. Cells undergoing apoptosis were analysed by flow cytometry. The apoptotic rate was determined for each condition as follows: Apoptotic rate = (early apoptotic rate +late apoptotic rate) × 100%.

2.8. *Transmission Electron Microscopy*

As described previously [24], cells (treated with ATO at 2.5 μmol/L or 5 μmol/L for 48 h) were fixed with 2.5% glutaraldehyde and 2% paraformaldehyde in 0.1 mol/L phosphate buffer, followed by 1% osmium tetroxide. After dehydration, thin sections were stained with uranyl acetate and lead citrate for observation. The ultrastructure of the cells was examined with a JEMI1230 transmission electron microscope (JEOL, Tokyo, Japan).

2.9. *Mitochondrial Staining*

Cells were grown on coverslips in six-well plates and cultured overnight. After treatment, the cells were washed with PBS thrice and added to 3.7% paraformaldehyde to stand for 10 min. Cells were then washed with PBS and stained with MitoTracher®Red CMXRos (100 nM) for 30 min or DAPI (5 μg/mL) for 15 min in the dark, at room temperature. Co-stained cells were washed and immediately observed using a fluorescence microscope (Olympus, Japan) equipped with a 20-objective lens. The area of red and green fluorescence was measured using software Image-Pro Plus 6.0.

2.10. *Reactive Oxygen Species Detection*

After treatment, the cells were rinsed with PBS and the reactive oxygen species (ROS) level was measured with the Fluorometric intracellular ROS Kit. The fluorescence microscope was used to observe the changes in ROS. Fluorescence intensity was analyzed quantitatively using Image-Pro Plus 6.0.

2.11. Western Blot Analysis

After treatment, cells were collected in a 70 μL RIPA buffer containing 1 mM PMSF (Beyotime, Haimen, China). The protein concentration was determined using the BCA kit (Solarbio, Bejing, China). Protein extracts (30 μg) were separated by 8–12% SDS-PAGE and electrophoretically transferred onto polyacrylamide difluoride membranes. After blocking with 5% non-fat milk in PBST for 1 h, the membranes were incubated with primary antibodies and secondary antibodies and visualized using ECL with film or CLINX Hemiscope (QinXiang, China). The relative signal intensity of bands was determined with Image J software. Protein expression levels were standardized by normalizing to GAPDH.

2.12. Statistical Analysis

Statistical analyses were performed using the SPSS16.0 software. All of the experiments were repeated three times in duplicate. The results were expressed as the mean ± SD. One-way ANOVA followed by Tukey's least significant difference post hoc test was used to analyze statistical differences between groups under different conditions. Statistical significance was set at $p < 0.05$.

3. Result

3.1. Parkin Aggravates the Proliferation Inhibition of HeLa Cells by ATO Treatment

To examine the function of Parkin in modulating cell proliferation, we used a HeLa cell line stably expressing YFP-Parkin to investigate the effects of Parkin on the viability of HeLa cells treated with ATO. The viability of the cells following treatment with various concentrations of ATO (0, 1, 2, 4, 6, and 8 μmol/L) for 24 h and 48 h was examined with MTT assays. As shown in Figure 1A, cell viability was markedly decreased in a dose- and time-dependent manner in both ATO-treated HeLa cells and YFP-Parkin HeLa cells, and the IC_{50} value of ATO in YFP-Parkin HeLa cells was much lower than that in HeLa cells (Figure 1B). The examination of the colony formation ability of the cells revealed, in the same way, that the overexpression of YFP-Parkin significantly decreased the colony number of YFP-Parkin HeLa cells compared to that of control HeLa cells after ATO administration (Figure 1C,D).

To further explore whether Parkin participates in the ATO-induced proliferation suppression of HeLa cells, we constructed a HeLa cell line with a loss-of-function mutation in Parkin (Parkin C431S). The mutation of Parkin C431S significantly weakened the proliferation-inhibitory effects of ATO on the YFP-Parkin HeLa cells, which were restored to close to the level of the control HeLa cells, as shown by MTT colorimetry and colony formation assays (Figure 1E–G). These findings indicate that the overexpression of Parkin could promote, in part, the sensitivity of HeLa cells to ATO.

3.2. Parkin Promotes ATO-Induced Apoptosis of HeLa Cells

To examine whether Parkin is a regulator of apoptosis in response to ATO treatment, the cells were double-stained with Annexin V-PE/7-AAD and PI. As shown in Figure 2A,B, ATO showed a much stronger apoptosis-inducing effect in YFP-Parkin HeLa cells than in HeLa cells, and the mutation of Parkin to Parkin C431S markedly impaired the apoptosis of ATO-treated YFP-Parkin HeLa cells. As an example, after treatment with 5 μmol/L ATO for 24 h, the apoptosis rates of HeLa cells, YFP-Parkin HeLa cells and Parkin C431S cells were 13.15 ± 1.67%, 27.48 ± 1.9% and 21.01 ± 3.1%, respectively. In addition, the analysis of the cell cycle distribution in ATO-treated HeLa cells showed that the overexpression of YFP-Parkin had no significant effects on the ATO-induced cell cycle changes that manifested as G2/M arrest (Figure 2C). As shown in Figure 2D, the expression of apoptosis-related proteins revealed that ATO produced a dose- and time-dependent increase in Cleaved-Caspase-9, Cleaved-Caspase-3 and Bax and a decrease in Bcl-2 proteins in both HeLa and YFP-Parkin HeLa cells, and the latter was much more pronounced. All of the above evidence indicated that Parkin expression greatly improved the sensitivity of HeLa cells to ATO-stimulated apoptosis.

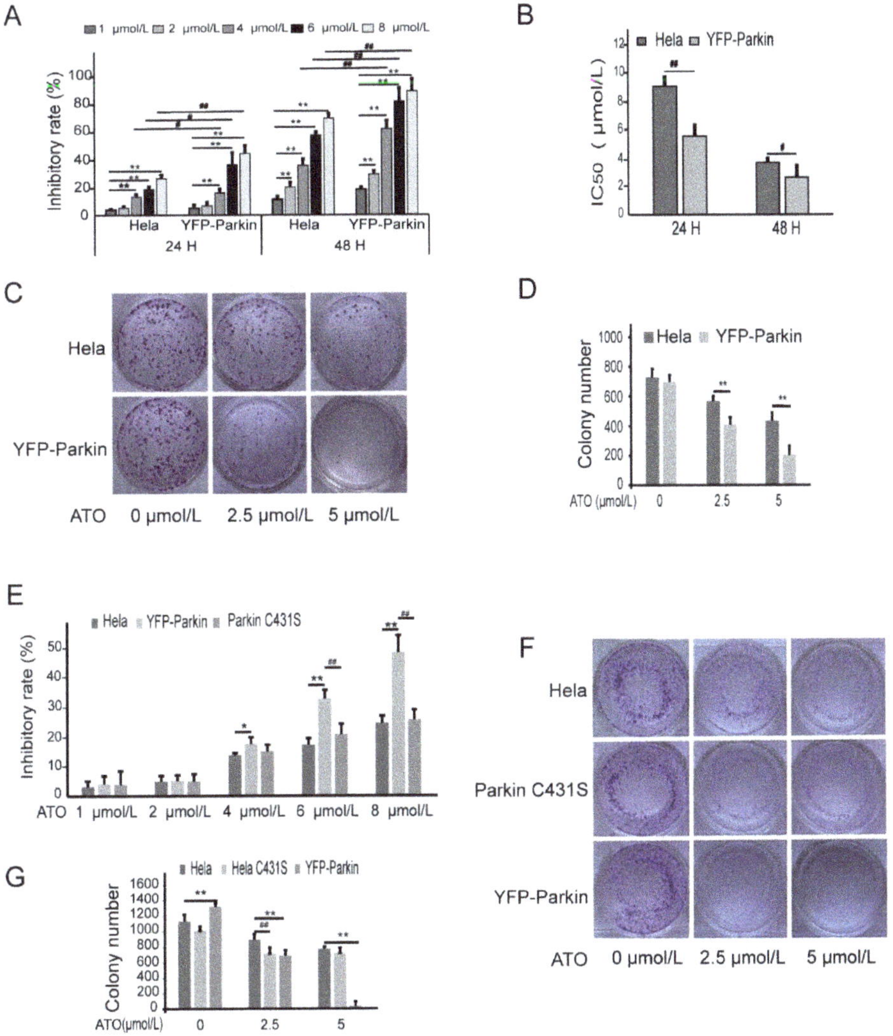

Figure 1. ATO modulates proliferation of HeLa cells. (**A**) MTT assays revealed that overexpression of YFP-Parkin significantly decreased the growth rate of the indicated cells. (**B**) The percentage of IC_{50}. (**C**) YFP-Parkin decreased the mean colony number in the colony formation assay. (**D**) Quantification of C. (**E**) MTT assays revealed that Parkin C431S rescued the growth rate after ATO treatment for 24 h. (**F**) Parkin C431S rescued the mean colony number in the colony formation assay. (**G**) Quantification of F. The histograms represent the mean ± SD of triplicate experiments. * $p < 0.05$ vs. 0 μmol/L group, ** $p < 0.01$ vs. 0 μmol/L group; # $p < 0.05$ vs. without Parkin group, ## $p < 0.01$ vs. without Parkin group.

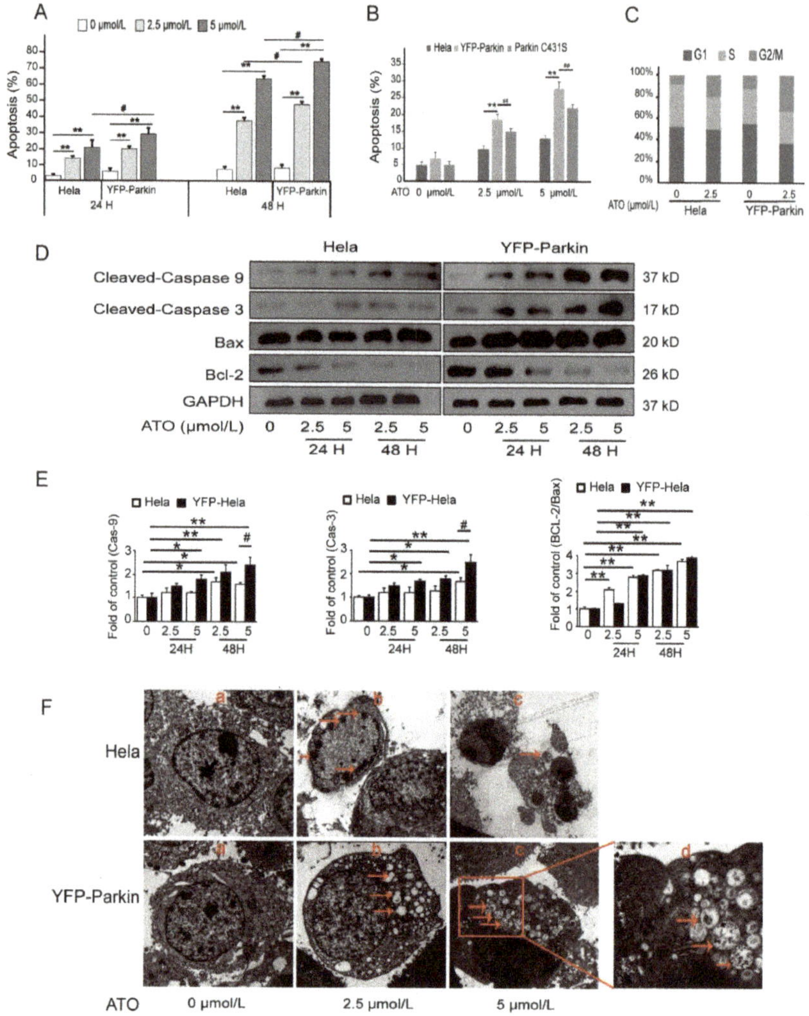

Figure 2. ATO induces the apoptosis of HeLa cells. (**A**) The apoptotic rate. (**B**) Parkin C431S rescued the apoptosis rate after ATO treatment for 24 h. (**C**) The percentage of cells in each phase of the cell cycle was analyzed after ATO treatment for 24 h by flow cytometry. (**D**) Western blot analysis of Cleaved-Caspase-3, Cleaved-Caspase-9, Bax, and Bcl-2 expression. (**E**) Quantification of D. (**F**) Morphology of apoptosis and autophagy was observed under a transmission electron microscope after ATO treatment for 24 h (a, b and c 5000×; d 15,000×). The arrowhead (↓) indicates the chromatin condensation, apoptotic bodies, autophagic vacuoles and mitophagosomes in apoptotic cells. The histograms represent the mean ± SD of triplicate experiments. * $p < 0.05$ vs. 0 μmol/L group, ** $p < 0.01$ vs. 0 μmol/L group, # $p < 0.05$ vs. without Parkin group; ## $p < 0.01$ vs. without Parkin group; GAPDH served as an internal control.

Transmission electron microscopy revealed that, as shown in Figure 2F, after exposure to ATO, both HeLa cells and YFP-Parkin HeLa cells exhibited morphological changes typical of apoptosis features, such as shrinkage and apoptotic bodies, which were more obvious in YFP-Parkin HeLa cells. At the same time, we also observed a large number of autophagic vacuoles and mitophagosomes in the cytoplasm of ATO-exposed YFP-Parkin

HeLa cells. These phenomena mean that Parkin strengthened ATO-triggered apoptosis in HeLa cells, which was likely achieved via the induction of autophagy or mitophagy.

3.3. *PINK1/Parkin Pathway Was Involved in the ATO-Induced Mitochondrial Damage and Mitophagy*

To ascertain whether the PINK1/Parkin pathway mediated mitophagy in the apoptosis of ATO-treated HeLa cells, we investigated the effects of Parkin on ATO-induced mitochondrial dysfunction in YFP-Parkin HeLa cells. To assess whether ATO exposure disturbed mitochondrial fission and fusion, we detected the expressions of fission and fusion-related proteins in both HeLa and YFP-Parkin HeLa cells. Our results showed that the expression of Drp1 increased, and the expression of Mfn1/2 decreased in the ATO-treated cells compared to the control cells (Figure 3A–D). We also observed the decreased expression of COX IV, a common mitochondrial marker (Figure 3E,F), and reduced Mito-Tracker fluorescence of mitochondria (Figure 4A–D) in ATO-treated HeLa and YFP-Parkin HeLa cells, which suggested that ATO might cause mitochondrial damage.

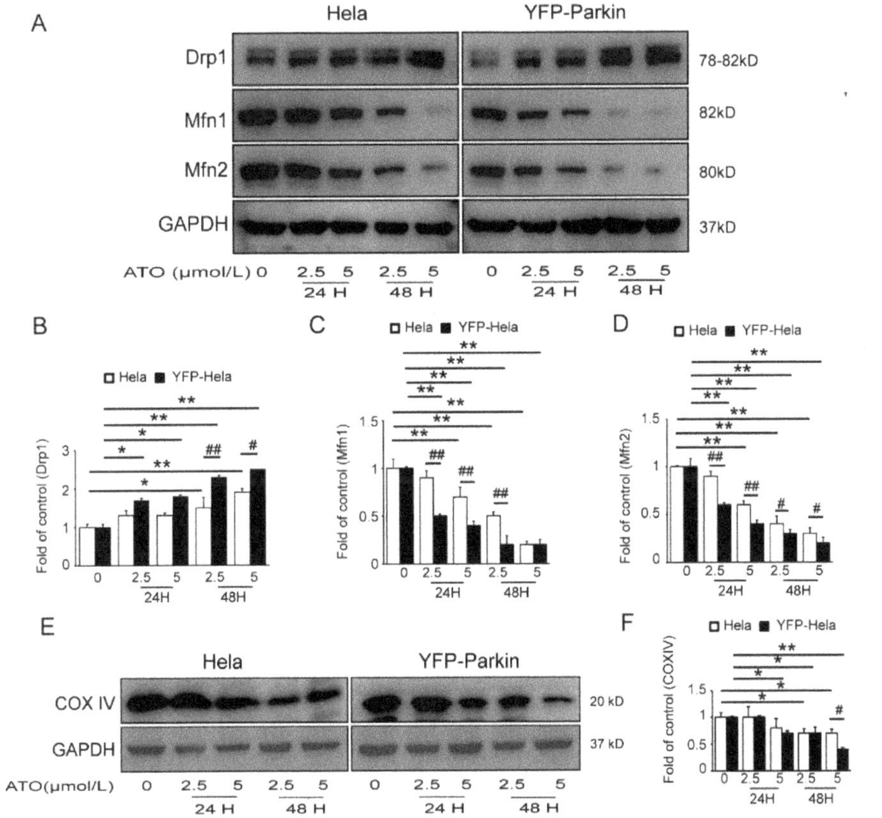

Figure 3. Effects of ATO exposure on mitochondrial dynamics in HeLa cells. (**A**) Western blot analysis of Drp1, Mfn1 and Mfn2 expression. (**B–D**) Quantification of A. (**E**) Western blot analysis of COX IV expression in the indicated cells. (**F**) Quantification of E. The histograms represent the mean ± SD of triplicate experiments. * $p < 0.05$ vs. 0 μmol/L group, ** $p < 0.01$ vs. 0 μmol/L group, # $p < 0.05$ vs. without Parkin group, ## $p < 0.01$ vs. without Parkin group; GAPDH served as an internal control.

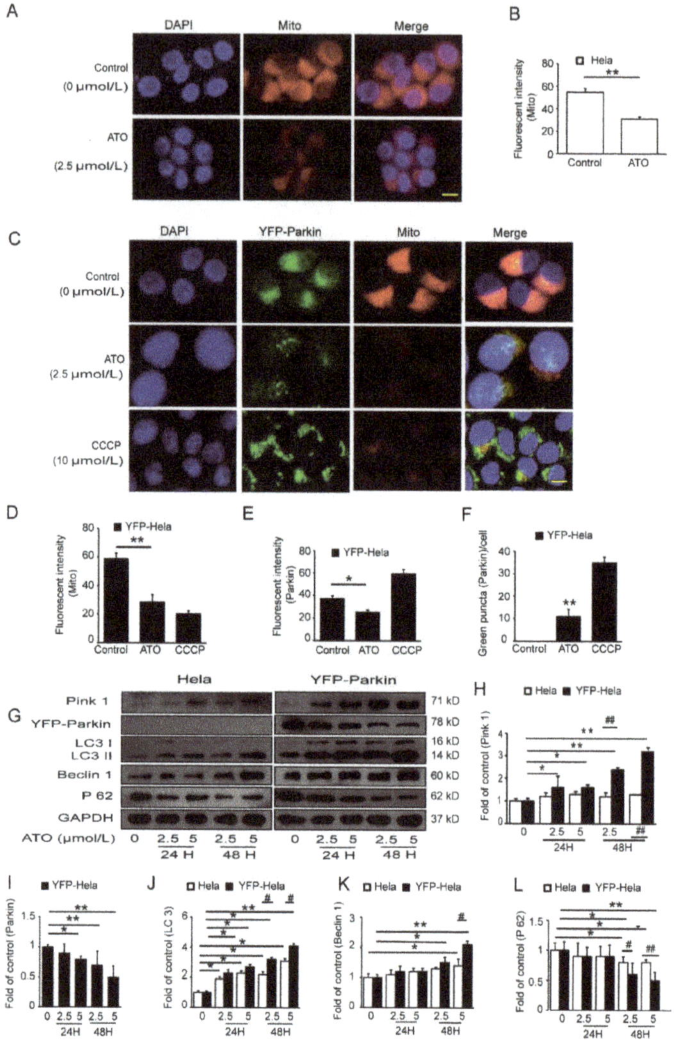

Figure 4. Regulation of PINK1/Parkin-mediated mitophagy by ATO. (**A**) HeLa cells were fixed and immunostained for nuclei (blue) and mitochondria (red) after ATO treatment for 24 h. The red fluorescence indicated the normal mitochondrial membrane potential. The samples were analyzed using a fluorescence microscopy. Scale bar, 10 μm. (**B**) shows the quantification of the red fluorescence, which was performed using Image J software. (**C**) YFP-Parkin HeLa cells were fixed and immunostained for nuclei (blue) and mitochondria (red) after ATO treatment for 24 h. The red fluorescence indicated the normal mitochondrial membrane potential. The green fluorescence was the marker for YFP-Parkin proteins. The level of mitophagy was evaluated by the number of green puncta. The samples were analyzed using a fluorescence microscopy. Scale bar, 10 μm. (**D**) Quantification of the red fluorescence, which was performed using Image J software. (**E**) Quantification of the green fluorescence, which was performed using Image J software. (**F**) Quantification of green puncta, which was performed using Image J software. (**G**) Western blot analysis of Parkin, PINK1, LC3, Beclin 1 and P62 expression in the indicated cells. (**H–L**) Quantification of G. The histograms represent the mean ± SD of triplicate experiments. * $p < 0.05$ vs. 0 μmol/L group, ** $p < 0.01$ vs. 0 μmol/L group, # $p < 0.05$ vs. without Parkin group; ## $p < 0.01$ vs. without Parkin group; GAPDH served as an internal control.

Usually, Parkin is recruited by PINK 1 to damaged mitochondria and then they are degraded via the activation of mitophagy. To test whether mitochondria labeled by Parkin display decreased membrane potential, we pulsed the cells with MitoTracker red, a potentiometric mitochondrial dye, before fixation. YFP-Parkin selectively accumulated on those mitochondria with MitoTracker staining (Figure 4C–F). These results also indicated that ATO resulted in mitochondrial damage and that Parkin might be recruited to impaired mitochondria to trigger their degradation by PINK1/Parkin-mediated mitophagy. To confirm this possibility, we assessed the expression of PINK1 and Parkin, as well as the autophagy-related proteins LC3, Beclin1 and P62, in ATO-exposed YFP-Parkin HeLa cells. The results of the Western blotting examination showed that after treatment with 2.5 and 5 μmol/L ATO for 24 h and 48 h, the increased expression of PINK1 and decreased expression of Parkin were accompanied by the increased transformation of LC3-I to LC3-II, increased expression of Beclin1 and decreased expression of P62 in YFP-Parkin HeLa cells compared to control HeLa cells (Figure 4G–L). These findings provide extremely strong evidence that Parkin is critically involved in the mitophagy of HeLa cells in response to ATO.

3.4. ROS Activate the PINK1/Parkin Pathway to Mediate Mitophagy and Mitophagic Death in ATO-Exposed HeLa Cells

Intracellular ROS production is one of the antitumoral mechanisms of ATO. To investigate the effect of the overexpression of Parkin on ATO-simulated ROS generation and whether ROS initiate the Parkin-mediated mitophagic death of HeLa cells by ATO, we incubated YFP-Parkin HeLa cells with 2.5 μmol/L ATO for 6 h. As shown in Figure 5, intracellular ROS overproduction (Figure 5A,B) occurred, accompanied by mitochondrial damage (Figure 5D–G), Parkin degradation and enhanced apoptosis (Figure 5C), and all of these effects could be reversed by the ROS scavenger acetylcysteine (NAC) (Figure 5D–I). These data suggested that the generation of ROS gave rise to mitochondrial dysfunction and activated the PINK1/Parkin pathway to trigger mitophagy and mediate mitophagic apoptosis in ATO-treated HeLa cells.

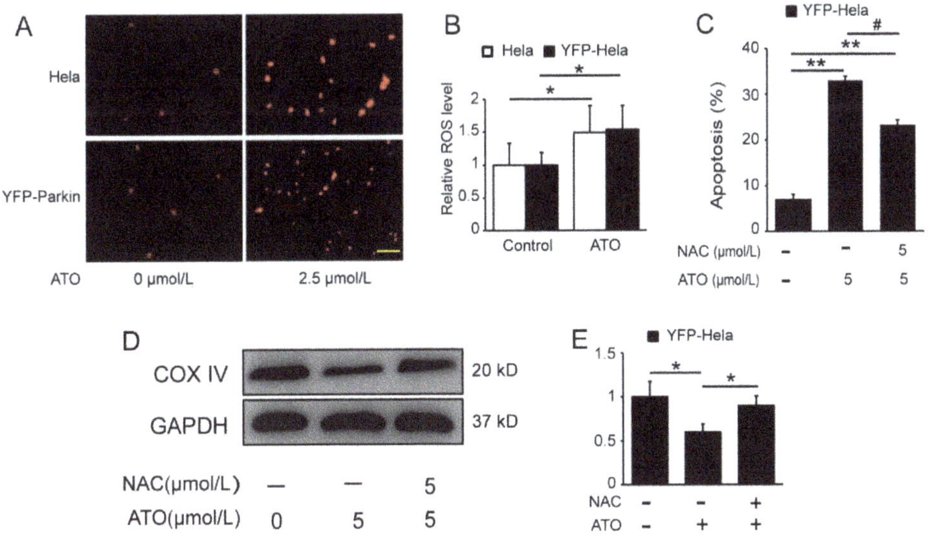

Figure 5. *Cont.*

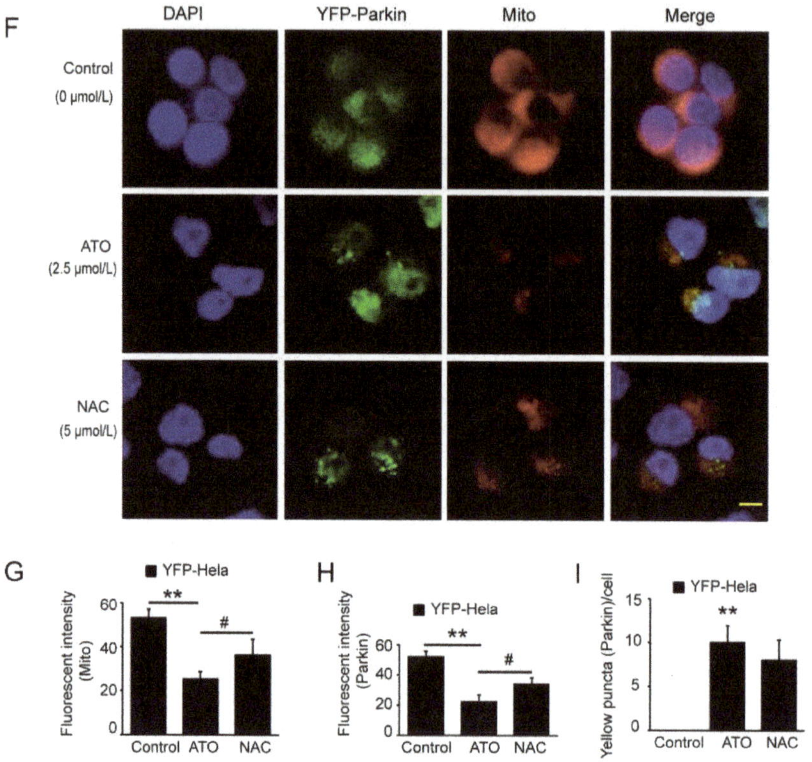

Figure 5. Potentiating effect of ATO on ROS production. (**A**) Red fluorescence indicates the production of ROS after cell treatment for 6 h, and the images were analyzed using a fluorescence microscopy. Scale bar, 10 μm. (**B**) shows the quantification of the red fluorescence, which was performed using Image J software. (**C**) YFP-Parkin HeLa cells were treated with the indicated concentrations of ATO for 24 h before detection of cell apoptosis by flow cytometry. (**D**) Western blot analysis of COX IV expression rescued after ATO treatment for 24 h in YFP-Parkin HeLa cells. (**E**) Quantification of D. (**F**) YFP-Parkin HeLa cells were fixed and immunostained for nuclei (blue) and mitochondria (red) after ATO treatment for 24 h. The red fluorescence indicated the normal mitochondrial membrane potential. The green fluorescence was the marker of YFP-Parkin proteins. The level of mitophagy was evaluated by the number of green puncta. The samples were analyzed using a fluorescence microscopy. Scale bar, 10 μm. (**G**) Quantification of the red fluorescence, which was performed using Image J software. (**H**) Quantification of the green fluorescence, which was performed using Image J software. (**I**) Quantification of green puncta, which was performed using Image J software. The histograms represent the mean ± S.D. of triplicate experiments. * $p < 0.05$ vs. 0 μmol/L group, ** $p < 0.01$ vs. 0 μmol/L group, # $p < 0.05$ vs. without Parkin group. Without NAC group; GAPDH served as an internal control.

4. Discussion

Parkin plays an essential role in mitochondrial quality control and mitochondria are one of the primary targets of ATO cytotoxicity [17,25,26]. To date, few studies have directly investigated the changes in Parkin in response to ATO cytotoxicity. In this study, we investigated the possible role of Parkin in promoting mitophagy in YFP-Parkin HeLa cells in response to ATO. Our data showed that the overexpression of Parkin significantly inhibited cell viability and induced mitochondrial damage as well as apoptosis in HeLa cells. During this process, ATO induced PINK1/Parkin-mediated mitophagy. Furthermore, we demonstrated that ROS may represent a key factor that promotes mitophagy following Parkin recruitment to mitochondria.

The molecular mechanism of ATO-induced apoptosis in tumor cells has been widely reported [16,17,21]. ATO activates different signaling pathways in different tumor cells. In this study, ATO caused mitochondrial impairment, including cytochrome c release and subsequent caspase-3 and -9 activation. ATO induced apoptosis via a mitochondria-dependent pathway in HeLa cells. The typical morphological characteristics of apoptosis were observed by electron microscopy after treatment with ATO. We generated HeLa cells stably expressing YFP-Parkin and found that Parkin overexpression can significantly increase the apoptosis of HeLa cells and inhibit cell proliferation. A large number of autophagic vacuoles and mitophagosomes appeared in the cytoplasm in YFP-Parkin HeLa cells exposed to ATO. These results suggested that ATO-induced mitochondrial damage activated PINK1/Parkin-mediated mitophagy. Furthermore, the mutation of the Parkin C431S alleviated ATO-induced cell apoptosis and proliferation. These findings showed that Parkin plays an important role in the activation of mitophagy under ATO treatment.

Previous studies indicated that the PINK1/Parkin pathway was widely accepted as the chief mechanism of mitophagy, closely related to the regulation of mitochondrial membrane potential, mitochondrial fusion/fission and mitochondrial function [27]. Narendra et al. demonstrated that Parkin is selectively recruited to dysfunctional mitochondria in mammalian cells and that after recruitment, Parkin mediates the engulfment of mitochondria by autophagosomes and their subsequent degradation [7]. Okatsu et al. showed that the mitophagy mechanism is controlled by PINK1/Parkin [28]. We hypothesized that ATO increased the clearance of impaired mitochondria by activating Parkin-mediated mitophagy. The Western blot results showed a significant reduction in Mfn1/2 in cells. The immunofluorescence assays showed that the intensity of the red fluorescence clearly decreased by the ATO treatment of the two cell lines. In YFP-Parkin HeLa cells, the intensity of the red fluorescence was much lower than that in HeLa cells. We next determined whether Parkin was recruited to mitochondria. As shown in Figure 4C, we found that YFP-Parkin colocalized with mitochondria with lower MitoTracker staining. These data indicate that the completion of mitophagy requires the recruitment of Parkin to damaged mitochondria post-ATO treatment, which is consistent with a previous study. PINK1 is upstream of Parkin and functions to recruit Parkin to mitochondria to trigger the process of mitophagy. The Western blot analysis also showed that the level of PINK1 was markedly increased in YFP-Parkin HeLa cells after treatment with ATO compared with HeLa control cells. These results suggested that the PINK1/Parkin pathway was activated in YFP-Parkin HeLa cells by ATO.

Subsequent experiments revealed that the ROS content was increased in ATO-induced cells. Xu et al. demonstrated that ROS functions as an upstream signal in the PINK1/Parkin pathway to mediate mitophagy progression [29]. Xiao et al. confirmed that ROS promotes mitophagy following Parkin translocation to mitochondria [30]. Our results are similar to those conclusions. ROS was clearly observed as early as 6 h post-ATO treatment. A marked overlay of Parkin and mitochondria could be observed 24 h after adding ATO in our study. Considering that ROS is upstream activators of Parkin, we determined whether ROS is involved in Parkin regulation under ATO. As shown in Figure 4D, after pretreatment with NAC for 1 h, exposure to ATO not only enhanced the green (Parkin) and red (mitochondria) fluorescence but also increased the visible colocalization of Parkin and mitochondria. It is speculated that the execution of mitophagy may be mainly propelled by ROS. Consistent with this, the protein expression of COX IV and the apoptosis ratio were markedly rescued by NAC compared with cells treated with ATO alone. Our data indicate that ROS acts downstream of Parkin recruitment to impaired mitochondria to drive mitophagy forward, while the activation of the PINK1/Parkin signaling pathway may contribute to the completion of mitophagy.

5. Conclusions

In conclusion, we found that cell viability decreased, the apoptotic rate increased, and PINK1/Parkin-mediated mitophagy was activated in YFP-Parkin-overexpressing cells treated with ATO. Further investigation indicated that ROS might represent a key factor that promotes mitophagy following Parkin recruitment to mitochondria. Our findings indicate that Parkin is critically involved in ATO-triggered mitophagy and it functions as a potential antiproliferative target in cancer cells.

Author Contributions: Conceptualization, H.W.and Z.Z.; methodology, J.Y., B.X. and Z.Z.; software, L.W.; J.C., J.W. and J.H. performed the research; Z.Z., X.Z. and J.Y. analyzed the data; H.W. and Z.Z. wrote the paper. All authors have read and agreed to the published version of the manuscript.

Funding: This work was supported by the Fundamental Research Funds for the Central Universities (No. lzujbky-2017-138); the Natural Science Foundation of Gansu Province (No. 18JR3RA291; No. 20JR5RA290; No. 17JR5A214; No. 21JR7RA448).

Data Availability Statement: Not applicable.

Conflicts of Interest: The authors declare no conflict of interest.

References

1. Bernardini, J.P.; Lazarou, M.; Dewson, G. Parkin and mitophagy in cancer. *Oncogene* **2017**, *36*, 1313–1327. [CrossRef] [PubMed]
2. Gao, B.L.; Yu, W.J.; Lv, P.; Liang, X.Y.; Sun, S.Q.; Zhang, Y.M. Parkin overexpression alleviates cardiac aging through facilitating K63-polyubiquitination of TBK1 to facilitate mitophagy. *Mol. Basis Dis.* **2021**, *1867*, 165997. [CrossRef] [PubMed]
3. Shiiba, I.; Takeda, K.; Nagashima, S.; Ito, N.; Tokuyama, T.; Yamashita, S.I.; Kanki, T.; Komatse, Y.; Urano, Y.; Fujikawa, Y.; et al. MITOL promotes cell survival by degrading Parkin during mitophagy. *EMBO Rep.* **2021**, *22*, e49097. [CrossRef] [PubMed]
4. Lin, C.Y.; Tsai, C.W. PINK1/Parkin-mediated mitophagy pathway is related to neuroprotection by carnosic acid in SH-SY5Y cells. *Food Chem. Toxicol.* **2019**, *125*, 430–437. [CrossRef]
5. Sun, D.Z.; Song, C.Q.; Xu, Y.M.; Wang, R.; Liu, W.; Liu, Z.; Dong, X.S. Involement of PINK1/Parkin-mediated mitophagy in paraquat-induced apoptosis in human lung epithelial-like A549 cells. *Toxicol. Vitr.* **2019**, *53*, 148–159. [CrossRef]
6. Wang, B.; Yin, X.Q.; Gan, W.D.; Pan, F.; Li, A.Y.; Xiang, Z.; Han, X.D.; Li, D.M. PRCC-TFE3 fusion-mediated PRKN/parkin-dependent mitophagy promotes cell survival and proliferation in PRCC-TFE3 translocation renal cell carcinoma. *Autophagy* **2021**, *17*, 2475–2493. [CrossRef]
7. Narendra, D.; Tanaka, A.; Suen, D.F.; Youle, R.J. Parkin is recruited selectively to impaired mitochondria and promotes their autophagy. *J. Cell Biol.* **2008**, *5*, 795–803. [CrossRef]
8. Wang, J.; Hu, T.; Wang, Q.; Chen, R.W.; Xie, Y.X.; Chang, H.Y.; Cheng, J. Repression of the AURKA-CXCL5 axis induces autophagic cell death and promotes radiosensitivity in non-amsll-cell lung cancer. *Cancer Lett.* **2021**, *509*, 89–104. [CrossRef]
9. Berger, A.K.; Cortese, G.P.; Amodeo, K.D.; Weihofen, A.; Letai, A.; LaVoie, M.J. Parkin selectively alters the intrinsic threshold for mitochondrial cytochrome c release. *Hum. Mol. Genet.* **2009**, *18*, 4317–4328. [CrossRef]
10. Carroll, R.G.; Hollville, E.; Martin, S.J. Parkin sensitizes toward apoptosis induced by mitochondrial depolarization through Promoting degradation of Mcl-1. *Cell Rep.* **2014**, *9*, 1538–1553. [CrossRef]
11. Bin, J.G.; Bai, T.Z.; Zhao, Q.X.; Duan, X.H.; Deng, S.X.; Xu, Y.J. Parkin overexpression reduces inflammation-mediated cardiomyocyte apoptosis through activating Nrf2/ARE signaling pathway. *J. Recept. Signal Transduct.* **2020**, *41*, 451–456. [CrossRef] [PubMed]
12. Johnson, B.N.; Berger, A.K.; Cortese, G.P.; LaVoie, M.J. The ubiquitin E3 ligase parkin regulated the proapoptotic function of Bax. *Proc. Natl. Acad. Sci. USA* **2012**, *16*, 6283–6288. [CrossRef] [PubMed]
13. Hovllville, E.; Carroll, R.G.; Cullen, S.P.; Martin, S.M. Bcl-2 family proteins participate in mitochondrial quality control by regulating Parkin/PINK1-Dependent mitophagy. *Mol. Cell* **2014**, *55*, 451–466. [CrossRef] [PubMed]
14. Kapadia, M.; Snoo, M.L.D.; Kalia, L.V.; Kalia, S.K. Regulation of Parkin-dependent mitophagy by Bcl-2-associated athanogene (BAG) family members. *Neural Regen. Res.* **2021**, *16*, 684–685.
15. Cheng, Y.H.; Chang, L.W.; Tsou, T.C. Mitogen-activated protein kinases mediate arsenic-induced down-regulation of surviving in human lung adenocarcinoma cells. *Mol. Toxicol.* **2006**, *80*, 310–318.
16. Gu, S.Y.; Chen, C.Z.; Jiang, X.J.; Zhang, Z.Z. ROS-mediated endoplasmic reticulum stress and mitochondrial dysfunction underlie apoptosis induced by resveratrol and arsenic trioxide in A549 cells. *Chemico-Biol. Interact.* **2016**, *245*, 100–109. [CrossRef]
17. Laka, K.; Makgoo, L.L.; Mbita, Z.K. Survivin splice variants in arsenic trioxide (As2O3)-induced deactivation of PI3K and MAPK cell signaling pathways in MCF-7 cells. *Genes* **2019**, *10*, 41. [CrossRef]
18. Cheng, J.; Wei, H.L.; Chen, J.; Xie, B. Antitumor effect of arsenic trioxide in human K562 and K562/ADM cells by autophagy. *Toxicol. Mech. Methods* **2012**, *22*, 512–519. [CrossRef]

19. Li, C.L.; Wei, H.L.; Chen, J.; Wang, B.; Xie, B.; Fan, L.L.; Li, L.L. Arsenic trioxide induces autophagy and antitumoreffects in Burkitt's lymphoma Raji cells. *Oncol. Rep.* **2014**, *32*, 1557–1563. [CrossRef]
20. Li, C.L.; Wei, H.L.; Chen, J.; Wang, B.; Xie, B.; Fan, L.L.; Li, L.L. Ebb-and-flow of macroautophagy in Raji cells induced by starvation and arsenic trioxide. *Asian Pac. J. Cancer Prev.* **2014**, *15*, 5715–5719. [CrossRef]
21. Zhang, P.Y.; Zhao, Z.X.; Zhang, W.J.; He, A.L.; Lei, B.; Zhang, W.G.; Chen, Y.X. Leukemia-associated gene MLAA-34 reduces arsenic trioxide-induced apoptosis in Hela cells via activation of the Wnt/β-catenin signaling pathway. *PLoS ONE* **2017**, *12*, e0186868. [CrossRef] [PubMed]
22. Niu, Z.D.; Zhang, W.Y.; Gu, X.Y.; Zhang, X.N.; Qi, Y.M.; Zhang, Y.M. Mitophagy inhibits proliferation by decreasing cyclooxygenase-2 (COX-2) in arsenic trioxide-treated HepG2 cells. *Environ. Toxicol. Pharmacol.* **2016**, *45*, 212–221. [CrossRef] [PubMed]
23. Watanabe, M.; Funakoshi, T.; Unuma, K.; Aki, T.; Uemura, K. Activation of the ubiquitin-proteasome system against arsenic trioxide cardiotoxicity involves ubiquitin ligase Parkin for mitochondrial homeostasis. *Toxicology* **2014**, *322*, 43–50. [CrossRef] [PubMed]
24. Chen, J.; Zhou, C.M.; Yi, J.; Sun, J.J.; Xie, B.; Zhang, Z.W.; Wang, Q.F.; Chen, G.; Jin, S.Y.; Hou, J.X.; et al. Metformin and arsenic trioxide synergize to trigger Parkin/pink1 -dependent mitophagic cell death in human cervical cancer Hela cells. *J. Cancer* **2021**, *12*, 6310–6319. [CrossRef]
25. Zhang, F.; Peng, W.X.; Zhang, J.; Dong, W.T.; Wu, J.H.; Wang, T.; Xie, Z.H. P53 and Parkin co-regulate mitophagy in bone marrw mesenchymal stem cells to promote the repair of early steroid-induced osteonecrosis of the femoral head. *Cell Death Dis.* **2020**, *11*, 42. [CrossRef] [PubMed]
26. Shires, S.E.; Quiles, J.M.; Najor, R.H.; Leon, L.J.; Cortez, M.Q.; Lampert, M.A.; Mark, A.; Gustafsson, A.B. Nuclear Parkin activates the ERRα transcriptional program and drices widespread changes in gene expression following hypoxia. *Sci. Rep.* **2020**, *10*, 8499. [CrossRef]
27. Chen, Y.P.; Shih, P.C.; Feng, C.W.; Wu, C.C.; Tsui, K.H.; Lin, Y.H.; Kuo, H.M.; Wen, Z.H. Paraxin activates excessive mitophagy and mitochondria-mediated apoptosis in human ovarian cancer by inducing reactive oxygen species. *Antioxidants* **2021**, *10*, 1883. [CrossRef]
28. Okatsu, K.; Koyano, F.; Kimura, M.; Kosako, H.; Saeki, Y.; Tanaka, K.; Matsuda, N. Phosphorylated ubiquitin chain is the genuine Parkin receptor. *J. Cell Biol.* **2015**, *1*, 111–128. [CrossRef]
29. Xu, J.; Wang, L.M.; Zhang, L.H.; Zheng, F.; Wang, F.; Leng, J.H.; Wang, K.Y.; Héroux, P.; Shen, H.M.; Wu, Y.H.; et al. Mon-2-ethylhexyl phthalate drives progression of PINK1-parkin-mediated mitophagy via increasing mitochondrial ROS to exacerbate cytotoxicity. *Redox Biol.* **2021**, *38*, 101776. [CrossRef]
30. Xiao, B.; Deng, X.; Grace, G.L.; Xie, S.P.; Zhou, Z.D.; Lim, K.L.; Tan, E.K. Superoxide drives progression of Parkin/PINK1-dependent mitophagy following translocation of Parkin to mitochondria. *Cell Death Dis.* **2017**, *8*, e3097. [CrossRef]

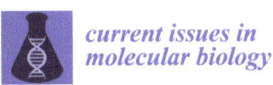

current issues in molecular biology

MDPI

Article

Expression of *GOT2* Is Epigenetically Regulated by DNA Methylation and Correlates with Immune Infiltrates in Clear-Cell Renal Cell Carcinoma

Wallax Augusto Silva Ferreira [1,*] and Edivaldo Herculano Correa de Oliveira [1,2]

1 Laboratory of Cytogenomics and Environmental Mutagenesis, Environment Section (SAMAM), Evandro Chagas Institute (IEC), BR 316, KM 7, s/n, Levilândia, Ananindeua 67030-000, PA, Brazil; ehco@ufpa.br

2 Faculty of Natural Sciences, Institute of Exact and Natural Sciences, Federal University of Pará (UFPA), Rua Augusto Correa, 01, Belém 66075-990, PA, Brazil

* Correspondence: wallaxaugusto@gmail.com or wallax.ferreira@icb.ufpa.br

Abstract: Clear cell renal cell carcinoma (KIRC) is the most common and highly malignant pathological type of kidney cancer, characterized by a profound metabolism dysregulation. As part of aspartate biosynthesis, mitochondrial *GOT2* (glutamic-oxaloacetic transaminase 2) is essential for regulating cellular energy production and biosynthesis, linking multiple pathways. Nevertheless, the expression profile and prognostic significance of *GOT2* in KIRC remain unclear. This study comprehensively analyzed the transcriptional levels, epigenetic regulation, correlation with immune infiltration, and prognosis of *GOT2* in KIRC using rigorous bioinformatics analysis. We discovered that the expression levels of both mRNA and protein of *GOT2* were remarkably decreased in KIRC tissues in comparison with normal tissues and were also significantly related to the clinical features and prognosis of KIRC. Remarkably, low *GOT2* expression was positively associated with poorer overall survival (OS) and disease-free survival (DFS). Further analysis revealed that *GOT2* downregulation is driven by DNA methylation in the promoter-related CpG islands. Finally, we also shed light on the influence of *GOT2* expression in immune cell infiltration, suggesting that *GOT2* may be a potential prognostic marker and therapeutic target for KIRC patients.

Keywords: KIRC; *GOT2*; multi-omics; epigenetics; immune cell infiltration

Citation: Ferreira, W.A.S.; de Oliveira, E.H.C. Expression of *GOT2* Is Epigenetically Regulated by DNA Methylation and Correlates with Immune Infiltrates in Clear-Cell Renal Cell Carcinoma. *Curr. Issues Mol. Biol.* **2022**, *44*, 2472–2489. https://doi.org/10.3390/cimb44060169

Academic Editor: Reza Alaghehbandan

Received: 7 March 2022
Accepted: 23 April 2022
Published: 25 May 2022

1. Introduction

Renal cell carcinoma (RCC) is one of the most common malignancies of the genitourinary system, representing 3% of all malignancies in adults [1]. In the subtypes of kidney cancer, kidney renal clear cell carcinoma (KIRC) accounts for about 75% of all RCC [2–4]. Indeed, due to the evident complexity from both morphological and molecular points of view [5–9], KIRC patients have heterogeneous clinical outcomes [10,11]. Most KIRC tumors are radiotherapy and chemotherapy-resistant, and ~30% of patients eventually develop metastases [3,12,13]. Thus, it is critical to identify new sensitive tumor biomarkers to advance the prognosis of KIRC.

Current shreds of evidence have demonstrated that KIRC is a metabolic disease [13–18]. This metabolic reprogramming is mainly related to loss-of-function mutation (or, less commonly, hypermethylation) in the von Hippel–Lindau (*VHL*) gene. VLH inactivation results in constitutive activation of hypoxia-inducible factors (HIF-1α and HIF-2α), thereby altering many genes involved in angiogenesis, metabolism, chromatin remodeling, extracellular matrix (ECM) and DNA repair [9,15,19–21]. Accordingly, mutations in several other genes (e.g., *PBRM1*, *SETD2*, *BAP1*, *KDM5C*, and *MTOR*) have been continuously identified to contribute to the pathogenesis (used to classify tumors into subgroups) and metabolic remodeling process of KIRC [22,23].

The metabolic shift in KIRC tumors covers different pathways and specific intermediates (e.g., amino acids, aerobic glycolysis, and fatty acid metabolism) [16,24], allowing cancer cells to rapidly proliferate, and survive during nutrient depletion and hypoxia, and evade the immune system [14]. Supporting this notion, aberrant tumor growth is promoted by an enhanced supply of specific metabolites, and some of them, such as aspartate (Asp), are limiting in some tumors [25–27]. Asp is usually synthesized in the mitochondrial matrix through the sequential actions of *MDH2* and glutamic-oxaloacetic transaminase 2 (*GOT2*) and then transported to the cytosol for use by *GOT1* and other enzymes [28].

GOT2, situated on chromosome 16q21, is a crucial enzyme for cancer cell metabolism, (i) mediating the reversible interconversion of oxaloacetate and glutamate into aspartate and α-ketoglutarate, providing energy for tumor cells (Krebs cycle) [29]; (ii) being a key transfer enzyme in the malate-aspartate NADH shuttle activity and oxidative protection [30], maintaining glycolysis, and (iii) participating in the amino acid metabolism of tumor cells [31]. Increasing evidence has shown that dysregulation of *GOT2* expression significantly influences tumor growth and the prognosis of several human neoplasms [30,32–38]. However, the role of *GOT2* in the development and prognosis of KIRC has not been reported. To address these issues, this study aims to evaluate the expression levels of *GOT2* in KIRC and determine its epigenetic modulation, prognostic value, and correlation with tumor-infiltrating immune cells in KIRC patients through multiple databases.

2. Materials and Methods

2.1. Differential Expression of GOT2 mRNA and Protein

Initially, pan-cancer analysis of *GOT2* transcription levels was performed via the TIMER2.0 database (Tumor Immune Estimation Resource, http://timer.cistrome.org/, accessed on 1 December 2021) [39], using the differential expression module across all TCGA tumors. The statistical significance computed by the Wilcoxon test was annotated by the number of stars (* p-value < 0.05; ** p-value < 0.01; *** p-value < 0.001).

To keep the focus of our analyses on KIRC tumors, the GEPIA2 database (Gene Expression Profiling Interactive Analysis 2) (http://gepia.cancer-pku.cn/index.html, accessed on 1 December 2021) [40] was used to confirm the differential expression found in the TIMER analysis by comparing the TCGA-KIRC (523 samples) with normal kidney samples from GTEx (Genotype-Tissue Expression project, http://www.gtexportal.org/home/index.html, accessed on 1 December 2021) (100 samples). The differential threshold of log2FC was 1 and the value cutoff of 0.05.

At the protein level, we used the UALCAN platform (http://ualcan.path.uab.edu/index.html, accessed on 1 December 2021) [41,42] to mine *GOT2* expression in the high-throughput mass spectrometry data, obtained from the Clinical Proteomic Tumor Analysis Consortium (CPTAC) of normal kidney tissues (n = 84) and primary KIRC tumors (N = 110) [43]. Integration and analysis of these data were described elsewhere [44,45]. Briefly, protein expression values (Log2 Spectral count ratio values) from CPTAC were first normalized within each sample profile, then normalized across samples. Then Z-values for each sample for GOT2 protein were calculated as standard deviations from the median across samples.

In this study, we checked the expression of GOT2 in the protein expression module of the HPA database (Human Protein Atlas, https://www.proteinatlas.org/, accessed on 1 December 2021) [46–48], and we analyzed the immunohistochemical results of GOT2 in tumor tissue (ID: 2176) and normal tissue (ID: 2067). The antibody used in both samples was HPA018139. All images of tissues stained by immunohistochemistry were manually annotated by a specialist, followed by verification by a second specialist. Protein expression score was based on immunohistochemical data manually scored concerning staining intensity (negative, weak, moderate, or strong) and the fraction of stained cells (<25%, 25–75% or >75%). Each combination of intensity and fractions was automatically converted into an protein expression level score as follows: negative—not detected; weak <25%—not detected; weak combined with either 25–75% or 75%—low; moderate <25%—low; moderate combined with either 25–75% or 75%—medium; strong <25%—medium, strong combined with either 25–75% or 75%—high (For more details, see https://www.proteinatlas.org/about/assays+annotation#ihk, accessed on 1 December 2021).

2.2. Clinical Correlations & Survival Analysis

Associations between clinicopathological parameters and mRNA expression of *GOT2* were analyzed using the UALCAN (http://ualcan.path.uab.edu/analysis.html, accessed on 5 December 2021) [41]. For these analyses, we included the following clinical features: cancer stage (stages 1, 2, 3, and 4), gender, tumor grade (1, 2, 3 and 4), KIRC subtypes (good risk: ccA; poor risk: ccB) [49], nodal metastasis status (N0: no regional lymph node metastasis; N1: metastases in 1 to 3 axillary lymph nodes).

The prognosis analysis was estimated by Kaplan–Meier (KM) survival curves generated by the Kaplan Meier (KM) Plotter (http://kmplot.com/analysis/, accessed on 5 December 2021) [50], GEPIA2 (http://gepia.cancer-pku.cn/index.html, accessed on 1 December 2021) [40] and HPA database (http://www.proteinatlas.org/, accessed on 5 December 2021) [46–48]. In this study, KIRC patients were split into high and low expression groups based on the median expression levels of *GOT2*, and then these two groups were compared in terms of relapse-free survival. Moreover, the hazard ratio (HR) with a 95% confidence interval (CI) and the *p*-value of the log-rank test were obtained. For all survival analyses, $p < 0.05$ was considered statistically significant.

2.3. GOT2 Methylation Analysis

To explore the DNA methylation level of all CpG islands located in GOT2 of KIRC-TCGA samples, we used the MethSurv database (https://biit.cs.ut.ee/methsurv/, accessed on 10 December 2021) [51]. Next, Shiny Methylation Analysis Resource Tool (SMART) (http://www.bioinfo-zs.com/smartapp/, accessed on 10 December 2021) [52] was used for differential methylation analysis of each *GOT2* probe and Spearman's correlation between methylation level (β-values, 450 k array) and mRNA level (Log2-scaled, TPM+1). CpG-aggregated methylation values were determined by mean (β-values).

2.4. Analysis of Immune Cell Infiltration

We calculated and compared the *GOT2* gene expression contributed by different immune cell types in kidney samples (TCGA tumor/normal and GTEx normal) by the GEPIA2021 (http://gepia2021.cancer-pku.cn/, accessed on 1 December 2021) [53]. For each GTEx/KIRC-TCGA sample, we run the CIBERSORT algorithm (absolute mode) with the default parameters to obtain the absolute proportions of 22 immune cell subtypes. The 22 immune cells included: memory B cells, naïve B cells, activated memory $CD4^+$ T cells, resting memory $CD4^+$ T cells, naïve $CD4^+$ T cells, $CD8^+$ T cells, follicular helper T cells (Tfh), regulatory T cells (Tregs), and gamma/delta T cells, activated dendritic cells (DC), resting dendritic cells, eosinophils, macrophages (M0–M2), activated mast cells, resting mast cells, monocytes, resting NK cells, activated NK cells, neutrophils, and plasma cells. ANOVA (analysis of variance) was used for quantitative comparison. Sidak's multiple comparisons test was used for the post-test, and $p < 0.05$ was considered significant.

2.5. Association between GOT2 and Tumor Microenvironment Exploration

The Tumor Immune Single-cell Hub (TISCH, http://tisch.comp-genomics.org/home/, accessed on 1 April 2022) is an online single-cell RNA-seq database focused on the tumor microenvironment (TME) [54]. In our analyses, two human KIRC scRNA-seq datasets [55,56] were used to obtain the *GOT2* average expression at the single-cell level. The expression of *GOT2* was collapsed by the mean value. The gene expression level displayed using UMAP and violin plots was quantified by the normalized values.

3. Results

3.1. Expression Level of GOT2 mRNA in Pan-Cancer

To determine whether *GOT2* expression correlates with cancer, we surveyed *GOT2* expression in multiple cancer types and adjacent normal tissues through the TIMER database. As shown in Figure 1, compared with normal tissues, *GOT2* expression was higher in BLCA (Bladder Urothelial Carcinoma), CESC (Cervical squamous cell carcinoma and endocervical adenocarcinoma), COAD (Colon adenocarcinoma), KICH (Kidney Chromophobe), ESCA (Esophageal carcinoma), LUAD (Lung adenocarcinoma), LUSC (Lung squamous cell carcinoma), STAD (Stomach adenocarcinoma) and UCEC (Uterine Corpus Endometrial Carcinoma). Conversely, *GOT2* had markedly lower expression in CHOL (Cholangiocarcinoma), GBM (Glioblastoma), KIRC (Kidney renal clear cell carcinoma), LIHC (Liver hepatocellular carcinoma), PRAD (Prostate adenocarcinoma) and THCA (Thyroid carcinoma). All these data indicated that the dysregulation of this glutamic-oxaloacetic transaminase was common across several tumors, including KIRC.

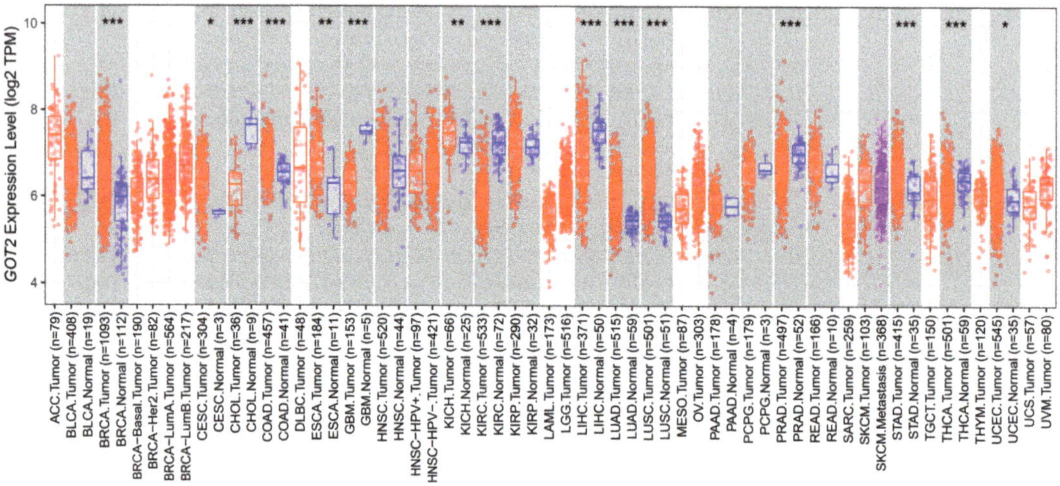

Figure 1. *GOT2* expression levels in pan-cancer (TCGA dataset). The box plot comparing specific *GOT2* expression in tumor samples (red plot) and paired normal tissues (blue plot) was derived from the TIMER database (* $p < 0.05$, ** $p < 0.01$, *** $p < 0.001$). TPM: transcripts per million.

3.2. GOT2 mRNA and Protein Are Downregulated and Correlated with Clinicopathological Parameters in KIRC

Next, to focus our analysis on KIRC, we investigated the transcription levels of *GOT2* performing a single-gene differential analysis using RNA-seq data from the TCGA database (KIRC-TCGA), compared with non-tumor tissues from the GTEx database by GEPIA2. Our results showed that the mRNA expression levels of *GOT2* in KIRC tissues ($n = 523$) were significantly lower than in adjacent normal tissues ($n = 100$) (Figure 2A). Correspondingly, in the CPTAC KIRC cohort, there was a significant downregulation of GOT2 in the tumors (Figure 2B), consistent with the immunohistochemical (IHC) staining images from the Human Protein Atlas (HPA) (Figure 2C). This further confirmed that the expression of *GOT2* in tumor tissues was significantly lower than that in normal tissues.

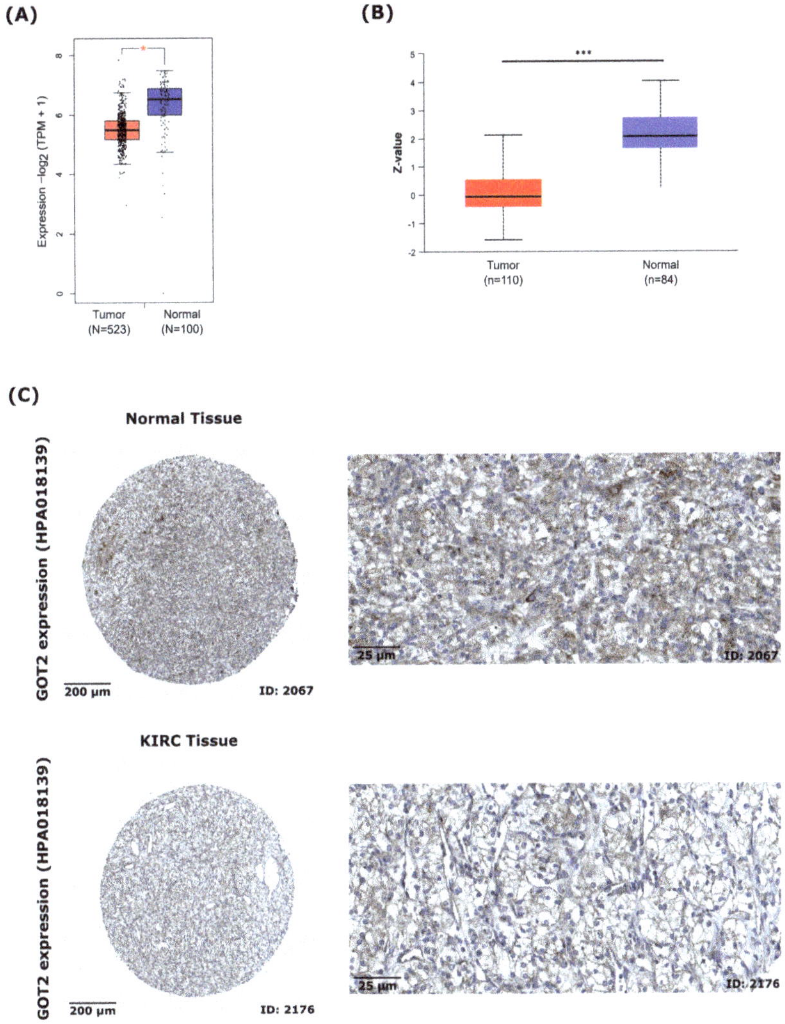

Figure 2. Expression of *GOT2* in KIRC and normal patients. (**A**) Differential expression of *GOT2* between KIRC samples from TCGA database (Red, n = 523 samples) and normal human kidney samples from GTEx database (Blue, n = 100 samples). (* $p < 0.05$, *** $p < 0.001$) (**B**) Significant downregulation of GOT2 protein level in the CPTAC KIRC cohort, analyzed by UALCAN. KIRC: n = 110; Normal samples: n = 84. Z-values represent standard deviations from the median across samples. (**C**) Representative images of immunohistochemical (IHC) staining of GOT2 protein in normal kidney tissue (Patient ID: 2067; Staining: medium; Intensity: moderate; Quantity: 75–25%; Location: cytoplasmic/membranous) and KIRC tissue (Patient ID: 2176; Staining: low; Intensity: weak; Quantity: >75%; Location: cytoplasmic/membranous) from the HPA database. Scale bars: left, 100 μm; right, 25 μm. Antibody used in both samples: HPA018139.

3.3. *Relationship between GOT2 Expression and Clinical Pathological Parameters of Patients with KIRC*

We next investigated the correlation between clinical parameters and the *GOT2* expression in KIRC. Data showed that *GOT2* expression levels were significantly associated with stage, gender, grade, KIRC subtypes, and nodal metastasis status (Figure 3A–G). Lastly, concerning the most commonly mutated genes in KIRC, patients harboring *VHL*, *PBRM1*, and *SETD2* mutations under-expressed *GOT2* (Figure 3H–J). Thus, it is likely that *GOT2* expression may serve as a potential diagnostic biomarker for KIRC patients.

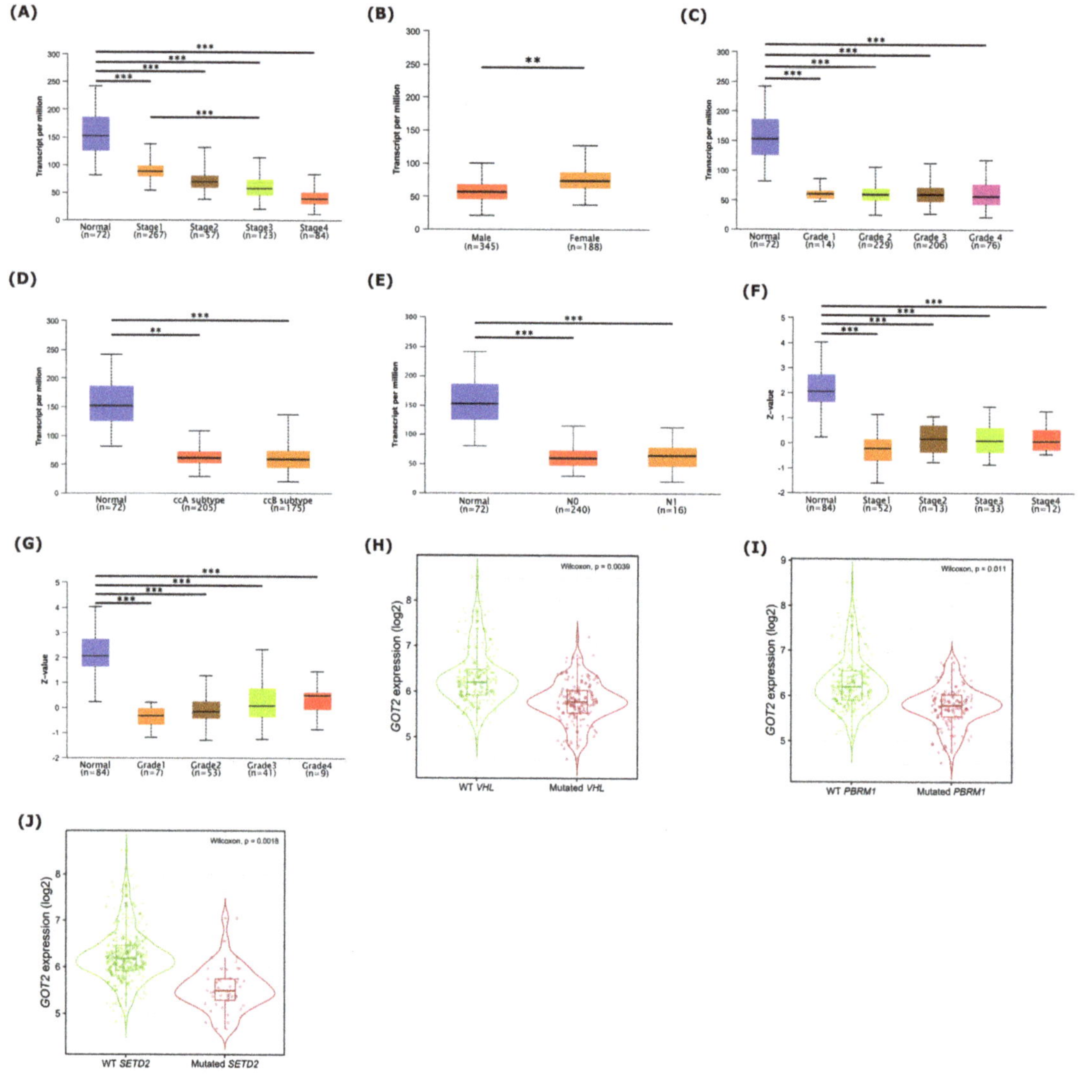

Figure 3. Association between *GOT2* expression and the clinicopathological features of KIRC patients. Box plots of *GOT2* mRNA expression according to: (**A**) KIRC stages (Stages 1, 2, 3 and 4). (**B**) gender (male, female). (**C**) KIRC grades (1, 2, 3 and 4). (**D**) clear cell renal cell carcinoma (ccRCC) good risk (ccA) and poor risk (ccB) subtype classification. (**E**) nodal metastasis status. GOT2 protein expression was differentially expressed in (**F**) clinical stages and (**G**) tumor grade ** $p < 0.01$, *** $p < 0.001$. *GOT2* expression in the KIRC cohort (TCGA) according to (**H**) VHL mutation status, (**I**) PBRM1 mutation status and (**J**) SETD2 mutation status, ** $p < 0.01$, *** $p < 0.001$.

3.4. Low Expression of GOT2 Is Associated with Poor Outcome in KIRC Patients

Initially, to explore the influence of *GOT2* expression on KIRC outcomes, we conducted a Kaplan–Meier test and Cox regression analysis to delve into the associations with overall survival (OS) and disease-free survival (DFS). As shown in Figure 4A,B, the OS and DFS of KIRC patients with low expression of *GOT2* were significantly shorter than those with high expression. At the same time, we also noticed that the low level of GOT2 protein was significantly related to the worse OS ($p = 0.023$) (Figure 4C). Additionally, we investigated the relationship between *GOT2* expression and clinicopathological features of KIRC patients in the Kaplan–Meier plotter database. Surprisingly, low *GOT2* mRNA expression was correlated with worse OS in KIRC patients with stage 4 (HR = 0.56, $p = 3.50 \times 10^{-2}$), grade 3 (HR = 0.53, $p = 7.90 \times 10^{-3}$), and low mutation burden (HR = 2.28, $p = 3.49 \times 10^{-2}$) (Table 1). Here, the differences in the clinical characteristics suggest that the use of *GOT2* as an indicator gene should be carefully combined with the patient's condition.

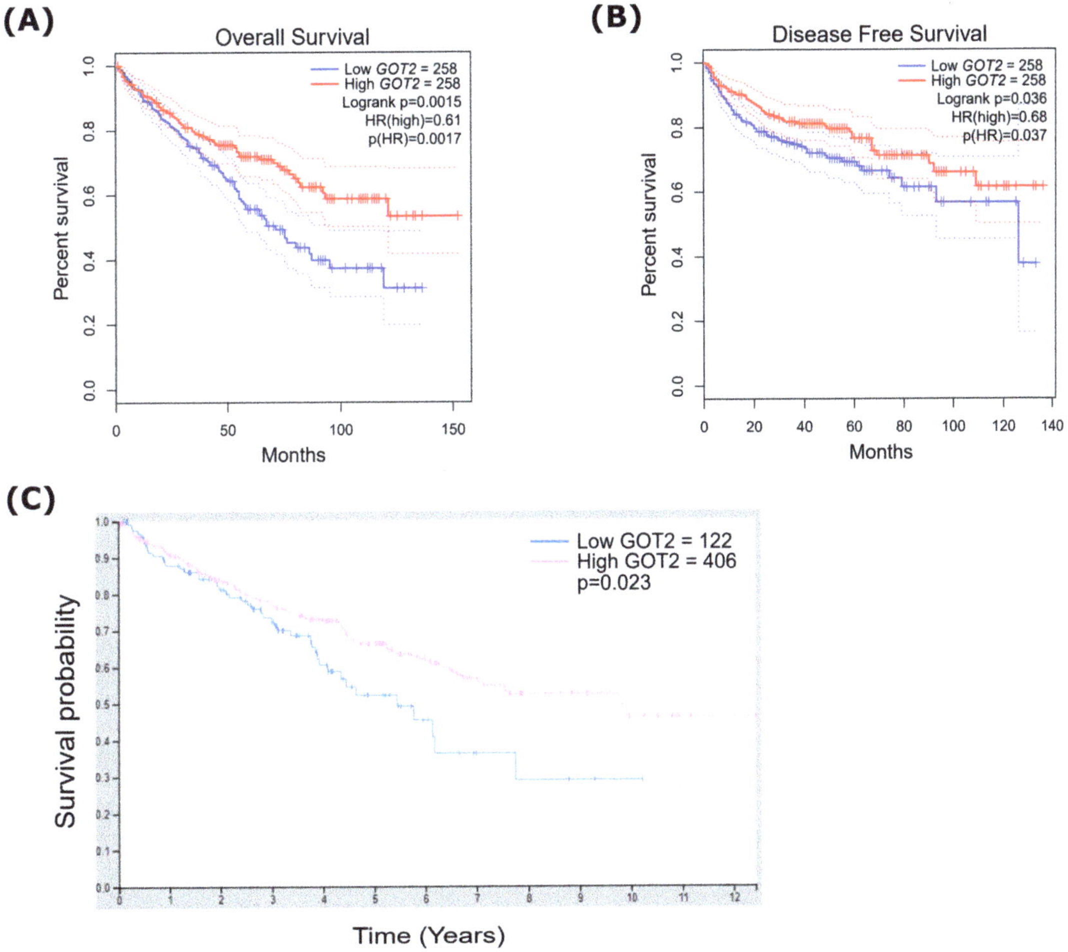

Figure 4. Kaplan–Meier survival analysis demonstrating the relationship between *GOT2* expression and prognosis in KIRC patients. Overexpression of *GOT2* mRNA prolonged (**A**) OS ($n = 258$) and (**B**) DFS (Disease-Free Survival; $n = 258$) of KIRC patients. (**C**) High expression of GOT2 protein prolonged OS of KIRC patients ($n = 528$) ($p = 0.023$). HR, hazard ratio; OS, overall survival; *GOT2*, Glutamic-Oxaloacetic Transaminase 2; KIRC, Kidney Renal Clear Cell Carcinoma.

Table 1. Correlation of *GOT2* mRNA expression and clinical outcomes in KIRC from TCGA database.

Clinicopathological Characteristics	*n*	Hazard Ratio (95% CI)	Logrank *p*
Stage			
1	265	1.67 (0.92–3.03)	8.94×10^{-2}
2	57	0.29 (0.06–1.31)	8.56×10^{-2}
3	123	0.35 (0.74–2.39)	3.47×10^{-1}
4	82	0.56 (0.32–0.97)	$\mathbf{3.50 \times 10^{-2}}$
Gender			
Female	186	0.62 (0.36–1.05)	7.45×10^{-2}
Male	344	0.71 (0.47–1.06)	9.56×10^{-2}
Grade			
1	14	-	-
2	227	1.51 (0.8–2.84)	2.04×10^{-1}
3	206	0.53 (0.33–0.85)	$\mathbf{7.90 \times 10^{-3}}$
4	75	0.64 (0.35–1.19)	1.60×10^{-1}
Mutation burden			
High	168	1.34 (0.77–2.34)	2.94×10^{-1}
Low	164	2.28 (1.04–4.99)	$\mathbf{3.49 \times 10^{-2}}$
Race			
White	459	0.72 (0.5–1.02)	6.17×10^{-2}
Asian	8	-	-
Black/African American	56	0.39 (0.12–1.29)	1.10×10^{-1}
Hemoglobin result			
Elevated	5	1.73 (0.1076–27.8905)	6.98×10^{-1}
Normal	184	0.70 (0.3784–1.3129)	2.70×10^{-1}
Low	261	1.22 (0.8484–1.7628)	2.81×10^{-1}
Laterality			
Right	280	1.12 (0.7181–1.7561)	6.11×10^{-1}
Left	248	0.71 (0.4721–1.0682)	1.00×10^{-1}
Bilateral	4	-	-
Serum calcium result			
Elevated	10	0.65 (0.1599–2.6446)	5.48×10^{-1}
Low	203	0.79 (0.5749–1.5257)	7.92×10^{-1}
Normal	150	0.62 (0.376–1.0547)	7.88×10^{-2}

Bold numbers indicate a statistically significant correlation with a *p*-value less than 0.05. Abbreviations: CI = confidence interval.

3.5. Hypermethylation of DNA in the Promoter Region Leads to Low Expression of GOT2 in KIRC

To further explore the epigenetic mechanism underlying *GOT2* underexpression, we analyzed the methylation level of seventeen probes covering the island (promoter region), N Shelf, S Shore, and Open Sea regions of *GOT2*, chosen through the UCSC Genome Browser (Table 2; Figure 5). Notably, the results showed that lower methylation levels for *GOT2* lay on probes at the promoter (island). At the same time, most hypermethylated sites fell in the open sea, N Shelf, and S Shore regions (Figure 6). Given that methylation of CpG sites within the gene promoter is a common mechanism in gene silencing, we next compared the methylation level of the probes that covered the *GOT2* promoter between normal vs. KIRC-TCGA samples (Figure 5). Interestingly, we found that the average methylation of all CpG sites (probes) near the TSS (transcription start site) of *GOT2* was significantly higher in tumor tissues than in the normal counterpart (Aggregation, $p = 0.00022$) (Figure 7A). Further analysis revealed a negative correlation between the methylation level and the mRNA of *GOT2* (Aggregation: R = −0.3, p = 0.0071) (Figure 7B), thus indicating that upregulation of DNA methylation level of CpGs island-associated promoter region may contribute to the downregulation of *GOT2* in KIRC patients.

Table 2. List of the 17 probes analyzed. CGI: CpG islands.

Probe	Chromosome	Start	End	CGI Position
cg04793118	chr16	58707974	58707975	Open Sea
cg06070269	chr16	58718235	58718236	Open Sea
cg06624121	chr16	58718247	58718248	Open Sea
cg04471375	chr16	58718261	58718262	Open Sea
cg06302295	chr16	58730982	58730983	N Shelf
cg08348831	chr16	58733959	58733960	Island
cg13626907	chr16	58733963	58733964	Island
cg14863484	chr16	58734106	58734107	Island
cg09082840	chr16	58734264	58734265	Island
cg16406345	chr16	58734350	58734351	Island
cg08578141	chr16	58734361	58734362	Island
cg10055227	chr16	58734416	58734417	Island
cg08950929	chr16	58734423	58734424	Island
cg00028829	chr16	58734507	58734508	S Shore
cg10140957	chr16	58734573	58734574	S Shore
cg13883681	chr16	58734661	58734662	S Shore
cg08987251	chr16	58735200	58735201	S Shore

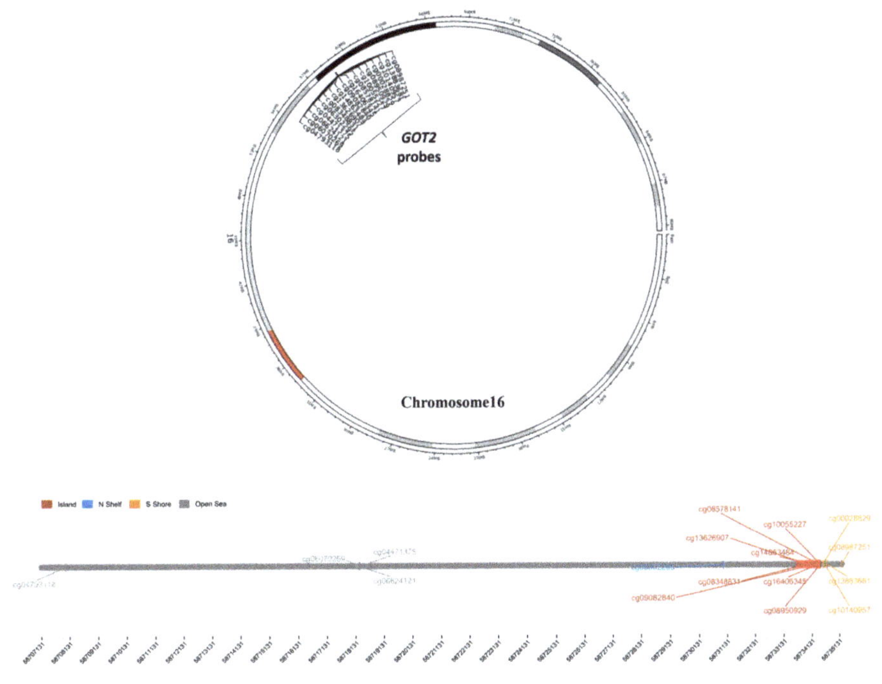

Figure 5. Chromosomal distribution of the methylation probes associated with *GOT2*. Upper panel: Circos plot depicting the genomic information of *GOT2* (16q21) and the probes used in this study. Lower panel: Segment plot showing the detailed information of genomic locations of each probe of *GOT2*, highlighting CpG island, N shelf, S Shore and Open Sea. The coverage of the promoter region is displayed as the red region (red box), which includes eight probes (cg08348831, cg13626907, cg14863484, cg09082840, cg16406345, cg08578141, cg10055227 and cg08950929). Numbers below represent the genomic length scale (1 kb).

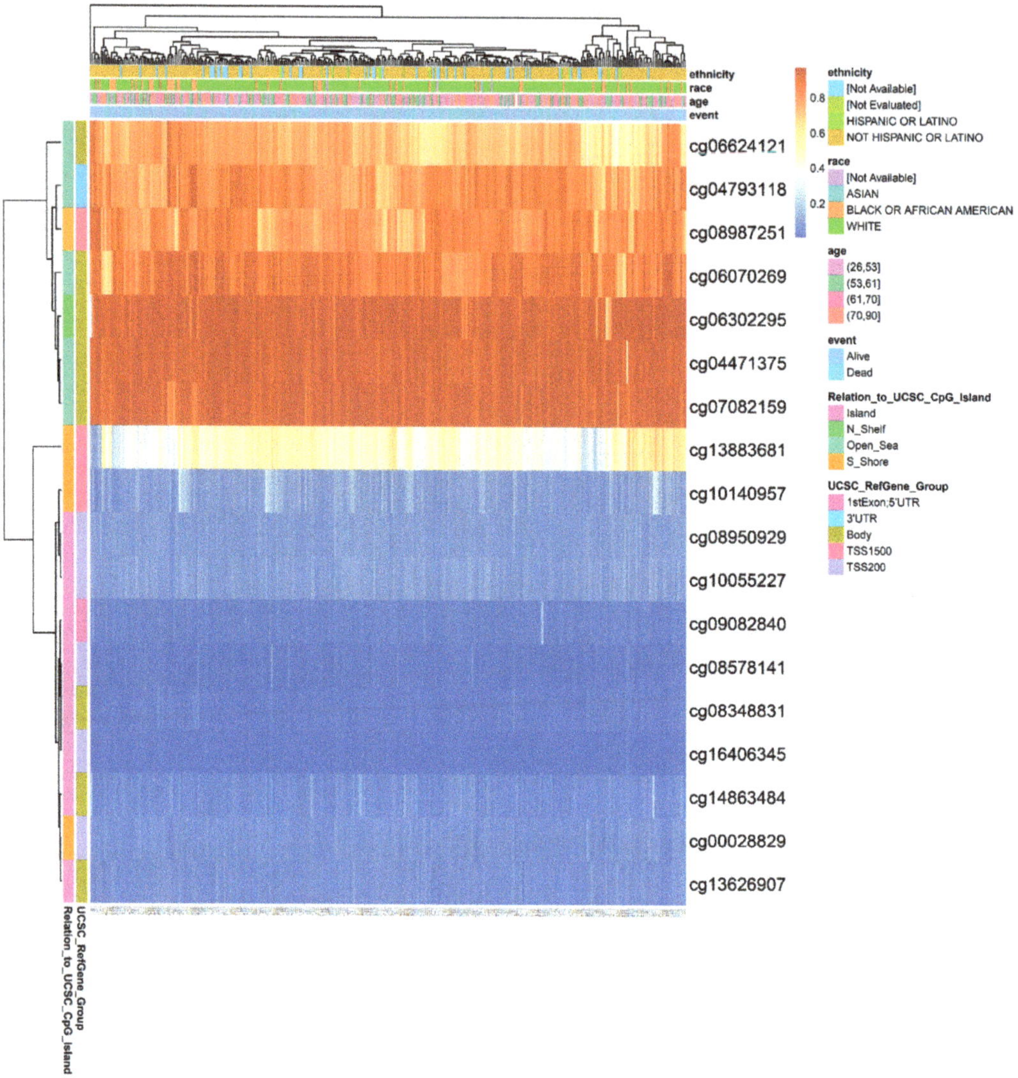

Figure 6. Dynamics of DNA methylation across all probes of *GOT2* in KIRC. Heat map showing the methylation levels of *GOT2* among different CpGs sites (probes) integrating ethnicity, race, age, vital status, and genomic regions of CpG sites (UCSC) from KIRC. Red to blue scale indicates high to low methylation levels.

3.6. *GOT2* Expression Correlates with Immune Cell Infiltration in KIRC

Tumor-infiltrating immune cells are essential for immune response and prognosis in KIRC patients [57,58]. To determine whether *GOT2* could potentially impact immune cell infiltration in KIRC, we first examined the differences in *GOT2* expression across the six immune subtypes proposed by Thorsson et al. [59]. We observed that *GOT2* expression was highest in patients harboring the C5 subtype (immunologically quiet) and lowest in patients exhibiting the C2 subtype (IFN-gamma dominant), indicating that *GOT2* can be used as a marker for immunophenotyping of patients with clear-cell renal cell carcinoma (Figure 8A).

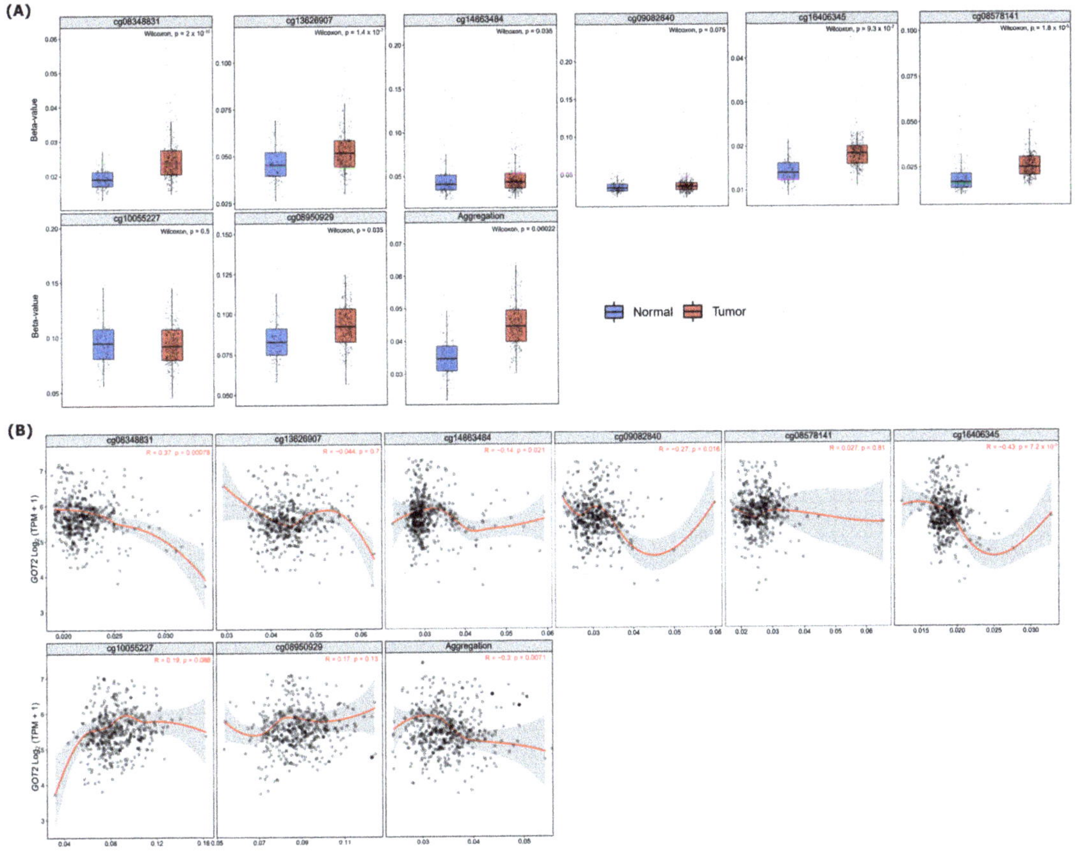

Figure 7. Hypermethylation of *GOT2* leads to downregulated expression in KIRC. (**A**) Differential methylation level of eighth *GOT2* probes (cg08348831, cg13626907, cg14863484, cg09082840, cg16406345, cg08578141, cg10055227 and cg08950929) between KIRC patients ($n = 313$) and normal samples ($n = 157$) from TCGA. (**B**) Spearman's correlation between methylation level (β-values, 450 k array) and mRNA level (Log2-scaled, TPM + 1) of *GOT2* in KIRC samples from TCGA.

To better understand the role of *GOT2* in the infiltration of immune cells in KIRC, we used the CIBERSORT deconvolution analysis [53] for rough correlation analysis. The immune-related signatures revealed that *GOT2* was higher in $CD8^+$ T cells, follicular helper $CD4^+$ T (Tfh) cells, M1 and M2 Macrophages in KIRC-TCGA tumors than in normal tissues (Figure 8B). To further expand and strengthen these results, the analysis of two independent single-cell RNA sequencing (scRNA-seq) datasets [55,56] showed that *GOT2* was mainly expressed within endothelial cells, followed by proliferative T cells (Tprolif), plasmacytoid dendritic cells (pDCs), exhausted $CD8^+$ T Cells (CD8Tex), Treg cells and conventional dendritic cells 2 (cDC2) (Figure 9A,B). These results imply that *GOT2* may play an essential role in the tumor microenvironment of the clear-cell renal cell carcinoma, affecting both stroma and immune cells. Interestingly, *GOT2* was broadly expressed within some clusters of immune cells (e.g., CD8ex and Tprolif) that also co-expressed some immune checkpoint inhibitors (e.g., *CTLA4*, *TIGIT*, *TOX*, *EOMES*, *LAG3*, *PDCD1*, *HAVCR2*, and *CD96*) (Figure 10), thus strongly suggesting that *GOT2* is involved in the dynamic regulation of immune homeostasis and is particularly relevant to T cell functionality.

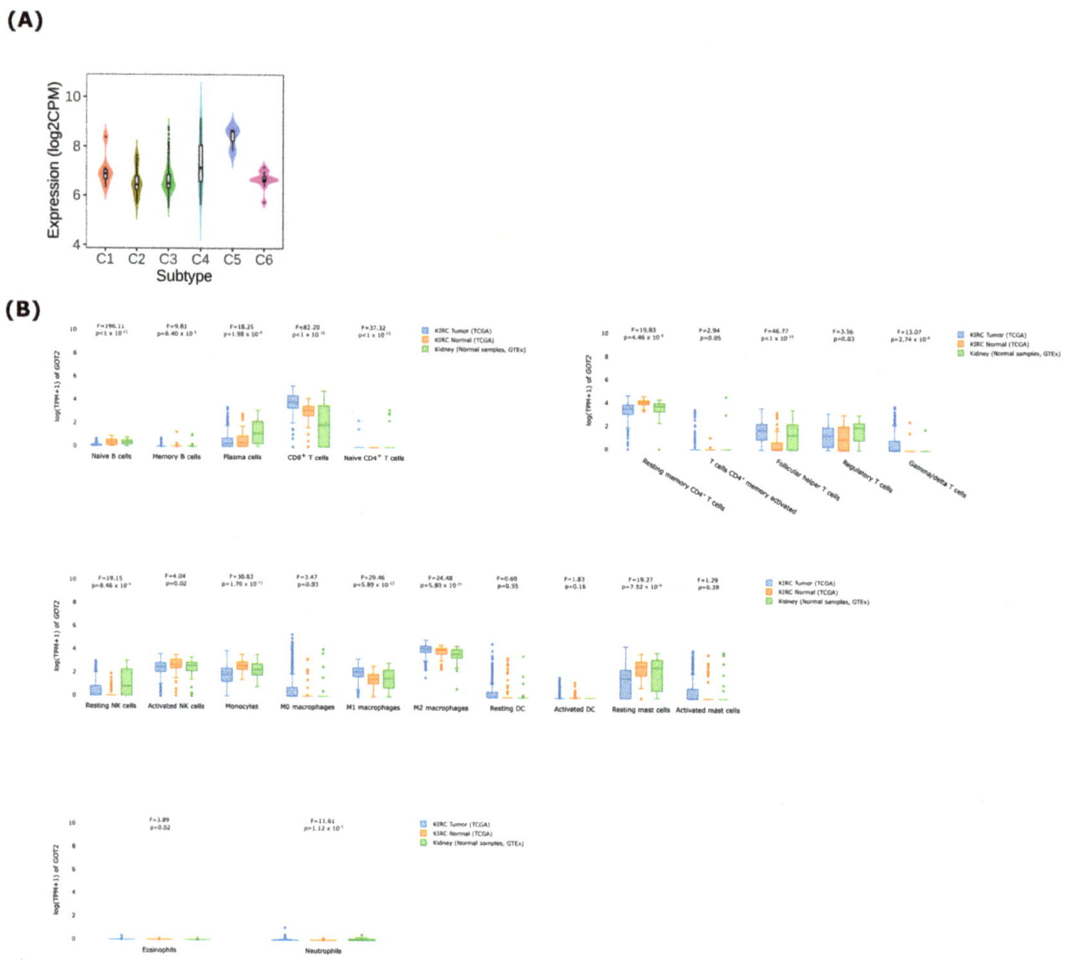

Figure 8. Association of *GOT2* expression, with immune subtypes and immune cell infiltration. (**A**) *GOT2* mRNA levels in TCGA-KIRC immune subtypes. C1: wound healing subtype (n = 7), C2: INF-γ dominant (n = 20), C3: inflammatory (n = 445), C4: lymphocyte depleted (n = 27), C5: immunologically quiet (n = 3), C6: TGF-β dominant (n = 16). One-way ANOVA p-value = 1.2×10^{-4}. (**B**) *GOT2* expression in different immune cells types in KIRC samples from TCGA and normal samples from TCGA and GTEx.

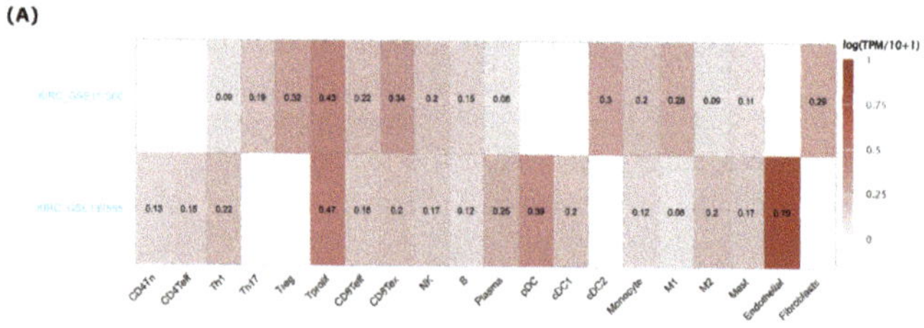

Figure 9. *Cont.*

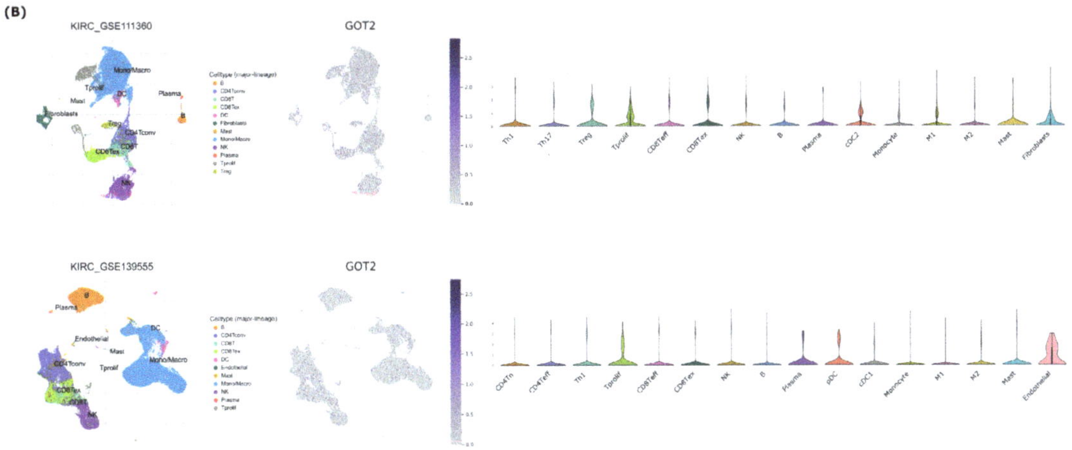

Figure 9. Expression of *GOT2* in scRNA-seq landscapes. (**A**) Heatmap of *GOT2* expression displayed heterogeneity in different clusters of cells in KIRC_GSE111360 [55] and KIRC_GSE139555 datasets [56]. (**B**) Expression of *GOT2* in GSE111360 (upper panel) and in GSE139555 (lower panel) datasets after Uniform Manifold Approximation and Projection (UMAP) processing. Violin diagrams depict the *GOT2* expression in different immune cells across each dataset analyzed.

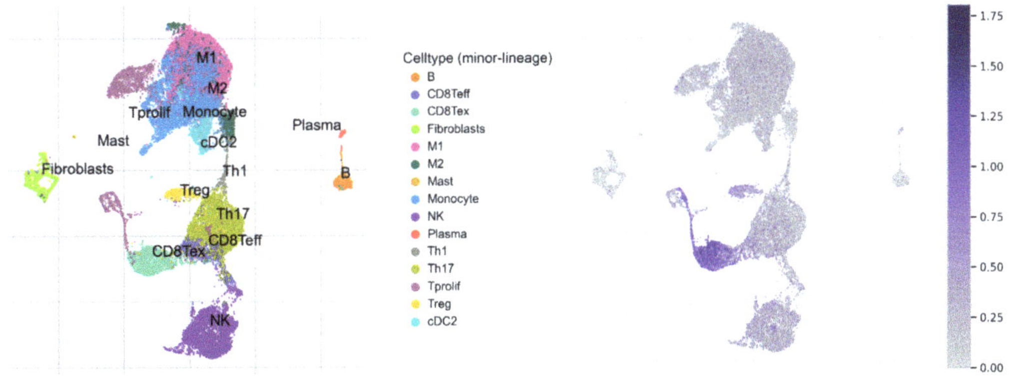

Figure 10. The expression source of the signature genes was revealed by single-cell analysis (GSE111360 dataset). The signature was composed of *GOT2*, *LAG3*, *CTLA4*, *EOMES*, *LGALS9*, *CD96*, *HAVCR2*, *PDCD1*, *TIGIT*, and *TOX*.

4. Discussion

KIRC is characterized by profound metabolic reprogramming that involves multiple pathways [14,15]. Current evidence suggests that changes in the supply of specific metabolites, such as aspartate, which is essential for nucleotide and protein synthesis in proliferating cells and maintains the reducing potential [28,60,61], can function as opportunistic fuel sources for high proliferation and tumor growth [25,27,62]. As part of the malate–aspartate shuttle, mitochondrial *GOT2* generates aspartate from oxaloacetate and glutamate [63]. Additionally, this enzyme is involved in energy transduction, specifically amino acid metabolism and the urea and TCA cycles. Thus far, there is no available information about the detailed roles of *GOT2* in KIRC. Herein, we elucidated the most comprehensive insights into understanding the epigenetic regulation and the potential association of *GOT2* with the clinical and immunity of KIRC.

Based on a pan-cancer perspective, we initially demonstrated that *GOT2* is differentially expressed in 18 tumor types, thus potentially being a therapeutic target. Further interrogating KIRC, we showed that the *GOT2* mRNA and protein levels were markedly decreased in KIRC patients than in normal tissues. Besides, we observed that this transaminase was markedly lower as the pathological stage increased and was also strongly impacted by other clinicopathological characteristics, which conferred a worse outcome. Our findings are consistent with Zhao et al. [64], who also reported the decreased expression and prognostic value of *GOT2* in hepatocellular carcinoma (HCC). The results from a recent study support a scenario in which in *VHL*-deficient KIRC, but not in non-clear renal cell carcinomas (NCRCC), the simultaneous suppression of *GOT1* and *GOT2* is *HIF1*α-dependent, which impairs oxidative and reductive aspartate biogenesis [61]. Hence, to compensate for the low levels of aspartate seen in the KIRC [65], glutamine metabolism has a dominant role in sustaining KIRC growth [66]. This conceivably explains the down-expression of *GOT2* in KIRC patients harboring *VHL*, *PBRM1*, and *SETD2* mutations seen in our study, thus suggesting that *GOT2* repression represents a specific metabolic feature of KIRC.

DNA methylation of specific CpG sites in the promoter region is tightly linked with transcription repression. In the last few years, its role in carcinogenesis has been of considerable interest [67–69]. It is currently well known that KIRC is characterized by many epigenome aberrations [70,71]. Furthermore, many studies have pointed out the occurrence of a pattern known as CpG island methylator phenotype (CIMP) in 20% of KIRC [71,72]. However, no study has previously been carried out to analyze the role of DNA methylation in *GOT2* expression in KIRC. Here, for the first time, we provided evidence that the methylation of the *GOT2* promoter was increased in KIRC patients compared to normal samples. Additionally, the correlation analysis results revealed that promoter methylation was negatively correlated with the regulation of gene expression. According to these results, it can be speculated that the DNA hypermethylation in the promoter-associated CpG islands may be one of the mechanisms leading to *GOT2* down-expression in KIRC. However, additional efforts are necessary to determine the potential impact of additional events, such as chromatin structural modifications, miRNAs, and the influence of metabolites on patients exhibiting *GOT2* promoter hypermethylation.

In addition, another innovative aspect of this study clarified the significant correlations between *GOT2* expression and various tumor-infiltrating immune cells in KIRC. Previous studies have found that T cells and macrophages represented the dominant populations in most KIRC cases [73,74], consistent with our findings, which indicates that *GOT2* expression was more likely to affect the tumor infiltration of subtypes of T cells, especially $CD8^+$ T cells and follicular helper $CD4^+$ T (Tfh) cells, and M1 and M2 macrophages compared to normal renal tissue. Our deep-dissection of individual cell subsets from scRNA-seq data revealed that *GOT2* was broadly expressed within exhausted $CD8^+$ T Cells (CD8Tex) and in the proliferative T cells (Tprolif). Unlike all solid tumors, high tumor-infiltrating $CD8^+$ T-cells predicted poor overall survival and inferior therapeutic responsiveness in patients with KIRC [75–77]. However, a comprehensive characterization of immune cells from KIRC patients using scRNA-seq along with T-cell-receptor (TCR) sequencing revealed that $CD8^+$ T-cells exhibited four distinct groups that may represent transcriptional states upon tumor infiltration with distinct prognostic significance: two of them were associated with a *PD-1*$^+$ *TIM-3*$^+$ exhausted subcluster, one with a proliferative subcluster, and a fourth with the higher levels of cytokine signaling [73]. Moreover, a correlation observed between increased clusters with the signature CD8_6 ($CD8^+$T-cells) and TAM_3 (macrophages) showed a better prognosis. In another study, a first-in-class CAR T-cell therapy co-expressing GOT2 enhanced T-cell metabolic function for treating GPC3-positive solid tumors, supporting the progress of a future first-in-human trial in subjects with GPC3-positive tumors [78]. Considering this context, we argue that *GOT2* is likely to play distinct roles at different stages of T-cell exhaustion and might potentially be modulated by the spectrum of changes in TME conditions of KIRC patients, including tumor metabolism, hypoxia, nutrient restriction, and exhaustion driven by chronic stimulation, thus strengthening the potential application of synergic modulation of the *GOT2* and T cell exhaustion markers in non-responsive KIRC patients to boost antitumor and immune responses.

In conclusion, using a series of rigorous bioinformatics analyses, we showed that the mRNA and the expression levels of the *GOT2* protein were significantly decreased in KIRC patients compared to normal ones. This low expression was positively associated with clinicopathological features, culminating in poor clinical outcomes for KIRC patients. Notably, we provide the first mechanism insights into the epigenetic-mediated regulation of *GOT2*, which is driven by the DNA methylation in the promoter-related CpG islands. Finally, we also shed light on the influence of *GOT2* expression in immune cell infiltration, suggesting that *GOT2* may be a potential prognostic marker and therapeutic target for KIRC patients.

Author Contributions: W.A.S.F. and E.H.C.d.O.: conceived and designed the study; W.A.S.F.: performed data mining; W.A.S.F.: writing—original draft preparation; E.H.C.d.O.: reviewed and editing; E.H.C.d.O.: funding acquisition. All authors have read and agreed to the published version of the manuscript.

Funding: This research was funded by the Pró-Reitoria de Pesquisa e Pós-Graduação (PROPESP/UFPA, Brazil) (PAPQ-02/2022).

Institutional Review Board Statement: Not applicable.

Informed Consent Statement: Not applicable.

Data Availability Statement: The datasets analyzed for this study can be found in the GEPIA2 dababase (http://gepia.cancer-pku.cn/index.html, accessed on 1 December 2021), GTEx (http://www.gtexportal.org/home/index.html, accessed on 1 December 2021), HPA database (Human Protein Atlas) (https://www.proteinatlas.org/, accessed on 1 December 2021), UALCAN (http://ualcan.path.uab.edu/analysis.html, accessed on 1 December 2021), Kaplan Meier (KM) Plotter (http://kmplot.com/analysis/, accessed on 5 December 2021), MethSurv database (https://biit.cs.ut.ee/methsurv/, accessed on 10 December 2021), Shiny Methylation Analysis Resource Tool (SMART) (http://www.bioinfo-zs.com/smartapp/, accessed on 10 December 2021), GEPIA2021 (http://gepia2021.cancer-pku.cn/, accessed on 1 December 2021) and Tumor Immune Estimation Resource (TIMER, https://cistrome.shinyapps.io/timer, accessed on 1 December 2021), Tumor Immune Single-cell Hub (TISCH, http://tisch.comp-genomics.org/home/, accessed on 1 April 2022). Further inquiries can be directed to the corresponding authors.

Acknowledgments: The authors are grateful to Evandro Chagas Institute (Brazil) and Pró-Reitoria de Pesquisa e Pós-Graduação (PROPESP/UFPA, Brazil) for financial and technical support.

Conflicts of Interest: The authors declare no conflict of interest.

References

1. Bray, F.; Ferlay, J.; Soerjomataram, I.; Siegel, R.L.; Torre, L.A.; Jemal, A. Global cancer statistics 2018: GLOBOCAN estimates of incidence and mortality worldwide for 36 cancers in 185 countries. *CA Cancer J. Clin.* **2018**, *68*, 394–424. [CrossRef] [PubMed]
2. Lopez-Beltran, A.; Carrasco, J.C.; Cheng, L.; Scarpelli, M.; Kirkali, Z.; Montironi, R. 2009 update on the classification of renal epithelial tumors in adults. *Int. J. Urol.* **2009**, *16*, 432–443. [CrossRef] [PubMed]
3. Hsieh, J.J.; Purdue, M.P.; Signoretti, S.; Swanton, C.; Albiges, L.; Schmidinger, M.; Heng, D.Y.; Larkin, J.; Ficarra, V. Renal cell carcinoma. *Nat. Rev. Dis. Primers* **2017**, *3*, 17009. [CrossRef] [PubMed]
4. Rini, B.I.; Campbell, S.C.; Escudier, B. Renal cell carcinoma. *Lancet* **2009**, *373*, 1119–1132. [CrossRef]
5. Gerlinger, M.; Rowan, A.J.; Horswell, S.; Math, M.; Larkin, J.; Endesfelder, D.; Gronroos, E.; Martinez, P.; Matthews, N.; Stewart, A.; et al. Intratumor heterogeneity and branched evolution revealed by multiregion sequencing. *N. Engl. J. Med.* **2012**, *366*, 883–892. [CrossRef]
6. Gerlinger, M.; Horswell, S.; Larkin, J.; Rowan, A.J.; Salm, M.; Varela, I.; Fisher, R.; McGranahan, N.; Matthews, N.; Santos, C.R.; et al. Genomic architecture and evolution of clear cell renal cell carcinomas defined by multiregion sequencing. *Nat. Genet.* **2014**, *46*, 225–233. [CrossRef]
7. Sankin, A.; Hakimi, A.A.; Mikkilineni, N.; Ostrovnaya, I.; Silk, M.T.; Liang, Y.; Mano, R.; Chevinsky, M.; Motzer, R.; Solomon, S.B.; et al. The impact of genetic heterogeneity on biomarker development in kidney cancer assessed by multiregional sampling. *Cancer Med.* **2014**, *3*, 1485–1492. [CrossRef] [PubMed]

8. Hsieh, J.J.; Chen, D.; Wang, P.I.; Marker, M.; Redzematovic, A.; Chen, Y.-B.; Selcuklu, S.D.; Weinhold, N.; Bouvier, N.; Huberman, K.H.; et al. Genomic Biomarkers of a Randomized Trial Comparing First-line Everolimus and Sunitinib in Patients with Metastatic Renal Cell Carcinoma. *Eur. Urol.* **2016**, *71*, 405–414. [CrossRef]
9. Jonasch, E.; Walker, C.L.; Rathmell, W.K. Clear cell renal cell carcinoma ontogeny and mechanisms of lethality. *Nat. Rev. Nephrol.* **2020**, *17*, 245–261. [CrossRef]
10. Molina, A.; Lin, X.; Korytowsky, B.; Matczak, E.; Lechuga, M.; Wiltshire, R.; Motzer, R. Sunitinib objective response in metastatic renal cell carcinoma: Analysis of 1059 patients treated on clinical trials. *Eur. J. Cancer* **2013**, *50*, 351–358. [CrossRef]
11. Motzer, R.J.; Barrios, C.H.; Kim, T.M.; Falcon, S.; Cosgriff, T.; Harker, W.G.; Srimuninnimit, V.; Pittman, K.; Sabbatini, R.; Rha, S.Y.; et al. Phase II Randomized Trial Comparing Sequential First-Line Everolimus and Second-Line Sunitinib Versus First-Line Sunitinib and Second-Line Everolimus in Patients With Metastatic Renal Cell Carcinoma. *J. Clin. Oncol.* **2014**, *32*, 2765–2772. [CrossRef]
12. Cui, H.; Shan, H.; Miao, M.Z.; Jiang, Z.; Meng, Y.; Chen, R.; Zhang, L.; Liu, Y. Identification of the key genes and pathways involved in the tumorigenesis and prognosis of kidney renal clear cell carcinoma. *Sci. Rep.* **2020**, *10*, 4271. [CrossRef]
13. Ragone, R.; Sallustio, F.; Piccinonna, S.; Rutigliano, M.; Vanessa, G.; Palazzo, S.; Lucarelli, G.; Ditonno, P.; Battaglia, M.; Fanizzi, F.P.; et al. Renal Cell Carcinoma: A Study through NMR-Based Metabolomics Combined with Transcriptomics. *Diseases* **2016**, *4*, 7. [CrossRef]
14. Wettersten, H.I.; Aboud, O.A.; Lara, P.N., Jr.; Weiss, R.H. Metabolic reprogramming in clear cell renal cell carcinoma. Nature reviews. *Nephrology* **2017**, *13*, 410–419.
15. Bianchi, C.; Meregalli, C.; Bombelli, S.; Di Stefano, V.; Salerno, F.; Torsello, B.; De Marco, S.; Bovo, G.; Cifola, I.; Mangano, E.; et al. The glucose and lipid metabolism reprogramming is grade-dependent in clear cell renal cell carcinoma primary cultures and is targetable to modulate cell viability and proliferation. *Oncotarget* **2017**, *8*, 113502–113515. [CrossRef]
16. Lucarelli, G.; Loizzo, D.; Franzin, R.; Battaglia, S.; Ferro, M.; Cantiello, F.; Castellano, G.; Bettocchi, C.; Ditonno, P.; Battaglia, M. Metabolomic insights into pathophysiological mechanisms and biomarker discovery in clear cell renal cell carcinoma. *Expert Rev. Mol. Diagn.* **2019**, *19*, 397–407. [CrossRef]
17. Lucarelli, G.; Rutigliano, M.; Sallustio, F.; Ribatti, D.; Giglio, A.; Signorile, M.L.; Grossi, V.; Sanese, P.; Napoli, A.; Maiorano, E.; et al. Integrated multi-omics characterization reveals a distinctive metabolic signature and the role of NDUFA4L2 in promoting angiogenesis, chemoresistance, and mitochondrial dysfunction in clear cell renal cell carcinoma. *Aging* **2018**, *10*, 3957–3985. [CrossRef]
18. Bombelli, S.; Torsello, B.; De Marco, S.; Lucarelli, G.; Cifola, I.; Grasselli, C.; Strada, G.; Bovo, G.; Perego, R.A.; Bianchi, C. 36-kDa Annexin A3 Isoform Negatively Modulates Lipid Storage in Clear Cell Renal Cell Carcinoma Cells. *Am. J. Pathol.* **2020**, *190*, 2317–2326. [CrossRef]
19. Maxwell, P.H.; Dachs, G.U.; Gleadle, J.M.; Nicholls, L.G.; Harris, A.L.; Stratford, I.J.; Hankinson, O.; Pugh, C.W.; Ratcliffe, P.J. Hypoxia-inducible factor-1 modulates gene expression in solid tumors and influences both angiogenesis and tumor growth. *Proc. Natl. Acad. Sci. USA* **1997**, *94*, 8104–8109. [CrossRef]
20. Flamme, I.; Krieg, M.; Plate, K.H. Up-Regulation of Vascular Endothelial Growth Factor in Stromal Cells of Hemangioblastomas Is Correlated with Up-Regulation of the Transcription Factor HRF/HIF-2α. *Am. J. Pathol.* **1998**, *153*, 25–29. [CrossRef]
21. Krieg, M.; Haas, R.; Brauch, H.; Acker, T.; Flamme, I.; Plate, K.H. Up-regulation of hypoxia-inducible factors HIF-1α and HIF-2α under normoxic conditions in renal carcinoma cells by von Hippel-Lindau tumor suppressor gene loss of function. *Oncogene* **2000**, *19*, 5435–5443. [CrossRef] [PubMed]
22. Díaz-Montero, C.M.; Rini, B.I.; Finke, J.H. The immunology of renal cell carcinoma. *Nat. Rev. Nephrol.* **2020**, *16*, 721–735. [CrossRef] [PubMed]
23. D'Avella, C.; Abbosh, P.; Pal, S.K.; Geynisman, D.M. Mutations in renal cell carcinoma. *Urol. Oncol. Semin. Orig. Investig.* **2018**, *38*, 763–773. [CrossRef]
24. Pandey, N.; Lanke, V.; Vinod, P.K. Network-based metabolic characterization of renal cell carcinoma. *Sci. Rep.* **2020**, *10*, 5955. [CrossRef] [PubMed]
25. Garcia-Bermudez, J.; Baudrier, L.; La, K.; Zhu, X.G.; Fidelin, J.; Sviderskiy, V.O.; Papagiannakopoulos, T.; Molina, H.; Snuderl, M.; Lewis, C.A.; et al. Publisher Correction: Aspartate is a limiting metabolite for cancer cell proliferation under hypoxia and in tumours. *Nat. Cell Biol.* **2018**, *20*, 775–781. [CrossRef]
26. Sullivan, L.B.; Luengo, A.; Danai, L.V.; Bush, L.N.; Diehl, F.F.; Hosios, A.M.; Lau, A.N.; Elmiligy, S.; Malstrom, S.; Lewis, C.A.; et al. Aspartate is an endogenous metabolic limitation for tumour growth. *Nat. Cell Biol.* **2018**, *20*, 782–788. [CrossRef] [PubMed]
27. Rabinovich, S.; Adler, L.; Yizhak, K.; Sarver, A.; Silberman, A.; Agron, S.; Stettner, N.; Sun, Q.; Brandis, A.; Helbling, D.; et al. Diversion of aspartate in ASS1-deficient tumours fosters de novo pyrimidine synthesis. *Nature* **2015**, *527*, 379–383. [CrossRef]
28. Birsoy, K.; Wang, T.; Chen, W.W.; Freinkman, E.; Abu-Remaileh, M.; Sabatini, D.M. An Essential Role of the Mitochondrial Electron Transport Chain in Cell Proliferation Is to Enable Aspartate Synthesis. *Cell* **2015**, *162*, 540–551. [CrossRef]
29. Van Karnebeek, C.D.M.; Ramos, R.J.; Wen, X.-Y.; Tarailo-Graovac, M.; Gleeson, J.G.; Skrypnyk, C.; Brand-Arzamendi, K.; Karbassi, F.; Issa, M.Y.; van der Lee, R.; et al. Bi-allelic GOT2 Mutations Cause a Treatable Malate-Aspartate Shuttle-Related Encephalopathy. *Am. J. Hum. Genet.* **2019**, *105*, 534–548. [CrossRef]

30. Yang, H.; Zhou, L.; Shi, Q.; Zhao, Y.; Lin, H.; Zhang, M.; Zhao, S.; Yang, Y.; Ling, Z.-Q.; Guan, K.-L.; et al. SIRT 3-dependent GOT2 acetylation status affects the malate–aspartate NADH shuttle activity and pancreatic tumor growth. *EMBO J.* **2015**, *34*, 1110–1125. [CrossRef]
31. Hsu, P.P.; Sabatini, D.M. Cancer Cell Metabolism: Warburg and Beyond. *Cell* **2008**, *134*, 703–707. [CrossRef]
32. Yang, S.; Hwang, S.; Kim, M.; Seo, S.B.; Lee, J.-H.; Jeong, S.M. Mitochondrial glutamine metabolism via GOT2 supports pancreatic cancer growth through senescence inhibition. *Cell Death Dis.* **2018**, *9*, 55. [CrossRef]
33. Guan, H.; Sun, C.; Gu, Y.; Li, J.; Ji, J.; Zhu, Y. Circular RNA circ_0003028 contributes to tumorigenesis by regulating GOT2 via miR-1298-5p in non-small cell lung cancer. *Bioengineered* **2021**, *12*, 2326–2340. [CrossRef]
34. Jin, M.; Shi, C.; Hua, Q.; Li, T.; Yang, C.; Wu, Y.; Zhao, L.; Yang, H.; Zhang, J.; Hu, C.; et al. High circ-SEC31A expression predicts unfavorable prognoses in non-small cell lung cancer by regulating the miR-520a-5p/GOT-2 axis. *Aging* **2020**, *12*, 10381–10397. [CrossRef]
35. Liu, M.; Liu, X.; Liu, S.; Xiao, F.; Guo, E.; Qin, X.; Wu, L.; Liang, Q.; Liang, Z.; Li, K.; et al. Big Data-Based Identification of Multi-Gene Prognostic Signatures in Liver Cancer. *Front. Oncol.* **2020**, *10*, 847. [CrossRef]
36. Du, F.; Chen, J.; Liu, H.; Cai, Y.; Cao, T.; Han, W.; Yi, X.; Qian, M.; Tian, D.; Nie, Y.; et al. SOX12 promotes colorectal cancer cell proliferation and metastasis by regulating asparagine synthesis. *Cell Death Dis.* **2019**, *10*, 239. [CrossRef]
37. Feist, M.; Schwarzfischer, P.; Heinrich, P.; Sun, X.; Kemper, J.; von Bonin, F.; Perez-Rubio, P.; Taruttis, F.; Rehberg, T.; Dettmer, K.; et al. Cooperative stat/nf-kappab signaling regulates lymphoma metabolic reprogramming and aberrant got2 expression. *Nat. Commun.* **2018**, *9*, 1514. [CrossRef]
38. Minchenko, O.H.; Riabovol, O.O.; Tsymbal, D.O.; Minchenko, D.O.; Ratushna, O.O. Effect of hypoxia on the expression of nuclear genes encoding mitochondrial proteins in U87 glioma cells. *Ukr. Biochem. J.* **2016**, *88*, 54–65. [CrossRef]
39. Li, T.; Fu, J.; Zeng, Z.; Cohen, D.; Li, J.; Chen, Q.; Li, B.; Liu, X.S. TIMER2.0 for analysis of tumor-infiltrating immune cells. *Nucleic Acids Res.* **2020**, *48*, W509–W514. [CrossRef]
40. Tang, Z.; Kang, B.; Li, C.; Chen, T.; Zhang, Z. GEPIA2: An enhanced web server for large-scale expression profiling and interactive analysis. *Nucleic Acids Res.* **2019**, *47*, W556–W560. [CrossRef] [PubMed]
41. Chandrashekar, D.S.; Bashel, B.; Balasubramanya, S.A.H.; Creighton, C.J.; Ponce-Rodriguez, I.; Chakravarthi, B.V.S.K.; Varambally, S. UALCAN: A portal for facilitating tumor subgroup gene expression and survival analyses. *Neoplasia* **2017**, *19*, 649–658. [CrossRef] [PubMed]
42. Chandrashekar, D.S.; Karthikeyan, S.K.; Korla, P.K.; Patel, H.; Shovon, A.R.; Athar, M.; Netto, G.J.; Qin, Z.S.; Kumar, S.; Manne, U.; et al. UALCAN: An update to the integrated cancer data analysis platform. *Neoplasia* **2022**, *25*, 18–27. [CrossRef] [PubMed]
43. Edwards, N.J.; Oberti, M.; Thangudu, R.R.; Cai, S.; McGarvey, P.B.; Jacob, S.; Madhavan, S.; Ketchum, K.A. The CPTAC Data Portal: A Resource for Cancer Proteomics Research. *J. Proteome Res.* **2015**, *14*, 2707–2713. [CrossRef]
44. Chen, F.; Chandrashekar, D.S.; Varambally, S.; Creighton, C.J. Pan-cancer molecular subtypes revealed by mass-spectrometry-based proteomic characterization of more than 500 human cancers. *Nat. Commun.* **2019**, *10*, 5679. [CrossRef]
45. Monsivais, D.; Vasquez, Y.M.; Chen, F.; Zhang, Y.; Chandrashekar, D.S.; Faver, J.C.; Masand, R.P.; Scheurer, M.E.; Varambally, S.; Matzuk, M.M.; et al. Mass-spectrometry-based proteomic correlates of grade and stage reveal pathways and kinases associated with aggressive human cancers. *Oncogene* **2021**, *40*, 2081–2095. [CrossRef]
46. Uhlén, M.; Zhang, C.; Lee, S.; Sjöstedt, E.; Fagerberg, L.; Bidkhori, G.; Benfeitas, R.; Arif, M.; Liu, Z.; Edfors, F.; et al. A pathology atlas of the human cancer transcriptome. *Science* **2017**, *357*, 2507. [CrossRef]
47. Thul, P.J.; Åkesson, L.; Wiking, M.; Mahdessian, D.; Geladaki, A.; Ait Blal, H.; Alm, T.; Asplund, A.; Björk, L.; Breckels, L.M.; et al. A subcellular map of the human proteome. *Science* **2017**, *356*, eaal3321. [CrossRef]
48. Uhlén, M.; Fagerberg, L.; Hallström, B.M.; Lindskog, C.; Oksvold, P.; Mardinoglu, A.; Sivertsson, Å.; Kampf, C.; Sjöstedt, E.; Asplund, A.; et al. Proteomics. Tissue-Based Map of the Human Proteome. *Science* **2015**, *347*, 1260419. [CrossRef]
49. Brooks, S.A.; Brannon, A.R.; Parker, J.S.; Fisher, J.C.; Sen, O.; Kattan, M.W.; Hakimi, A.A.; Hsieh, J.J.; Choueiri, T.K.; Tamboli, P.; et al. ClearCode34: A Prognostic Risk Predictor for Localized Clear Cell Renal Cell Carcinoma. *Eur. Urol.* **2014**, *66*, 77–84. [CrossRef]
50. Győrffy, B.; Lanczky, A.; Eklund, A.C.; Denkert, C.; Budczies, J.; Li, Q.; Szallasi, Z. An online survival analysis tool to rapidly assess the effect of 22,277 genes on breast cancer prognosis using microarray data of 1809 patients. *Breast Cancer Res. Treat.* **2009**, *123*, 725–731. [CrossRef]
51. Modhukur, V.; Iljasenko, T.; Metsalu, T.; Lokk, K.; Laisk-Podar, T.; Vilo, J. MethSurv: A web tool to perform multivariable survival analysis using DNA methylation data. *Epigenomics* **2018**, *10*, 277–288. [CrossRef] [PubMed]
52. Li, Y.; Ge, D.; Lu, C. The SMART App: An interactive web application for comprehensive DNA methylation analysis and visualization. *Epigenet. Chromatin* **2019**, *12*, 71. [CrossRef] [PubMed]
53. Li, C.; Tang, Z.; Zhang, W.; Ye, Z.; Liu, F. GEPIA2021: Integrating multiple deconvolution-based analysis into GEPIA. *Nucleic Acids Res.* **2021**, *49*, W242–W246. [CrossRef] [PubMed]
54. Sun, D.; Wang, J.; Han, Y.; Dong, X.; Ge, J.; Zheng, R.; Shi, X.; Wang, B.; Li, Z.; Ren, P.; et al. TISCH: A comprehensive web resource enabling interactive single-cell transcriptome visualization of tumor microenvironment. *Nucleic Acids Res.* **2020**, *49*, D1420–D1430. [CrossRef]

55. Neal, J.T.; Li, X.; Zhu, J.; Giangarra, V.; Grzeskowiak, C.L.; Ju, J.; Liu, I.H.; Chiou, S.-H.; Salahudeen, A.A.; Smith, A.R.; et al. Organoid Modeling of the Tumor Immune Microenvironment. *Cell* **2018**, *175*, 1972–1988.e16. [CrossRef]
56. Wu, T.D.; Madireddi, S.; de Almeida, P.E.; Banchereau, R.; Chen, Y.-J.J.; Chitre, A.S.; Chiang, E.Y.; Iftikhar, H.; O'Gorman, W.E.; Au-Yeung, A.; et al. Peripheral T cell expansion predicts tumour infiltration and clinical response. *Nature* **2020**, *579*, 274–278. [CrossRef]
57. Liu, Z.; Liu, C.; Xiao, M.; Han, Y.; Zhang, S.; Xu, B. Bioinformatics Analysis of the Prognostic and Biological Significance of ZDHHC-Protein Acyltransferases in Kidney Renal Clear Cell Carcinoma. *Front. Oncol.* **2020**, *10*, 565414. [CrossRef]
58. Ghatalia, P.; Gordetsky, J.; Kuo, F.; Dulaimi, E.; Cai, K.Q.; Devarajan, K.; Bae, S.; Naik, G.; Chan, T.A.; Uzzo, R.; et al. Prognostic impact of immune gene expression signature and tumor infiltrating immune cells in localized clear cell renal cell carcinoma. *J. Immunother. Cancer* **2019**, *7*, 139. [CrossRef]
59. Thorsson, V.; Gibbs, D.L.; Brown, S.D.; Wolf, D.; Bortone, D.S.; Ou Yang, T.-H.; Porta-Pardo, E.; Gao, G.F.; Plaisier, C.L.; Eddy, J.A.; et al. The Immune Landscape of Cancer. *Immunity* **2018**, *48*, 812–830.e14. [CrossRef]
60. Lane, A.N.; Fan, T.W.-M. Regulation of mammalian nucleotide metabolism and biosynthesis. *Nucleic Acids Res.* **2015**, *43*, 2466–2485. [CrossRef]
61. Meléndez-Rodríguez, F.; Urrutia, A.A.; Lorendeau, D.; Rinaldi, G.; Roche, O.; Böğürcü-Seidel, N.; Muelas, M.O.; Ciller, C.M.; Turiel, G.; Bouthelier, A.; et al. HIF1α Suppresses Tumor Cell Proliferation through Inhibition of Aspartate Biosynthesis. *Cell Rep.* **2019**, *26*, 2257–2265.e4. [CrossRef]
62. Sullivan, L.B.; Gui, D.Y.; Hosios, A.M.; Bush, L.N.; Freinkman, E.; Vander Heiden, M.G. Supporting Aspartate Biosynthesis Is an Essential Function of Respiration in Proliferating Cells. *Cell* **2015**, *162*, 552–563. [CrossRef]
63. Lieu, E.L.; Nguyen, T.; Rhyne, S.; Kim, J. Amino acids in cancer. *Exp. Mol. Med.* **2020**, *52*, 15–30. [CrossRef]
64. Zhao, Y.; Zhang, J.; Wang, S.; Jiang, Q.; Xu, K. Identification and Validation of a Nine-Gene Amino Acid Metabolism-Related Risk Signature in HCC. *Front. Cell Dev. Biol.* **2021**, *9*, 731790. [CrossRef]
65. Hakimi, A.A.; Reznik, E.; Lee, C.-H.; Creighton, C.J.; Brannon, A.R.; Luna, A.; Aksoy, B.A.; Liu, E.M.; Shen, R.; Lee, W.; et al. An Integrated Metabolic Atlas of Clear Cell Renal Cell Carcinoma. *Cancer Cell* **2016**, *29*, 104–116. [CrossRef]
66. Shroff, E.H.; Eberlin, L.S.; Dang, V.M.; Gouw, A.M.; Gabay, M.; Adam, S.J.; Bellovin, D.I.; Tran, P.T.; Philbrick, W.M.; Garcia-Ocana, A.; et al. MYC oncogene overexpression drives renal cell carcinoma in a mouse model through glutamine metabolism. *Proc. Natl. Acad. Sci. USA* **2015**, *112*, 6539–6544. [CrossRef]
67. Sharma, S.; Kelly, T.K.; Jones, P.A. Epigenetics in cancer. *Carcinogenesis* **2010**, *31*, 27–36. [CrossRef]
68. Sandoval, J.; Esteller, M. Cancer epigenomics: Beyond genomics. *Curr. Opin. Genet. Dev.* **2012**, *22*, 50–55. [CrossRef]
69. Das, P.M.; Singal, R. DNA methylation and cancer. *J. Clin. Oncol. Off. J. Am. Soc. Clin. Oncol.* **2004**, *22*, 4632–4642. [CrossRef]
70. Lasseigne, B.N.; Brooks, J.D. The Role of DNA Methylation in Renal Cell Carcinoma. *Mol. Diagn. Ther.* **2018**, *22*, 431–442. [CrossRef]
71. Mehdi, A.; Riazalhosseini, Y. Epigenome Aberrations: Emerging Driving Factors of the Clear Cell Renal Cell Carcinoma. *Int. J. Mol. Sci.* **2017**, *18*, 1774. [CrossRef] [PubMed]
72. Cancer Genome Atlas Research, N. Comprehensive molecular characterization of clear cell renal cell carcinoma. *Nature* **2013**, *499*, 43–49. [CrossRef]
73. Borcherding, N.; Vishwakarma, A.; Voigt, A.P.; Bellizzi, A.; Kaplan, J.; Nepple, K.; Salem, A.K.; Jenkins, R.W.; Zakharia, Y.; Zhang, W. Mapping the immune environment in clear cell renal carcinoma by single-cell genomics. *Commun. Biol.* **2021**, *4*, 122. [CrossRef] [PubMed]
74. Wang, Q.; Hu, J.; Kang, W.; Wang, J.; Xiang, Y.; Fu, M.; Gao, H.; Huang, Z. Tumor microenvironment immune subtypes for classification of novel clear cell renal cell carcinoma profiles with prognostic and therapeutic implications. *Medicine* **2021**, *100*, e24949. [CrossRef] [PubMed]
75. Nakano, O.; Sato, M.; Naito, Y.; Suzuki, K.; Orikasa, S.; Aizawa, M.; Suzuki, Y.; Shintaku, I.; Nagura, H.; Ohtani, H. Proliferative activity of intratumoral CD8(+) T-lymphocytes as a prognostic factor in human renal cell carcinoma: Clinicopathologic demonstration of antitumor immunity. *Cancer Res.* **2001**, *61*, 5132–5136. [PubMed]
76. Qi, Y.; Xia, Y.; Lin, Z.; Qu, Y.; Qi, Y.; Chen, Y.; Zhou, Q.; Zeng, H.; Wang, J.; Chang, Y.; et al. Tumor-infiltrating CD39+CD8+ T cells determine poor prognosis and immune evasion in clear cell renal cell carcinoma patients. *Cancer Immunol. Immunother.* **2020**, *69*, 1565–1576. [CrossRef] [PubMed]
77. Giraldo, N.; Becht, E.; Pagès, F.; Skliris, G.P.; Verkarre, V.; Vano, Y.A.; Mejean, A.; Saint-Aubert, N.; Lacroix, L.; Natario, I.; et al. Orchestration and Prognostic Significance of Immune Checkpoints in the Microenvironment of Primary and Metastatic Renal Cell Cancer. *Clin. Cancer Res.* **2015**, *21*, 3031–3040. [CrossRef] [PubMed]
78. Choi, E.; Whiteman, K.; Pai, T.; Hickman, T.; Johnson, T.; Friedman, T.; Parikh, A.; Gilbert, M.; Shen, B.; Weiss, G.J.; et al. Abstract 2184: BOXR1030: A first-in-class CAR T-cell therapy co-expressing GOT2 enhances T-cell metabolic function for the treatment of GPC3-positive solid tumors. *Immunology* **2020**, *80*, 2184. [CrossRef]

current issues in molecular biology

Article

NEAT1–SOD2 Axis Confers Sorafenib and Lenvatinib Resistance by Activating AKT in Liver Cancer Cell Lines

Hiroyuki Tsuchiya [1,*], Ririko Shinonaga [1,†], Hiromi Sakaguchi [2], Yutaka Kitagawa [2] and Kenji Yoshida [2]

1 Division of Molecular and Genetic Medicine, Graduate School of Medicine, Tottori University, 86 Nishi-cho, Yonago 683-8503, Japan
2 Department of Radiation Oncology, Tottori University Hospital, 86 Nishi-cho, Yonago 683-8503, Japan
* Correspondence: tsuchiyah@tottori-u.ac.jp; Tel.: +81-859-38-6435
† Current address: Cell Fate Dynamics and Therapeutics, Institute for Life and Medical Science (LiMe), Kyoto University, Shogoin Kawaharacho 53, Sakyo, Kyoto 606-8507, Japan.

Abstract: This study investigated the effects of a long noncoding RNA, nuclear paraspeckle assembly transcript 1 (NEAT1) variant 1 (NEAT1v1) on drug resistance in liver cancer cell lines. NEAT1 knockdown activated mitogen-activated protein kinase (MAPK) signaling pathways, including MAPK kinase (MEK)/extracellular signal-regulated kinase (ERK), but suppressed AKT. Moreover, NEAT1 knockdown sensitized liver cancer cells to sorafenib and lenvatinib, both clinically used for treating hepatocellular carcinoma, whereas it conferred resistance to an AKT-targeted drug, capivasertib. NEAT1v1 overexpression suppressed MEK/ERK and activated AKT, resulting in resistance to sorafenib and lenvatinib and sensitization to capivasertib. Superoxide dismutase 2 (SOD2) knockdown reverted the effects of NEAT1v1 overexpression on the sensitivity to the molecular-targeted drugs. Although NEAT1 or SOD2 knockdown enhanced endoplasmic reticulum (ER) stress, concomitant with the suppression of AKT, taurodeoxycholate, an ER stress suppressor, did not restore AKT activity. Although further in vivo and clinical studies are needed, these results suggested that NEAT1v1 switches the growth modality of liver cancer cell lines from MEK/ERK-dependent to AKT-dependent mode via SOD2 and regulates sensitivity to the molecular-targeted drugs independent of ER stress.

Keywords: NEAT1; liver cancer cell line; drug resistance; SOD2; MAPK; AKT

Citation: Tsuchiya, H.; Shinonaga, R.; Sakaguchi, H.; Kitagawa, Y.; Yoshida, K. NEAT1–SOD2 Axis Confers Sorafenib and Lenvatinib Resistance by Activating AKT in Liver Cancer Cell Lines. *Curr. Issues Mol. Biol.* **2023**, *45*, 1073–1085. https://doi.org/10.3390/cimb45020071

Academic Editor: Asita Elengoe

Received: 6 December 2022
Revised: 25 January 2023
Accepted: 27 January 2023
Published: 29 January 2023

1. Introduction

Liver cancer is one of the malignant tumors with high mortality, making it the third leading cause of cancer-related deaths worldwide [1]. Hepatocellular carcinoma (HCC) is the most prevalent subtype of liver cancer. HCC in the early to middle stages can be treated with surgical resection or locoregional therapies, including radiofrequency ablation and transarterial chemoembolization [2]. In contrast, systemic chemotherapy is indicated for patients with advanced HCC who are not eligible for surgery or locoregional therapies [2].

Sorafenib and lenvatinib are multityrosine kinase inhibitors clinically used to treat HCC, although their preferential targets differ. The primary target kinases of sorafenib are vascular endothelial growth factor receptor (VEGFR), platelet-derived growth factor receptor (PDGFR), FMS-like tyrosine kinase 3 (FLT3), c-KIT, RAF1, and B-RAF, thereby suppressing mainly the mitogen-activated protein kinase (MAPK) cascade, including MAPK kinase (MEK) and extracellular signal-regulated kinase (ERK) [3,4]. Lenvatinib preferentially inhibits the tyrosine kinase activities of VEGFR, PDGFR, RET, c-KIT, and fibroblast growth factor receptor (FGFR), leading to the suppression of their downstream signaling pathways, including the MEK/ERK and phosphatidylinositol 3-kinase (PI3K)/AKT pathways [4–6]. By virtue of their inhibitory properties, both suppress tumor cell proliferation and neoangiogenesis and eventually prolong the survival of patients with advanced HCC [2]. However, the efficacy is limited in part by drug resistance [7].

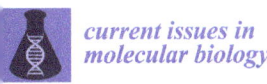

It was demonstrated that alternative activation of P38MAPK supported RAF-independent activation of the MEK/ERK pathway in the presence of sorafenib and was required to acquire sorafenib resistance in the mouse HCC model [8]. AKT activation also induced sorafenib resistance in HCC by suppressing sorafenib-induced autophagic cell death [9,10] or by inducing forkhead box M1 expression via a transcription factor, activator protein 1 [11]. AKT-targeted drugs, MK2206 and ipatasertib, reversed sorafenib resistance in HCC cells [9,12]. Hepatocyte growth factor activated AKT through c-MET and concomitantly attenuated the antitumor effects of lenvatinib in HCC cells highly expressing c-MET [13]. Likewise, the suppression of phosphatase and tensin homologs by proprotein convertase subtilisin/kexin type 9 activated AKT, leading to the acquisition of sorafenib resistance in HCC [14].

A long noncoding RNA, nuclear paraspeckle assembly transcript 1 (NEAT1), is required for the formation of paraspeckle [15]. The NEAT1 gene is expressed as two variant isoforms: NEAT1v1 (3.8 kb in length in humans) and NEAT1v2 (22.7 kb). Both of these are transcribed from the same nucleotide position but have different sites of transcriptional termination [15]. NEAT1v2 is required for the formation of paraspeckle, which is a nuclear substructure found in most cultured cells. NEAT1v1 is also incorporated into paraspeckles, but it also exists in "microspeckle", outside of paraspeckles [16], suggesting that NEAT1v1 has intrinsic functions independent of NEAT1v2. We have previously demonstrated that NEAT1v1, but not NEAT1v2, is involved in the maintenance of liver cancer stem cells and confers radioresistance to liver cancer cell lines [17,18]. It confers radioresistance to liver cancer cell lines by inducing mitophagy via superoxide dismutase 2 (SOD2) and γ-aminobutyric acid A receptor-associated protein (GABARAP) [19]. Moreover, NEAT1 induced sorafenib resistance in HCC cells by activating the c-MET/AKT pathway via microRNA (miR)-335 [20] or promoting autophagy via the miR-204/autophagy-related 3 axis [21]. However, the relationship between NEAT1 and lenvatinib has not been investigated.

This study investigated the effects of NEAT1v1 on the sensitivity of liver cancer cell lines to sorafenib and lenvatinib and found that NEAT1v1 confers resistance to these drugs by activating AKT via superoxide dismutase 2 (SOD2) on the one hand, and NEAT1v1 concomitantly sensitizes cells to an AKT-targeted drug, capivasertib, on the other hand. In agreement with these findings, NEAT1v1 activates AKT while suppressing MAPK signaling molecules, including MEK/ERK, P38MAPK, and c-Jun N-terminal kinase (JNK). Although NEAT1 or SOD2 knockdown increased ER stress, concomitant with AKT suppression, taurodeoxycholate (TUDC), an ER stress suppressor, did not restore AKT activity. These results suggested that the NEAT1v1–SOD2 axis promotes AKT-dependent growth independent of ER stress. Moreover, AKT-targeted drugs are promising as another therapeutic option for treating advanced HCC.

2. Materials and Methods

All resources used in the present study are summarized in Table S1.

2.1. Cell Culture

Human liver cancer cell lines HLE and HuH6 were purchased from the Japanese Collection of Research Bioresources Cell Bank (Osaka, Japan) and maintained in Dulbecco's Modified Eagle's Medium (Nissui Pharmaceutical, Tokyo, Japan) supplemented with 10% inactivated fetal bovine serum (Sigma-Aldrich, St. Louis, MO, USA). HLF and HuH6 cells overexpressing human NEAT1v1 were reported previously [16,17]. In brief, HLF and HuH6 cells were stably transfected with pcDNA6-hNEAT1v1-AcGFP [18]. Following blasticidin (Kaken Pharmaceutical, Tokyo, Japan) selection, AcGFP-positive cells were sorted by flow cytometry.

2.2. Adenovirus Construction

The construction of adenovirus vectors was previously reported [17–19]. In brief, shNT, shNEAT1a/b, or shSOD2a/b were ligated into BsaI-digested pENTR/U6-AmCyan1 with Ligation High version 2 (Toyobo, Osaka, Japan). These oligo DNAs are shown in Table S1. The shRNA and AmCyan1-expressing cassettes were transferred by the LR reaction to pAd/BLOCK-iT-DEST (Thermo Fisher Scientific, Waltham, MA, USA). Adenovirus vectors were constructed by transfecting adenovirus plasmid DNA with Lipofec-tAMINE2000 into 293A cells (Thermo Fisher Scientific) according to the manufacturer's protocol. Adenovirus titer was determined by the infectious genome titration protocol [22]. Adenovirus transduction was performed at 200 multiplicities of infection 24 h after seeding.

2.3. Drug Treatment and WST Assay

Cells were treated with sorafenib (Adipogen Life Sciences, San Diego, CA, USA), lenvatinib (Toronto Research Chemicals, Toronto, ON, Canada), and capivasertib (Adooq Bioscience, Irvine, CA, USA) at the concentrations indicated in the figures, or DMSO as the control for 48 h in a 96-well plate. In the knockdown experiments, adenovirus vectors were transduced 48 h before drug treatment. After treatment, the WST assay was performed with Cell Counting Kit-8 (Dojindo, Kumamoto, Japan) according to the manufacturer's protocol.

2.4. TUDC Treatment

Cells were seeded in a 3.5 cm dish for 24 h. TUDC (Nacalai Tesque, Kyoto, Japan) was added to cells at a concentration of 200 mM. The adenovirus vectors were transduced at the same time as the TUDC treatment. After 48 h incubation, mRNA or protein was recovered from cells.

2.5. Reverse Transcription-Quantitative PCR (RT-qPCR) and Western Blot Analysis

RT-qPCR and Western blot analysis were performed as reported previously [17–19]. mRNA and protein samples were prepared 48 h after seeding, drug treatment, or adenovirus transduction. The primers used for RT-qPCR are summarized in Table S2. An amount of 0.2–1 μg of total RNA was used for the RT reaction, while an amount of 20–100 μg of protein was used for Western blot analysis. β-Actin was used as an internal control for calculating the relative mRNA expression levels. The antibodies for Western blot analysis were as follows: AKT (#9272), P-AKT (S473; #9271), P-AMP-activated protein kinase α (AMPKα) (T172; D79.5E; #4188), P-eukaryotic translation initiation factor 2α (EIF2α) (S51; #9721), P-ERK1/2 (Y202/204; #9101), inositol-requiring enzyme 1α (IRE1α) (14C10; #3294), JNK (#9252), P-JNK (T183/Y185; #9251), P-MEK1/2 (Ser217/221; 41G9;#9154), P-mammalian target of rapamycin (mTOR) (S2448; #2971), P38 (#9212), and P-P38 (T180/Y182; #9211) from Cell Signaling Technology (Danvers, MA, USA); AMPKα1/2 (D-6; sc-74461), activating transcription factor 4 (ATF4) (B-3; sc-390063), ATF6α (F-7; sc-166659), EIF2α (D-3; sc-133132), ERK1/2 (C-9; sc-514302), glyceraldehyde 3-phosphate dehydrogenase (GAPDH) (G-9; sc-365062), MEK1/2 (9G3; sc-81504), PRKR-like ER kinase (PERK) (B-5; sc-377400), β-tubulin (βTUB) (G-8; sc-55529), and X-box binding protein 1 (XBP1) (F-4; sc-8015) from Santa Cruz Biotechnology (Santa Cruz, CA, USA); and P-IRE1(S724; EPR5253; ab124945) from Abcam (Cambridge, MA, USA). After transferring proteins to polyvinylidene difluoride membranes, the membranes were horizontally cut and probed with the antibodies. GAPDH and βTUB (for total and unphosphorylated proteins) and total proteins (for corresponding phosphorylated proteins) were used for internal control.

2.6. Statistical Analysis

Three or more independent samples for each experiment were analyzed, and all experimental values were expressed as the mean ± standard deviation. The differences between the two groups were assessed by Student's *t*-test. Multiple comparisons were made by Dunnett's or Tukey's tests as indicated. $p < 0.05$ was considered statistically significant.

3. Results

3.1. NEAT1 Knockdown Sensitizes Liver Cancer Cells to Sorafenib and Lenvatinib

Two NEAT1-specific short hairpin RNAs (shRNAs; shNEAT1a and shNEAT1b) previously constructed [17,18] were used in this study. Both shRNAs activated MEK and ERK in liver cancer cell lines (HLF and HuH6; Figure 1A). Moreover, NEAT1 knockdown also activated P38MAPK and JNK (Figure S1A). After treatment with sorafenib and lenvatinib, the viability of cells knocked down for NEAT1 decreased significantly more than that of cells transduced with nontargeting shRNA (shNT; Figure 1B), suggesting that NEAT1 knockdown sensitized cells to these drugs. Although the activation of MEK and ERK by NEAT1 knockdown was higher in HuH6 cells than in HLF cells, the sensitization was similar between the cell lines.

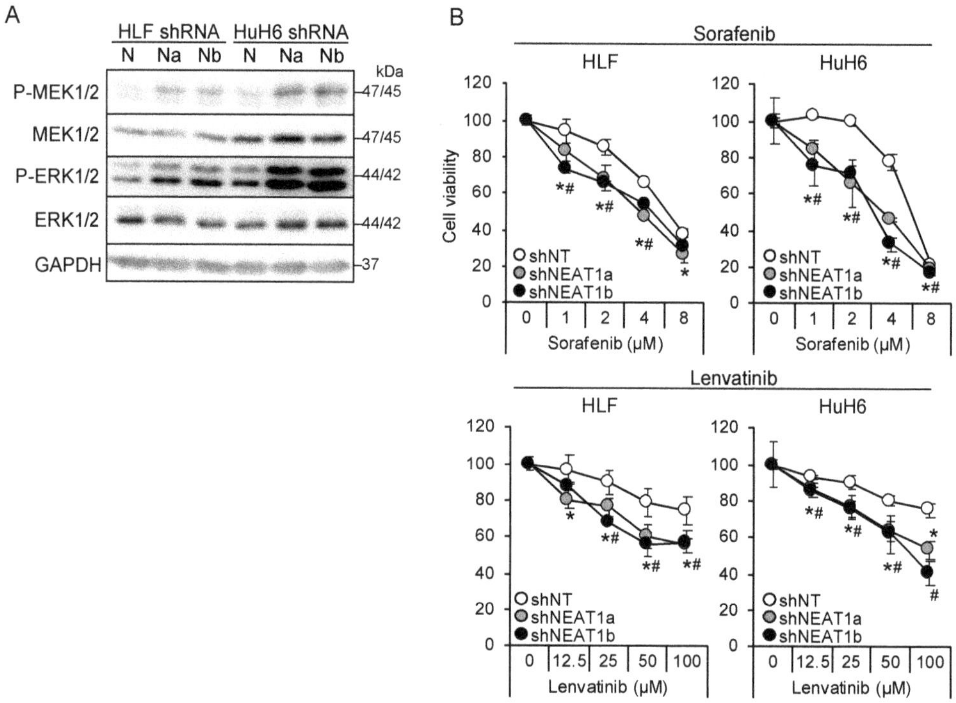

Figure 1. NEAT1 knockdown induces sorafenib and lenvatinib resistance. (**A**) Representative Western blot images for the indicated proteins. GAPDH is shown as an internal control. HLF and HuH6 cell lines were transduced with adenoviruses expressing nontargeting shRNA [shNT (N)] or NEAT1-specific shRNAs [shNEAT1a (Na) and shNEAT1b (Nb)] for 48 h. (**B**) Viabilities of HLF and HuH6 cells treated with sorafenib or lenvatinib at the concentrations indicated in the figure for 48 h relative to cells treated with dimethyl sulfoxide (DMSO; 100%). Cells were transduced with adenoviruses expressing shNT, shNEAT1a, and shNEAT1b 48 h before drug treatment. * $p < 0.05$ vs. shNT vs. shNEAT1a; # $p < 0.05$ shNT vs. shNEAT1b (Dunnett's test; $n = 4$).

3.2. NEAT1 Knockdown Confers Resistance against an AKT-Targeted Drug, Capivasertib

It is suggested that AKT activation is one of the mechanisms underlying sorafenib resistance in HCC [9,10]. Thus, this study investigated whether AKT activity was affected by NEAT1 knockdown. AKT phosphorylation decreased in liver cancer cell lines knocked down for NEAT1 (Figure 2A). In contrast to MEK and ERK, the activation of AKT by NEAT1 knockdown was similar between HLF and HuH6 cell lines. Representative targets of AKT, mTOR, and AMPK were also examined, but their phosphorylation statuses were not changed (Figure S1A). In contrast to sorafenib and lenvatinib, cells showed resistance to an ATP-competitive AKT inhibitor, capivasertib (Figure 2B). These results suggested that NEAT1 knockdown endows liver cancer cells with resistance to AKT-targeted drugs.

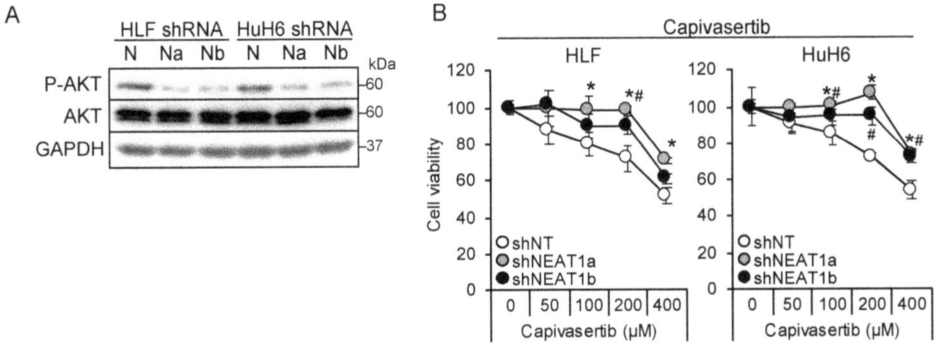

Figure 2. NEAT1 knockdown induces resistance against an AKT-targeted drug, capivasertib. (**A**) Representative Western blot images for the indicated proteins. GAPDH is shown as an internal control. HLF and HuH6 cell lines were transduced with adenoviruses expressing non-targeting shRNA [shNT (N)] or NEAT1-specific shRNAs [shNEAT1a (Na) and shNEAT1b (Nb)] for 48 h. (**B**) Viabilities of HLF and HuH6 cells treated with capivasertib at the concentrations indicated in the figure for 48 h relative to cells treated with DMSO (100%). Cells were transduced with adenoviruses expressing shNT, shNEAT1a, and shNEAT1b 48 h before drug treatment. * $p < 0.05$ vs. shNT vs. shNEAT1a; # $p < 0.05$ shNT vs. shNEAT1b (Dunnett's test; $n = 4$).

3.3. NEAT1v1 Plays a Role as a Molecular Switch of Cell Growth Modality

The shorter isoform, NEAT1v1, but not the longer one, NEAT1v2, is sufficient to induce cancer stemness and radioresistance [17–19]. Thus, liver cancer cell lines overexpressing NEAT1v1 [18,19] were used to examine whether NEAT1v1 could determine drug sensitivity. NEAT1v1 overexpression suppressed MEK and ERK, whereas P38MAPK and JNK phosphorylation was not affected (Figure 3A and Figure S1B). In contrast to knockdown, the activation of MEK and ERK by NEAT1 overexpression was similar between HLF and HuH6 cell lines. In addition, sensitivity to sorafenib and lenvatinib significantly decreased (Figure 3B). In contrast, NEAT1v1 activated AKT and sensitized cells to capivasertib (Figure 3C,D), whereas mTOR and AMPK were not activated (Figure S1B). AKT activation and sensitization to capivasertibe were more prominent in HuH6 cells than in HLF cells. These results suggested that NEAT1v1 is a molecular switch of growth modalities from MEK/ERK- to AKT-dependent growth in liver cancer cells.

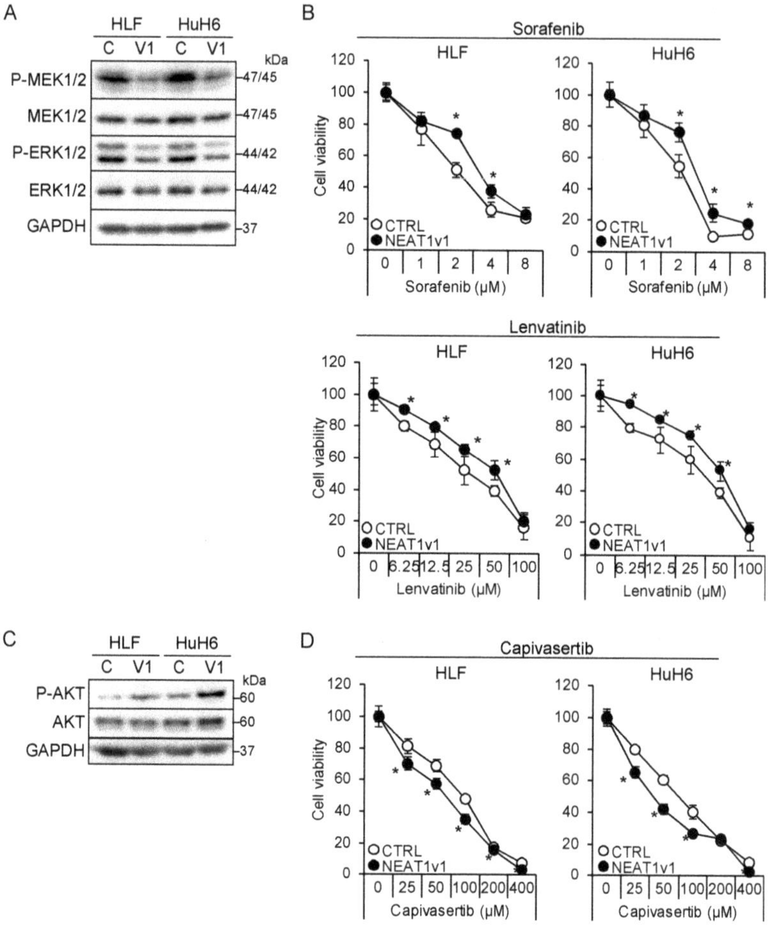

Figure 3. NEAT1v1 plays a role as a molecular switch of cell growth modality. (**A**,**C**) Representative Western blot images for the indicated proteins in control (**C**) or NEAT1v1-overexpressing (V1) cells. GAPDH is shown as an internal control. (**B**,**D**) Viabilities of control (CTRL) or NEAT1v1-overexpressing (NEAT1v1) cells treated with sorafenib (**B**), lenvatinib (**B**), or capivasertib (**D**) at the concentrations indicated in the figure for 48 h relative to cells treated with DMSO (100%). * $p < 0.05$ vs. CTRL (Student's *t*-test; $n = 4$).

3.4. NEAT1v1 Regulates the Growth Modality of Liver Cancer Cells through SOD2

We found that NEAT1 knockdown and NEAT1v1 overexpression resulted in the downregulation and upregulation, respectively, of SOD2 in liver cancer cells (Figure 4A,B). SOD2 knockdown in liver cancer cell lines overexpressing NEAT1v1 activated MEK and ERK as well as P38MAPK and JNK and sensitized cells to sorafenib and lenvatinib (Figure 4C,D and Figure S1C). The activation of MEK and ERK by SOD2 knockdown was higher in HuH6 cells than in HLF cells, similar to NEAT1 knockdown. In addition, it concomitantly suppressed AKT and conferred resistance to capivasertib (Figure 4E,F). However, mTOR and AMPK were not affected by SOD2 knockdown (Figure S1C). These results suggested that SOD2 switches the growth modalities of liver cancer cells downstream of NEAT1v1.

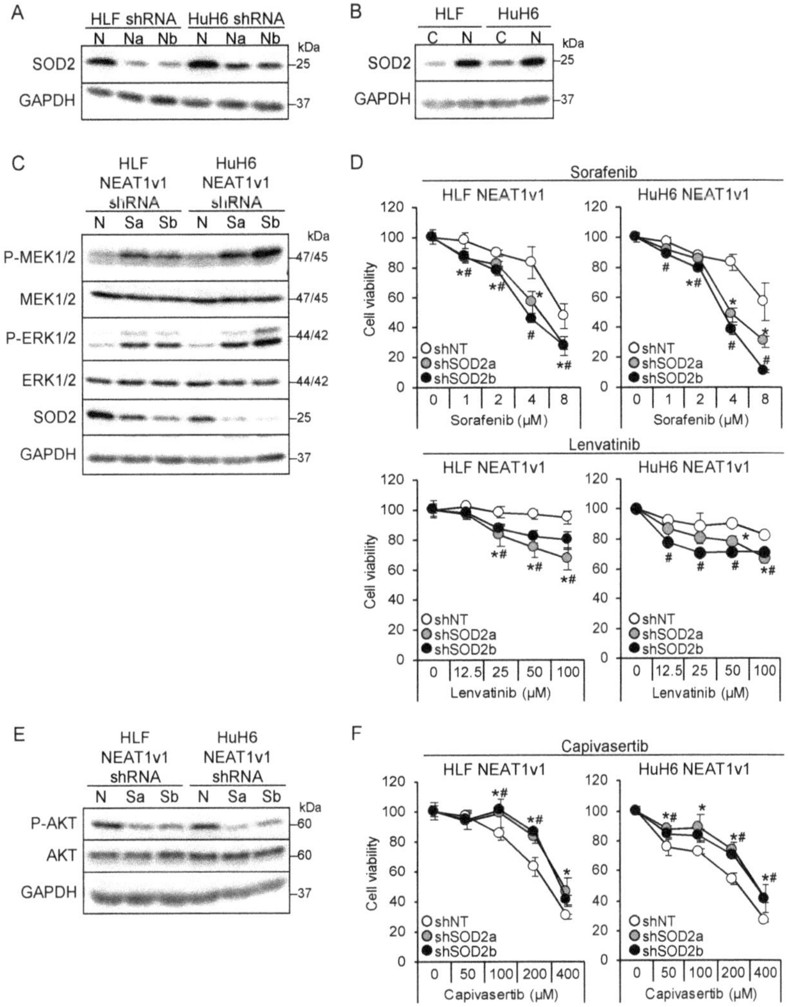

Figure 4. NEAT1v1 regulates cell growth modality through SOD2. (**A**) Representative Western blot images for SOD2 expression in HLF and HuH6 cells transduced with adenoviruses expressing shNT (N), shNEAT1a (Na), or shNEAT1b (Nb) for 48 h. GAPDH is shown as an internal control. (**B**) Representative Western blot images for SOD2 expression in control (**C**) or NEAT1v1-overexpressing (V1) HLF and HuH6 cells. (**C**,**E**) Representative Western blot images for the indicated proteins in NEAT1v1-overexpressing cells transduced with adenoviruses expressing shNT (N) or SOD2-specific shRNAs [shSOD2a (Sa) and shSOD2b (Sb)] for 48 h. (**D**,**F**) Viabilities of cells treated with sorafenib (**D**), lenvatinib (**D**), or capivasertib (**F**) at the concentrations indicated in the figure for 48 h relative to cells treated with DMSO (100%). NEAT1v1-overexpressing cells were transduced with adenoviruses expressing shNT, shSOD2a, and shSOD2b 48 h before drug treatment. * $p < 0.05$ vs. shNT vs. shSOD2a; # $p < 0.05$ shNT vs. shSOD2b (Dunnett's test; n = 4).

3.5. NEAT1v1 or SOD2 Knockdown Suppresses AKT Activity Independent of ER Stress

Unfolded protein accumulation in the endoplasmic reticulum (ER) causes ER stress, which triggers several pathways, including PERK/EIF2α, IRE1α/XBP1, and ATF6, to adapt to stress [23]. Upon increased ER stress, PERK is activated by self-phosphorylation and phosphorylates EIF2α. IRE1α, also activated by self-phosphorylation, executes XBP1 mRNA splicing, leading to XBP1 protein translation. ATF6 is activated by processing, and the p50ATF6 fragment translocates to the nucleus to initiate its target gene transcription. ER stress is also associated with the mTORC1/PI3K/AKT pathway [23]. Whereas, it was reported that SOD2 suppression by anticancer drugs increases oxidative stress, further aggravating ER stress [24]. These results implicate that SOD2 may regulate AKT activity through ER stress.

ER stress induces the expression of its target genes, including binding-immunoglobulin protein (BIP), CCAAT/enhancer-binding protein homologous protein (CHOP), and ER oxidoreductase 1α (ERO1α), to ameliorate stress or induce apoptosis [23]. In liver cancer cell lines overexpressing NEAT1v1, these target genes, except for ERO1α, were downregulated (Figure S2A, Table S3). As previously observed [18], NEAT1v2 was upregulated only in HuH6 cells overexpressing NEAT1v1 (Figure S2A, Table S3). NEAT1 or SOD2 knockdown significantly increased BIP, CHOP, and ERO1α expression, whereas an ER stress inhibitor, TUDC, significantly suppressed their expression (Figures 5A and S2B, Table S3). This result suggested that NEAT1 or SOD2 knockdown enhanced ER stress and that TUDC effectively counteracted it. In agreement with this, PERK was activated by NEAT1 or SOD2 knockdown, as indicated by an autophosphorylation-induced mobility shift and increased EIF2α phosphorylation (Figure 5B). Concomitantly, AKT activity was inhibited, as expected (Figure 5B). In contrast, the IRE1α/XBP1 pathway was suppressed by SOD2 knockdown, and ATF6 was unaffected by NEAT1 and SOD2 knockdown (Figure 5B). In agreement with the expression of ER stress target genes (Figure 5A), TUDC treatment suppressed PERK activation and EIF2α phosphorylation induced by NEAT1 and SOD2 knockdown (Figure 5C). However, the inhibition of AKT activity was not mitigated by TUDC (Figure 5C). These results suggested that NEAT1v1 or SOD2 knockdown suppresses AKT activity independent of ER stress.

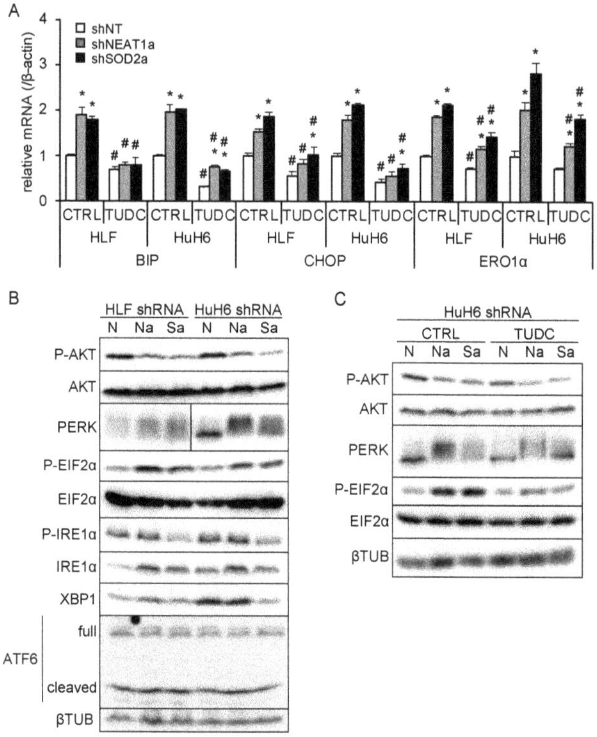

Figure 5. NEAT1v1 or SOD2 knockdown suppresses AKT activity independent of ER stress. (**A**) mRNA expression of ER stress target genes (BIP, CHOP, and ERO1α). HLF and HuH6 cell

lines were transduced with adenoviruses expressing shNT, shNEAT1a, or shSOD2a in the presence of 0 mM (H_2O; CTRL) or 2 mM TUDC for 48 h. * $p < 0.05$ vs. shNT; # $p < 0.05$ vs. CTRL (Tukey's test; $n = 3$). (**B**) Representative Western blot images for the indicated proteins. βTUB is shown as an internal control. HLF and HuH6 cell lines were transduced with adenoviruses expressing shNT (N), shNEAT1a (Na), or shSOD2a (Sa) for 48 h. (**C**) Representative Western blot images for the indicated proteins. GAPDH is shown as an internal control. HuH6 cell lines were transduced with adenoviruses expressing shNT (N), shNEAT1a (Na), or shSOD2a (Sa) in the presence of 0 mM (H_2O; CTRL) or 2 mM TUDC for 48 h.

4. Discussion

Although the precise function of NEAT1 in tumors is not fully clarified yet, NEAT1v1 plays important roles in HCC progression, such as the maintenance of liver CSCs and the acquisition of radioresistance through autophagy [17–19]. This study further elucidated that NEAT1v1 induces sorafenib and lenvatinib resistance, concomitantly with the suppression of MAPK signaling pathways and AKT activation through SOD2. Moreover, as a consequence of the AKT activation, NEAT1v1 sensitizes liver cancer cell lines to an AKT-targeted drug, capivasertib, suggesting that NEAT1v1 induces AKT addiction [25]. Although sorafenib and lenvatinib are clinically used for treating advanced HCC [2], their clinical efficacy is limited partly by the acquisition of drug resistance [7,26]. These results suggest that the NEAT1v1–SOD2 axis is one of the mechanisms underlying resistance to sorafenib and lenvatinib, as well as radiotherapy, and can be a therapeutic target and diagnostic marker for improving their clinical efficacy (Figure 6). Nonetheless, there are some limitations in our study that should be noted. First, all experiments in this study were performed in liver cancer cell lines; thus, the results must be further validated by in vivo studies. Second, HLF and HuH6 cell lines were established from HCC with mutations in the *TP53* gene and *telomerase reverse transcriptase* gene promoter of a 68-year-old male patient and hepatoblastoma with mutations in the *TP53* and *β-catenin* genes of a 1-year-old male patient, respectively [27,28]; thus, the pathological and pathogenic differences, and sex bias must be taken into consideration. Lastly, although the modulated expression of NEAT1v1 significantly changed cell viability, the changes are unlikely to be clinically significant under the current experimental conditions. This might be due to several reasons, such as a single-exposure and one-endpoint assessment, and an insufficient knockdown/overexpression efficiency. These weaknesses should be addressed by optimizing drug treatment conditions and employing knockout/rescue cell lines. Moreover, repeated exposure and continuous observation and assessment of tumors in in vivo models would show a more clinically significant effect than in vitro. However, a more important finding of this study is that NEAT1v1 is involved in the regulation of drug sensitivity in liver cancer cell lines by modulating growth signaling pathways.

We demonstrated that NEAT1 or SOD2 knockdown concomitantly activates P38MAPK and JNK in addition to MEK and ERK, while it was reported that P38MAPK can directly activate MEK in a RAF-independent manner [8]. However, in contrast to our results, this P38MAPK-induced MEK activation rendered HCC cells resistant to sorafenib [8]. Moreover, P38MAPK and JNK phosphorylation was not affected by NEAT1v1 overexpression, suggesting that these MAPK signaling molecules are unlikely to be involved in the mechanism underlying the drug resistance induced by the NEAT1v1–SOD2 axis. Based on these results, it is postulated that the NEAT1v1–SOD2 axis endows liver cancer cells with MEK/ERK-independent and AKT-dependent cell growth. Although this notion likely explains how NEAT1v1 lowered sensitivity to sorafenib and lenvatinib, more precise studies are needed.

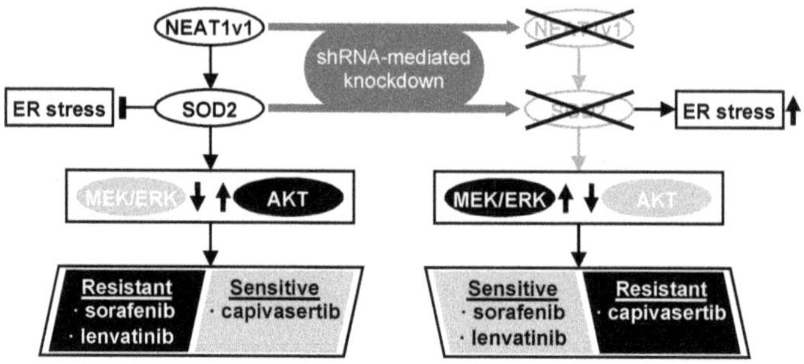

Figure 6. NEAT1v1 activates the AKT pathway through SOD2, thereby conferring sorafenib and lenvatinib resistance in liver cancer cells, which are concomitantly sensitized to capivasertib. This result suggests that NEAT1v1 switches the growth modality of liver cancer cells from MEK/ERK-dependent to AKT-dependent mode via SOD2. Consistently, NEAT1 or SOD2 knockdown results in MEK/ERK activation, thereby sensitizing liver cancer cells to sorafenib and lenvatinib and conferring capivasertib resistance. NEAT1v1 or SOD2 knockdown also exacerbates ER stress; however, AKT is suppressed in an ER stress-independent manner.

The targets of sorafenib are VEGFR, PDGFR, FLT3, c-KIT, RAF1, and B-RAF [3,4], whereas lenvatinib inhibits VEGFR, PDGFR, RET, c-KIT, and FGFR [4–6]. Because RAF1 and B-RAF heterodimers phosphorylate MEK [29], it is thought that sorafenib preferentially inhibits the MAPK pathway [30]. In agreement with these findings, AKT activation is one of the mechanisms underlying the acquisition of sorafenib resistance in HCC [9–14]. Moreover, another group also reported that NEAT1 activates AKT via c-MET in HCC [20]. Although the relation between SOD2 and c-MET remains unclear, it may be worth studying it from the viewpoint of oxidative stress. In contrast, because receptor-type tyrosine kinase members transduce an extracellular signal to the MAPK and PI3K/AKT pathways, lenvatinib inhibits both pathways [4,6,31]. However, the mechanisms underlying lenvatinib resistance are currently not well understood. It was reported that the activities of AKT and ERK decreased and increased, respectively, in thyroid cancer cells treated with lenvatinib [32]. A MEK inhibitor, selumetinib, enhanced the cytotoxic effects of lenvatinib [32], suggesting that thyroid cancer cells switch cell growth modalities from AKT-dependent to MEK/ERK-dependent mode to acquire resistance against lenvatinib. Therefore, molecular switches between the two modes would provide important insights into lenvatinib resistance.

It is suggested that ER stress is involved in the regulation of AKT [23]. The suppression of SOD2 activity by anticancer drugs increased ER stress via oxidative stress, thereby inhibiting AKT, leading to apoptosis in HeLa cells [24]. Interestingly, SOD2 suppression concomitantly activates ERK and inhibits P38MAPK and JNK [24]. In contrast, an herbicide, paraquat, increases ER stress and activates AKT via PERK to promote epithelial-to-mesenchymal transition in pulmonary epithelial cells [33]. A VEGFR2-targeted drug, apatinib, was also shown to activate AKT via IRE1α activated by ER stress in esophageal squamous cell carcinoma [34]. Thus, AKT regulation by ER stress likely depends on the cellular context [23]. In liver cancer cell lines, NEAT1 or SOD2 knockdown activates the PERK/EIF2α pathway. However, amelioration of ER stress by TUDC fails to restore AKT activity. These results indicate that the NEAT1v1–SOD2 axis regulates AKT activity independent of ER stress.

It was demonstrated that a hydrogen peroxide-producing enzyme, NADPH oxidase 4, activates AKT to promote the growth and metastasis of lung cancer [35]. Therefore, hydrogen peroxide produced by SOD2 could be involved in the activation of AKT in liver cancer cells overexpressing NEAT1v1. Moreover, a genome-wide screen using a CRISPR/Cas9 library identified that Kelch-like ECH-associated protein 1 deficiency conferred resistance to sorafenib and lenvatinib in HCC cells [36]. Mechanistically, KEAP1 deficiency induced the activation of nuclear factor erythroid 2-related factor 2, which decreased sorafenib-

and lenvatinib-induced oxidative stress through the upregulation of antioxidative stress factors [36]. These findings suggest that oxidative stress might play a central role in NEAT1v1-induced chemoresistance. The clarification of the underlying mechanism may provide a novel target for treating advanced HCC.

The recurrence of tumors is still a serious clinical problem, especially for systemic chemotherapy for advanced HCC. The data based on the WST assay demonstrated significant but modest sensitizing effects of NEAT1v1 knockdown to capivasertib, suggesting that the efficiency of NEAT1v1 knockdown must be improved to achieve clinically significant efficacy. In terms of this viewpoint, the recent successes of clinical trials using siRNAs or antisense oligonucleotides that target hepatocyte RNAs [37–39] make it attractive to establish NEAT1v1-targeting therapy in combination with capivasertib. Pre-clinical studies using in vivo models will provide more clinically important information for the development of a next-generation therapy for advanced HCC.

5. Conclusions

The NEAT1v1–SOD2 axis switches the growth modality from MEK/ERK- to AKT-dependent mode in male HCC and hepatoma cell lines and confers sorafenib and lenvatinib resistance. NEAT1v1 concomitantly sensitizes the liver cancer cell lines to an AKT-targeted drug, capivasertib. These findings would provide valuable clues to enhance the efficacy of sorafenib and lenvatinib treatment. Moreover, AKT would be a promising target for novel drugs for advanced HCC treatment.

Supplementary Materials: The following are available online at https://www.mdpi.com/article/10.3390/cimb45020071/s1.

Author Contributions: Conceptualization, H.T., H.S. and Y.K.; methodology, H.T. and H.S.; investigation, H.T. and R.S.; data curation, H.T., R.S., H.S. and Y.K.; writing—original draft preparation, H.T.; writing—review and editing, R.S., H.S., Y.K. and K.Y.; supervision, H.T. and K.Y.; project administration, H.T. and K.Y.; funding acquisition, H.T. and H.S. All authors have read and agreed to the published version of the manuscript.

Funding: This work was funded by JSPS KAKENHI Grant Number JP19K08469 (H.T.), and JP19K17460 (H.S.).

Institutional Review Board Statement: This article does not contain any studies involving human participants performed by any of the authors.

Informed Consent Statement: Not applicable.

Data Availability Statement: The raw data are available upon request, please contact the corresponding author.

Acknowledgments: The authors are sincerely grateful to A. Fox (University of Western Australia) for the gift of pCRII_TOPO_hNEAT1.

Conflicts of Interest: The authors have no competing interests.

References

1. Sung, H.; Ferlay, J.; Siegel, R.L.; Laversanne, M.; Soerjomataram, I.; Jemal, A.; Bray, F. Global Cancer Statistics 2020: GLOBOCAN Estimates of Incidence and Mortality Worldwide for 36 Cancers in 185 Countries. *CA Cancer J. Clin.* **2021**, *71*, 209–249. [CrossRef]
2. Llovet, J.M.; Pinyol, R.; Kelley, R.K.; El-Khoueiry, A.; Reeves, H.L.; Wang, X.W.; Gores, G.J.; Villanueva, A. Molecular pathogenesis and systemic therapies for hepatocellular carcinoma. *Nat. Cancer* **2022**, *3*, 386–401. [CrossRef] [PubMed]
3. Wilhelm, S.M.; Carter, C.; Tang, L.; Wilkie, D.; McNabola, A.; Rong, H.; Chen, C.; Zhang, X.; Vincent, P.; McHugh, M.; et al. BAY 43-9006 Exhibits Broad Spectrum Oral Antitumor Activity and Targets the RAF/MEK/ERK Pathway and Receptor Tyrosine Kinases Involved in Tumor Progression and Angiogenesis. *Cancer Res.* **2004**, *64*, 7099–7109. [CrossRef] [PubMed]
4. Tohyama, O.; Matsui, J.; Kodama, K.; Hata-Sugi, N.; Kimura, T.; Okamoto, K.; Minoshima, Y.; Iwata, M.; Funahashi, Y. Antitumor Activity of Lenvatinib (E7080): An Angiogenesis Inhibitor That Targets Multiple Receptor Tyrosine Kinases in Preclinical Human Thyroid Cancer Models. *J. Thyroid Res.* **2014**, *2014*, 638747. [CrossRef] [PubMed]

5. Matsuki, M.; Hoshi, T.; Yamamoto, Y.; Ikemori-Kawada, M.; Minoshima, Y.; Funahashi, Y.; Matsui, J. Lenvatinib inhibits angiogenesis and tumor fibroblast growth factor signaling pathways in human hepatocellular carcinoma models. *Cancer Med.* **2018**, *7*, 2641–2653. [CrossRef] [PubMed]
6. Ferrari, S.M.; Bocci, G.; Di Desidero, T.; Elia, G.; Ruffilli, I.; Ragusa, F.; Orlandi, P.; Paparo, S.R.; Patrizio, A.; Piaggi, S.; et al. Lenvatinib exhibits antineoplastic activity in anaplastic thyroid cancer in vitro and in vivo. *Oncol. Rep.* **2018**, *39*, 2225–2234. [CrossRef] [PubMed]
7. Jindal, A.; Thadi, A.; Shailubhai, K. Hepatocellular Carcinoma: Etiology and Current and Future Drugs. *J. Clin. Exp. Hepatol.* **2019**, *9*, 221–232. [CrossRef]
8. Rudalska, R.; Dauch, D.; Longerich, T.; McJunkin, K.; Wuestefeld, T.; Kang, T.-W.; Hohmeyer, A.; Pesic, M.; Leibold, J.; von Thun, A.; et al. In vivo RNAi screening identifies a mechanism of sorafenib resistance in liver cancer. *Nat. Med.* **2014**, *20*, 1138–1146. [CrossRef]
9. Zhai, B.; Hu, F.; Jiang, X.; Xu, J.; Zhao, D.; Liu, B.; Pan, S.; Dong, X.; Tan, G.; Wei, Z.; et al. Inhibition of Akt Reverses the Acquired Resistance to Sorafenib by Switching Protective Autophagy to Autophagic Cell Death in Hepatocellular Carcinoma. *Mol. Cancer Ther.* **2014**, *13*, 1589–1598. [CrossRef]
10. He, C.; Dong, X.; Zhai, B.; Jiang, X.; Dong, D.; Li, B.; Jiang, H.; Xu, S.; Sun, X. MiR-21 mediates sorafenib resistance of hepatocellular carcinoma cells by inhibiting autophagy via the PTEN/Akt pathway. *Oncotarget* **2015**, *6*, 28867–28881. [CrossRef]
11. Yan, D.; Yan, X.; Dai, X.; Chen, L.; Sun, L.; Li, T.; He, F.; Lian, J.; Cai, W. Activation of AKT/AP1/FoxM1 signaling confers sorafenib resistance to liver cancer cells. *Oncol. Rep.* **2019**, *42*, 785–796. [CrossRef] [PubMed]
12. Zhai, B.; Zhang, X.; Sun, B.; Cao, L.; Zhao, L.; Li, J.; Ge, N.; Chen, L.; Qian, H.; Yin, Z. MK2206 overcomes the resistance of human liver cancer stem cells to sorafenib by inhibition of pAkt and upregulation of pERK. *Tumor Biol.* **2016**, *37*, 8047–8055. [CrossRef] [PubMed]
13. Fu, R.; Jiang, S.; Li, J.; Chen, H.; Zhang, X. Activation of the HGF/c-MET axis promotes lenvatinib resistance in hepatocellular carcinoma cells with high c-MET expression. *Med. Oncol.* **2020**, *37*, 24. [CrossRef]
14. Sun, Y.; Zhang, H.; Meng, J.; Guo, F.; Ren, D.; Wu, H.; Jin, X. S-palmitoylation of PCSK9 induces sorafenib resistance in liver cancer by activating the PI3K/AKT pathway. *Cell Rep.* **2022**, *40*, 111194. [CrossRef] [PubMed]
15. Nakagawa, S.; Naganuma, T.; Shioi, G.; Hirose, T. Paraspeckles are subpopulation-specific nuclear bodies that are not essential in mice. *J. Cell Biol.* **2011**, *193*, 31–39. [CrossRef] [PubMed]
16. Li, R.; Harvey, A.R.; Hodgetts, S.I.; Fox, A.H. Functional dissection of NEAT1 using genome editing reveals substantial localization of the NEAT1_1 isoform outside paraspeckles. *RNA* **2017**, *23*, 872–881. [CrossRef]
17. Koyama, S.; Tsuchiya, H.; Amisaki, M.; Sakaguchi, H.; Honjo, S.; Fujiwara, Y.; Shiota, G. NEAT1 is Required for the Expression of the Liver Cancer Stem Cell Marker CD44. *Int. J. Mol. Sci.* **2020**, *21*, 1927. [CrossRef]
18. Sakaguchi, H.; Tsuchiya, H.; Kitagawa, Y.; Tanino, T.; Yoshida, K.; Uchida, N.; Shiota, G. *NEAT1* Confers Radioresistance to Hepatocellular Carcinoma Cells by Inducing Autophagy through GABARAP. *Int. J. Mol. Sci.* **2022**, *23*, 711. [CrossRef]
19. Tsuchiya, H.; Shinonaga, R.; Sakaguchi, H.; Kitagawa, Y.; Yoshida, K.; Shiota, G. NEAT1 Confers Radioresistance to Hepatocellular Carcinoma Cells by Inducing PINK1/Parkin-Mediated Mitophagy. *Int. J. Mol. Sci.* **2022**, *23*, 14397. [CrossRef]
20. Chen, S.; Xia, X. Long noncoding RNA NEAT1 suppresses sorafenib sensitivity of hepatocellular carcinoma cells via regulating miR-335–c-Met. *J. Cell. Physiol.* **2019**, *234*, 14999–15009. [CrossRef]
21. Li, X.; Zhou, Y.; Yang, L.; Ma, Y.; Peng, X.; Yang, S.; Li, H.; Liu, J. LncRNA NEAT1 promotes autophagy via regulating miR-204/ATG3 and enhanced cell resistance to sorafenib in hepatocellular carcinoma. *J. Cell. Physiol.* **2020**, *235*, 3402–3413. [CrossRef] [PubMed]
22. Gallaher, S.D.; Berk, A.J. A rapid Q-PCR titration protocol for adenovirus and helper-dependent adenovirus vectors that produces biologically relevant results. *J. Virol. Methods* **2013**, *192*, 28–38. [CrossRef]
23. Appenzeller-Herzog, C.; Hall, M.N. Bidirectional crosstalk between endoplasmic reticulum stress and mTOR signaling. *Trends Cell Biol.* **2012**, *22*, 274–282. [CrossRef] [PubMed]
24. Li, M.; Wu, C.; Muhammad, J.S.; Yan, D.; Tsuneyama, K.; Hatta, H.; Cui, Z.-G.; Inadera, H. Melatonin sensitises shikonin-induced cancer cell death mediated by oxidative stress via inhibition of the SIRT3/SOD2-AKT pathway. *Redox Biol.* **2020**, *36*, 101632. [CrossRef] [PubMed]
25. Pfeifer, M.; Grau, M.; Lenze, D.; Wenzel, S.-S.; Wolf, A.; Wollert-Wulf, B.; Dietze, K.; Nogai, H.; Storek, B.; Madle, H.; et al. PTEN loss defines a PI3K/AKT pathway-dependent germinal center subtype of diffuse large B-cell lymphoma. *Proc. Natl. Acad. Sci. USA* **2013**, *110*, 12420–12425. [CrossRef] [PubMed]
26. Chidambaranathan-Reghupaty, S.; Fisher, P.B.; Sarkar, D. Hepatocellular carcinoma (HCC): Epidemiology, etiology and molecular classification. *Adv. Cancer Res.* **2021**, *149*, 1–61. [CrossRef]
27. Doi, I.; Namba, M.; Sato, J. Establishment and some biological characteristics of human hepatoma cell lines. *Gan* **1975**, *66*, 385–392.
28. Doi, I. Establishment of a cell line and its clonal sublines from a patient with hepatoblastoma. *Gan* **1976**, *67*, 1–10.
29. Rushworth, L.K.; Hindley, A.D.; O'Neill, E.; Kolch, W. Regulation and Role of Raf-1/B-Raf Heterodimerization. *Mol. Cell. Biol.* **2006**, *26*, 2262–2272. [CrossRef]
30. Gedaly, R.; Angulo, P.; Hundley, J.; Daily, M.; Chen, C.; Evers, B.M. PKI-587 and Sorafenib Targeting PI3K/AKT/mTOR and Ras/Raf/MAPK Pathways Synergistically Inhibit HCC Cell Proliferation. *J. Surg. Res.* **2012**, *176*, 542–548. [CrossRef]

31. Okamoto, K.; Kodama, K.; Takase, K.; Sugi, N.H.; Yamamoto, Y.; Iwata, M.; Tsuruoka, A. Antitumor activities of the targeted multi-tyrosine kinase inhibitor lenvatinib (E7080) against RET gene fusion-driven tumor models. *Cancer Lett.* **2013**, *340*, 97–103. [CrossRef] [PubMed]
32. Enomoto, K.; Hirayama, S.; Kumashiro, N.; Jing, X.; Kimura, T.; Tamagawa, S.; Matsuzaki, I.; Murata, S.-I.; Hotomi, M. Synergistic Effects of Lenvatinib (E7080) and MEK Inhibitors against Anaplastic Thyroid Cancer in Preclinical Models. *Cancers* **2021**, *13*, 862. [CrossRef] [PubMed]
33. Meng, X.; Liu, K.; Xie, H.; Zhu, Y.; Jin, W.; Lu, J.; Wang, R. Endoplasmic reticulum stress promotes epithelial-mesenchymal transition via the PERK signaling pathway in paraquat-induced pulmonary fibrosis. *Mol. Med. Rep.* **2021**, *24*, 525. [CrossRef]
34. Wang, Y.-M.; Xu, X.; Tang, J.; Sun, Z.-Y.; Fu, Y.-J.; Zhao, X.-J.; Ma, X.-M.; Ye, Q. Apatinib induces endoplasmic reticulum stress-mediated apoptosis and autophagy and potentiates cell sensitivity to paclitaxel via the IRE-1α–AKT–mTOR pathway in esophageal squamous cell carcinoma. *Cell Biosci.* **2021**, *11*, 124. [CrossRef] [PubMed]
35. Chen, B.; Song, Y.; Zhan, Y.; Zhou, S.; Ke, J.; Ao, W.; Zhang, Y.; Liang, Q.; He, M.; Li, S.; et al. Fangchinoline inhibits non-small cell lung cancer metastasis by reversing epithelial-mesenchymal transition and suppressing the cytosolic ROS-related Akt-mTOR signaling pathway. *Cancer Lett.* **2022**, *543*, 215783. [CrossRef] [PubMed]
36. Zheng, A.; Chevalier, N.; Calderoni, M.; Dubuis, G.; Dormond, O.; Ziros, P.G.; Sykiotis, G.P.; Widmann, C. CRISPR/Cas9 genome-wide screening identifies KEAP1 as a sorafenib, lenvatinib, and regorafenib sensitivity gene in hepatocellular carcinoma. *Oncotarget* **2019**, *10*, 7058–7070. [CrossRef]
37. Ray, K.K.; Wright, R.S.; Kallend, D.; Koenig, W.; Leiter, L.A.; Raal, F.J.; Bisch, J.A.; Richardson, T.; Jaros, M.; Wijngaard, P.L.; et al. Two Phase 3 Trials of Inclisiran in Patients with Elevated LDL Cholesterol. *N. Engl. J. Med.* **2020**, *382*, 1507–1519. [CrossRef]
38. Strnad, P.; Mandorfer, M.; Choudhury, G.; Griffiths, W.; Trautwein, C.; Loomba, R.; Schluep, T.; Chang, T.; Yi, M.; Given, B.D.; et al. Fazirsiran for Liver Disease Associated with Alpha$_1$-Antitrypsin Deficiency. *N. Engl. J. Med.* **2022**, *387*, 514–524. [CrossRef]
39. Yuen, M.-F.; Lim, S.-G.; Plesniak, R.; Tsuji, K.; Janssen, H.L.; Pojoga, C.; Gadano, A.; Popescu, C.P.; Stepanova, T.; Asselah, T.; et al. Efficacy and Safety of Bepirovirsen in Chronic Hepatitis B Infection. *N. Engl. J. Med.* **2022**, *387*, 1957–1968. [CrossRef]

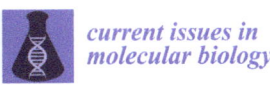

current issues in molecular biology

Article

Cellular, Molecular and Proteomic Characteristics of Early Hepatocellular Carcinoma

Athanasios Armakolas [1,2], Vasiliki Dimopoulou [1], Adrianos Nezos [1], George Stamatakis [3], Martina Samiotaki [3], George Panayotou [3], Maria Tampaki [2], Martha Stathaki [1], Spyridon Dourakis [2] and John Koskinas [2,*]

1 Physiology Laboratory, Athens Medical School, National and Kapodistrian University of Athens, 115 27 Athens, Greece
2 B' Department of Medicine, Hippokration Hospital, National and Kapodistrian University of Athens, 115 27 Athens, Greece
3 Institute for Bioinnovation, Biomedical Sciences Research Center "Alexander Fleming", 166 72 Vari, Greece
* Correspondence: jkoskinas@med.uoa.gr

Abstract: Hepatocellular carcinoma (HCC) accounts for the majority of primary liver cancers. Early detection/diagnosis is vital for the prognosis of HCC, whereas diagnosis at late stages is associated with very low survival rate. Early diagnosis is based on 6-month surveillance of the patient and the use of at least two imaging modalities. The aim of this study was to investigate diagnostic markers for the detection of early HCC based on proteome analysis, microRNAs (miRNAs) and circulating tumor cells (CTCs) in the blood of patients with cirrhosis or early or advanced HCC. We studied 89 patients with HCC, of whom 33 had early HCC and 28 were cirrhotic. CTCs were detected by real-time quantitative reverse transcription PCR and immunofluorescence using the markers epithelial cell adhesion molecule (EPCAM), vimentin, alpha fetoprotein (aFP) and surface major vault protein (sMVP). Expression of the five most common HCC-involved miRNAs (miR-122, miR-200a, miR-200b, miR-221, miR-222) was examined in serum using quantitative real time PCR (qRT-PCR). Finally, patient serum was analyzed via whole proteome analysis (LC/MS). Of 53 patients with advanced HCC, 27 (51%) had detectable CTCs. Among these, 10/27 (37%) presented evidence of mesenchymal or intermediate stage cells (vimentin and/or sMVP positive). Moreover, 5/17 (29%) patients with early HCC and 2/28 (7%) cirrhotic patients had detectable CTCs. Patients with early or advanced HCC exhibited a significant increase in miR-200b when compared to cirrhotic patients. Our proteome analysis indicated that early HCC patients present a significant upregulation of APOA2, APOC3 proteins when compared to cirrhotic patients. When taken in combination, this covers the 100% of the patients with early HCC. miR-200b, APOA2 and APOC3 proteins are sensitive markers and can be potentially useful in combination for the early diagnosis of HCC.

Keywords: HCC; CTCs; miRNAs; proteomics

Citation: Armakolas, A.; Dimopoulou, V.; Nezos, A.; Stamatakis, G.; Samiotaki, M.; Panayotou, G.; Tampaki, M.; Stathaki, M.; Dourakis, S.; Koskinas, J. Cellular, Molecular and Proteomic Characteristics of Early Hepatocellular Carcinoma. *Curr. Issues Mol. Biol.* **2022**, *44*, 4714–4734. https://doi.org/10.3390/cimb44100322

Academic Editor: Dumitru A. Iacobas

Received: 22 August 2022
Accepted: 3 October 2022
Published: 10 October 2022

1. Introduction

Liver cancer is the second-most lethal form of cancer, with a 5-year survival rate of 18% [1]. It is the sixth-most common cancer in terms of incidence rate, with nearly 800,000 cases reported annually. Recent evidence suggests that liver cancer incidence is increasing, with a mortality rate estimate of more than 1 million in 2030 [1,2]. Hepatocellular carcinoma (HCC) accounts for the majority of primary liver cancers. Early detection and diagnosis are vital for the prognosis of the disease, whereas in patients with HCC that are diagnosed at late stages, the odds are dismayingly low. Survival rates for advanced HCC vary by country, with the one-year survival rate being approximately 20% after diagnosis [1–3]. The majority of HCCs occur in patients with underlying liver disease, usually as a result of chronic hepatitis B or C virus (HBV or HCV) infection, alcohol abuse or metabolic syndrome [4].

Chronic hepatitis B virus (HBV) and hepatitis C virus (HCV) infections are major global public health problems, with an estimated 1 million and 450,000 deaths yearly worldwide, respectively [5]. Up to 40% of patients with chronic hepatitis B virus infection and approximately 20–30% of subjects infected with HCV are estimated to develop liver cirrhosis. Most commonly, deaths occur from complications of liver cirrhosis, namely liver failure and HCC [5].

In the absence of antiviral therapy, the prognosis for patients with HBV cirrhosis is poor, with a 5-year survival of only 14%, compared to 84% in patients with compensated HBV cirrhosis. Similarly, patients with HCV cirrhosis in the absence of anti-viral therapy have a 67%–91% mortality rate due to liver-related causes, including HCC or hepatic failure [6]. However, patients with HCV cirrhosis, even if cured of HCV infection, maintain a significant risk of HCC development [6,7].

Non-alcoholic steatohepatitis (NASH) as a liver manifestation of metabolic syndrome and obesity is rapidly increasing and is expected to become the predominant risk factor for HCC in high-income regions in the coming years [7].

It is estimated that 2–5% of patients with cirrhosis will develop HCC annually [1,2]. For early detection of HCC, it is recommended that all patients with cirrhosis should be under 6-month surveillance with ultrasound by an experienced radiologist. In the presence of a nodule, a contrast-enhanced CT or MRI should be performed. A distinctive pattern of hyper-enhancement in the arterial phase and washout in venous or delayed phases is diagnostic for HCC. MRI screening for small HCC ($\leq$3 cm) has 78.82% sensitivity, 78.46% specificity, 78.67% accuracy, 82.72% positive predictive value and 73.91% negative predictive value [8]. For patients with nodules that are less than 1 cm in diameter, follow-up imaging is mandatory [2,9].

Since early diagnosis is vital for the prognosis of HCC, it is easy to understand the necessity of the development of diagnostic tools for the early diagnosis of HCC in cirrhotic patients. In this study, we investigated various biological markers in the blood of patients with early and advanced HCC and cirrhosis without HCC. The markers examined were: circulating tumor cells (CTCs) of epithelial (epithelial cell adhesion molecule, (EPCAM), mesenchymal (vimentin) and intermediate stage (surface major vault protein) subtypes and a set of microRNAs (miRNAs) that are involved in HCC [10–13]: miR-122, which is associated with metastasis and poor prognosis [2]; miR-200a and miR-200b, which have been found as early markers of HCC development [14,15]; and miR-221 and miR-222, which have been shown to target the cell-cycle-dependent kinase inhibitor p27 and induce cell proliferation, metastasis and infiltration [15,16]. The blood of all these patients was also examined for protein signatures using whole proteome analysis.

Previous evidence indicated the possibility of the use of multi-omics techniques prior to the establisment of more accurate diagnostic tools in HCC [17,18]. Our aim was to determine useful markers to predict the development of HCC at very early stages in cirrhotic patients.

2. Materials and Methods

2.1. Patient Selection

A total of 117 patients were included in the study—mean age 65 $\pm$ 11 years, 87 (74%) males. With respect to the etiology of the underlying liver disease, 41 (35%) cases were viral-related, 39 (33%) alcohol- and/or metabolic-syndrome-related and 37 (31.6%) related to other etiology (Table 1).

Concerning the presence of HCC, 89 patients had non-treated HCC (56 patients with advanced and 33 with early HCC) and 28 were age-matched cirrhotic patients with no evidence of HCC, based on imaging studies and normal serum aFP levels. Five healthy controls were also recruited in the study. Total cellular RNA, serum miRNA and serum proteins were extracted from the five healthy individuals, and the average value obtained in each case was used as the reference value.

Early HCC was defined as a nodule $\leq$3 cm. Diagnosis of HCC was based on MRI-characteristic hemodynamic pattern and/or liver biopsy. Whole blood, plasma and serum samples were obtained from all patients.

Table 1. Patient Characteristics.

	HCC Patients (n = 89)	Cirrhotic Patients (n = 28)
Median age, years (range)	65 ± 11 years	60 ± 7 years
Male sex, n (%)	67, (75%)	20, (71%)
Liver Disease, n (%)		
Viral hepatitis	32, (36%)	9, (32%)
Non-viral hepatitis	57, (64%)	19, (68%)
HCC stage, n (%)		
BCLC A	33, (37%)	n/a
BCLC C	56, (63%)	
Microvascular invasion		
(Available data in 55 patients)	23, (26%)	n/a
Serum aFP values >20 ng/ml		
Early HCC	7/33 (21%)	-
Advanced HCC	42/56 (75%)	-
Cirrhotic	-	0/28 (0%)

HCC, hepatocelllar carcinoma; BCLC, Barcelona Clinic liver cancer staging system; AFP, alpha fetoprotein; n/a, not available.

The study was approved by the National and Kapodistrian University of Athens Medical School Ethics Committee (Approval code: 029/19.11.18). All subjects gave informed consent in accordance with the Declaration of Helsinki. Written informed consent was obtained from every patient and healthy individual involved in this study. All methods were performed in accordance with the relevant guidelines and regulations. The relatively small number of patients included in the study is mainly due to the prospective nature of the study, the availability of patients in the out-patient clinic and the restricted sample collection time frame in order to obtain a meaningful follow-up period.

2.2. CTCs

Cells were isolated from 10 mL peripheral blood, and after treatment with erythrocytes lysis buffer, the pellet was treated with TRIZOL reagent according to the manufacturer's instructions. Briefly, 2.5 mL freshly isolated peripheral whole blood samples was mixed with 7.5 mL erythrocyte lysis buffer (ELB) and incubated on ice for 30 min, followed by centrifugation 400 g/10 min/4 °C and lysis of the cell pellet using TRIzol Reagent (Cat # 15596026, Thermo Scientific, Waltham, MA, USA). RNA extraction was performed according to the manufacturer's instructions and stored at −80 °C. The quantity and quality of RNA samples was spectrophotometrically recorded (Biospec Nano, Kyoto, Japan).

2.3. MicroRNAs

Plasma miRNA was extracted with a Nucleospin miRNA plasma kit (Cat # 740971, MACHEREY-NAGEL GmbH & Co. KG, Duren, Germany) according to manufacturer instructions.

2.4. cDNA Synthesis and Real-Time PCR

A Mir-X miRNA First-Strand Synthesis kit (Cat # 638316, Takara Bio Inc., Otsu, Shiga, Japan) was used for the synthesis of first-strand cDNA from our purified miRNA samples according to the manufacturer's instructions. Complementary DNA samples were diluted 1:5 with nuclease-free water (Qiagen, Hilden, Germany) immediately after synthesis and stored at −20 °C.

Total RNA (0.5μg) obtained from peripheral whole blood samples was reverse transcribed using PrimeScript RT Reagent Kit (Perfect Real Time) (Cat # RR037A, Takara Bio Inc., Otsu, Shiga, Japan). Complementary DNA samples were diluted 1:10 with nuclease-free water (Qiagen, Germany) immediately after synthesis and stored at −20 °C.

Quantitative real-time polymerase chain reaction (qRT-PCR) was carried out using the Bio-Rad IQ5 thermocycler and the Kapa Biosystems SYBR Green (Cat # KR0389, Kapa Biosystems, Cape Town, South Africa).

Specific 5′ primer to amplify only miR-122, miR-221, miR-222, miR-200a and miR-200b was synthesized according to Mir-X miRNA First-Strand Synthesis kit instructions and the miRBase Sequence Database (University of Manchester: http://www.mirbase.org assessed on 20 May 2018). Primer sequences are shown in Supplementary Table S1. As a normalization miRNA, U6 snRNA control (reference) was used, supplied by the Mir-X miRNA First-Strand Synthesis kit. Despite the fact that the quantitative determinations of the miRNAs examined were not obtained using probes, the specificity of our assay was induced by the isolation of only small-size RNAs from plasma; the primers were selected under stringent conditions, and they were specific for only one target; and the PCR conditions were set prior to favor the amplification of small molecules.

Specific primers to amplify only cDNA (exon-intron spanning) for circulating-tumor-cell-related genes: MVP, EPCAM, vimentin, aFP and the normalization (reference) gene GAPDH were designed using the Beacon Designer software. The sequences of each primer set are shown in Supplementary Table S1.

The reaction was performed in a total volume of 20 μL per reaction and constituted of 2 μL of template cDNA, 0.4 μM of each primer, 10 μL 2× KAPA SYBR Green Mix (Kapa Biosystems, Cape Town, South Africa) and ultra-pure water. A two-step amplification protocol was applied, starting with step 1, with one cycle at 95 °C for 4 min followed step 2, with 40 cycles at 95 °C for 5 s and 63 °C for 30 s. The specificity of the amplified products was determined via melting curve analysis.

The threshold cycles (Ct) generated by the qPCR system were used to calculate relative gene expression levels between different samples [13,19]. Briefly, the Ct of the target gene was subtracted from the Ct of the reference gene for the two groups, and the relative expression of each sample was determined using the $2^{-\Delta\Delta Ct}$ method. All reactions were performed in duplicate. The average value of the samples from the HC group in each case was used as the reference sample.

The average calibrator ΔCq value in EPCAM was obtained from these samples, and individual ΔCqs of the calibrator were within 2 SDs of the average calibrator ΔCq value. Samples were classified as positive for a particular gene if the $2^{-\Delta\Delta Cq}$ was 2.0 or more (i.e., 100% or more than what is found in healthy blood) [19]. Additionally, prior to the determination of the cutoff points that would define positive from negative patients in the case of CTC and miRNA markers' ROC, curve analysis was carried out using biomarkers' continuous variables (Figure S1).

2.5. *Immunofluorescence Staining (IF)*

Peripheral blood (10 mL) was centrifuged, and the cells were resuspended in RPMI 10% FBS. The cells were then cultured on chamber slides for 6 hrs, and they were stained by a direct immunofluorescence method. Cells were rinsed in PBS and fixed with ice-cold 80% methanol for 10 min at room temperature. Firstly, they were stained with anti-EPCAM and anti-MVP fluorescent primary antibodies for 1 hr at room temperature, and they were then permeabilized with 0.5% Triton X-100 (Cat # 9002-93-1, Sigma Aldrich, St. Louis, MO, USA) for 10 min. They were then incubated with the anti-vimentin fluorescent primary antibody again for 1 h at room temperature. The antibodies used were Texas Red conjugated anti-EPCAM (1:100), Alexa fluor 480 anti-Vimentin (1:100) and Alexa fluor 647 anti-MVP (1:100) (Cat # ab286811, ab195877, ab208627, respectively, all from Abcam, Cambridge, UK). After 3 washes, samples were stained with 4′,6-diamidino-2-phenylindole (DAPI) (1 μg/mL) for viewing with a microscope (Olympus BX40, Tokyo, Japan). Patients were considered positive for the existence of CTCs when more than 10 EPCAM-positive cells were detected in 5 mL of blood.

2.6. Proteome Analysis

Liquid Chromatography Mass Spectrometry (LC MS)

For each patient and normal sample, 30 μL of serum was analyzed. Peptides were separated with an Acclaim Pepmap 75 μm × 50 cm on an Ultimate 3000 nano LC system set at 40 °C. The HPLC was performed using 350 nl/min flow with 0.1% formic acid in water as solvent A and as 0.1% formic acid in acetonitrile as solvent B. For each injection, 500 ng was loaded and eluted with a linear gradient from 8% to 24% buffer B over 50 min; then, the gradient was ramped up to 36% buffer B over 10 min. The column was washed with 100% buffer B for 5 min and was left to equilibrate to 8% buffer B for 15 min.

The gas phase fractionation (GFP) of the pooled sample was performed using 12 × 50 m/z windows ranging from 400–1000 m/z at 60,000 MS1 with an AGC of 3 e6 for 60 ms and MS2 at 30,000 with an AGC of 1×10^6 for 60 ms; NCE was set to 27, and +2H was assumed as the default charge state. The GPF-DIA acquisitions used 4 m/z precursor isolation windows in a staggered window pattern with optimized window placements.

The DIA conditions for the sample analysis were (a) measured at 390–1010 m/z with 60,000 resolution and an AGC target of 3e6, IT of 60 ms for MS1, and (b) 76 × 8 m/z DIA isolation windows were measured at 15,000 resolution in a staggered-window pattern with optimized window placements from 400 to 1000 m/z using an NCE 0f 27 at a +2 default charge state.

Data processing proteomics:

The raw files from GFP were overlap demultiplexed with 10 ppm accuracy after peak picking in ProteoWizard (version 3.0.18299). Searches were performed using EncyclopeDIA (version 0.9), which was configured to use default settings: 10 ppm precursor, fragment and library tolerances. EncyclopeDIA was allowed to consider both B and Y ions, and trypsin digestion was assumed.

The created library was imported in skyline (version 20.1) by removing repeated peptides. The raw files were imported and reintegrated by an m-prophet model with 1:1 decoys, and the peptides were filtered for dotp >0.7. The total fragment areas for each peptide were exported, and all peptides corresponding to a protein were summed. The statistical analysis of the samples was performed in Perseus version (1.6.10.0). The total fragment area of each protein from the previous step was normalized to the total area of the sample.

The mass spectrometry proteomics data have been deposited to the ProteomeXchange Consortium via the PRIDE [1] partner repository with the dataset identifier PXD033306.

2.7. Statistical Analysis

Statistical analysis was performed by SPSS V23 (SPSS software; SPSS Inc, Chicago, IL, USA) and ROC curves were performed using the MedCalc® Statistical Software, version 20.113 (MedCalc Software Ltd., Ostend, Belgium; https://www.medcalc.org; accessed on 21 August 2022). Data were expressed as frequencies, mean ± SD, or median with interquartile range (IQR), as appropriate. Quantitative variables were compared using Student's *t*-test or Mann–Whitney test for normally distributed and non-normally distributed variables, respectively. Qualitative variables were compared with Chi-squared test or Fisher's exact-test, as appropriate. The relationships between quantitative variables were assessed using the Spearman's correlation coefficient. All tests were two-sided, and p values < 0.05 were considered to be significant.

3. Results

3.1. CTCs

Circulating tumor cells of all types were detected in the serum of patients with HCC of all stages. Detection was obtained by qRT-PCR in total RNA isolated from peripheral blood and by immune-fluorescence in blood cells. Cells and cellular RNA were evaluated in 53 out of 56 patients with advanced HCC, in 17 out of 33 patients with early HCC and in all 28 patients with cirrhosis (Figure 1). Only patients that were positive both for IF and qRT-PCR were considered positive for the existence of CTCs (Figure 1).

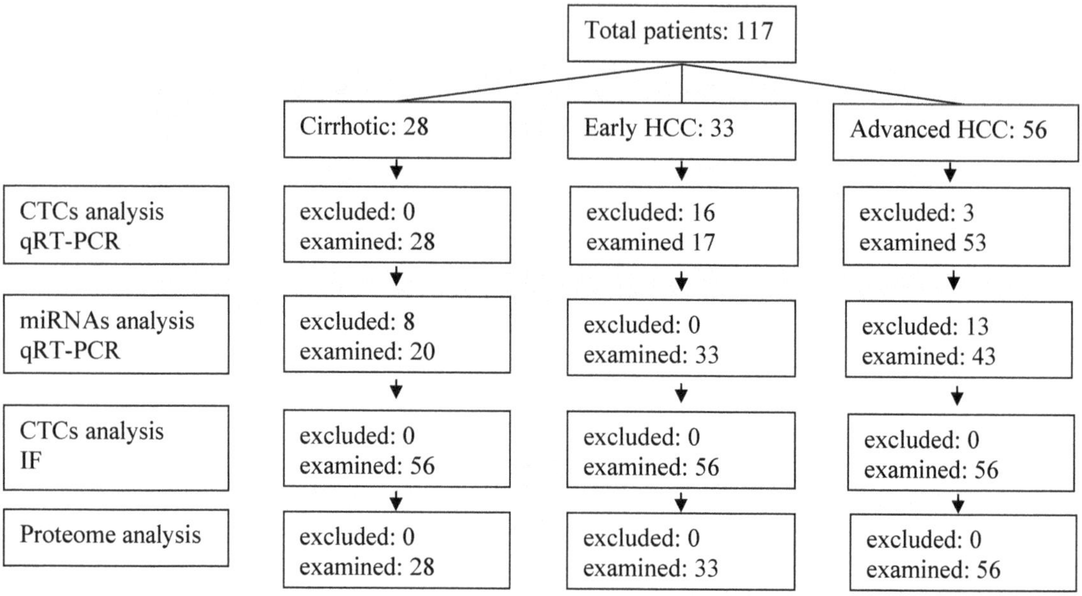

Figure 1. Flow cart of the patients involved in the experimental procedure.

Out of 53 patients with advanced HCC who had been evaluated, 27 (~51%) presented evidence of CTCs (Table 2). Ten patients out of twenty-seven, exhibited mesenchymal-type CTCs, presenting high vimentin and low EPCAM expression. Six of the patients with CTCs of mesenchymal subtype also presented evidence of sMVP too, as was observed by IF, whereas four patients showed sMVP expression (intermediate differentiation) without expressing vimentin. The remaining patients presented CTCs of epithelial subtype expressing only EPCAM (Figure 2a,b).

Table 2. Synopsis of the results indicating the presence of CTCs in each group.

	EPCAM	Vimentin	AFP	sMVP
Advanced HCC (53 pts)	27/53	10/53	12/53	6/53
Early HCC (17 pts)	5/17	3/17	1/17	1/17
Cirrhosis (28 pts)	2/28	1/28	1/28	0/28

Pts: patients; aFP: alpha fetoprotein; EPCAM: epithelial cellular adhesion molecule; sMVP: surface major vault protein.

In patients with early HCC, 5 (29%) out of 17 patients who had been evaluated presented evidence of CTCs. Three out of five showed high vimentin expression, and one out of three additionally presented sMVP expression. Comparison of the actual numbers of CTCs of any type suggested that individuals with advanced cancer presented significantly higher number of CTCs compared to individuals with cirrhosis ($p = 0.009$) (Figure 1a,b). In patients with cirrhosis, 2 (7%) out of 28 presented evidence of CTCs, and 1 of them also presented high vimentin levels (Table 2) (Figure 2a,b). During a 2-year follow-up period, none of these patients developed HCC.

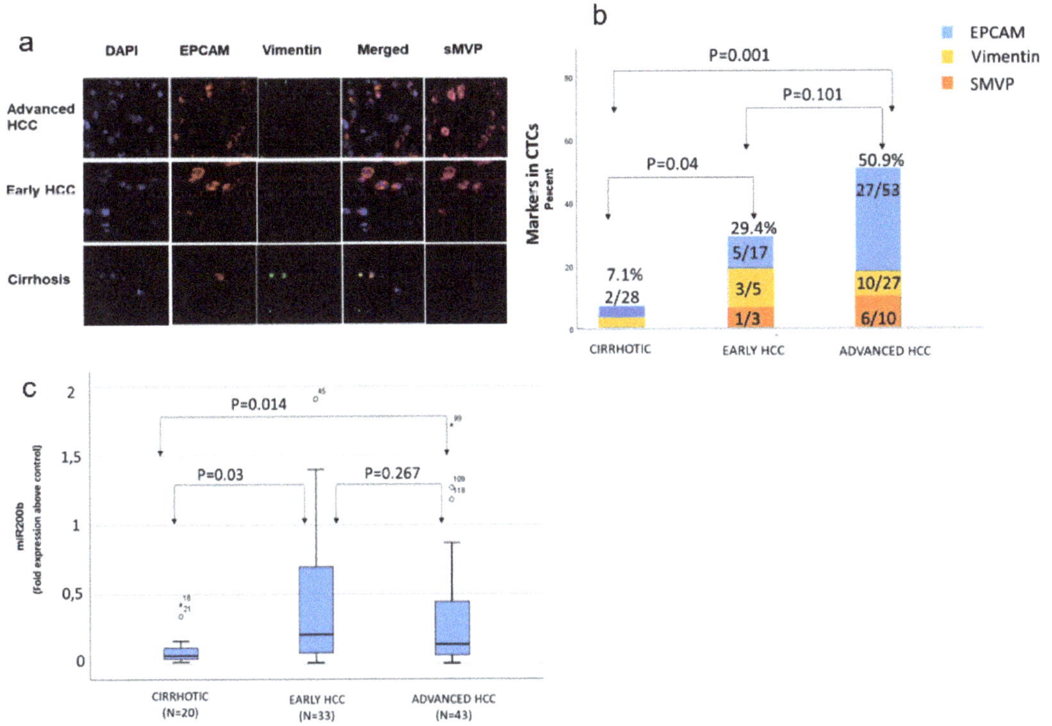

Figure 2. Detection of circulating tumor cells (CTCs) in the blood of cirrhotic and cancer patients: (**a**) detection of CTCs via immunofluorescence in the three different groups of patients.; (**b**) real-time quantitative reverse transcription PCR analysis for the detection of circulating cells expressing epithelial cellular adhesion molecule (EPCAM), vimentin and sMVP (Pearson chi-square test). DAPI: 4′,6-diamidino-2-phenylindole; sMVP: surface major vault protein; HCC: hepatocellular carcinoma. (**c**) Detection of microRNAs (miR) in the blood of hepatocellular carcinoma (HCC) and cirrhotic patients indicated that miR-200b presented significant difference between cirrhotic and early/advanced HCC (Mann–Whitney test), (* $p < 0.05$).

aFP was measured in the serum and in the blood isolated mRNA of patients. aFP expression in CTCs was detected in 12 out of the 27 EPCAM-positive patients with advanced HCC and in the majority of the cases was associated with the existence of vimentin. Expression of aFP was observed only in one out of five early HCC patients with detectable CTCs who also had elevated serum aFP levels (56 ng/mL). Finally, in cirrhotic patients, aFP expression (qRT-PCR) was observed in one out of two patients with CTCs, and one of these two patients exhibited vimentin expression. Regarding serum aFP levels, only 7 out of 33 (21%) patients with early HCC had elevated levels (>20 ng/mL), in contrast to 42 out of 56 (75%) patients with advanced HCC. None of the cirrhotic patients had elevated serum aFP (Table 1).

The presence and subtypes of CTCs in all groups studied did not correlate to the type of underlying liver disease (viral/non-viral).

3.2. MiRNAs

MiRNAs from the plasma of patients were evaluated in 43 out of 56 patients with advanced HCC, in all 33 patients with early HCC and in 20 out of 28 cirrhotic patients.

Quantitative analysis of the five most important miRNAs associated to diagnosis and prognosis of HCC (miR-122, miR-200a, miR-200b, miR-221 and miR-222) suggested that the only marker that was significantly associated with early HCC was miR-200b. Patients with either early or advanced HCC presented a significant upregulation in miR-200b levels compared to the cirrhotic patients (Figure 1c). Furthermore, miR-200b upregulation was observed in all patients that presented CTCs, and it was proportional to EPCAM elevation as was monitored by qRT-PCR (Table 3). miR-200b level did not correlate to the type of underlying liver disease (viral/non-viral).

Table 3. miRNA upregulation in comparison with CTCs.

(i)	EPCAM	Vimentin	AFP	sMVP
Cirrhotic				
miR-200b	* 2/2	* 1/1	* 1/1	0/2
miR-122	* 2/2	* 1/1	0/1	0/2
miR-221	0/2	0/1	0/1	0/2
miR-222	0/2	0/1	0/1	0/2
miR-200a	0/2	0/1	0/1	0/2
Early HCC				
miR-200b	* 4/5	* 3/3	* 1/1	0/1
miR-122	* 3/5	* 1/3	0/1	0/1
miR-221	* 2/5	0/3	0/1	0/1
miR-222	* 3/5	* 1/3	0/1	0/1
miR-200a	0/5	0/3	0/1	0/1
Advanced HCC				
miR-200b	* 7/27	* 7/10	* 3/12	* 2/6
miR-122	* 11/27	* 9/10	* 7/12	* 4/6
miR-221	* 5/27	* 5/10	* 3/12	* 2/6
miR-222	* 12/27	* 8/10	* 10/12	* 4/6
miR-200a	* 3/27	* 3/10	* 3/12	0/6

(ii)			
miR-200b	Cirrhotic	Early HCC	Advanced HCC
CTCs	2	4	7
No CTCs	1	2	3
miR-122	Cirrhotic	Early HCC	Advanced HCC
CTCs	2	3	11
No CTCs	7	7	6
miR-221	Cirrhotic	Early HCC	Advanced HCC
CTCs	0	2	5
No CTCs	13	8	9
miR-222	Cirrhotic	Early HCC	Advanced HCC
CTCs	0	3	12
No CTCs	13	7	8
miR-200a	Cirrhotic	Early HCC	Advanced HCC
CTCs	0	0	3
No CTCs	7	4	6

* The patients that presented miRNA upregulation also presented EPCAM, vimentin, AFP and sMVP upregulation. (i) Pathological expression of different miRNAs in patients that presented CTCs of various subtypes (EPCAM, vimentin, sMVP). (ii) Correlation of CTCs with the pathological expression of miRNAs.

3.3. Proteomics

Proteomics were evaluated in all patients of this study; 56 with advanced HCC, 33 with early HCC and 28 with cirrhosis.

By defining high-stringency criteria, a total of 228 proteins that show a statistically significant difference in expression between patients with early HCC, cirrhotic patients and patients with advanced HCC were identified (Figure 3a). To determine markers that will be useful for determining early HCC, a proteomic comparison was performed between patients with cirrhosis and patients with early HCC with Student's *t*-test.

The results were displayed on a volcano plot. A total of 53 proteins were found that can be used as potential biomarkers to determine early HCC from cirrhosis. These proteins were further tested using the Metascape online platform enrichment tool to determine their function, the biological processes involved, their involvement in carcinogenesis and possible protein interactions. Out of a total of 53 proteins, 31 were upregulated and 22 were downregulated in patients with early HCC compared to cirrhotic patients (Figure 3b). The proteins that were up regulated in early HCC compared to cirrhotic were: AGT, APCS, APOA1, APOA2, APOA4, APOC3, APOM, ARFIP1, C1RL, CFHR3, CLU, CPN1, CPN2, DMRT2, F5, GC, HBA1, ITIH1, ITIH2, KLKB1, LRG1, OGT, PON3, PROS1, SAA4, SERPINA4, SERPINC1, SERPIND1, SPG11, TTR and VTN. The significance of these proteins is also presented in a heatmap in comparison to cirrhotic and HCC patients (early and advanced) (Figure 3c).

These proteins were further analyzed in five central mechanisms involved in carcinogenesis: growth (Figure 4a), immune response (Figure 4b), angiogenesis (Figure 4c), proliferation (Figure 4d) and metastases (Figure 4e). Seven proteins were which showed the greatest statistical significance in terms of expression intensity as presented in the thermograms in patients with early HCC as compared to cirrhotic patients were selected. These proteins were APOA2, APOC3, CLU, OGT, APOD, VTN and HRG. CLU protein (apoliporotein J) seems to be a very potent marker for HCC since it presented high statistical significance, and it seems to be involved in many biological processes that are central to HCC (Figure 4b,c,e,f). Examination of each of these markers and the remaining 24 of all the 31 markers that were found to be up regulated in the early HCC patients indicated that none of these markers alone present the sensitivity or the specificity to stand as a diagnostic or prognostic marker. However, the combination of APOA2 and APOC3 upregulation covered 100% of the patients with early HCC. In detail, APOC3 was upregulated in 29/33 (88%) and APOA2 in 28/33 (85%) in early HCC patients. Both markers were also found to be upregulated and in 2/28 cirrhotic patients.

Protein profile was not affected by the etiology of underlying liver disease. Data are available via ProteomeXchange with identifier PXD033306.

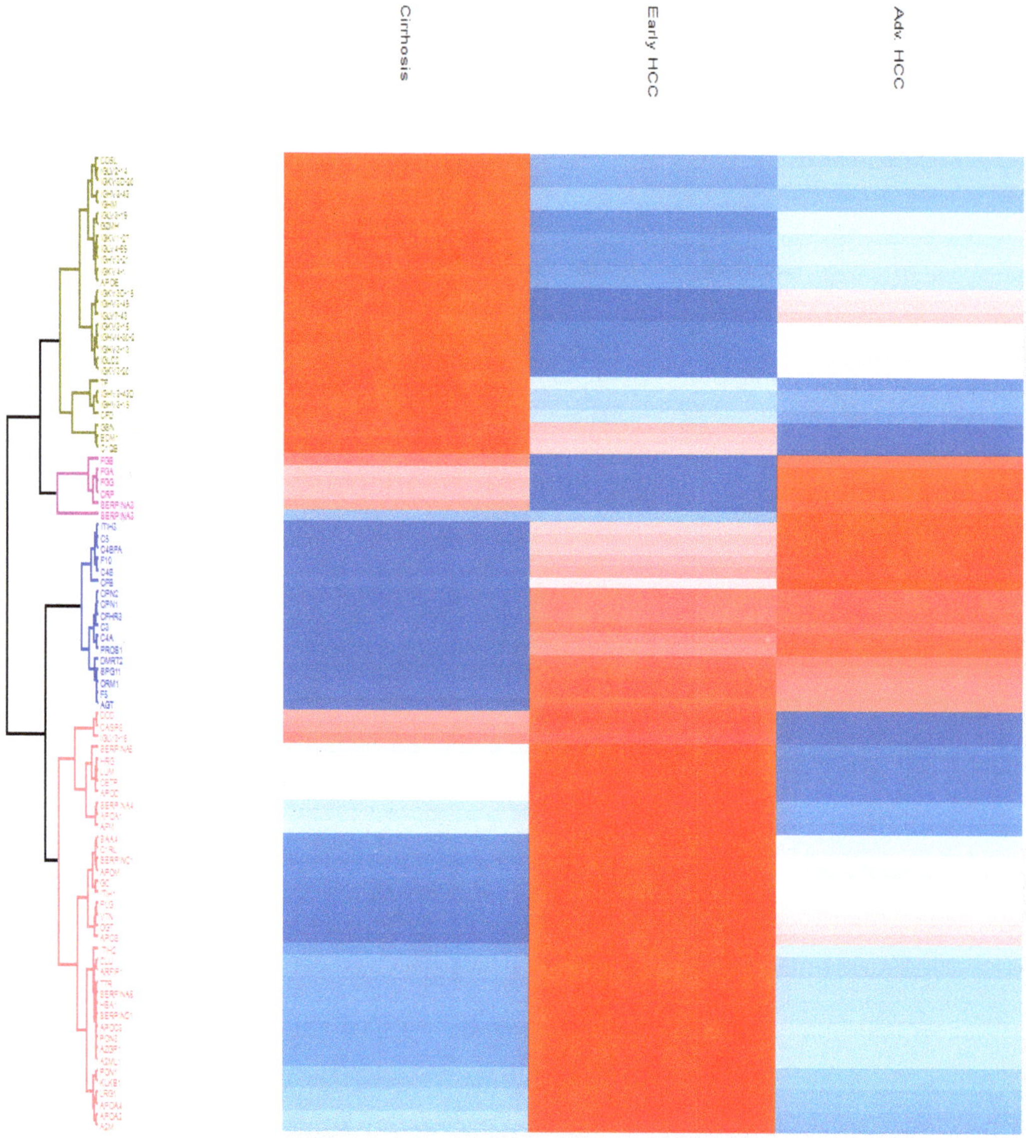

(**a**)

Figure 3. *Cont.*

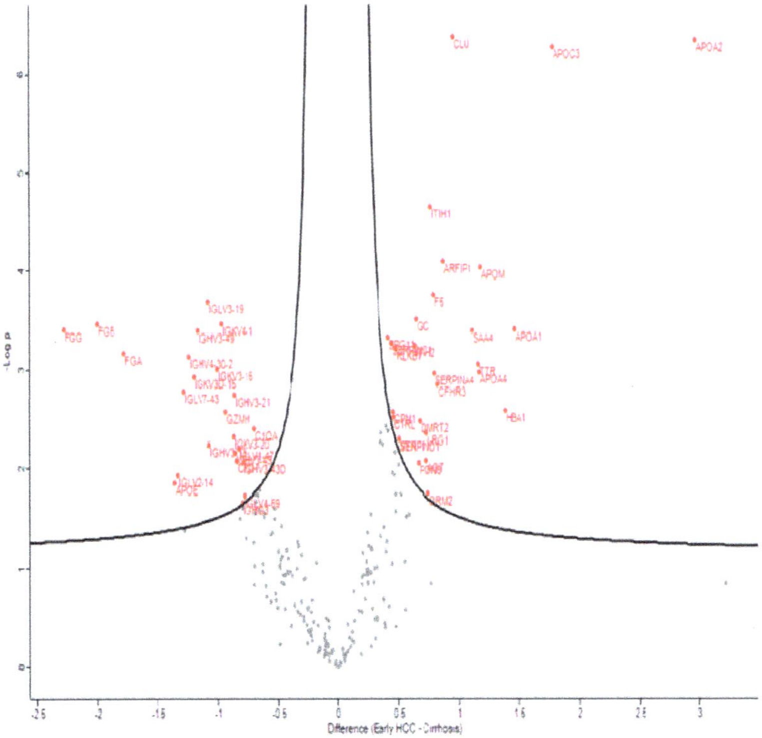

(**b**)

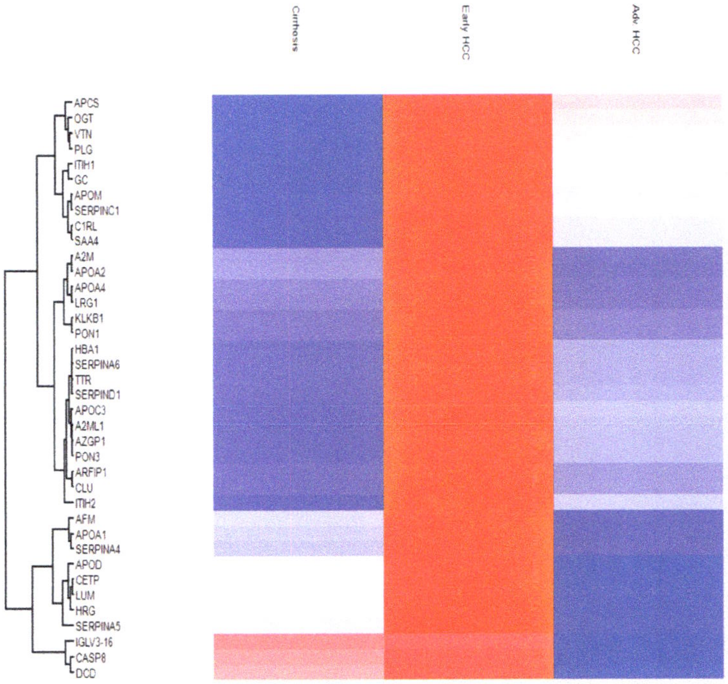

(**c**)

Figure 3. Proteomics comparison of early hepatocellular carcinoma (HCC) vs cirrhotic patients: (**a**) heatmap of the proteomes in all three patient categories; (**b**) volcano plot produced after the comparison of patients with early HCC and cirrhotic patients. At the right side of the plot, the proteins that were upregulated in early HCC patients are observed. The farther away the protein from the lines of the plot, the greater the significance; (**c**) comparison of the 3 patient categories for the 31 proteins determined by the volcano plot.

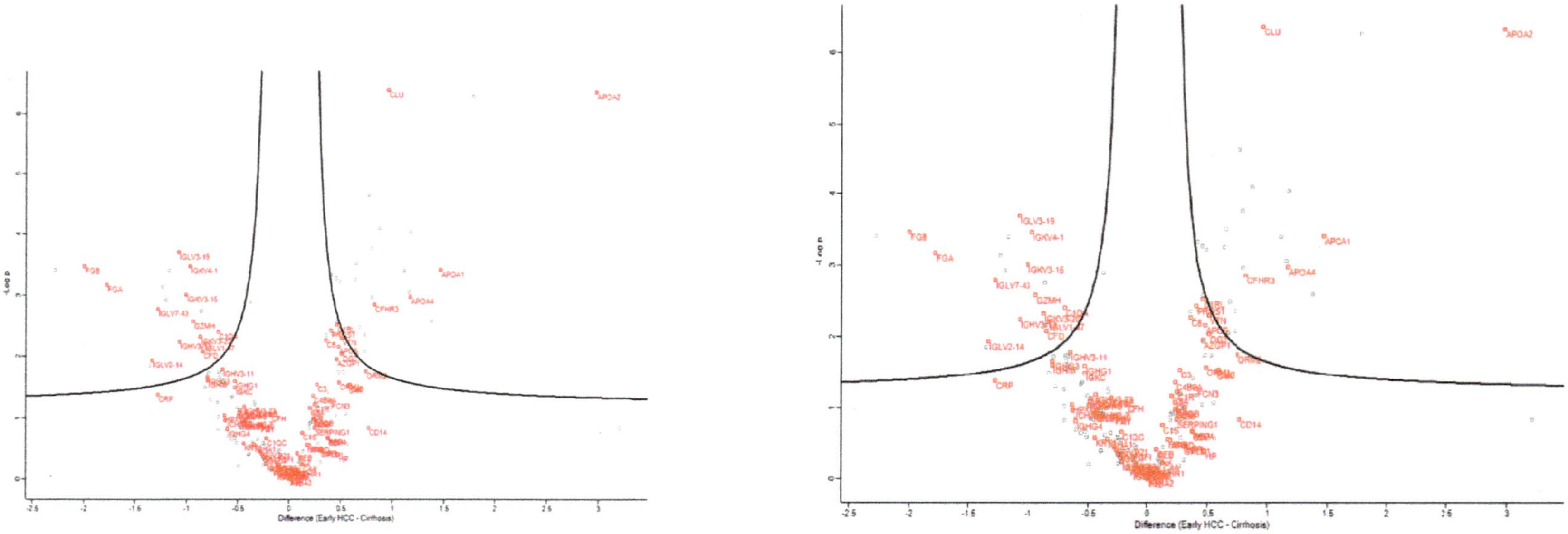

(a) (b)

Figure 4. *Cont.*

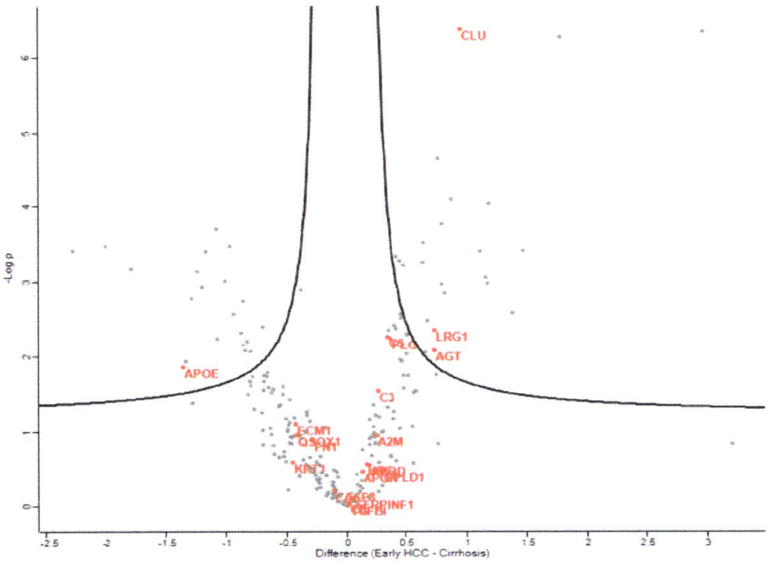

(c)

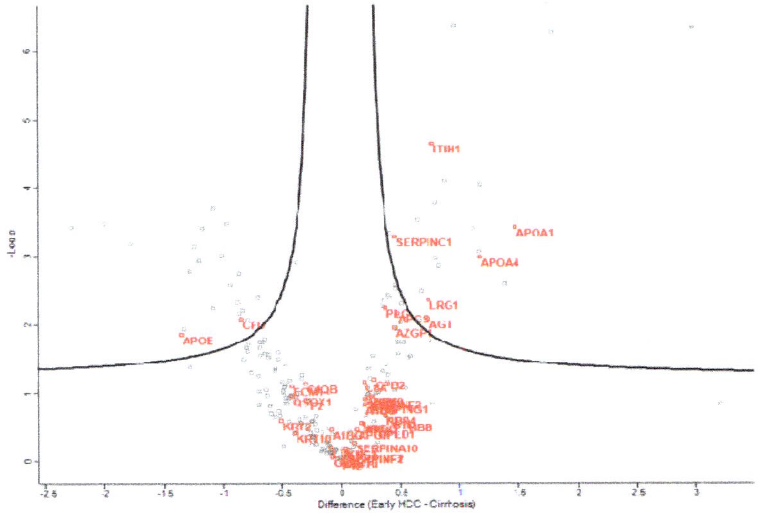

(d)

Figure 4. *Cont.*

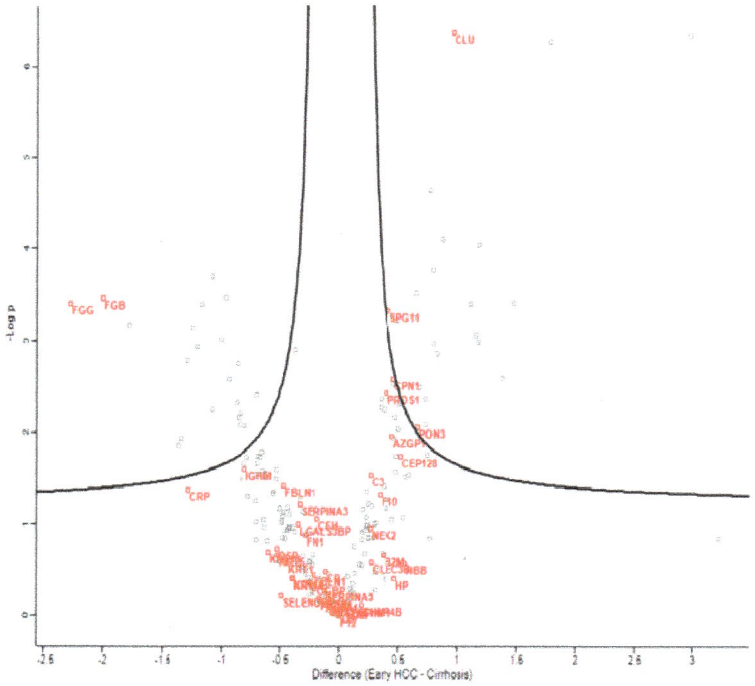

(e)

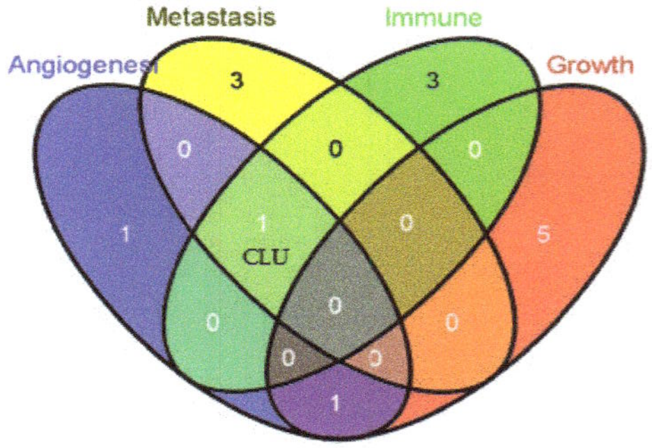

(f)

Figure 4. Volcano analyses of cirrhotic vs early HCC patients. The patients are compared for the expression of the proteins involved in: (**a**) growth; (**b**) immune response; (**c**) angiogenesis; (**d**) proliferation; (**e**) metastasis. (**f**) CLU protein (apolipoprotein J) seems to be central in metastases, angiogenesis and immune response.

3.4. Clinical Significance of ApoA2, APOC3 and mir-200b

Receiver operating characteristic (ROC) analysis indicated the significance of both protein markers APOA2 and APOC3. APOA2 presented sensitivity of 84.95% and specificity of 89.29% with an area under the curve (AUC) of 0.81, and APOC3 presented 93.94% sensitivity and 89.29% specificity, AUC= 0.916 ($p < 0.001$ for both markers) (Figure 5a,b,d). miR-200b presented sensitivity of 63.6% and specificity of 82.1%, AUC= 0.729 ($p < 0.001$) (Figure 5c,d).

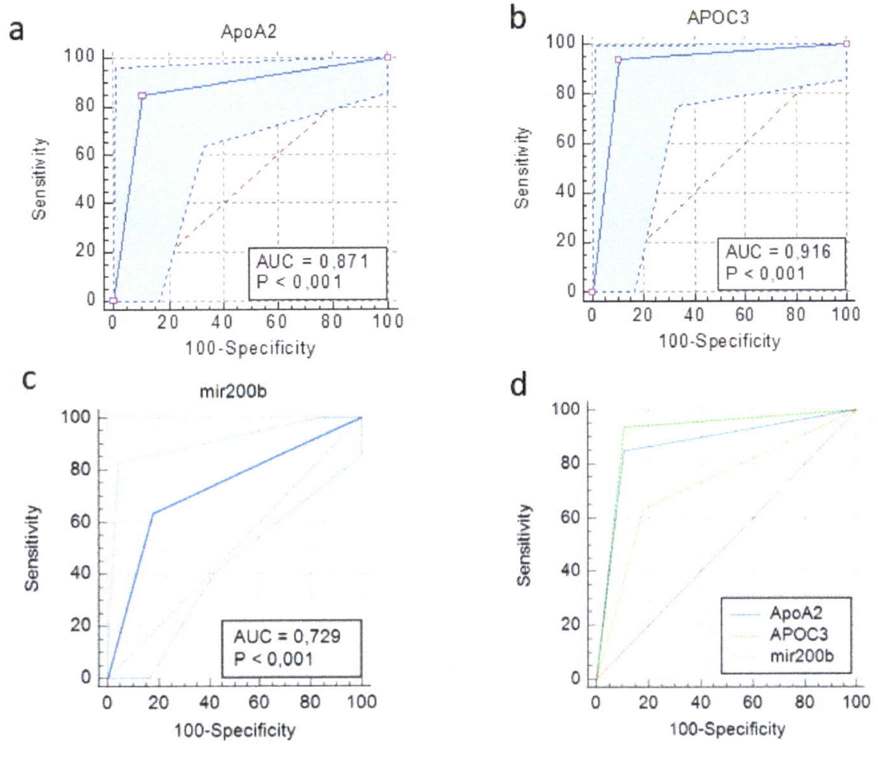

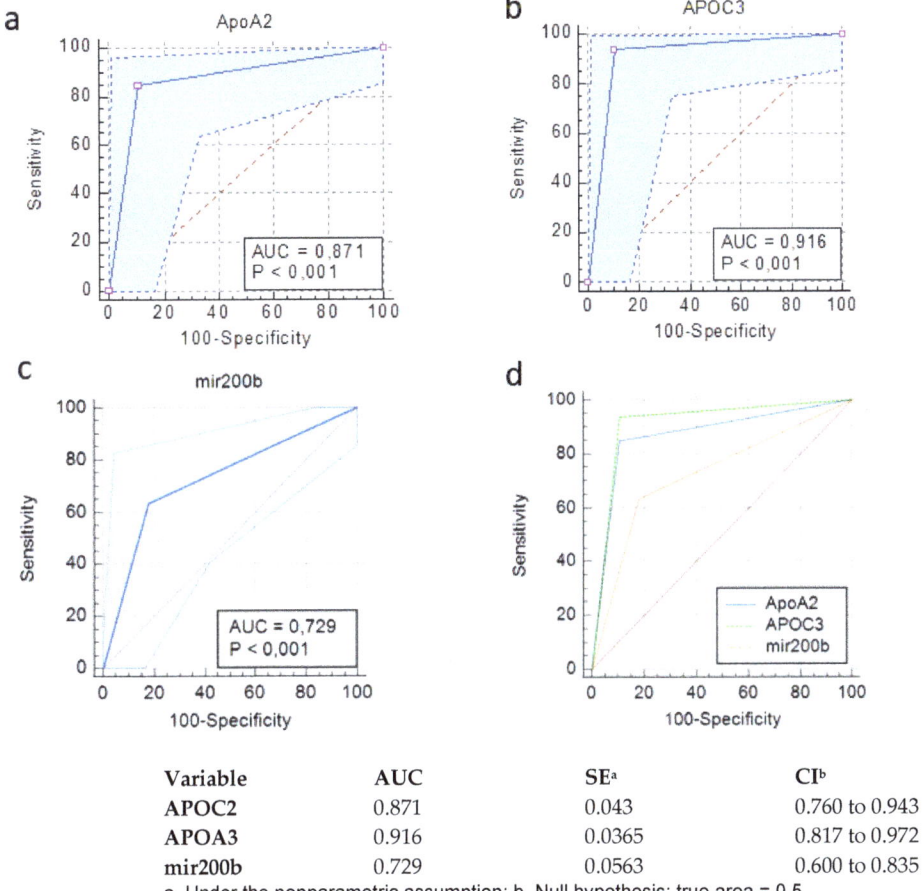

Variable	AUC	SE[a]	CI[b]
APOC2	0.871	0.043	0.760 to 0.943
APOA3	0.916	0.0365	0.817 to 0.972
mir200b	0.729	0.0563	0.600 to 0.835

a. Under the nonparametric assumption; b. Null hypothesis: true area = 0.5

Figure 5. ROC analysis for the determination of specificity and sensitivity of selected markers to define early HCC from cirrhotic patients. (**a**) APOA2, (**b**) APOC3, (**c**) miR-200b, (**d**) the three markers combined.

4. Discussion

Early detection of HCC is associated with better prognosis and a five-year survival of 40–70%, which drops dramatically in intermediate (3-year survival 10–40%) and advanced stages (<12 months) [1–3]. Therefore, early detection of HCC is crucial for the survival of these patients. In clinical practice surveillance for HCC, ultrasound (U/S) every 6 months in cirrhotic patients is mandatory for early detection, followed by contrast enhancing images by computed tomography/magnetic resonance imaging (CT/MRI) for confirmation. However, their sensitivity and specificity depend on size and type of HCC. Liver biopsy is an invasive method that is not always feasible and sometimes not diagnostic for various reasons. Having said that, detection of HCC at very early stages is still a challenging task. Recent sophisticated omics techniques have been applied for HCC and have reported possible prognostic models based on molecular signatures, but with low sensitivity and accuracy [17–19]. Consequently, the necessity of the development of more accurate and sensitive tools guiding early diagnosis and prognostic assessments in HCC patients remains.

In this study, we combined the power of proteomics, CTCs and HCC-associated miRNAs in the blood of patients with early HCC, advanced HCC and cirrhosis with no obvious HCC, aiming towards the determination of novel and accurate HCC diagnostic markers which could be combined for the development of a novel tool for the diagnosis and/or prognosis of early HCC.

To our knowledge, the vast majority of studies that aim towards the determination of novel biomarkers in HCC compare cirrhotic patients or healthy controls to HCC patients [17–20]. In this study, our aim was to investigate possible cellular and molecular differences by analyzing the proteome and miRNA profiles of patients with specifically early HCC in comparison to cirrhotic patients without HCC.

CTC analysis enables early cancer detection, prognosis prediction and therapy response monitoring in patients with HCC. Despite this, translation of CTC analysis from bench to patient is a challenging task mainly due to the fact of the heterogeneity of the studies in respect to detection techniques/technologies and the lack of standardized assays [16]. In this study, we determined epithelial, mesenchymal and intermediate-stage CTCs (EPCAM, vimentin and sMVP, respectively) via qRT PCR and IF. Furthermore, we examined the expression of aFP by qRT-PCR.

Changes in cytoplasmic aFP expression have been found to be associated with the expression of several metastasis-related mesenchymal proteins: keratin 19 (K19), EPCAM, matrix metalloproteinase 2/9 (MMP2/9) and C-X-C motif chemokine receptor 4 (CXCR4). According to this, aFP plays a critical role in promoting the metastasis of HCC. Furthermore, aFP modulates the expression of PD-L1 and B7-H4, resulting in the immune escape of HCC and consequently strengthening its ability to metastasize [20,21].

The notion that cancer epithelial cells are responsible for metastases tends to be abolished [21]. On these grounds, the target of this study—in respect to CTC detection—was to determine the tumor cells that switch or are prone to switching towards a mesenchymal phenotype rather than to detect tumor epithelial cells. This may also explain the lower number of advanced HCC patients that present CTCs in their periphery.

EPCAM expression is observed in epithelial cells and in cancer cells at early stages of the epithelial to mesenchymal transition (EMT) [20–22]. Vimentin is a mesenchymal marker, and sMVP is a marker of intermediate differentiation towards mesenchymal phenotype [10–13].

CTCs were detected in 51% of patients with advanced HCC, in 29% of patients with early HCC and in 7% of cirrhotic patients. These patients presented evidence of epithelial CTCs and of mesenchymal subtypes as was determined by both techniques. From our findings, we can conclude that despite the fact that CTCs are considered to be associated with advanced cancer, are predictors of prognosis [10–13,23,24] and are suitable for patient monitoring, they are not suitable as a diagnostic tool for defining cirrhotic patients prone to develop early HCC or for defining early HCC patients. For the markers examined, CTCs seem to be a very sensitive prognostic marker but with rather low specificity since they are also observed in some cirrhotic patients. Our results concur with several other studies which demonstrated the presence of CTCs in at least 12% of cirrhotic patients [19,23].

With respect to miRNA analysis, miR-200a and miR-200b repression have been proposed as early markers of HCC development [16,25]. It has been demonstrated that miR-200b was highly expressed in the plasma-derived exosome of ovarian cancer patients, promoting the proliferation and invasion of cancer cells. The proposed mechanism is the induction of macrophage M2 polarization by suppressing KLF6 expression [25]. Furthermore, miR-200b was significantly upregulated in patients with early HCC compared to cirrhotic patients, and miR-200b abnormal expression in serum was related to tumor occurrence and development [26].

MiR-200b inhibits the epithelial to mesenchymal transition through ZEB1 and ZEB2 inhibition, allowing the expression of E-cadherin and leading to EMT inhibition. It also reduces angiogenesis and inhibits the Notch 1 signaling pathway [27]. The Notch 1 mode of action is controversial since initial evidence suggested it as a tumor suppressor. On the other hand, there is growing evidence indicating that Notch 1 overexpression may also exert an oncogenic effect [28].

In this study we found that patients with either early or advanced HCC presented a significant upregulation in miR-200b levels compared to the cirrhotic patients. HCC patients with CTCs exhibited significant upregulation of miR-200b that was proportional to the elevation of EPCAM.

Taking into account the stability of miRNAs, a major question that arises is if the levels of miR-200b in the blood remain constantly high so they can be detected and if they remain high despite the absence of CTCs in the blood. Our results are rather controversial when compared to other studies regarding the use of the miRNAs as useful prognostic and diagnostic markers for HCC [24–29]. It seems that the majority of the mRNAs used in this study, with the exception of miR-200b, apart from being present in a number of HCC patients independently from the presence of CTCs were also expressed in a number of cirrhotic patients (Table 3). Therefore, based on our evidence, they presented low specificity to be used as diagnostic biomarkers.

In most proteomics studies in the literature, the comparison of biomarkers is amongst cirrhotic, and generally, HCC patients at any stage rather than cirrhotic and early HCC patients [30,31]. Our evidence suggests that advanced HCC patients present significantly different proteome profiles when compared to early HCC patients. Therefore, we compared the proteome profiles of non-HCC cirrhotic patients to those of patients with early HCC.

Proteomics analysis of these individuals led to the identification of a total of 31 markers, which were upregulated in patients with early HCC compared to cirrhotic patients. These markers were selected in respect to their significance in respect to upregulation in patients with early HCC when compared to cirrhotic patients and to their significant involvement in the five major cancer-related functions examined (growth, immune response, angiogenesis, proliferation and metastases) as found in our study. Examination of the protein profile of each cirrhotic and early HCC patient suggests that most of the markers, despite the initial significance in the comparison between groups, may not be suitable as diagnostic markers since they seem to lack specificity in many instances. Ideally, a marker should be capable of identifying early HCC from cirrhotic patients. In this study, APOC3 protein—which was found to have one of the highest significant differences in the comparison between groups—was upregulated in 88% of patients with early HCC and in 2/28 (7%) cirrhotic patients, whereas APOA2 was found to be expressed in 85% of early HCC patients and in 2/29 (7%) cirrhotic patients. It is worth mentioning that when we examined both markers, we covered 100% of the patients with early HCC. Apolipoproteins (APOs) have been increasingly reported for their relationships with tumors [31]. In this study, several apolipoproteins seem to be related to the development of early HCC (Figure 6). It has been previously suggested that APOA2 can be used as a biomarker in HCC and in prostate cancer [30,32]. Normally, APOA2 lipoprotein is mainly produced in the liver and is found in plasma as a monomer, homodimer, or heterodimer with apolipoprotein D. These are all involved to the catalysis of lipoproteins towards high-density lipoprotein (HDL) [31].

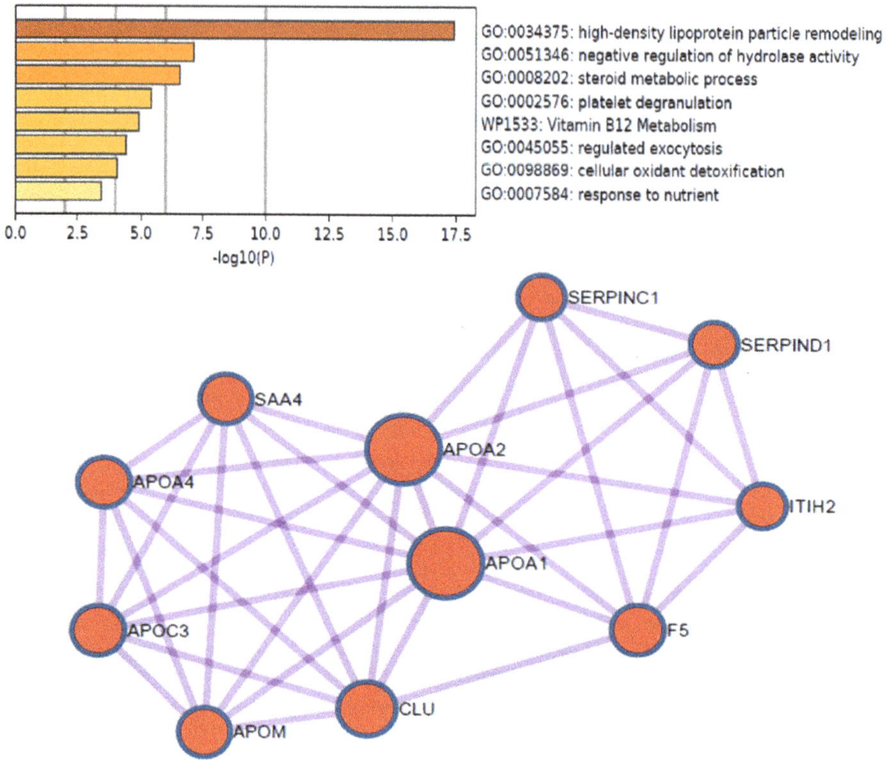

Figure 6. Enrichment analysis of the biological processes related to the proteins found from the proteomics analysis, where apolipoproteins seem to play a central role in hepatocellular carcinoma (Metascape.org, accessed on 21 August 2022).

APOC3 is an inhibitor of lipoprotein and hepatic lipases, and it has been proposed to inhibit the liver's absorption of triglyceride-rich particles. At the cellular level, APOC3 appears to promote the assembly and secretion of triglyceride-rich low-density lipoprotein (VLDL) particles from hepatocytes under lipid-rich conditions [33]. In cancer, APOC3 induces the NLRP3 inflammasome via caspase 8 and toll-like receptors 2 and 4 [33]. Cytosolic caspase 8 is a mediator of death receptor signaling. Caspase 8 depletion induces G2/M arrest, p53 stabilization and induction of p53-dependent intrinsic apoptosis in tumor cells [34]. From the above, it can be understood that APOC3 is directly involved in cancer progression, whereas APOA2 is a newly recognized biomarker in HCC [30–32].

In our quest towards the determination of novel biomarkers that will define HCC at very early stages, our evidence suggests that a combined tool of three markers (miR-200b, APOA2 and APOC3) could diagnose HCC at very early stages accurately and with high sensitivity. Each of these markers separately presents greater sensitivity and specificity than aFP, the traditionally used marker for HCC [35]. The detection of elevated serum aFP levels in only 21% of early HCC observed in this study supports the above results.

We acknowledge that this study has limitations, which are mainly the small sample size and the lack of a validation cohort. However, we believe that these preliminary data are significant and prompt further investigation to elucidate the role of the proposed molecules as a useful diagnostic tool for patients with cirrhosis prone to developing HCC.

5. Conclusions

Our evidence suggests that patients with early HCC present with a distinct proteome profile when compared to cirrhotic patients. The combined upregulation of APOA2 and APOC3 can be used as an accurate and sensitive diagnostic tool for the diagnosis of early HCC. Moreover, based on our findings, the increased levels of miR-200b in serum could also contribute to the identification of early HCC development. Finally, CTCs, although significantly increased in patients with advanced HCC in our study, showed low specificity in the distinction between HCC and non-HCC patients. Further confirmation of the above results in larger studies may highlight the utility of combined proteomics and molecular biomarkers in the timely diagnosis of early HCC and in individualized cancer management.

Supplementary Materials: The following supporting information can be downloaded at: https://www.mdpi.com/article/10.3390/cimb44100322/s1. Table S1: Specific gene and primer sequence for the gene expression analysis. Figure S1: Determination of the significance of EPCAM, Vimentin and miR200b as biological markers for the determination of early HCC, Considering the area under the curve only miR200b seems to be a suitable molecule as a biomarker for early HCC diagnosis.

Author Contributions: A.A and J.K.: study design. J.K: provision of patient material. A.A., V.D., A.N., G.S., M.S. (Martina Samiotaki) and M.S. (Martha Stathaki): methodology (sample collection, CTC isolation, miRNA extraction, RNA extraction, protein extraction, qRT-PCR, immunofluorescense, LC/MS). M.S. (Martina Samiotaki) and M.T.: statistical analysis. G.P., S.D. and J.K.: supervision. A.A. and J.K.: writing—original draft preparation. All authors have read and agreed to the published version of the manuscript.

Funding: We acknowledge the support of part of this work (proteome analysis) by the project "The Greek Research Infrastructure for Personalised Medicine (pMED-GR)" (MIS 5002802), which is implemented under the action "Reinforcement of the Research and Innovation Infrastructure", funded by the Operational Programme "Competitiveness, Entrepreneurship and Innovation" (NSRF 2014-2020) and co-financed by Greece and the European Union (European Regional Development Fund). The rest of this research received no external funding.

Institutional Review Board Statement: The study was approved by the National and Kapodistrian University of Athens Medical School Ethics Committee (Approval code: 029/19.11.18).

Informed Consent Statement: All subjects gave informed consent in accordance with the Declaration of Helsinki. All methods were performed in accordance with the relevant guidelines and regulations.

Data Availability Statement: The mass spectrometry proteomics data have been deposited to the ProteomeXchange Consortium via the PRIDE [1] partner repository with the dataset identifier PXD033306.

Conflicts of Interest: The authors declare no conflict of interest.

Abbreviations

Hepatocellular carcinoma	(HCC)
microRNAs	(miRNAs)
Circulating tumor cells	(CTCs)
Epithelial cell adhesion molecule	(EPCAM)
Alpha fetoprotein	(aFP)
Surface major vault protein	(sMVP)
Chronic hepatitis B virus	(HBV)
Hepatitis C virus	(HCV)
Non-alcoholic steatohepatitis	(NASH)
Erythrocyte lysis buffer	(ELB)
Quantitative real-time polymerase chain reaction	(qRT-PCR)
Immunofluorescence staining	(IF)
4′,6-diamidino-2-phenylindole	(DAPI)
Liquid chromatography mass spectrometry	(LC MS)
Gas phase fractionation	(GFP)

Computed tomography/magnetic resonance imaging	(CT/MRI)
Matrix metalloproteinase 2/9	(MMP2/9)
C-X-C motif chemokine receptor 4	(CXCR4)
Apolipoprotein	(APO)
High-density lipoprotein	(HDL)
Triglyceride-rich low-density lipoprotein	(VLDL)

References

1. World Health Organisation. Projections of Mortality Causes of Death, 2016 to 2060. Available online: http://www.who.int/healthinfo/global_burden_disease/projection/en/ (accessed on 2 February 2022).
2. Villanueva, A. Hepatocellular Carcinoma. *The New Engl. J. Med.* **2019**, *380*, 1450–1462. [CrossRef]
3. Llovet, J.M.; Kelley, R.K.; Villanueva, A.; Singal, A.G.; Pikarsky, E.; Roayaie, S.; Lencioni, R.; Koike, K.; Zucman-Rossi, J.; Finn, R.S. Hepatocellular carcinoma. *Nat. Rev. Dis. Prim.* **2021**, *7*, 6. [CrossRef]
4. Akinyemiju, T.; Abera, S.; Ahmed, M.; Alam, N.; Alemayohu, M.A.; Allen, C.; Al-Raddadi, R.; Alvis-Guzman, N.; Amoako, Y.; Artaman, A.; et al. The Burden of Primary Liver Cancer and Underlying Etiologies From 1990 to 2015 at the Global, Regional and National Level: Results From the Global Burden of Disease Study 2015. *JAMA Oncol.* **2017**, *3*, 1683–1691. [PubMed]
5. Xie, Y. Hepatitis B Virus-Associated Hepatocellular Carcinoma. *Adv. Exp. Med. Biol.* **2017**, *1018*, 11–21. [PubMed]
6. Harrod, E.; Moctezuma-Velazquez, C.; Gurakar, A.; Ala, A.; Dieterich, D.; Saberi, B. Management of concomitant hepato-cellular carcinoma and chronic hepatitis C: A review. *Hepatoma Res.* **2019**, *5*, 28.
7. Méndez-Sánchez, N.; Valencia-Rodríguez, A.; Coronel-Castillo, C.E.; Qi, X. Narrative review of hepatocellular carcinoma: From molecular bases to therapeutic approach. *Dig. Med. Res.* **2021**, *4*, 21037. [CrossRef]
8. Wang, G.; Zhu, S.; Li, X. Comparison of values of CT and MRI imaging in the diagnosis of hepatocellular carcinoma and analysis of prognostic factors. *Oncol. Lett.* **2018**, *17*, 1184–1188. [CrossRef]
9. Tang, Y.; Xiao, G.; Shen, Z.; Zhuang, C.; Xie, Y.; Zhang, X.; Yang, Z.; Guan, J.; Shen, Y.; Chen, Y.; et al. Noninvasive Detection of Extracellular pH in Human Benign and Malignant Liver Tumors Using CEST MRI. *Front. Oncol.* **2020**, *10*, 578985. [CrossRef]
10. Lee, H.M.; Joh, J.W.; Seong, C.G.; Kim, W.-T.; Kim, M.K.; Choi, H.S.; Kim, S.Y.; Jang, Y.-J.; Sinn, D.H.; Choi, G.S.; et al. Cell-surface major vault protein promotes cancer progression through harboring mesenchymal and intermediate circulating tumor cells in hepatocellular carcinomas. *Sci. Rep.* **2017**, *7*, 1–15. [CrossRef] [PubMed]
11. Liu, X.-N.; Cui, D.-N.; Li, Y.-F.; Liu, Y.-H.; Liu, G.; Liu, L. Multiple "Omics" data-based biomarker screening for hepatocellular carcinoma diagnosis. *World J. Gastroenterol.* **2019**, *25*, 4199–4212. [CrossRef]
12. Schulze, K.; Gasch, C.; Staufer, K.; Nashan, B.; Lohse, A.W.; Pantel, K.; Riethdorf, S.; Wege, H. Presence of EpCAM-positive circulating tumor cells as biomarker for systemic disease strongly correlates to survival in patients with hepatocellular carci-noma. *Int. J. Cancer* **2013**, *133*, 2165–2171. [CrossRef]
13. Guo, W.; Yang, X.R.; Sun, Y.F.; Shen, M.N.; Ma, X.L.; Wu, J.; Zhang, C.Y.; Zhou, Y.; Xu, Y.; Hu, B.; et al. Clinical significance of EpCAM mRNA-positive circulating tumor cells in hepatocellular carcinoma by an optimized negative enrichment and qRT-PCR-based platform. *Clin. Cancer Res. Off. J. Am. Assoc. Cancer Res.* **2014**, *20*, 4794–4805. [CrossRef] [PubMed]
14. Bandiera, S.; Pfeffer, S.; BaumertT, F.; ZeiselM, B. miR-122—A key factor and therapeutic target in liver disease. *J. Hepatol.* **2015**, *62*, 448–457. [CrossRef] [PubMed]
15. Dhayat, S.A.; Mardin, W.A.; Kφhler, G.; Bahde, R.; Vowinkel, T.; Wolters, H.; Senninger, N.; Haier, J.; Mees, S.T. The microRNA-200 family—A potential diagnostic marker in hepatocellular carcinoma? *J. Surg. Oncol.* **2014**, *110*, 430–438. [CrossRef]
16. Mizuguchi, Y.; Takizawa, T.; Yoshida, H.; Uchida, E. Dysregulated miRNA in progression of hepatocellular carcinoma: A systematic review. *Hepatol. Res.* **2015**, *46*, 391–406. [CrossRef]
17. Kaur, H.; Lathwal, A.; Raghava, G.P.S. Integrative multi-omics approach for stratification of tumor recurrence risk groups of Hepatocellular Carcinoma patients. *bioRxiv* **2021**. [CrossRef]
18. Cui, D.; Li, W.; Jiang, D.; Wu, J.; Xie, J.; Wu, Y. Advances in Multi-Omics Applications in HBV-Associated Hepatocellular Carcinoma. *Front. Med.* **2021**, *8*, 8. [CrossRef]
19. Wang, P.-X.; Cheng, J.-W.; Yang, X.-R. Detection of circulating tumor cells in hepatocellular carcinoma: Applications in di-agnosis prognosis prediction and personalized treatment. *Hepatoma Res.* **2020**, *6*, 61. [CrossRef]
20. Di Bisceglie, A.M.; Hoofnagle, J.H. Elevations in serum alpha-fetoprotein levels in patients with chronic hepatitis B. *Cancer* **1989**, *64*, 2117–2120. [CrossRef]
21. Li, Y.; Tang, Z.-Y.; Ye, S.-L.; Liu, Y.-K.; Chen, J.; Xue, Q.; Chen, J.; Gao, D.-M.; Bao, W.-H. Establishment of cell clones with different metastatic potential from the metastatic hepatocellular carcinoma cell line MHCC97. *World J. Gastroenterol.* **2001**, *7*, 630–636. [CrossRef]
22. Genna, A.; Vanwynsberghe, A.M.; Villard, A.V.; Pottier, C.; Ancel, J.; Polette, M.; Gilles, C. EMT-Associated Heterogeneity in Circulating Tumor Cells: Sticky Friends on the Road to Metastasis. *Cancers* **2020**, *12*, 1632. [CrossRef] [PubMed]
23. Espejo-Cruz, M.L.; Gonzαlez-Rubio, S.; Zamora-Olaya, J.; Amado-Torres, V.; Alejandre, R.; Sαnchez-Frvas, M.; Ciria, R.; De la Mata, M.; Rodrvguez-Perαlvarez, M.; Ferrvn, G. Circulating Tumor Cells in Hepatocellular Carcinoma: A Comprehensive Re-view and Critical Appraisal. *Int. J. Mol. Sci.* **2021**, *22*, 13073. [CrossRef]

24. Fan, J.-L.; Yang, Y.-F.; Yuan, C.-H.; Chen, H.; Wang, F.-B. Circulating tumor cells for predicting the prognostic of patients with hepatocellular carcinoma: A meta analysis. *Cell. Physiol. Biochem.* **2015**, *37*, 629–640. [CrossRef] [PubMed]
25. Xiong, J.; He, X.; Xu, Y.; Zhang, W. FuF MiR-200b is upregulated in plasma-derived exosomes and functions as an oncogene by promoting macrophage M2 polarization in ovarian cancer. *J. Ovarian Res.* **2021**, *14*, 1–10. [CrossRef]
26. Wang, X.; Liao, X.; Huang, K.; Zeng, X.; Liu, Z.; Zhou, X.; Yu, T.; Yang, C.; Yu, L.; Wang, Q.; et al. Clustered microRNAs hsa-miR-221-3p/hsa-miR-222-3p and their targeted genes might be prognostic predictors for hepato-cellular carcinoma. *J. Cancer* **2019**, *10*, 2520–2533. [CrossRef]
27. Tsai, S.-C.; Lin, C.-C.; Shih, T.-C.; Tseng, R.-J.; Yu, M.-C.; Lin, Y.-J.; Hsieh, S.-Y. The miR-200b-ZEB1 circuit regulates diverse stemness of human hepatocellular carcinoma. *Mol. Carcinog.* **2017**, *56*, 2035–2047. [CrossRef]
28. Sokolowski, K.M.; Balamurugan, M.; Kunnimalaiyaan, S.; Gamblin, T.C.; Kunnimalaiyaan, M. Notch signaling in hepato-cellular carcinoma: Molecular targeting in an advanced disease. *Hepatoma Res.* **2015**, *1*, 11–18.
29. Wong, Q.W.-L.; Ching, A.K.-K.; Chan, A.W.-H.; Choy, K.-W.; To, K.-F.; Lai, P.B.-S.; Wong, N. MiR-222 overexpression confers cell migratory advantages in hepatocellular carcinoma through enhancing AKT signaling. *Clin. Cancer Res.* **2010**, *16*, 867–875. [CrossRef] [PubMed]
30. Malik, G.; Ward, M.D.; Gupta, S.K.; Trosset, M.W.; EGrizzle, W.; Adam, B.-L.; IDiaz, J.; Semmes, O.J. Serum levels of an isoform of apolipoprotein A-II as a potential marker for prostate cancer. *Clin. Cancer Res.* **2005**, *11*, 1073–1085. [CrossRef]
31. Mendivil, C.O.; Zheng, C.; Furtado, J.; Lel, J.; Sacks, F.M. Metabolism of very-low-density lipoprotein and low-density lipoprotein containing apolipoprotein C-III and not other small apolipoproteins. *Arterioscler. Thromb. Vasc. Biol.* **2010**, *30*, 239–245. [CrossRef] [PubMed]
32. Liu, Y.; Sogawa, K.; Sunaga, M.; Umemura, H.; Satoh, M.; Kazami, T.; Yoshikawa, M.; Tomonaga, T.; Yokosuka, O.; Nomura, F. Increased Concentrations of Apo A-I and Apo A-II Fragments in the Serum of Patients With Hepatocellular Carcinoma by Magnetic Beads–Assisted MALDI-TOF Mass Spectrometry. *Am. J. Clin. Pathol.* **2014**, *141*, 52–61. [CrossRef] [PubMed]
33. Zewinger, S.; Reiser, J.; Jankowski, V.; Alansary, D.; Hahm, E.; Triem, S.; Klug, M.; Schunk, S.J.; Schmit, D.; Kramann, R.; et al. Apolipoprotein C3 induces inflammation and organ damage by al-ternative inflammasome activation. *Nat. Immunol.* **2020**, *21*, 30–41. [CrossRef] [PubMed]
34. Móller, I.; Strozyk, E.; Schindler, S.; Beissert, S.; Oo, H.Z.; Sauter, T.; Lucarelli, P.; Raeth, S.; Hausser, A.; Al Nakouzi, N.; et al. Cancer Cells Employ Nuclear Caspase-8 to Overcome the p53-Dependent G2/M Checkpoint through Cleavage of USP28. *Mol. Cell* **2020**, *77*, 970–984.e7. [CrossRef]
35. Luo, P.; Wu, S.; Yu, Y.; Ming, X.; Li, S.; Zuo, X.; Tu, J. Current Status and Perspective Biomarkers in AFP Negative HCC: Towards Screening for and Diagnosing Hepatocellular Carcinoma at an Earlier Stage. *Pathol. Oncol. Res.* **2020**, *26*, 599–603. [CrossRef]

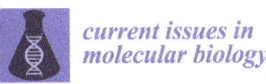

current issues in molecular biology

MDPI

Article

Single Circulating-Tumor-Cell-Targeted Sequencing to Identify Somatic Variants in Liquid Biopsies in Non-Small-Cell Lung Cancer Patients

Mouadh Barbirou [1], Amanda Miller [1], Yariswamy Manjunath [2,3], Arturo B. Ramirez [4], Nolan G. Ericson [4], Kevin F. Staveley-O'Carroll [2,3,5], Jonathan B. Mitchem [2,3,5], Wesley C. Warren [5,6], Aadel A. Chaudhuri [5,7], Yi Huang [5,7], Guangfu Li [2,3,5], Peter J. Tonellato [1] and Jussuf T. Kaifi [1,2,3,5,*,†]

1 Center for Biomedical Informatics, Department of Health Management and Informatics, School of Medicine, University of Missouri, Columbia, MO 65212, USA; barbiroum@missouri.edu (M.B.); milleraa@health.missouri.edu (A.M.); tonellatop@health.missouri.edu (P.J.T.)
2 Department of Surgery, University of Missouri School of Medicine, Columbia, MO 65212, USA; yariswamym@health.missouri.edu (Y.M.); ocarrollk@health.missouri.edu (K.F.S.-O.); mitchemj@health.missouri.edu (J.B.M.); liguan@health.missouri.edu (G.L.)
3 Harry S. Truman Memorial Veterans' Hospital, Columbia, MO 65201, USA
4 RareCyte®, Inc., Seattle, WA 98121, USA; aramirez@rarecyte.com (A.B.R.); nericson@rarecyte.com (N.G.E.)
5 Siteman Cancer Center, St. Louis, MO 63110, USA; warrenwc@missouri.edu (W.C.W.); aadel@wustl.edu (A.A.C.); huangyi1@wustl.edu (Y.H.)
6 Department of Animal Sciences and Surgery, Informatics and Data Sciences Institute, Bond Life Sciences Center, University of Missouri, Columbia, MO 65211, USA
7 Department of Radiation Oncology, Washington University School of Medicine, St. Louis, MO 63110, USA
* Correspondence: kaifij@health.missouri.edu
† Current address: Department of Surgery, Ellis Fischel Cancer Center, University of Missouri, 1 Hospital Drive, Columbia, MO 65212, USA.

Citation: Barbirou, M.; Miller, A.; Manjunath, Y.; Ramirez, A.B.; Ericson, N.G.; Staveley-O'Carroll, K.F.; Mitchem, J.B.; Warren, W.C.; Chaudhuri, A.A.; Huang, Y.; et al. Single Circulating-Tumor-Cell-Targeted Sequencing to Identify Somatic Variants in Liquid Biopsies in Non-Small-Cell Lung Cancer Patients. *Curr. Issues Mol. Biol.* **2022**, *44*, 750–763. https://doi.org/10.3390/cimb44020052

Academic Editor: Anna Kawiak

Received: 12 January 2022
Accepted: 31 January 2022
Published: 2 February 2022

Abstract: Non-small-cell lung cancer (NSCLC) accounts for most cancer-related deaths worldwide. Liquid biopsy by a blood draw to detect circulating tumor cells (CTCs) is a tool for molecular profiling of cancer using single-cell and next-generation sequencing (NGS) technologies. The aim of the study was to identify somatic variants in single CTCs isolated from NSCLC patients by targeted NGS. Thirty-one subjects (20 NSCLC patients, 11 smokers without cancer) were enrolled for blood draws (7.5 mL). CTCs were identified by immunofluorescence, individually retrieved, and DNA-extracted. Targeted NGS was performed to detect somatic variants (single-nucleotide variants (SNVs) and insertions/deletions (Indels)) across 65 oncogenes and tumor suppressor genes. Cancer-associated variants were classified using OncoKB database. NSCLC patients had significantly higher CTC counts than control smokers ($p = 0.0132$; Mann–Whitney test). Analyzing 23 CTCs and 13 white blood cells across seven patients revealed a total of 644 somatic variants that occurred in all CTCs within the same subject, ranging from 1 to 137 per patient. The highest number of variants detected in ≥1 CTC within a patient was 441. A total of 18/65 (27.7%) genes were highly mutated. Mutations with oncogenic impact were identified in functional domains of seven oncogenes/tumor suppressor genes (*NF1, PTCH1, TP53, SMARCB1, SMAD4, KRAS*, and *ERBB2*). Single CTC-targeted NGS detects heterogeneous and shared mutational signatures within and between NSCLC patients. CTC single-cell genomics have potential for integration in NSCLC precision oncology.

Keywords: circulating tumor cells; non-small-cell lung cancer; single cell next generation sequencing

1. Introduction

Lung cancer is by far the leading cause of cancer-related deaths worldwide. Non-small-cell lung cancer (NSCLC) accounts for >80% of all lung cancer subtypes [1]. Although lung cancer screening of long-term smokers by low-dose computed tomography (LDCT) significantly increases detection at curable stages and improves survival [2], 75% of NSCLC patients are diagnosed at advanced stages III–IV with 5-year survival rates of <25% [3]. The eligibility of NSCLC patients with advanced disease to receive targeted therapy relies on profiling of driver oncogenes and tumor suppressor genes mutation analyses performed on invasive tumor tissue biopsies [4]. However, these tumor tissue biopsies are associated with significant morbidities and costs. Due to these limitations, invasive biopsies are typically only performed once and consequently do not reflect tumor evolution over time and development of resistant clones during therapies [5,6]. Therefore, developing non-invasive and repeatable, real-time diagnostic tests for NSCLC patients appears critical to improve management [7].

Liquid biopsy approaches by a simple blood draw are minimally invasive and easily repeatable alternatives to tissue biopsies [8]. For example, cell-free circulating tumor (ct)DNA has evolved as a blood-based option to identify genetic tumor alterations for NSCLC and other cancer patients [9,10]. Whereas ctDNA release is thought to be related to tumor cell turnover [11,12], circulating tumor cells (CTCs) are shed into the blood from the primary tumor [13] and may be reflective of tumor resistance to treatment. ctDNA is a useful biomarker to predict disease recurrence following surgical tumor resections and is a predictor of treatment responses in solid cancers, such as in patients suffering from malignant melanoma [14]. CTC-derived DNA may also offer mutational insights into future metastatic recurrences, as CTCs represent a whole cancer and—in some cases—exhibit tumorigenic properties [15]. Importantly, recent studies suggest that CTCs exhibit unique genetic alterations that are not detected in ctDNA, whereas ctDNA can reveal genomic alterations not detected in CTCs [16]. These findings of unique mutations detected by different liquid biopsy modalities provide a strong rationale for further exploration of CTC single-cell sequencing assays. Novel CTC sequencing liquid biomarker assays are likely to provide novel information for NSCLC patients to investigate resistance mechanisms for personalized medicine [8,17].

In this study, a CTC detection platform that integrates detection and single-cell retrieval for targeted NGS of individual CTCs was applied to NSCLC patients. In a prospective pilot trial, CTCs were enumerated and single CTCs (and control white blood cells (WBCs)) underwent targeted NGS to detect somatic variants in oncogenes and tumor suppressor genes in liquid biopsies. To serve as risk-matched controls, long-term smokers without lung cancer were recruited from a lung cancer screening program. Single CTC-targeted NGS could detect heterogeneous and shared mutational signatures within and between NSCLC patients. In addition to other liquid biomarkers, CTC single-cell genomics have potential for integration in NSCLC precision oncology.

2. Materials and Methods

2.1. Enrollment of Subjects

Subjects were prospectively recruited at the Ellis Fischel Cancer Center at the University of Missouri (MU), consisting of 20 patients with pathologically confirmed NSCLC and 11 high-risk chronic smokers without cancer determined by screening LDCT. Clinicopathologic data were obtained from chart review. The TNM staging manual of the American Joint Committee on Cancer (AJCC, Chicago, IL, USA; 8th edition) was applied. A healthy volunteer blood donor was included for validation of the platform by spiking with a known number of human NSCLC adenocarcinoma cell line A549 cells. Studies involving human subjects were approved by the University of Missouri Institutional Review Board (MU IRB Number 2010166; approved 16 April 2016) and were performed according to the Helsinki Declaration.

2.2. CTC Enumeration with NSCLC Cell Line Cells Spiked into Healthy Human Blood and Study Subjects' Blood Samples

In alignment with the traditional definition of a CTC of the FDA-approved CellSearch® platform, a multi-parameter immunofluorescence staining pattern analysis was performed and identified a CTC as CK/EpCAM+ (epithelial markers) and CD45- (WBC marker) with a DAPI+ nucleus (Supplementary Figure S1A). The mean fluorescent intensity values of the whole cell were used to distinguish strong from weak staining for each biomarker. A comparison of multiple slides of cancer cells and WBCs showed that tumor cells consistently displayed strong CK/EpCAM, and weak CD45 staining compared with WBCs in the same sample that had weak CK/EpCAM and strong CD45 expression (Figure 1A). Based on the explicit and distinct pattern of tumor cells, a cut-off was used to define tumor cells with mean fluorescent intensity for CK > 500, EpCAM > 100, and CD45 < 100. For initial validation of the technology (AccuCyte, RareCyte, Seattle, WA, USA) [18] before testing clinical samples, blood samples of a healthy control donor were spiked by single-cell micropipetting with a known number (N = 0, 100, 200, 1000) of human NSCLC adenocarcinoma cancer cells (A549; ATCC) with the analytic personnel being blinded. Results showed a similar immunofluorescence staining profile between the reference spiked A549 cell line in healthy donors' blood and detected CTCs from study subjects (spiked/retrieved: 0/0; 100/76; 200/208; 1000/1223; linear regression $r^2 = 0.999$) (Supplementary Figure S1B). Following validation of correlation with a variety of spiked cancer cells, the same protocols were applied to the clinical samples of NSCLC and screening subjects.

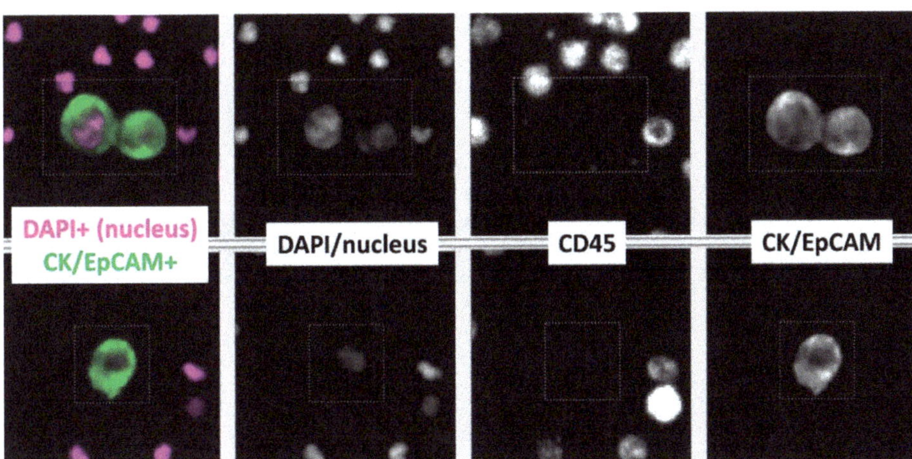

Figure 1. Four-channel fluorescent images of circulating tumor cells (CTCs) detected in NSCLC patients' blood (7.5 mL). CTCs from two different NSCLC patients are shown, identified as cytokeratin (CK)/EpCAM+ and CD45- cells with DAPI+ nuclei. (Magnification ×10).

Phlebotomies were performed and blood (7.5 mL) was collected in AccuCyte BCT tubes and shipped overnight to RareCyte Inc. (Seattle, WA, USA) for CTC enumeration and single-cell retrieval of CTCs and WBCs in NSCLC patients. Processing was performed using the AccuCyte sample preparation system to isolate nucleated cells and spread them evenly onto SuperFrost™ Plus Microscope Slides (Fisherbrand™, Fisher Scientific, Hampton, NH, USA). The slides were air-dried at room temperature and banked for later staining (stored at −20 °C). Enumeration and retrieval of CTCs and WBCs were performed using CyteFinder® instrument based on CF405, Sytox Orange, CF647, and QD800 tags to target the Pre-label, Nucleus, CK/EpCAM, and CD45, respectively. Slide images were analyzed by CyteMapper® software. Then, cells were individually retrieved and dispensed in PCR tubes for downstream NGS. CTCs were defined by nuclear size ≥8 µm in diameter, presence of a well-defined and visible cytoplasm, and immunofluorescence staining in the corresponding channels of predicted biomarkers (CK+ and/or EpCAM+, CD45-, DAPI+ nucleus).

2.3. Targeted NGS of Single CTCs

For targeted NGS sequencing of single CTCs, the CleanPlex OncoZoom Panel kit (Paragon Genomics, Inc., Hayward, CA, USA) was used, with a modified protocol to sequence single cells. Standard bioinformatic and visualization workflows for dissemination of sequence data were applied. DNA was extracted from a sorted single cell using a Single Cell Lysis Kit (cat. 4458235; Thermo Fisher Scientific, Indianapolis, IN, USA). Six microliters of total reaction volume was used for first-step amplification. To establish the single-cell retrieval protocol, we used human lung cancer cells (A549) spiked into healthy blood. We retrieved three A549 cells and two WBCs from the healthy donor blood. For subsequent analysis of clinical samples, a targeted-genome amplification and sequencing of 2–6 CTCs and 1–2 WBCs per patient was performed covering 601 amplicons in 65 genes. Concentration and quality of prepared libraries were assessed via fragment analysis using the Advanced Analytics High Sensitivity NGS Fragment Analysis Kit (cat. DNF-474-0500; Agilent, Santa Clara, CA, USA). An individual library quality ratio score was determined by dividing the fragment analysis trace concentration (ng/µL) (250–350 bp peak concentration) by the fragment peak concentration (ng/µL) (150–190 bp). The intent was to remove samples with higher concentration of primer dimers and to remove samples with low concentration of library fragments, presenting poor quality with ratio scores less than 1. Passing libraries were denatured, pooled and diluted to a final loading concentration of 1.5 pMol prior to sequencing on the NextSeq 500 system at 2 × 151 bp using the NextSeq Mid Output v2 (300 cycle) kit (cat. 15057939; Illumina, San Diego, CA, USA). FASTQ files were pre-processed for adapter trimming using cutadapt version 1.18 [19] and then assessed using FastQC [20] and MultiQC [21]. The paired-end reads were aligned to the GRCh37 human reference genome with bwa version 0.7.17 [22]. Subsequent analysis was restricted to the targeted regions of the panel; variant calling was performed in the targeted regions of the OncoZoom panel with 100 bp of padding. The resulting BAM files were cleaned using the base quality score recalibration provided by GATK v. 4.1.9.0 [23].

2.4. Somatic Variant (SNVs and Indels) Analysis

Somatic single-nucleotide variants (SNVs) and insertions/deletions (Indels) were called using Mutect2 in GATK v. 4.1.9.0 for each subject individually with the gnomAD database as a "germline-resource" from the GATK resource bundle (https://console.cloud.google.com/storage/browser/genomics-public-data/resources/broad/hg38/v0, accessed on 20 January 2021). Multi-sample mode of Mutect2 [24] was run for a joint analysis that included all CTCs to determine the shared variants among the CTCs within a subject. Then, Mutect2 was run for each individual CTC separately to determine variants in ≥1 CTCs within a subject. Initial variant filtering used FilterMutectCalls with default parameters followed by annotation using ANNOVAR allele frequency and gene information [25]. A second filtering according to the Minor Allele Frequency (MAF) ≥1% in the 1000 Genomes [22] and ExAC databases [26] was performed. Additional annotation by RefSeq Gene definition was performed to predict the variant's genomic region and corresponding gene name [27]. The shared variants in all CTCs across the samples were grouped by gene. The variants in genes shared by at least two subjects were matched with a potential oncogenic impact according to the open access OncoKB database [28] that annotates biological and oncogenic effects and the prognostic significance of somatic molecular alterations, including the ones predictive of drug responses based on US Food and Drug Administration (FDA) labeling. They were visualized in lollipop plots illustrating the genomic position and functional impact of these variants using cBioPortal MutationMapper [29,30].

2.5. Statistical Analysis

Statistical analyses were performed with R version 4.0.2 [31] and Prism v8.0.1 (GraphPad Software, San Diego, CA, USA). To compare non-parametric CTC counts between two groups, the Mann–Whitney test was applied. A p value of <0.05 considered statistically significant.

3. Results

3.1. Clinical Characteristics of Subjects

A total of 31 subjects were prospectively enrolled. Out of these, 20 patients were diagnosed with NSCLC and 11 subjects were risk-matched controls consisting of long-term smokers (all ≥30 pack years) without evidence of lung cancer on screening LDCT scans of the chest. Clinical characteristics of all 31 study subjects are described in Table 1. NSCLC patients were staged by AJCC classification as localized/loco-regional disease stage I-III in N = 9 (45%) and metastatic stage IV in N = 11 (55%). No significant differences were observed between the two study groups of cancer patients and controls with regard to relevant clinical parameters, such as age and extent of smoking history (defined by pack years).

Table 1. Subjects' characteristics.

	N
Total number of of subjects	31
NSCLC patients	20 (65%)
Median age (range)	66 (55–75)
Gender	
• Females	14 (70%)
• Males	6 (30%)
Smoking history	
Never-smokers	3 (15%)
Smokers	17 (85%)
• Pack years <5	1 (6%)
• Pack years ≥5–30	2 (12%)
• Pack years ≥30	14 (82%)
Tumor stage (TNM/AJCC 8th ed.)	
• I–III (localized/locoregional disease)	9 (45%)
• IV (metastatic)	11 (55%)
Histologic subtype	
• Adenocarcinoma	13 (65%)
• Squamous cell carcinoma	7 (35%)
Smokers without cancer	11 (35%)
Median age (range)	67 (52–76)
Gender	
• Females (%)	7 (64%)
• Males (%)	4 (36%)
Smoking history	
• Pack years ≥30	11 (100%)

3.2. CTC Enumeration in NSCLC Patients and Chronic Smokers without Cancer

A multiplex immunofluorescence approach identified a CTC as CK/EpCAM+ (epithelial markers) and CD45- (WBC marker) with a DAPI+ nucleus (Figure 1). Following validation of protocols and the CTC detection technology in blinded spiking experiments with A549 lung cancer cells into healthy human blood (supplementary Figure S1A,B), CTCs were detected in 12/20 (60%) of NSCLC patients at a mean of 13.4 ± 1.78 SEM with a median of 1 (range 0–237) (Table 2). In long-term smokers without cancer, CTCs were detected in 2/11 (18%) subjects with a mean of 0.18 ± 0.12 SEM and a median of 0 with a range of 0–1. A statistically significantly higher number of CTCs were found in NSCLC patients compared to control long-term smokers without lung cancer as determined by

LDCT screening (p = 0.0132; Mann–Whitney test) (Figure 2A). Subsequently, we compared the CTC counts between NSCLC patients according to tumor stage, grouping patients into two categories: patients with localized or loco-regional cancer disease (stage I–III) and patients with metastatic disease (stage IV). The mean CTC count was clearly higher in metastatic/stage IV NSCLC patients ((mean 23.45 ± 21.36 SEM; median 2 (range 0–237)) than non-metastatic stage I–III patients ((mean 1.11 ± 0.70 SEM); median 0 (range 0–6)), although not reaching level of statistical significance (p = 0.0651) (Figure 2B).

Table 2. CTC enumeration in control high-risk subjects without cancer and NSCLC patients.

	N	Circulating Tumor Cells/7.5 mL Blood			*p* Value
		Present (%)	Mean (±SEM)	Median (Range)	
Smokers without cancer	11	2 (18%)	0.18 (±0.12)	0 (0–1)	
NSCLC patients	20	12 (60%)	13.40 (± 11.78)	1 (0–237)	0.0132 *
NSCLC tumor stage					
• I–III (non-metastatic)	9	3 (33%)	1.11 (±0.70)	0 (0–6)	0.0651 †
• IV (metastatic)	11	9 (82%)	23.45 (±21.36)	2 (0–237)	

* Comparing smokers without cancer vs. NSCLC, † comparing stage I–III vs. IV: Mann–Whitney test.

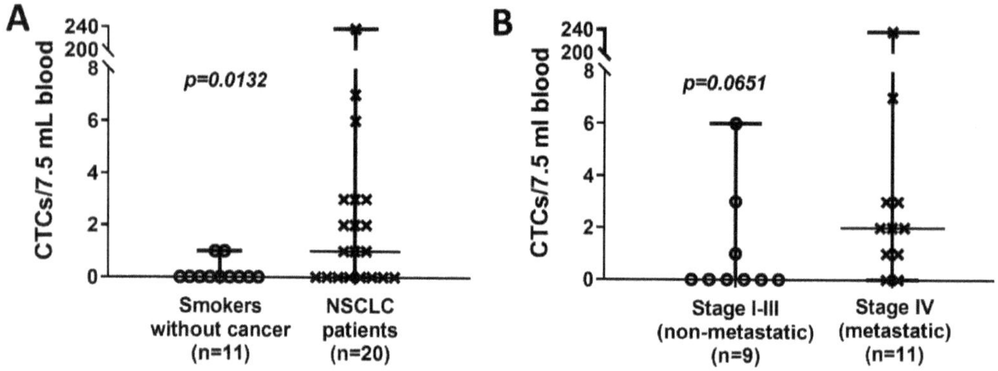

Figure 2. CTC counts in the study populations. (**A**). CTC counts in high-risk controls (long-term smokers) without cancer and patients diagnosed with NSCLC. (**B**). Distribution of CTC count for NSCLC patients by tumor stages, separating them in localized/loco-regional stage I–III versus advanced, metastatic stage IV. (Scatter dot plots; p values were calculated with Mann–Whitney test).

3.3. *Characterization of Single CTCs Somatic Variants in NSCLC Patients*

Seven NSCLC patients (stage I: N = 2; metastatic stage IV: N = 5) that had ≥2 CTCs detected were selected for single-cell sequencing (Table 3). A total of 36 single cells (23 CTCs and 13 WBCs) from these seven NSCLC patients underwent targeted NGS. As we processed an input with low library DNA concentration, only libraries with a library quality ratio score of greater than 1 with clean amplification of the targeted region in the fragment analysis (as defined in the methods) were combined and sequenced after quality-control examination of library DNA concentrations and fragment sizes (Supplementary Figure S2A). With a target of 500-fold coverage, we generated a total of 2,769 Mb data per sample, with a mean of 76.9 and a range of 26 to 129 Mb per sample. A Phred score of 30 for all FASTQ files was observed, indicating high base quality (supplementary Figure S2B).

Table 3. NSCLC patients (N = 7) selected for single CTC sequencing.

Patient ID	NSCLC Stage	Number of CTCs Detected	Number of CTCs Sequenced	Number of WBCs Sequenced	Total Number of Shared Variants Detected in All Sequenced CTCs within the Same Subject	Total Number of Variants Detected in ≥1 Sequenced CTC within the Same Subject
RL13	I	6	4	2	1	125
RL5	I	3	2	1	85	148
RL14	IV	3	3	2	72	121
RL16	IV	237	6	2	130	441
RL17	IV	7	3	2	91	147
RL19	IV	2	2	2	137	194
RL20	IV	3	3	2	101	245

CTC: circulating tumor cells, WBCs: white blood cells.

Somatic variants (the sum of all SNVs and Indels) were counted in sequenced CTCs according to the number of appearances (1) within each subject and (2) across all seven subjects to compare variant incidences and variant types in NSCLC patients. The number of shared variants that were detected in all sequenced CTCs within a subject was determined and is shown in Table 3. Adding up all variants shared by all sequenced CTCs within a subject, a total of 644 shared variants were identified in all seven NSCLC patients combined. After allele frequency filtering using a cutoff of <0.01 MAF based on the 1000 Genomes and ExAC databases and as outlined in the methods, a total of 617 variants remained, and these are summarized per patient in Table 3 (2nd last column). In the two stage I patients, one patient (RL13) had only one variant shared by all CTCs, whereas 85 shared variants were detected in all sequenced CTCs in the other stage I patient (RL5). In the five metastatic/stage IV NSCLC patients, an increased number of variants shared by all CTCs within the subject was found, ranging from 72 to 137.

Finally, variants detected in ≥1 CTCs (but not necessarily in all CTCs) within an NSCLC patient were determined (Table 3; last column). A higher number of variants in ≥1 CTCs within a subject was detected in stage IV/metastatic patients in comparison to the two stage I patients. The highest number of 441 shared variants in ≥1 CTCs was noted within a metastatic NSCLC patient (RL16). In all seven NSCLC patients, the total number of variants detected in ≥1 CTCs within the same subject ranged from 121 to 441.

3.4. Single CTC Somatic Variants Detected in Oncogenes and Tumor-Suppressor Genes

Variants detected in the 65 oncogenes and tumor-suppressor genes included in the targeted NGS panel were classified (Table 4). Since some of the 617 shared variants detected in all CTCs within a subjected listed in Table 3 appeared in ≥1 subject, the variants that appeared more than once across the seven patients were counted only once. This reduced the number of shared variants to a total of 598 in various genes. Analysis showed that 18 (27.7%) oncogenes/tumor-suppressor genes were highly mutated, defined as >10 somatic variants detected per gene. Out of these 18 genes, 14 (77.8%) showed a shared somatic variant by at least two patients (Table 5). The highest number of shared variants was detected in the TP53 gene (four variants)—a known tumor-suppressor gene described in multiple cancers, including NSCLC.

Shared somatic variants detected in CTCs among the seven NSCLC patients were also matched against variants described in OncoKB [26], a knowledge base containing somatic mutations in cancer-associated genes with diagnostic and therapeutic relevance. Visualization plots were then generated to highlight genomic alterations and their potential impact on specific functional domains of select genes. Variants in genes with known impact in cancer were found in functional domains in 7 (50%) out to 14 oncogenes/tumor suppressor genes (NF1, PTCH1, TP53, SMARCB1, SMAD4, KRAS, and ERBB2) (Figure 3). With the only exception of PTCH1, 6/7 (85.8%) of these cancer-associated genes have been described to be associated with NSCLC development.

Table 4. Variants per oncogene/tumor-suppressor gene that were detected in all sequenced CTCs within a subject (total number of variants per gene were combined from all seven subjects; variants detected in ≥1 subject were counted once only, adding up to a total of 598 variants).

Gene	Number of Variants
NF1	69
BRCA2	33
PTCH1	33
NF2	27
ATM	25
EGFR	24
ERBB3	24
PIK3CA	19
APC	17
BRCA1	17
RB1	13
SMARCB1	13
TP53	13
CDH1	12
CSF1R	11
NOTCH1	11
SMO	11
TERT	11
ABL1	10
MLH1	10
EZH2	9
FGFR2	9
SMAD4	9
DNMT3A	8
FBXW7	8
MTOR	8
CTNNB1	7
ERBB2	7
HNF1A	7
KIT	7
PIK3R1	7
PTEN	7
ALK	6
CDKN2A	6
PTPN11	6
ERBB4	5
JAK2	5
JAK3	5
RET	5
BRAF	4
KRAS	4
LOC100507346	4
STK11	4
VHL	4
FGFR3	3
FLT3	3
HRAS	3
IDH1	3
IDH2	3
KDR	3
MET	3
NPM1	3
TSC1	3
AKT1	2
FGFR1	2
GNAQ	2
GNAS	2

Table 4. *Cont.*

Gene	Number of Variants
NRAS	2
PDGFRA	2
ATM; C11orf65	1
FBXW7-AS1	1
GNA11	1
MAP2K1	1
MSH6	1

Table 5. Variants detected in oncogenes/tumor-suppressor genes in all sequenced CTCs within a subject that were shared by ≥2 NSCLC patients.

Gene	Number of Variants
TP53	4
NF1	2
SMARCB1	1
SMAD4	1
PTEN	1
PTCH1	1
MAP2K1	1
KRAS	1
JAK3	1
ERBB2	1
DNMT3A	1
CTNNB1	1
ABL1	1
ALK	1

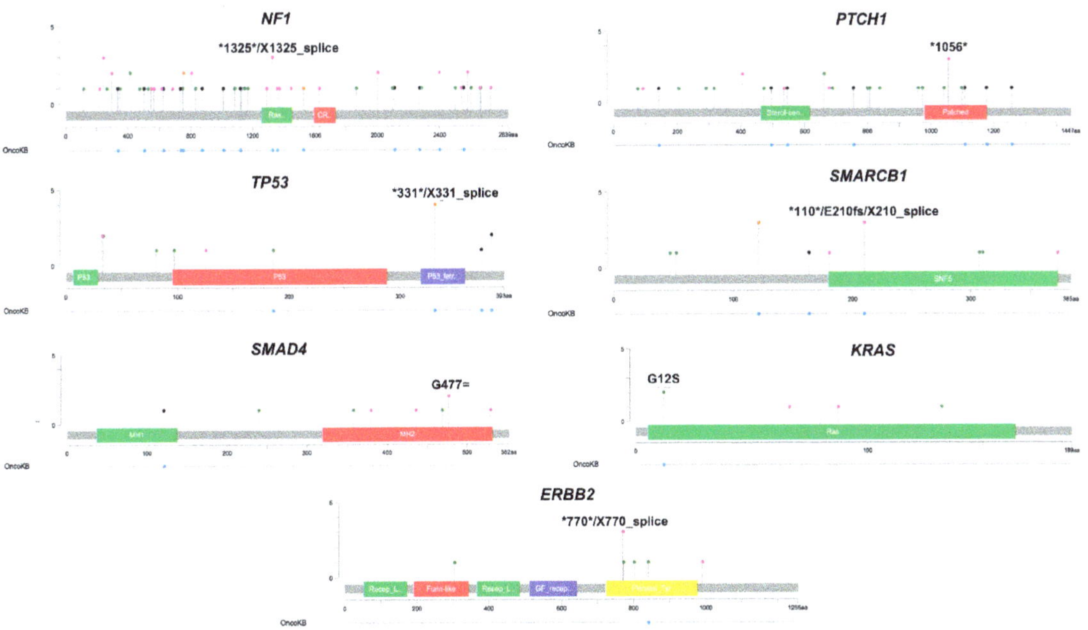

Figure 3. Gene locations of somatic variants in oncogenes and tumor-suppressor genes with predicted oncogenic impact, as per OncoKB database. Seven oncogenes/tumor suppressor genes were identified

with shared somatic variants detected in ≥2 CTCs. Lollipop plots show the variant gene locations and their predicted oncogenic impact, according to OncoKB database. In some genes, at the same base position different mutations are observed in multiple individual CTCs. For example, in NF1 we observed a base substitution in coding sequence position of 1325 that results in the introduction of a stop codon and, separately at the same base position, a base substitution occurs that alters the conserved splice acceptor/donor site for exon/intron splicing. Colored boxes represent mutations in specific functional domains (*: stop codon; mutation types: green: missense, black: truncating, orange: splice, pink: others).

4. Discussion

In contrast to invasive lung cancer tissue biopsies, which are associated with significant morbidities and costs, minimally invasive liquid biopsies by simple blood draws hold great promise to improve clinical management of NSCLC. Liquid biopsies in cancer patients can identify somatic variants and cancer-associated mutations at the time of diagnosis or later on in real-time to allow precise adjustments of therapy management or monitoring of disease progression [32]. Beyond CTC enumeration, the present study provides a technical assessment of single CTC-targeted NGS in NSCLC patients. We successfully detected and retrieved single CTCs using 7.5 mL of blood and then performed targeted NGS with a panel targeting more than 2900 hotspots in 65 genes with known cancer-associations. This approach led to identification of distinct and shared variants in and across NSCLC patients' single CTCs. Cancer-associated variants detected in single CTCs could be matched to known cancer mutations from an established oncology knowledge base. Analysis of a relatively low number of single CTCs per patient still allowed identification of key oncogene and tumor suppressor gene variants known and not yet known to be associated with NSCLC disease.

Current state-of-the-art molecular testing is performed by one-time invasive tumor tissue biopsy. In some cases, low tumor cellularity requires even an invasive repeat biopsy, which is again associated with morbidities, costs, and delay in care. In contrast to non-cellular liquid biomarkers (e.g., ctDNA or extracellular vesicles), a CTC in the blood represents a whole, morphologically intact tumor cell. Lung cancer patient-derived CTCs can be tumorigenic in vivo in immunodeficient mice [15,33], indicating that micrometastatic CTCs may carry mutations of future metastases. In our analysis of single CTCs retrieved from seven NSCLC patients, somatic variants with potential oncologic impact were detected in all CTCs analyzed. CTCs of at least two NSCLC patients shared variants in six oncogenes/tumor suppressor genes: *NF1*, *TP53*, *SMARCB1*, *SMAD4*, *KRAS*, and *ERBB2*. Variants in the *NF1* tumor suppressor gene have been previously described to be present in 10% of NSCLC tumor tissues, and they are frequently paired with *KRAS* and *ERBB* cancer driver variants [34]. Additionally, variants of the *NF1* gene have been found relatively frequently in male smokers and coexist with *TP53* variants [35]. Several studies have reported the *TP53* gene variants as a predictor of NSCLC patients' poor prognosis [34,36]. For instance, an analysis conducted using The Cancer Genome Atlas (TCGA) revealed that NSCLC patients with *TP53* variants had significantly shorter survival rates than those without [36]. With regard to *SMARCB1*, gene variants and loss of expression were reported in up to 5% of NSCLC cases and have been associated with poor clinical outcome [37]. The *SMAD4* pathway has been identified as a potential target for tumor treatment [38]. The findings of that study indicated that variants of the *SMAD4* gene play an important role for NSCLC metastasis as they were observed in advanced stage IV patients only. *SMAD4* serum concentration also correlated with a malignant NSCLC phenotype that was associated with metastasis and clinical progression [39]. In addition, *SMAD4* and its transcription factor play a regulatory function for many target genes, also increasing the risk of cell tumorigenesis in lung cancer [40]. *SMAD4* variants and related expression may regulate the signal transduction pathways involved in NSCLC tumorigenesis, such as the *TGF-β/SMAD4* pathway [41]. *KRAS* variants were also detected in the current study in single CTCs. *KRAS* variants have been detected in 51% of advanced NSCLC patients, which included older patients and current or former smokers [42]. This study also showed a higher prevalence of 51% of KRAS variants in NSCLC adenocarcinoma patients, in contrast to previously reported prevalence of 20–40% [42]. We also detected variants in the *ERBB2* gene in single CTCs in our cohort, a well-known driver oncogene in NSCLC [43]. *ERBB2* gene is a member of the *EGFR* family that is involved in several biological scenarios in malignant diseases [44]. It has been demonstrated that *ERBB2* is involved in a series of cancer-associated processes, such as cell proliferation, cell survival, and differentiation [45]. With regard to these oncogenes and tumor suppressor genes, previously published results in NSCLC are in concordance with our findings on cancer-associated genes that were found to be mutated in single CTCs of NSCLC patients. Findings support that single CTCs identified using the platform in our study with criteria also applied by the FDA-approved CellSearch® system can be analyzed individually by targeted NGS to identify variants in known NSCLC-associated oncogenes and tumor suppressor genes. As a liquid biopsy modality, single CTC-seq analyses may provide a complementary technology to assist clinicians for better management of NSCLC patients using a non-invasive protocol with a small sample of peripheral blood.

There are several limitations with the present pilot study. Most important is that the cohort is a small sample size, so the analysis has limited power. The study lacks mutational information on the matched tumor tissues that were not analyzed for comparison with the targeted sequencing performed on single CTCs. Matched-tumor tissues for comparative analyses were not available in our study. Additionally, we performed testing just at one timepoint, which did not allow us to study longitudinal changes of detected somatic variants over time. Longer follow-up times, particularly in screening subjects without cancer, and our focus on initial diagnosis resulted in determining mutations at a single time point only. Finally, we did not compare CTC-seq data with other genetic liquid biopsy modalities, such as ctDNA, to correlate findings on CTCs with other liquid biopsy technologies. In particular the integration of CTC-seq with ctDNA findings promises more comprehensive profiling of NSCLC-associated mutations by liquid biopsy.

In summary, our study presents a robust method of CTC detection in NSCLC patients and, as a critical addition, an option for genetic profiling of single CTCs. Distinct and shared variants within and across NSCLC patients can be identified in oncogenes and tumor suppressor genes in single CTCs, even in cases with low CTC numbers. Further investigations, including a larger prospective cohort of different tumor stages of NSCLC patients, will be required for further validation. Single CTC variant detection by sequencing may have potential clinical value for diagnosis and therapy management of NSCLC patients.

Supplementary Materials: The following supporting information can be downloaded at: https://www.mdpi.com/article/10.3390/cimb44020052/s1.

Author Contributions: M.B.: Participated in study design, carried out the study and managed all project study participants who aided with experiments, patient consenting and chart, data review, manuscript preparation, and data analysis. A.M.: Data analysis and processing, sequencing alignment, variant calling, and manuscript editing. Y.M.: Participated in patient consenting, statistical analyses, manuscript editing, and chart. A.B.R.: Liquid biopsy processing and analysis and manuscript editing. N.G.E.: Study design, sequencing analysis, and manuscript editing. K.F.S.-O.: Participated in the original idea. J.B.M.: Study design, data interpretation, and manuscript preparation. W.C.W.: Discussed study design, sequencing data, and interpretation. A.A.C.: Study design, data interpretation, and manuscript preparation. Y.H.: Statistical analysis and interpretation. G.L.: Reviewed methodology and manuscript. P.J.T.: Supervised the bioinformatics and statistical analysis data and interpretation, and final manuscript preparation. J.T.K.: Project principal investigator, original idea, study concept and design, study analysis, discussion of results, and manuscript preparation. All authors have read and agreed to the published version of the manuscript.

Funding: This work was supported in part by funding provided by the Center for Biomedical Informatics (P.J.T.) and Department of Surgery (J.T.K.), School of Medicine, University of Missouri, Columbia, MO, USA, an Ellis Fischel Cancer Center Pilot Grant (J.T.K.; A.A.C.), and RareCyte, Inc., Seattle, WA, USA.

Institutional Review Board Statement: All subject investigations conformed to the principles outlined in the Declaration of Helsinki and have been performed with permission of the study protocol approved by the Institutional Review Board (IRB), University of Missouri, Columbia, MO (MU IRB Number 2010166).

Informed Consent Statement: Informed consent was obtained from all subjects involved in the study.

Data Availability Statement: The data presented in this study are available on request from the corresponding author.

Acknowledgments: We thank all individuals who participated in the present study. We express our thanks to Eduardo J. Simoes and Nathan J. Bivens for their assistance with experiments, discussion of results, and suggested ideas for consideration.

Conflicts of Interest: Nolan G. Ericson and Arturo B. Ramirez are employees of RareCyte Inc. The remaining authors have no disclosures. The funders had no role in the design of the study; in the collection, analyses, or interpretation of data; in the writing of the manuscript; or in the decision to publish the results.

References

1. Siegel, R.L.; Miller, K.D.; Jemal, A. Cancer statistics, 2020. *CA Cancer J. Clin.* **2020**, *70*, 7–30. [CrossRef] [PubMed]
2. Aberle, D.R.; Adams, A.M.; Berg, C.D.; Black, W.C.; Clapp, J.D.; Fagerstrom, R.M.; Gareen, I.F.; Gatsonis, C.; Marcus, P.M.; Sicks, J.D. Reduced lung-cancer mortality with low-dose computed tomographic screening. *N. Engl. J. Med.* **2011**, *365*, 395–409. [CrossRef] [PubMed]
3. Detterbeck, F.C.; Chansky, K.; Groome, P.; Bolejack, V.; Crowley, J.; Shemanski, L.; Kennedy, C.; Krasnik, M.; Peake, M.; Rami-Porta, R.; et al. The IASLC Lung Cancer Staging Project: Methodology and Validation Used in the Development of Proposals for Revision of the Stage Classification of NSCLC in the Forthcoming (Eighth) Edition of the TNM Classification of Lung Cancer. *J. Thorac. Oncol.* **2016**, *11*, 1433–1446. [CrossRef]
4. National Comprehensive Cancer Network (NCCN) Clinical Practice Guidelines in Oncology: Non-Small Cell Lung Cancer (Version 6.2020). Available online: https://www.nccn.org/professionals/physician_gls/pdf/nscl.pdf (accessed on 1 July 2021).
5. Horn, L.; Whisenant, J.G.; Wakelee, H.; Reckamp, K.L.; Qiao, H.; Leal, T.A.; Du, L.; Hernandez, J.; Huang, V.; Blumenschein, G.R.; et al. Monitoring Therapeutic Response and Resistance: Analysis of Circulating Tumor DNA in Patients With ALK+ Lung Cancer. *J. Thorac. Oncol.* **2019**, *14*, 1901–1911. [CrossRef]
6. Dagogo-Jack, I.; Brannon, A.R.; Ferris, L.A.; Campbell, C.D.; Lin, J.J.; Schultz, K.R.; Ackil, J.; Stevens, S.; Dardaei, L.; Yoda, S.; et al. Tracking the Evolution of Resistance to ALK Tyrosine Kinase Inhibitors through Longitudinal Analysis of Circulating Tumor DNA. *JCO Precis. Oncol.* **2018**, *2018*, 1–14. [CrossRef] [PubMed]
7. Aggarwal, C.; Thompson, J.C.; Chien, A.L.; Quinn, K.J.; Hwang, W.T.; Black, T.A.; Yee, S.S.; Christensen, T.E.; LaRiviere, M.J.; Silva, B.A.; et al. Baseline Plasma Tumor Mutation Burden Predicts Response to Pembrolizumab-based Therapy in Patients with Metastatic Non-Small Cell Lung Cancer. *Clin. Cancer Res.* **2020**, *26*, 2354–2361. [CrossRef] [PubMed]
8. Alix-Panabieres, C.; Pantel, K. Liquid Biopsy: From Discovery to Clinical Application. *Cancer Discov.* **2021**, *11*, 858–873. [CrossRef]
9. Rolfo, C.; Mack, P.; Scagliotti, G.V.; Aggarwal, C.; Arcila, M.E.; Barlesi, F.; Bivona, T.; Diehn, M.; Dive, C.; Dziadziuszko, R.; et al. Liquid Biopsy for Advanced NSCLC: A Consensus Statement From the International Association for the Study of Lung Cancer. *J. Thorac. Oncol.* **2021**, *16*, 1647–1662. [CrossRef]
10. Powles, T.; Assaf, Z.J.; Davarpanah, N.; Banchereau, R.; Szabados, B.E.; Yuen, K.C.; Grivas, P.; Hussain, M.; Oudard, S.; Gschwend, J.E.; et al. ctDNA guiding adjuvant immunotherapy in urothelial carcinoma. *Nature* **2021**, *595*, 432–437. [CrossRef]
11. Rostami, A.; Lambie, M.; Yu, C.W.; Stambolic, V.; Waldron, J.N.; Bratman, S.V. Senescence, Necrosis, and Apoptosis Govern Circulating Cell-free DNA Release Kinetics. *Cell Rep.* **2020**, *31*, 107830. [CrossRef]
12. Jahr, S.; Hentze, H.; Englisch, S.; Hardt, D.; Fackelmayer, F.O.; Hesch, R.D.; Knippers, R. DNA fragments in the blood plasma of cancer patients: Quantitations and evidence for their origin from apoptotic and necrotic cells. *Cancer Res.* **2001**, *61*, 1659–1665. [PubMed]
13. Heitzer, E.; Haque, I.S.; Roberts, C.E.S.; Speicher, M.R. Current and future perspectives of liquid biopsies in genomics-driven oncology. *Nat. Rev. Genet.* **2019**, *20*, 71–88. [CrossRef] [PubMed]
14. Revythis, A.; Shah, S.; Kutka, M.; Moschetta, M.; Ozturk, M.A.; Pappas-Gogos, G.; Ioannidou, E.; Sheriff, M.; Rassy, E.; Boussios, S. Unraveling the Wide Spectrum of Melanoma Biomarkers. *Diagnostics* **2021**, *11*, 1341. [CrossRef] [PubMed]
15. Morrow, C.J.; Trapani, F.; Metcalf, R.L.; Bertolini, G.; Hodgkinson, C.L.; Khandelwal, G.; Kelly, P.; Galvin, M.; Carter, L.; Simpson, K.L.; et al. Tumourigenic non-small-cell lung cancer mesenchymal circulating tumour cells: A clinical case study. *Ann. Oncol.* **2016**, *27*, 1155–1160. [CrossRef] [PubMed]
16. Onidani, K.; Shoji, H.; Kakizaki, T.; Yoshimoto, S.; Okaya, S.; Miura, N.; Sekikawa, S.; Furuta, K.; Lim, C.T.; Shibahara, T.; et al. Monitoring of cancer patients via next-generation sequencing of patient-derived circulating tumor cells and tumor DNA. *Cancer Sci.* **2019**, *110*, 2590–2599. [CrossRef] [PubMed]

17. Alix-Panabieres, C.; Pantel, K. Clinical Applications of Circulating Tumor Cells and Circulating Tumor DNA as Liquid Biopsy. *Cancer Discov.* **2016**, *6*, 479–491. [CrossRef]
18. Campton, D.E.; Ramirez, A.B.; Nordberg, J.J.; Drovetto, N.; Clein, A.C.; Varshavskaya, P.; Friemel, B.H.; Quarre, S.; Breman, A.; Dorschner, M.; et al. High-recovery visual identification and single-cell retrieval of circulating tumor cells for genomic analysis using a dual-technology platform integrated with automated immunofluorescence staining. *BMC Cancer* **2015**, *15*, 360. [CrossRef]
19. Martin, M. Cutadapt Removes Adapter Sequences from High-Throughput Sequencing Reads. *EMBnet J.* **2011**, *17*, 10–12. [CrossRef]
20. Andrews, S. FastQC: A Quality Control Tool for High throughput Sequence Data. Available online: http://www.bioinformatics.babraham.ac.uk/proj (accessed on 15 January 2021).
21. Ewels, P.; Magnusson, M.; Lundin, S.; Kaller, M. MultiQC: Summarize analysis results for multiple tools and samples in a single report. *Bioinformatics* **2016**, *32*, 3047–3048. [CrossRef]
22. Genomes Project, C.; Auton, A.; Brooks, L.D.; Durbin, R.M.; Garrison, E.P.; Kang, H.M.; Korbel, J.O.; Marchini, J.L.; McCarthy, S.; McVean, G.A.; et al. A global reference for human genetic variation. *Nature* **2015**, *526*, 68–74. [CrossRef]
23. van der Auwera, G.A.; O'Connor, B.D. *Genomics in the Cloud: Using Docker, GATK, and WDL in Terra*, 1st ed.; O'Reilly Media: Sebastopol, VA, USA, 2020.
24. Benjamin, D.; Sato, T.; Cibulskis, K.; Getz, G.; Stewart, C.; Lichtenstein, L. Calling Somatic SNVs and Indels with Mutect2. *bioRxiv* **2019**. [CrossRef]
25. Wang, K.; Li, M.; Hakonarson, H. ANNOVAR: Functional annotation of genetic variants from high-throughput sequencing data. *Nucleic. Acids Res.* **2010**, *38*, e164. [CrossRef]
26. Karczewski, K.J.; Weisburd, B.; Thomas, B.; Solomonson, M.; Ruderfer, D.M.; Kavanagh, D.; Hamamsy, T.; Lek, M.; Samocha, K.E.; Cummings, B.B.; et al. The ExAC browser: Displaying reference data information from over 60,000 exomes. *Nucleic. Acids Res.* **2017**, *45*, D840–D845. [CrossRef] [PubMed]
27. Tatusova, T.; DiCuccio, M.; Badretdin, A.; Chetvernin, V.; Nawrocki, E.P.; Zaslavsky, L.; Lomsadze, A.; Pruitt, K.D.; Borodovsky, M.; Ostell, J. NCBI prokaryotic genome annotation pipeline. *Nucleic. Acids Res.* **2016**, *44*, 6614–6624. [CrossRef] [PubMed]
28. Chakravarty, D.; Gao, J.; Phillips, S.M.; Kundra, R.; Zhang, H.; Wang, J.; Rudolph, J.E.; Yaeger, R.; Soumerai, T.; Nissan, M.H.; et al. OncoKB: A Precision Oncology Knowledge Base. *JCO Precis. Oncol.* **2017**, *2017*, 1–16. [CrossRef] [PubMed]
29. Cerami, E.; Gao, J.; Dogrusoz, U.; Gross, B.E.; Sumer, S.O.; Aksoy, B.A.; Jacobsen, A.; Byrne, C.J.; Heuer, M.L.; Larsson, E.; et al. The cBio cancer genomics portal: An open platform for exploring multidimensional cancer genomics data. *Cancer Discov.* **2012**, *2*, 401–404. [CrossRef] [PubMed]
30. Gao, J.; Aksoy, B.A.; Dogrusoz, U.; Dresdner, G.; Gross, B.; Sumer, S.O.; Sun, Y.; Jacobsen, A.; Sinha, R.; Larsson, E.; et al. Integrative analysis of complex cancer genomics and clinical profiles using the cBioPortal. *Sci. Signal* **2013**, *6*, pl1. [CrossRef]
31. Team, R.C. *R: A Language and Environment for Statistical Computing*; R Foundation for Statistical Computing: Vienna, Austria, 2020; Available online: https://www.R-project.org/ (accessed on 30 January 2021).
32. Wan, J.C.M.; Massie, C.; Garcia-Corbacho, J.; Mouliere, F.; Brenton, J.D.; Caldas, C.; Pacey, S.; Baird, R.; Rosenfeld, N. Liquid biopsies come of age: Towards implementation of circulating tumour DNA. *Nat. Rev. Cancer* **2017**, *17*, 223–238. [CrossRef]
33. Hodgkinson, C.L.; Morrow, C.J.; Li, Y.; Metcalf, R.L.; Rothwell, D.G.; Trapani, F.; Polanski, R.; Burt, D.J.; Simpson, K.L.; Morris, K.; et al. Tumorigenicity and genetic profiling of circulating tumor cells in small-cell lung cancer. *Nat. Med.* **2014**, *20*, 897–903. [CrossRef]
34. Gray, P.N.; Dunlop, C.L.; Elliott, A.M. Not All Next Generation Sequencing Diagnostics are Created Equal: Understanding the Nuances of Solid Tumor Assay Design for Somatic Mutation Detection. *Cancers* **2015**, *7*, 1313–1332. [CrossRef]
35. Redig, A.J.; Capelletti, M.; Dahlberg, S.E.; Sholl, L.M.; Mach, S.; Fontes, C.; Shi, Y.; Chalasani, P.; Janne, P.A. Clinical and Molecular Characteristics of NF1-Mutant Lung Cancer. *Clin. Cancer Res.* **2016**, *22*, 3148–3156. [CrossRef] [PubMed]
36. Wang, X.; Sun, Q. TP53 mutations, expression and interaction networks in human cancers. *Oncotarget* **2017**, *8*, 624–643. [CrossRef] [PubMed]
37. Gandhi, J.S.; Alnoor, F.; Sadiq, Q.; Solares, J.; Gradowski, J.F. SMARCA4 (BRG1) and SMARCB1 (INI1) expression in TTF-1 negative neuroendocrine carcinomas including merkel cell carcinoma. *Pathol. Res. Pract.* **2021**, *219*, 153341. [CrossRef] [PubMed]
38. Ozawa, H.; Ranaweera, R.S.; Izumchenko, E.; Makarev, E.; Zhavoronkov, A.; Fertig, E.J.; Howard, J.D.; Markovic, A.; Bedi, A.; Ravi, R.; et al. SMAD4 Loss Is Associated with Cetuximab Resistance and Induction of MAPK/JNK Activation in Head and Neck Cancer Cells. *Clin. Cancer Res.* **2017**, *23*, 5162–5175. [CrossRef]
39. Guo, X.; Li, M.; Wang, X.; Pan, Y.; Li, J. Correlation between loss of Smad4 and clinical parameters of non-small cell lung cancer: An observational cohort study. *BMC Pulm. Med.* **2021**, *21*, 111. [CrossRef]
40. Haeger, S.M.; Thompson, J.J.; Kalra, S.; Cleaver, T.G.; Merrick, D.; Wang, X.J.; Malkoski, S.P. Smad4 loss promotes lung cancer formation but increases sensitivity to DNA topoisomerase inhibitors. *Oncogene* **2016**, *35*, 577–586. [CrossRef]
41. Wang, Y.; Tan, X.; Tang, Y.; Zhang, C.; Xu, J.; Zhou, J.; Cheng, X.; Hou, N.; Liu, W.; Yang, G.; et al. Dysregulated Tgfbr2/ERK-Smad4/SOX2 Signaling Promotes Lung Squamous Cell Carcinoma Formation. *Cancer Res.* **2019**, *79*, 4466–4479. [CrossRef]
42. Kartolo, A.; Feilotter, H.; Hopman, W.; Fung, A.S.; Robinson, A. A single institution study evaluating outcomes of PD-L1 high KRAS-mutant advanced non-small cell lung cancer (NSCLC) patients treated with first line immune checkpoint inhibitors. *Cancer Treat. Res. Commun.* **2021**, *27*, 100330. [CrossRef]

43. Byeon, S.; Lee, B.; Park, W.Y.; Choi, Y.L.; Jung, H.A.; Sun, J.M.; Ahn, J.S.; Ahn, M.J.; Park, K.; Lee, S.H. Benefit of Targeted DNA Sequencing in Advanced Non-Small-Cell Lung Cancer Patients without EGFR and ALK Alterations on Conventional Tests. *Clin. Lung Cancer* **2020**, *21*, e182–e190. [CrossRef]
44. Liberelle, M.; Jonckheere, N.; Melnyk, P.; Van Seuningen, I.; Lebegue, N. EGF-Containing Membrane-Bound Mucins: A Hidden ErbB2 Targeting Pathway? *J. Med. Chem.* **2020**, *63*, 5074–5088. [CrossRef]
45. Xu, Y.; Zhang, L.; Xia, L.; Zhu, X. MicroRNA-133a-3p suppresses malignant behavior of non-small cell lung cancer cells by negatively regulating ERBB2. *Oncol. Lett.* **2021**, *21*, 457. [CrossRef] [PubMed]

current issues in molecular biology

MDPI

Communication

Possibility of SARS-CoV-2 Infection in the Metastatic Microenvironment of Cancer

Takuma Hayashi [1,2,*], Kenji Sano [3] and Ikuo Konishi [1,4]

1 National Hospital Organization, Kyoto Medical Center, Kyoto 612-8555, Japan; ikuokonishi08@yahoo.co.jp
2 START-Program, Japan Science and Technology Agency (JST), Tokyo 102-8666, Japan
3 Shinsyu University Hospital, Matsumoto 390-8621, Japan; kenjisano12@yahoo.co.jp
4 Graduate School of Medicine, Kyoto University, Kyoto 606-8501, Japan
* Correspondence: takumah@shinshu-u.ac.jp

Abstract: According to a report from the World Health Organization (WHO), the mortality and disease severity induced by the severe acute respiratory syndrome coronavirus 2 (SARS-CoV-2) are significantly higher in cancer patients than those of individuals with no known condition. Common and cancer-specific risk factors might be involved in the mortality and severity rates observed in the coronavirus disease 2019 (COVID-19). Similarly, various factors might contribute to the aggravation of COVID-19 in patients with cancer. However, the factors involved in the aggravation of COVID-19 in cancer patients have not been fully investigated so far. The formation of metastases in other organs is common in cancer patients. Therefore, the present study investigated the association between lung metastatic lesion formation and SARS-CoV-2 infectivity. In the pulmonary micrometastatic niche of patients with ovarian cancer, alveolar epithelial stem-like cells were found adjacent to ovarian cancer. Moreover, angiotensin-converting enzyme 2, a host-side receptor for SARS-CoV-2, was expressed in these alveolar epithelial stem-like cells. Furthermore, the spike glycoprotein receptor-binding domain (RBD) of SARS-CoV-2 was bound to alveolar epithelial stem-like cells. Altogether, these data suggested that patients with cancer and pulmonary micrometastases are more susceptible to SARS-CoV-2. The prevention of de novo niche formation in metastatic diseases might constitute a new strategy for the clinical treatment of COVID-19 for patients with cancer.

Keywords: alveolar epithelial stem cells; ACE2; SARS-CoV-2; COVID-19; RBD of spike glycoprotein; metastatic microenvironment

Citation: Hayashi, T.; Sano, K.; Konishi, I. Possibility of SARS-CoV-2 Infection in the Metastatic Microenvironment of Cancer. *Curr. Issues Mol. Biol.* **2022**, *44*, 233–241. https://doi.org/10.3390/cimb44010017

Academic Editors: Dumitru A. Iacobas and Peter C. Hart

Received: 3 December 2021
Accepted: 4 January 2022
Published: 5 January 2022

1. Introduction

The United States and other countries have found it difficult to contain the coronavirus disease 2019 (COVID-19) pandemic due to the spread through respiratory droplets of the severe acute respiratory syndrome coronavirus 2 (SARS-CoV-2) and the inconsistent adherence to effective public health measures, including wearing masks and maintaining social distancing. According to the reports of the World Health Organization, the mortality rate of cancer patients infected with SARS-CoV-2 is 7.6%, which is fairly higher than the 1.4% mortality rate of individuals infected with SARS-CoV-2 without complications [1]. Among patients with cancer and COVID-19, the 30-day all-cause mortality is high, i.e., a mortality rate of 13.3%, and has been associated with general and cancer-specific risk factors [1,2]. Moreover, according to a report from the Japan Ministry of Health, Labor, and Welfare, as of August 2020, the severity rate of patients with solid cancer infected with SARS-CoV-2 was much higher than that of all patients infected with SARS-CoV-2 [3]. The reason why COVID-19 is more severe in cancer patients is not fully understood. However, immunity against the virus seems reduced in cancer patients receiving therapeutic anticancer agents [4].

Lung stem cells able to regenerate lung tissue are used in research on lung diseases, including cancer and infectious diseases. Furthermore, it has become possible to prepare a lung culture model using lung stem cells in the laboratory. However, the biological characteristics of human lung stem cells have not been clarified. Therefore, the development of a lung culture model, especially one modeling the most peripheral part of the lung where gas exchange occurs, has not progressed.

Human lung culture systems have been reported to model lung infections, including SARS-CoV-2 infections responsible for COVID-19-associated pneumonia [5,6]. Moreover, Kuo et al. succeeded in creating a human peripheral lung culture model using lung stem cells [5]. Epithelial cell adhesion molecule-positive alveolar epithelial type 2 (AT2) stem cells differentiate into AT2 progenitor cells that express angiotensin-converting enzyme 2 (ACE2). ACE2 is a receptor for SARS-CoV-2. In contrast, cytokeratin 5-positive lung stem cells differentiate into bronchial and/or bronchiolar epithelial progenitor cells. In other words, the AT2 progenitor cells used as human peripheral lung culture model express ACE2, a host cell receptor for SARS-CoV-2 [5]. Thus, SARS-CoV-2 can infect alveolar epithelial stem cells.

Previous clinical studies have shown that SARS-CoV-2 infection rates and COVID-19 severity are higher in cancer and in people with a history of cancer than they are in healthy individuals. The specific reasons for these observations have not been clarified. Previously, we have shown that, in lung metastases of ovarian cancer, alveolar epithelial cells adjacent to ovarian cancer cells transform into alveolar epithelial stem-like cells [7]. ACE2 is strongly expressed in alveolar epithelial stem-like cells [5]. The present study aims to confirm the expression of ACE2 in alveolar epithelial stem-like cells adjacent to ovarian cancer cells in lung metastases of ovarian cancer and to investigate the binding of SARS-CoV-2 spike glycoprotein to alveolar epithelial stem-like cells. We showed that ACE2 was indeed expressed in alveolar epithelial stem-like cells adjacent to ovarian cancer in the pulmonary micrometastatic niche. Furthermore, the receptor-binding domain (RBD) of SARS-CoV-2 spike glycoprotein bound to alveolar epithelial stem-like cells. Altogether, these data indicate that cancer patients with pulmonary micrometastases might be more susceptible to SARS-CoV-2. The prevention of de novo niche formation in metastatic diseases might constitute a new strategy for the clinical treatment of COVID-19 in patients with cancer.

2. Materials and Methods

2.1. Case Selection for Immunohistochemical Staining

To examine the biological and medical characteristics of the pulmonary micrometastatic niche, 4 cases with high-grade serous ovarian adenocarcinomas were selected from 69 primary epithelial ovarian cancers stained using an anti-human S100 calcium-binding protein A4 (S100A4) antibody. S100A4 is used as marker for ovarian cancer (Supplementary Table S1). Sixty-nine patients with ovarian carcinoma visited the Shinshu University Hospital (Matsumoto, Nagano, Japan) between 1994 and 2003. They underwent surgery followed by combination chemotherapy with taxane-based and platinum preparations. The Supplementary Materials show the detailed medical conditions of the patients.

2.2. Antibodies and Immunohistochemistry (IHC)

IHC staining of S100A4, cluster of differentiation 90 (CD90 or Thy1), ACE2, and the RBD of the SARS-CoV-2 spike glycoprotein was performed on tissue sections from pulmonary micrometastases of patients with high-grade serous ovarian cancer. Tumor tissue sections were incubated with the appropriate primary antibodies at 4 °C overnight. For the staining of the RBD of the SARS-CoV-2 spike glycoprotein, tumor tissue sections were incubated with 10 ng of recombinant RBD of the spike protein (Sino Biological Inc., Beijing, China) at 4 °C overnight. After this incubation, the sections were incubated with the mouse monoclonal antibody recognizing the RBD of SARS-CoV-2 spike glycoprotein at 4 °C overnight.

IHC-stained sections were visualized under a confocal microscope (Leica TCS SP8, Wetzlar, Germany) according to the manufacturer's procedure. Photographs of the healthy alveoli and bronchioles areas (Bron.) as well as of metastases areas (Met.) were taken from tissue sections of lung metastases resected from patients with ovarian cancer, as shown in the result section. Then, the expression levels of each factor were calculated using fluorescent color. These IHC experiments on human tissue sections were performed using standard procedures at Shinshu University (Matsumoto, Nagano, Japan) and the National Hospital Organization Kyoto Medical Center (Kyoto, Kyoto, Japan) in accordance with the institutional guidelines (approval no. M192). The Supplementary Materials section contains the list of primary or secondary antibodies used in the experiments and the detailed materials and methods.

2.3. Ethical Approval and Consent to Participate

This study was reviewed and approved by the Central Ethics Review Board of the National Hospital Organization Headquarters in Japan (Tokyo, Japan) and Shinshu University (Nagano, Japan). The ethical approval was obtained on 17 August 2019 (approval number NHO H31-02). The authors attended educational lectures supervised by the Japanese Government on medical ethics in 2020 and 2021. The completion numbers for the authors are AP0000151756, AP0000151757, AP0000151769, and AP000351128. Consent to participate was needed as this work was considered clinical research. All subjects signed informed consent when they were briefed about the experiments and agreed with the contents of the study. The authors attended a seminar on the ethics of experimental research using human materials on 2 July 2020 and 20 July 2021 to become familiar with the importance and ethics of clinical experiments (National Hospital Organization Kyoto Medical Center and Shinshu University School of Medicine). The experiments with human materials performed in the present study were approved by the ethics committee (approval number KMC R02-0702).

Details of materials and methods are described in the Supplementary Materials.

3. Results

Niches promoting metastatic colonization have been previously investigated using models with human-in-mouse ovarian cancer xenograft in immunodeficient mice. For example, CD34-positive lineage ovarian cancer stem-like cells sorted using the side population procedure were injected into the mammary fat pads of BALB/c *nu/nu* mice [8]. CD90, also known as Thy1, is used as a marker for several stem cells [9]. S100 calcium-binding protein A4 (S100A4), a member of the S100 calcium-binding protein family secreted by ovarian cancer cells, supports tumorigenesis by stimulating angiogenesis [9] (Supplementary Table S1). Pathological examinations have shown the existence of S100A4-negative and CD90-positive stem-like cells in vimentin-positive normal neighboring alveolar epithelial cells [9]. Similar to this previous observation, we found that the initialization of mimicry represented incomplete differentiation of normal alveolar epithelial cells toward the stem-like lineage in pulmonary micrometastases of patients with ovarian cancer (Figure 1A).

ACE2, a host-side receptor for SARS-CoV-2, expressed in CD90-positive alveolar epithelial stem-like cells in the pulmonary metastatic niches of patients with high-grade serous ovarian cancer, is essential (Figure 1A and Table 1). Furthermore, histopathological analyses showed that the RBD of the SARS-CoV-2 spike glycoprotein bound to ACE2-expressing CD90-positive alveolar epithelial stem-like cells (Figure 1A and Table 1). Based on these findings, SARS-CoV-2 is deemed to infect the alveolar epithelial stem-like cells in pulmonary micrometastases of patients with ovarian cancer.

The pathological examination with an anti-human CD90 monoclonal antibody revealed that CD90 was not expressed in the alveolar and bronchiolar epithelial cells (Figure 1B). However, analyses with an anti-human ACE2 monoclonal antibody showed that bronchiolar epithelial cells expressed ACE2 (Figure 1B, Table 1). Therefore, the binding of SARS-CoV-2 RBD to the ACE2-positive bronchiolar epithelial cells was confirmed in the normal tissue section (Figure 1B).

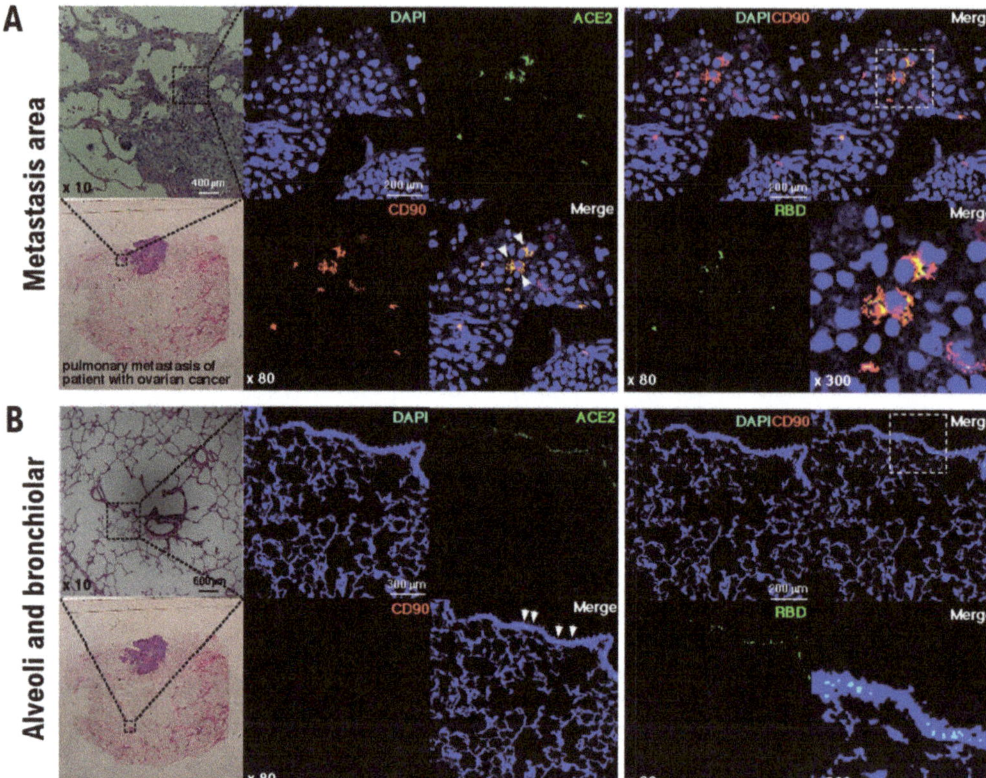

Figure 1. Binding of the RBD of the SARS-CoV-2 spike glycoprotein to the stem-like cells in normal neighboring alveolar epithelial cells. Immunohistochemical studies were performed using pulmonary metastatic tissue surgically excised from patients with high-grade serous ovarian carcinoma. The expression levels of CD90 and ACE2 and the binding of RBD in the normal alveolar and bronchiolar areas as well as in the metastases were investigated by pathological studies. In the photographs of the metastases (**A**) and normal alveolar and bronchiolar areas (**B**), the expression of each factor and binding activity are indicated by fluorescent color. Human ACE2-positive (green) and CD90-positive (red) stem-like cells, indicated by white arrowheads in human normal neighboring alveolar epithelial cells, were found in pulmonary micrometastases. Human CD90-positive cells (red) were not detected in the metastatic colonies of human serous ovarian carcinoma. Immunohistochemical studies were performed using an antibody recognizing human ACE2 (green), a monoclonal antibody detecting the spike glycoprotein of SARS-CoV-2 (green), and an antibody specific for human CD90 (red), which is a biomarker for stem-like cells. The binding of the RBD of SARS-CoV-2 spike glycoprotein (green) to CD90-positive (red) alveolar epithelial stem-like cells was observed and is indicated in yellow. Anti-human CD90 (Abcam ab133350), anti-human ACE2 (ORIGENE, Rockville, MD, USA), and anti-spike glycoprotein of SARS-CoV-2 (GeneTex, Inc., Irvine, CA, USA) antibodies, as well as recombinant spike glycoprotein of SARS-CoV-2 protein (BioVision, Milpitas, CA, USA), were used. The experiments were performed five times with similar results.

Table 1. Characteristics of patients with ovarian cancer and lung metastases as well as CD90 and ACE2 expression in the metastasis areas and the alveolar and bronchiolar areas.

Patient No.	Age Range	Age at Surgery (Years)	Histological Type	FIGO Stage	Grade	No. of Lung Metastatic Lesions	Pulmonary Metastatic Niche		Vital Status
							CD90* (%)	ACE2* (%)	
1	40 s	40–45	HG serous	IVA	3	Single	36.43	27.42	Alive
2	50 s	50–55	HG serous	IVA	3	Single	33.87	18.93	Alive
3	50 s	50–55	HG serous	IVA	3	Multiple	38.32	29.38	Deceased
4	40 s	45–50	HG serous	IVB	3	Multiple	32.67	28.05	Alive

Normal alveolar and bronchiolar areas

Patient No.	Normal Alveoli				Normal Bronchioles			
	CD90* (%)		ACE2* (%)		CD90* (%)		ACE2* (%)	
1	4.53	11.82	3.23	20.67	4.53	11.82	3.23	20.67
2	3.91	12.57	3.18	21.46	3.91	12.57	3.18	21.46
3	4.34	12.71	3.45	22.05	4.34	12.71	3.45	22.05
4	4.08	13.43	2.98	21.92	4.08	13.43	2.98	21.92

FIGO stage, the FIGO (International Federation of Gynecology and Obstetrics) staging system is commonly used for cancers of the female reproductive organs. High grade (HG) serous, high-grade serous ovarian adenocarcinoma. CD90*, proportion of CD90-positive alveolar epithelial stem-like cells in pulmonary metastatic niches, normal alveoli, and bronchioles assessed by immunohistochemical experiments using anti-human CD90 monoclonal antibody. ACE2*, proportion of ACE2-positive alveolar epithelial stem-like cells in pulmonary metastatic niches, normal alveoli, and bronchioles assessed by immunohistochemical experiments using anti-human ACE2 monoclonal antibodies. The expression levels of each factor were determined by measuring the fluorescence intensities. Percentages are the ratio of CD90 or ACE2-positive cells to the total cell counts.

In the photographs of the normal alveolar and bronchiolar areas as well as the metastases areas, the expression levels of each factor were determined by measuring the fluorescence intensity. Moreover, the ratios of CD90-positive, ACE2-positive, and RBD-positive cells were determined in these areas. Quantitative analysis showed that the average ratio of CD90-positive cells was higher in the metastatic areas (36.4%) than that in the normal alveoli and bronchioles areas (4.5%) (Figure 2A, Table 1, Supplementary Figure S1). Additionally, the average proportion of ACE2-positive cells in the normal alveolar and bronchiolar areas was 11.8%, and that in the metastatic areas was 27.4% (Figure 2A, Table 1, Supplementary Figure S2). The average ratio of bronchiolar epithelial cells binding to SARS-CoV-2 RBD in normal alveolar and bronchiolar areas was 7.3%, whereas it reached 15.5% in the metastatic areas (Figure 2A, Table 1). Figure 2B shows the normal alveoli and bronchioles areas and metastases areas.

The ratio of ACE2 and RBD double-positive cells to the total number of ACE2-positive cells was 39.3% in the normal alveolar and bronchiolar areas and 43.8% in the metastatic areas (Figure 2C), suggesting that the binding property of RBD to ACE2-positive cells did not significantly change between normal alveolar and bronchiolar areas and the metastatic areas.

Previous studies have demonstrated that, in the pulmonary micrometastatic niche of other cancer types, the reprogramming of mimicry probably represents the incomplete differentiation of normal alveolar epithelial cells into stem-like lineages [7,10]. Presumably, the pulmonary micrometastatic niche is a target for SARS-CoV-2 infection.

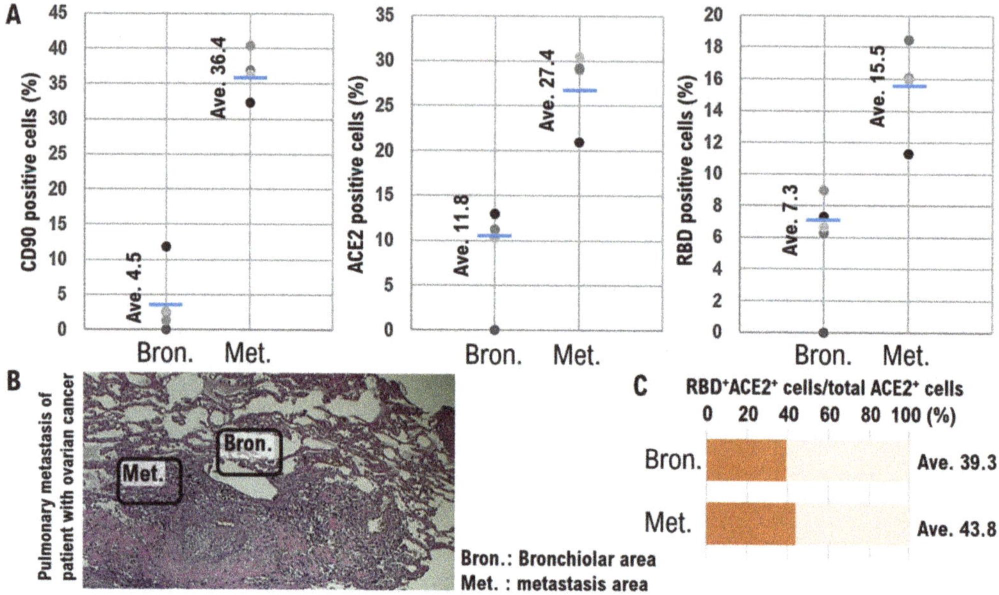

Figure 2. Quantification of immunofluorescence colocalization in imaging for CD90 and ACE2 expression levels and RBD binding. A quantitative analysis was conducted using ImageJ Version 1.53 m, a public domain software for image analysis (NIH ImageJ, Bethesda, MD, USA). The CD90 and ACE2 expression levels and the binding of RBD in normal alveolar and bronchiolar areas (Bron.) and metastases areas (Met.) were investigated through pathological studies. (**A**) The ratios of the number of target protein-positive cells to the total number of cells are shown in the dispersion diagram. (**B**) Photographs of the normal alveoli and bronchioles areas (Bron.) and metastases areas (Met.) from a tissue section of lung metastases resected from a patient with ovarian cancer. The expression levels of each factor were determined by measuring fluorescence intensities. (**C**) The proportion of bronchiolar epithelial cells associated with RBD of SARS-CoV-2 in normal alveoli and bronchiolar regions is indicated as the ratio of RBD-positive cells to total cell count.

4. Discussion

Previous clinical studies have shown that SARS-CoV-2 infection rates and COVID-19 severity rates are higher in cancer patients and in people with a history of cancer than in healthy individuals. The specific reasons for these observations have not been clarified. Previously, we showed that, in lung metastases of ovarian cancer, alveolar epithelial cells adjacent to ovarian cancer cells transform into alveolar epithelial stem-like cells [7]. Alveolar epithelial stem-like cells have been reported to strongly express ACE2 [5]. The present work aimed to confirm the expression of ACE2 in alveolar epithelial stem-like cells adjacent to ovarian cancer cells in lung metastases of ovarian cancer and to investigate the binding of the SARS-CoV-2 spike glycoprotein to alveolar epithelial stem-like cells. Our IHC analyses showed the expression of ACE2 in alveolar epithelial stem-like cells adjacent to ovarian cancer cells in lung metastases of ovarian cancer. We also demonstrated the binding of SARS-CoV-2 spike glycoprotein to alveolar epithelial stem-like cells.

A high 30-day all-cause mortality has been reported in patients with cancer and COVID-19. Various factors contributing to COVID-19 severity in cancer patients have been identified. A previous report has shown that SARS-CoV-2 infects alveolar and bronchiolar epithelial cells [11]. Within the pulmonary metastatic niche, alveolar epithelial cells adjacent to metastatic cancer cells are differentiated into alveolar epithelial stem-like cells. Our experiments showed that, although the expression of ACE2 was not strong in normal alveolar epithelial cells, ACE2 was clearly expressed in alveolar epithelial stem-like cells. We also observed the binding of the alveolar epithelial stem-like cells to the RBD of the SARS-CoV-2 spike glycoprotein. Therefore, SARS-CoV-2 infection of stem cells and/or epithelial progenitor cells present in the metastatic niche in patients with cancer is likely a factor contributing to COVID-19 severity.

We examined the environmental niche of lung metastases of ovarian cancer. However, the infection rate of SARS-CoV-2 and the severity of COVID-19 also increase in patients with other cancer types. The results showing that ovarian cancer cells form a metastatic niche near the alveolar stem cells are reminiscent of a previous finding demonstrating that prostate cancer cells metastasizing to the bone settle near the stem cells in the bone marrow, promoting the development of a metastatic environment that supports tumor growth [12]. A recent report also described cancer-associated parenchymal cells that show stem-cell-like characteristics, the expression of lung progenitor markers, multilineage differentiation potential, and self-renewal activity [13].

To obtain accurate histopathological information of positive or negative SARS-CoV-2 infection in tissues from COVID-19 cancer patients, histopathological experiments with metastatic tissues of cancer patients infected with SARS-CoV-2 or with COVID-19 symptoms must be performed. Thus far, treatment aimed at reducing lung metastases has been limited to surgical treatment. However, in clinical practice, lung metastases have been reduced using immune checkpoint inhibitors and/or poly ADP-ribose polymerase inhibitors. Furthermore, the efficacy of the anti-S100A4 antibody drug in suppressing the metastatic ability of malignant tumors in other organs, including ovarian cancer, has been investigated [9,14,15].

A longer follow-up is needed to better understand the effect of COVID-19 on the treatment outcomes of patients with cancer, including on the ability to continue specific cancer treatments. In such a significant intersection of cancer medicine and infectious diseases, the prevention of de novo niche formation of metastatic disease might constitute a novel strategy for the clinical treatment of COVID-19.

5. Conclusions

Previous clinical studies have shown that SARS-CoV-2 infection rates and COVID-19 severity rates are higher in cancer patients currently being treated and people with a history of cancer than in healthy individuals. The specific reasons for the high SARS-CoV-2 infection rate and COVID-19 aggravation rate in cancer patients have not been clarified. Infection of SARS-CoV-2 into the niche of metastatic lesions in cancer patients may be one of the reasons for the higher rate of SARS-CoV-2 infection and COVID-19 severity compared to healthy individuals.

Supplementary Materials: The following supporting information can be downloaded at: https://www.mdpi.com/article/10.3390/cimb44010017/s1. Table S1: Ovarian cancer cells express S100A4. In particular, in the case of high grade serous ovarian cancer, S100A4 is strongly expressed; Figure S1: IHC experiments with anti-human ACE2 monoclonal antibody do not provide the medical evidence, which demonstrate CD90 expressions in any cells of respiratory tissues, i.e., cells making up nasopharynx, bronchus, and lung tissues. −, 0–10% positive cells; +, 10–50% positive cells, more than 50% positive cells; Figure S2: IHC experiments with anti-human ACE2 monoclonal antibody demonstrate that in case of ciliated cells (ciliary rootlets) of nasopharynx and bronchus, ACE2 is markedly expressed. However, lung cells i.e., alveolar cells and macrophages unclearly express ACE2. −, 0–10% positive cells; +, 10–50% positive cells, more than 50% positive cells.

Author Contributions: T.H. and K.S. performed most of the clinical treatments and diagnostic pathological studies, coordinated the project, created the study, and wrote the manuscript. I.K. carefully reviewed the manuscript and commented on the aspects of clinical medicine. I.K. shared information on clinical medicine and oversaw the entirety of the study. All authors have read and agreed to the published version of the manuscript.

Funding: This clinical research was performed with the support of the following research funding: Japan Society for the Promotion of Science for T.H. (Grant No. 19K09840), the START-program Japan Science and Technology Agency (JST) for T.H. (Grant No. STSC20001), and the National Hospital Organization Multicenter clinical study for T.H. (Grant No. 2019- Cancer in general-02).

Institutional Review Board Statement: This study was reviewed and approved by the Central Ethics Review Board of the National Hospital Organization Headquarters in Japan (Tokyo, Japan) and Shinshu University (Nagano, Japan). The ethical approval was obtained on 17 August 2019 (approval number NHO H31-02). The authors attended educational lectures supervised by the Japanese Government on medical ethics in 2020 and 2021. The completion numbers for the authors are AP0000151756, AP0000151757, AP0000151769, and AP000351128. The authors attended a seminar on the ethics of experimental research using human materials on 2 July 2020 and 20 July 2021 to become familiar with the importance and ethics of clinical experiments (National Hospital Organization Kyoto Medical Center and Shinshu University School of Medicine). The experiments with human materials performed in the present study were approved by the ethics committee (approval number KMC R02-0702).

Informed Consent Statement: Consent to participate was needed as this work was considered clinical research. All subjects signed informed consent when they were briefed about the experiments and agreed with the contents of the study.

Data Availability Statement: This manuscript is an editorial and does not contain research data. Therefore, there are no research data or information to be published or opened.

Acknowledgments: We thank all the medical staff and co-medical staff for their contribution to the medical research at National Hospital Organization Kyoto Medical Center. We appreciate Crimson Interactive Japan Co., Ltd., for revising and polishing our manuscript.

Conflicts of Interest: The authors declare no potential conflict of interest.

References

1. Kuderer, N.M.; Choueiri, T.K.; Shah, D.P.; Shyr, Y.; Rubinstein, S.M.; Rivera, D.R.; Shete, S.; Hsu, C.Y.; Desai, A.; de Lima Lopes, G., Jr.; et al. Clinical impact of COVID-19 on patients with cancer (CCC19): A cohort study. *Lancet* **2020**, *395*, 1907–1918. [CrossRef]
2. Moris, D.; Tsilimigras, D.I.; Schizas, D. Cancer and COVID-19. *Lancet* **2020**, *396*, 1066. [CrossRef]
3. Kudo, K.; Ichihara, E. COVID-19 Pneumonia. *Gan Kagaku Ryoho* **2020**, *47*, 1657–1661.
4. Cesaro, S.; Giacchino, M.; Fioredda, F.; Barone, A.; Battisti, L.; Bezzio, S.; Frenos, S.; De Santis, R.; Livadiotti, S.; Marinello, S.; et al. Guidelines on vaccinations in paediatric haematology and oncology patients. *Biomed. Res. Int.* **2014**, *2014*, 707691. [CrossRef] [PubMed]
5. Salahudeen, A.A.; Choi, S.S.; Rustagi, A.; Zhu, J.; van Unen, V.; Sean, M.; Flynn, R.A.; Margalef-Català, M.; Santos, A.J.; Ju, J.; et al. Progenitor identification and SARS-CoV-2 infection in human distal lung organoids. *Nature* **2020**, *588*, 670–675. [CrossRef] [PubMed]
6. Katsura, H.; Sontake, V.; Tata, A.; Kobayashi, Y.; Edwards, C.E.; Heaton, B.E.; Konkimalla, A.; Asakura, T.; Mikami, Y.; Fritch, E.J.; et al. Human Lung Stem Cell-Based Alveolospheres Provide Insights into SARS-CoV-2-Mediated Interferon Responses and Pneumocyte Dysfunction. *Cell Stem Cell* **2020**, *27*, 890–904.e8. [CrossRef] [PubMed]
7. Hayashi, T.; Sano, K.; Aburatani, H.; Yaegashi, N.; Konishi, I. Initialization of epithelial cells by tumor cells in a metastatic microenvironment. *Oncogene* **2020**, *39*, 2638–2640. [CrossRef] [PubMed]
8. Hayashi, T.; Sano, K.; Ichimura, T.; Kanai, Y.; Zharhary, D.; Aburatani, H.; Yaegashi, N.; Konishi, I. Characteristics of Leiomyosarcoma: Induction of Hematogenous Metastasis by Isolated Uterine Mesenchymal Tumor Stem-like Cells. *Anticancer Res.* **2020**, *40*, 1255–1265. [CrossRef] [PubMed]
9. Horiuchi, A.; Hayashi, T.; Kikuchi, N.; Hayashi, A.; Fuseya, C.; Shiozawa, T.; Konishi, I. Hypoxia upregulates ovarian cancer invasiveness via the binding of HIF-1α to a hypoxia-induced, methylation-free hypoxia response element of S100A4 gene. *Int. J. Cancer* **2012**, *131*, 1755–1767. [CrossRef] [PubMed]
10. Ombrato, L.; Nolan, E.; Kurelac, I.; Mavousian, A.; Bridgeman, V.L.; Heinze, I.; Chakravarty, P.; Horswell, S.; Gonzalez-Gualda, E.; Matacchione, G.; et al. Metastatic-niche labelling reveals parenchymal cells with stem features. *Nature* **2019**, *572*, 603–608. [CrossRef] [PubMed]
11. Yao, X.H.; He, Z.C.; Li, T.Y.; Zhang, H.R.; Wang, Y.; Mou, H.; Guo, Q.; Yu, S.C.; Ding, Y.; Liu, X.; et al. Pathological evidence for residual SARS-CoV-2 in pulmonary tissues of a ready-for-discharge patient. *Cell Res.* **2020**, *30*, 541–543. [CrossRef] [PubMed]
12. Shiozawa, Y.; Pedersen, E.A.; Havens, A.M.; Jung, Y.; Mishra, A.; Joseph, J.; Kim, J.K.; Patel, L.R.; Ying, C.; Ziegler, A.M.; et al. Human prostate cancer metastases target the hematopoietic stem cell niche to establish footholds in mouse bone marrow. *J. Clin. Investig.* **2011**, *121*, 1298–1312. [CrossRef] [PubMed]

13. Asselin-Labat, M.L. Cells tagged near an early spread of cancer. *Nature* **2019**, *572*, 589–590. [CrossRef] [PubMed]
14. Link, T.; Kuhlmann, J.D.; Kobelt, D.; Herrmann, P.; Vassileva, Y.D.; Kramer, M.; Frank, K.; Göckenjan, M.; Wimberger, P.; Stein, U. Clinical relevance of circulating MACC1 and S100A4 transcripts for ovarian cancer. *Mol. Oncol.* **2019**, *13*, 1268–1279. [CrossRef] [PubMed]
15. Maděrka, M.; Pilka, R.; Neubert, D.; Hambálek, J. New serum tumor markers S100, TFF3 and AIF-1 and their possible use in oncogynecology. *Ceska Gynekol.* **2019**, *84*, 303–308. [PubMed]

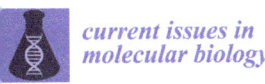

current issues in molecular biology

MDPI

Review

Exceptional Repositioning of Dog Dewormer: Fenbendazole Fever

Tania Sultana [1,†], Umair Jan [1], Hyunsu Lee [1], Hyejin Lee [1] and Jeong Ik Lee [1,2,*]

1 Regenerative Medicine Laboratory, Center for Stem Cell Research, Department of Biomedical Science and Technology, Institute of Biomedical Science and Technology, Konkuk University, Seoul 05029, Korea
2 Department of Veterinary Obstetrics and Theriogenology, College of Veterinary Medicine, Konkuk University, Seoul 05029, Korea
* Correspondence: jeongik@konkuk.ac.kr; Tel.: +82-2-2049-6234
† Current affiliation: Department of Biomedical Science, College of Medicine, Florida State University, Tallahassee, FL 32306, USA.

Abstract: Fenbendazole (FZ) is a benzimidazole carbamate drug with broad-spectrum antiparasitic activity in humans and animals. The mechanism of action of FZ is associated with microtubular polymerization inhibition and glucose uptake blockade resulting in reduced glycogen stores and decreased ATP formation in the adult stages of susceptible parasites. A completely cured case of lung cancer became known globally and greatly influenced the cancer community in South Korea. Desperate Korean patients with cancer began self-administering FZ without their physician's knowledge, which interfered with the outcome of the cancer treatment planned by their oncologists. On the basis of presented evidence, this review provides valuable information from PubMed, Naver, Google Scholar, and Social Network Services (SNS) on the effects of FZ in a broad range of preclinical studies on cancer. In addition, we suggest investigating the self-administration of products, including supplements, herbs, or bioactive compounds, by patients to circumvent waiting for long and costly FZ clinical trials.

Keywords: fenbendazole; microtubule polymerization; self-administration; cancer

Citation: Sultana, T.; Jan, U.; Lee, H.; Lee, H.; Lee, J.I. Exceptional Repositioning of Dog Dewormer: Fenbendazole Fever. *Curr. Issues Mol. Biol.* **2022**, *44*, 4977–4986. https://doi.org/10.3390/cimb44100338

Academic Editor: Dumitru A. Iacobas

Received: 24 September 2022
Accepted: 11 October 2022
Published: 17 October 2022

1. Introduction

Over the past few decades, a considerable amount of research has been conducted on novel oncological therapies; however, cancer remains a major global cause of morbidity and mortality [1]. Developing novel anticancer drugs requires considerable funding for large-scale investigations, experimentation, testing, corroboration, and the subsequent evaluation of efficacy, pharmacokinetics, and toxicity [2]. After this arduous process, only 5% of oncology drugs enter phase I clinical trials [3]. Chemotherapy is considered to be one of the most methodical and vigorous strategies to treat malignant tumors. However, the development of multidrug resistance in patients with cancer receiving traditional chemotherapeutics causes 90% of deaths [4]. Under these circumstances, new therapeutic alternatives are in demand. However, the conventional method of developing new anticancer drugs is onerous, stringent, and costly. The estimated time to discover a single drug candidate is 11.4–13.5 years, and it costs approximately USD 1–2.5 billion to take it through all the obligatory trials required by the U.S. Food and Drug Administration [5].

Drug repurposing or reprofiling has gained recognition and has enabled existing pharmaceutical products to be reconsidered for promising alternative applications, as their pharmacodynamics, pharmacokinetics, and toxicity profile are already well-known in animals and humans [6]. With repurposing, new drugs could be ready for clinical trials faster, reducing development time; this is also economically appealing by expediting integration into medical practice compared with other drug development strategies [7,8]. Examples of high-potential drugs recognized within the Repurposing Drugs in Oncology (ReDO) project include clarithromycin, cimetidine, diclofenac, mebendazole (MBZ), and nitroglycerine [8]. In addition, several antiparasitic drugs that have been in clinical use for decades have been investigated for repurposing in oncology [9]. The repositioning of anthelminthic pleiotropic benzimidazole carbamate (BZ) group drugs such as MBZ, albendazole (ABZ), and flubendazole has recently opened new avenues in cancer treatment owing to their easy access, low cost as generic drugs, and safety in human application [10,11]. Another potent and efficient pharmacological candidate from this group for repurposing as an anticancer drug is fenbendazole (FZ), which is widely used in veterinary medicine to treat parasitic worms including ascarids, whipworms, hookworms, and a single species of tapeworm, *Taenia pisiformis*, in humans and animals [12]. Although there are considerable research and successful cases regarding the anticancer activity and mechanism of action of FZ, there is ongoing social controversy concerning its application in cancer treatment [13,14]. In this review, we summarize the current evidence on the anticancer activity of FZ and the drawbacks of patient self-administration in the treatment regime.

2. Fenbendazole

Methyl N-(6-phenylsulfanyl-1H-benzimidazole-2-yl (FZ) is a safe broad-spectrum antiparasitic drug [12] with proven applications in treating different types of parasitic infections caused by helminths in livestock [15], companion animals [16,17], and laboratory animals [18]. It has a high safety margin with a low degree of toxicity in experimental animals [19] and is well-tolerated by most species, even at sixfold the prescribed dose and threefold the recommended duration. The FZ dosage for dogs is 50 mg/kg/day for 3 days, and it is also safely administered at specific doses in other livestock [20]. Orally administered FZ is poorly absorbed in the bloodstream; thus, it is necessary to retain FZ as long as possible in the rumen, where it is progressively absorbed for higher efficacy. Rarely found side effects are diarrhea and vomiting. Metabolism to its sulfoxide derivative is mainly in the liver, and excretion occurs mainly through feces with a very small amount excreted via urine.

3. Mechanism of Action

FZ primarily inhibits tubulin polymerization and promotes microtubular (MT) disruption in parasite cells (Figure 1) [21]. Tubulin, a structural protein of MTs, is the leading molecular target of benzimidazoles [22] and has prominent functions in cell proliferation, motility, division, the intercellular transport of organelles, the maintenance of the cell shape, and the secretion process of cells in all living organisms [23]. By binding with beta-tubulin, FZ blocks MT polymerization in worms and thus perturbs glucose uptake, eventually emptying glycogen reserves and adversely affecting energy management mechanisms. As a result, the whole process eventually contributes to the death of the parasites [20]. Additionally, the poor absorption of FZ from the intestine manifests as reduced levels of drugs and their active ingredients in tissues compared with those within the gut, where the targeted parasites are present [24].

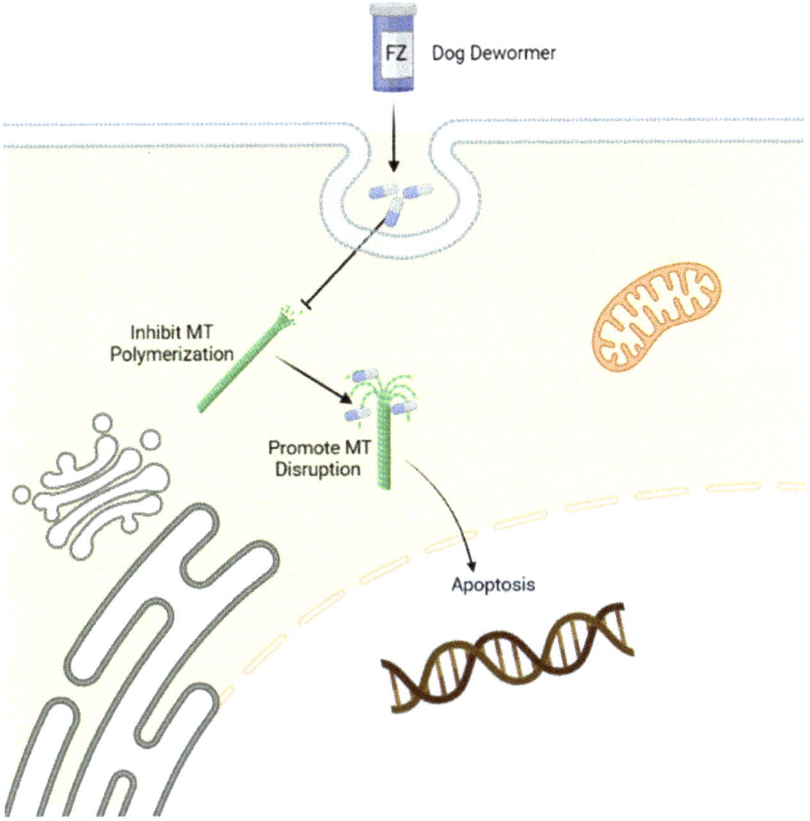

Figure 1. Mechanism of action of fenbendazole (FZ) targeting tubulin. Tubulin is the leading molecular target of FZ, which selectively binds to the β-tubulin of microtubules to disrupt microtubular polymerization, promoting immobilization and the death of parasites. The figure was created using Biorender (https://biorender.com/) (accessed on 15 May 2022).

4. Anticancer Activity of FZ

MTs are one of the major components of the cytoskeleton, and their role in cell division, the maintenance of cell shape and structure, motility, and intracellular trafficking renders them one of the most important targets for anticancer therapy. Several broadly used anticancer drugs induce their antineoplastic effects by perturbing MT dynamics. A class of anticancer drugs acts by inhibiting MT polymerization (vincristine, vinblastine), while another class stabilizes MTs (paclitaxel, docetaxel), suggesting that FZ could have potential anticancer effects [25,26]. The consequence of disrupting tubulin and dynamic MT stability with these classes of anticancer drugs in dividing cells is apoptosis and metaphase arrest. FZ exhibits moderate MT depolymerizing activity in human cancer cell lines, and has a potent anticancer effect in vitro and in vivo [27]. The antitumor effects of FZ are summarized in Table 1.

Table 1. Anticancer activity of fenbendazole.

Cell Source	Cell Lines	Species	Procedure of Study	Cancer Type	Target Pathway	Reference
Human	A549, H460, and H1299	-	In vitro	Lung cancer	Tubulin polymerization	[13]
Human	H460 and A549	Mice	In vivo and in vitro	NSCLC	Endoplasmic reticulum stress, ROS production, decreased mitochondrial membrane potential, and cytochrome c release	[27]
Human	P493-6 B	SCID mice	In vivo and in vitro	Lymphoma	Tubulin disruption	[28]
Mice	EMT6	BALB/c Rw mice	In vivo	Lung cancer	Tubulin disruption	[29]
Human	BMSC, HFF, and HL60	Mice	In vivo and in vitro	Leukemia	Granulocyte differentiation and PI3K/AKT, JAK/STAT, and MAPK pathways	[30]
-	-	Human	In vivo	NSCLC	-	[31]
Rat	H4IIE	Mice	In vitro	Hepatocellular Carcinoma	MAPKs, glucose generation, and reactive oxygen species (ROS)	[32]

BMSC, bone marrow-derived mesenchymal stem cells; HFF, human foreskin fibroblast cells; NSCLC, nonsmall-cell lung cancer.

5. Anticancer Activity of FZ in Preclinical Models

The anticancer activity of FZ has been investigated in different cell lines. FZ exhibits depolymerizing MT activity toward human cancer cell lines that manifests as a significant anticancer effect in vitro and in vivo. The mechanism of action of the FZ antitumor effect is predominantly the disruption of MT dynamics, p53 activation, and the regulation of genes associated with multiple biological pathways. FZ treatment also causes the depletion of glucose uptake in cancer cells by downregulating key glycolytic enzymes and GLUT transporters [24,27,33]. FZ selectively inhibits the growth of H4IIE cells by upregulating p21, and downregulating cyclins B and D at G1/S and G2/M phases, resulting in apoptosis exclusively in actively growing cells with low confluency, but not in quiescent cells. MAPKs, glucose generation, and ROS are unlikely targets of FZ in H4IIE rat hepatocellular carcinoma cells [32]. Treating human cancer cell lines with FZ induces apoptosis, whereas normal cells remain unaffected. Many apoptosis regulatory proteins such as cyclins, p53, and IκBα that are normally degraded by the ubiquitin–proteasome pathway accumulate in FZ-treated cells. Moreover, FZ produced distinct endoplasmic reticulum (ER) stress-associated genes, such as *ATF3*, *GADD153*, *GRP78*, *IRE1α*, and *NOXA*, in experimental cells. Thus, FZ treatment in human cancer cells induced decreased mitochondrial membrane potential, ROS production, ER stress, and cytochrome *c* release, eventually leading to cancer cell death [34]. FZ exhibits considerable affinity for mammalian tubulin in MT and is toxic in human cancer cells (H460, A549) at micromolar concentrations. Additionally, FZ exposure causes the mitochondrial translocation of p53, and effectively inhibits the expression of GLUT transporters, glucose uptake, and levels of hexokinase, which is a key glycolytic enzyme potentially linked to p53 activation and the alteration of MT dynamics. Orally administered FZ successfully blocked the growth of human xenografts in a *nu/nu* mice model [27]. Moreover, Qiwen et al. reported that FZ treatment is toxic to EMT6 mouse mammary tumor cells in vitro, with toxicity increasing after 24 h incubation with high FZ doses. However, FZ did not alter the dose–response curves for radiation on EMT6 cells under either aerobic or hypoxic conditions [24]. In contrast, Ping et al. reported that FZ or vitamins alone had no growth inhibitory effect on P493-6 human lymphoma cell lines in SCID mice. In combination with vitamin supplements, FZ significantly inhibited tumor growth through its antimicrotubular activity [28]. The effect of a therapeutic diet containing 150 ppm FZ for 6 weeks on the growth of EMT6 mouse mammary tumors in BALB/c mice injected intradermally was examined. The results revealed that the FZ diet did not alter tumor growth, metastasis, or invasion. Therefore, the authors suggested being cautious in applying FZ diets to mouse colonies used in cancer research. [29]. HL-60 cells, a human leukemia cell line, were treated with FZ to investigate the anticancer potential in the absence or presence of N-acetyl cysteine (NAC), an inhibitor of ROS production. NAC could significantly recover the decreased metabolic activity of HL-60 cells induced by 0.5–1 μM FZ treatments. The results proved that FZ manifests anticancer activity in HL-60 cells via ROS production [35]. Ji-Yun also reported the antitumor effect of FZ and paclitaxel via ROS on HL-60 cells at a certain concentration [36]. Moreover, FZ and its synthetic analog induced oxidative stress by accumulating ROS. In addition, FZ activated the p38-MAPK signaling pathway to inhibit the proliferation and increase the apoptosis of HeLa cells. FZ also impaired glucose metabolism, and prevented HeLa cell migration and invasion [37]. Koh et al. found that FZ, MBZ, and oxibendazole possess remarkable anticancer activities at the cellular level via tubulin depolymerization, but their poor pharmacokinetic parameters limit their use as systemic anticancer drugs. However, the same authors assured patients on the application of oxibendazole or FZ as a cancer treatment option [13]. Thus, FZ was investigated using a novel transcriptional drug repositioning approach based on both bioinformatic and cheminformatic components for the identification of new active compounds and modes of action. FZ induced the differentiation of leukemia cells, HL60, to granulocytes, at a low concentration of 0.1 μM within 3 days, causing apoptosis in cancer cells [30].

6. Complete Cure for Cancer by Self-Administering FZ with Supplements

In August 2016, a businessman from Oklahoma, Joe Tippens, was diagnosed with small-cell lung cancer and underwent a clinical trial under the supervision of his oncologist. He was informed of a short life expectancy from 3 months to 1 year. However, a veterinarian recommended to try FZ with vitamin E supplements, cannabidiol (CBD) oil, and bioavailable curcumin while going through the clinical trial. A positron emission tomography (PET) scan after 3 months did not detect any cancer cells anywhere in his body. Tippens was the only patient cured of cancer among the 1100 patients included in that clinical trial [38]. Tippens shared his success story through Social Network Services (SNS) via a closed group, 'my cancer story rocks' (33,900 members) [39], and also in his blog, Get Busy Living (Figure 2), and mentioned at least 60 known FZ success stories [40]. The blog has been read by thousands from 60 different countries, as mentioned in his blog (Figure 3). A protocol is also available on the website recommending FZ 222 mg (1 gm of Panacur™ or Safeguard™) daily for different types of cancer such as colorectal, colon, lung, pancreatic, and prostate cancers, melanoma, lymphoma, and glioblastoma.

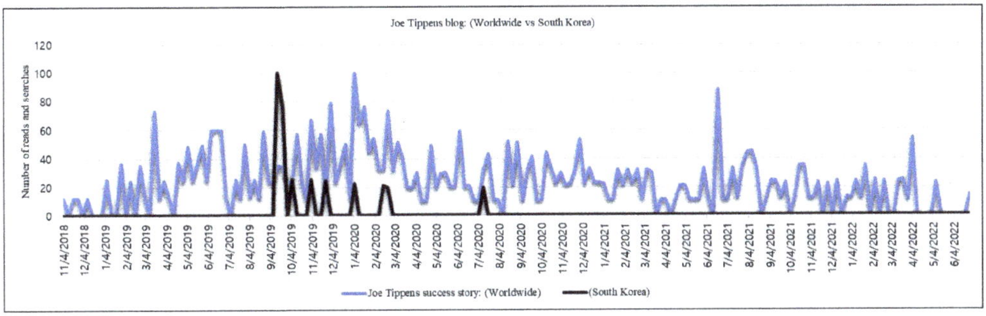

Figure 2. Comparative reading of Joe Tippens' blog. The number indicates the average number of reads and searches for the use of fenbendazole as an anticancer drug after the release of Tippens' story. A volume of 100 indicates the highest number, while 0 indicates the lowest. The data were collected using Google Trends and Naver Data Lab (South Korean search engine).

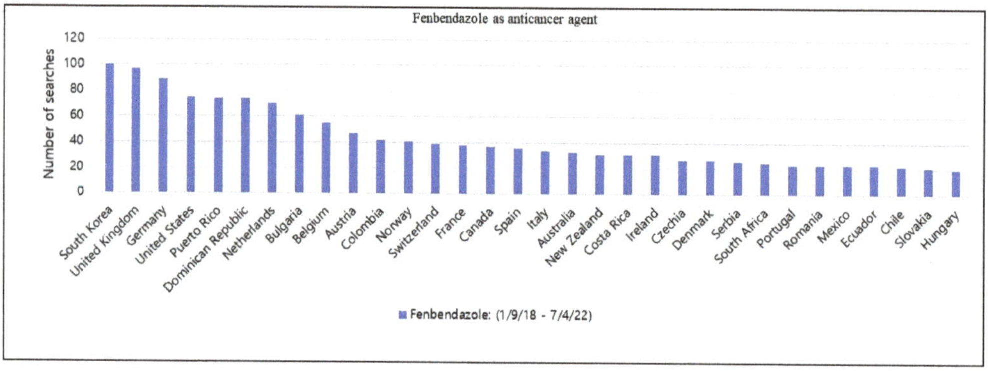

Figure 3. Dissemination of Joe Tippens' cure story. Thousands of people globally read, listened to, and followed his protocol after his success. The data represent the number of searches conducted by different countries. A value of 100 indicates the highest popularity with the highest number of searches, while 0 indicates the lowest popularity with the lowest number of searches. The data were collected using Google Trends and Naver Data Lab (South Korea search engine).

7. Liver Injury after Self-Administration of FZ

An 80-year-old patient with advanced nonsmall-cell lung cancer (NSCLC) was treated with pembrolizumab monotherapy. Nine months after treatment initiation, the patient began to present symptoms of severe liver injury. The physical findings and vital signs of the patient were unremarkable. A diligent interview with her and her family disclosed that she had been taking oral FZ for a month solely on the basis of SNS suggesting its effectiveness against cancer. After discontinuing self-administered FZ, the patient's liver injury gradually improved. Pembrolizumab monotherapy was suspended due to an enlarged tumor in the right upper lobe of her lung [31].

8. Dissemination of FZ in South Korea

Cancer has been the leading cause of death in Korea since 1983 [41]. The news of Joe Tippens disseminated rapidly among online South Korean cancer patient communities and on social media since September 2019, when a South Korean YouTube channel introduced Joe Tippens' story on their channel [42]. The video amassed more than 2.4 million views within 3 months. The Korean Medical Association, Korean Pharmaceutical Association, Korean Veterinary Association, and Ministry of Food and Drug Safety have warned patients not to take FZ, as no clinical trials have been conducted on humans. However, disregarding these warnings, dozens of patients with terminal cancer self-administered FZ, regularly uploading videos and reporting on the positive changes observed in their bodies. Vet pharmacies all over South Korea have reported shortages of FZ and frequent inquiries about the availability of this antiparasitic agent by desperate people hoping to cure cancer for themselves or their family members. A South Korean comedian and singer, Kim Chul-Min, who was suffering from Stage 4 lung cancer, joined the FZ bandwagon, reporting that his body pain was alleviated, and his blood test results improved after he had administered FZ in early October 2019 [43] (Figure 4). However, Kim stopped taking FZ after 8 months, mentioning that the drug was ineffective; indeed, FZ use led to serious side effects, recently causing his death [44]. Moreover, an internal medicine specialist in South Korea motivated many cancer patients to urge the government to conduct a clinical trial to determine the oncological efficacy of FZ in humans [45], and also expressed reluctance to use expensive cancer treatments in favor of FZ [46]. The controversy grew further when a Korean oncologist, Kim Ja-young, uploaded a YouTube video favoring FZ titled, 'Is dog dewormer safe for humans', depicting the appropriate dosage of the drug; this video reached 60,000–180,000 views and received more than 500 thankful comments within a very short period [47]. However, the medical community has labeled this video as a source of false and exaggerated information.

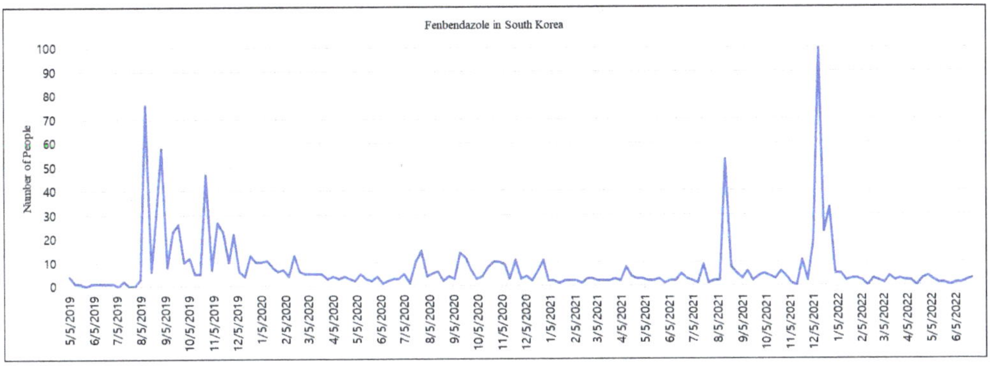

Figure 4. Dissemination of self-administration of fenbendazole. People started following Kim Chul-Min's story because of celebrity endorsement. The data represent the number of cancer patients or their relatives following Min. A volume of 100 indicates the highest number of searches. The data were collected using Naver Data Lab (a South Korean search engine).

Another SNS group directly connected to Joe Tippens had pet owners who self-administer FZ using autoprescriptions to treat canine/feline cancer without informing veterinarians [48]. Patients are selling these drugs to each other, and specialized drug sellers are dealing directly with patients, which is strictly prohibited by Korean healthcare laws and regulations. The members are showing eagerness to use FZ or other benzimidazole antiparasitic drugs, including ABZ, MBZ, and FLU, and drugs from different groups such as niclosamide [49], pyrvinium [50], and ivermectin [51] (which is effective against the COVID-19 virus [52]). An increase in the popularity of FZ in South Korea has been observed in online news, YouTube videos, social media, etc. We also discovered that a large number of Google search results regarding the use of FZ against cancer were from South Korea (70.3%; 104,000/148,000).

9. Discussion

To accelerate anticancer drug development, agents in clinical use for different indications are screened for repurposing. The approval of such repurposed drugs may be expedited owing to the availability of preclinical and clinical data on pharmacokinetics, toxicities, and regimens [53]. The anticancer activities of several anthelmintics, owing to their microtubule disruption ability, have generated considerable interest. The synergism of BZs, particularly FZ, with many clinically approved anticancer drugs is advantageous in repurposing these drugs. In the veterinary field, pets are often prescribed human medicines, since those available for companion animals are insufficient to cover all animal diseases. Likewise, commercially available animal drugs might be utilized in human medicine as long as human clinical trials have been conducted. However, globally, patients with cancer use SNS to repurpose veterinary medicines as anticancer drugs, following dosage regimens provided by self-cured patients. Sources of medical information on SNS are often unproven, and it is challenging for nonmedical professionals to precisely select and filter complex medical information. Considerable effort is required to set the effective dosages of FZ for humans. Moreover, if a patient self-administers a veterinary dewormer with an established anticancer drug while participating in a clinical trial, the outcome of the clinical trial could be altered completely, resulting in huge economic and time losses. Therefore, the reason for the therapeutic or adverse effects of the trialed drug in the clinical trial remains equivocal, and oncologists are confounded by their lack of knowledge regarding the self-administration of the dewormer (FZ) by patients.

FZ began to gain popularity as a human anticancer drug in South Korea in September 2019. The serious consequences of such events will likely emerge soon in South Korea. As the number of success stories being published online increases, the uninformed self-administration of FZ will also continue to increase in South Korea. Any prohibition by the government or medical doctors will likely not be followed by desperate patients with cancer. The resulting situation would lead to possible global FZ self-administration by patients with cancer. Therefore, an alternative must be offered to relieve patients from the long and expensive wait for FZ clinical trials in humans.

Indeed, patients with cancer and their families desperately seek remedies to treat the disease, but administering drugs with limited or no human safety profiles is a concern for clinicians. However, there are many exceptions for investigational drugs in clinical trials. There is a plethora of evidence that several BZ carbamates, particularly FZ, show anticancer potential in vivo, in vitro, and in silico. The role of MT is well-characterized, and the mechanism of action of MT disruptors such as FZ is similar to that of major chemotherapeutic agents such as vinblastine and vincristine. Nevertheless, nonmedical professionals cannot make appropriate medical judgments regarding the best usage of FZ. Therefore, we argue in favor of FZ owing to its well-established pharmacokinetics, excellent toxicity profile, and low cost, and suggest that an initial assessment of patients by interdisciplinary researchers such as veterinarians, oncologists, and pharmacologists be conducted to determine the best initial dosage and facilitate clinical trials on FZ in the near future.

10. Conclusions

Comprehensive verification through evidence-based medicine is crucial in reducing needless confusion in healthcare. Despite the potential anticancer capabilities of FZ, its pharmacokinetics, safety, and tolerance profiles in humans must be confirmed in extensive clinical trials before it can be used in any therapeutic context. Experts must further attempt to provide patients with reliable medical information.

11. Materials and Methods

Process of Article Selection

We used search engines PubMed, Naver, and Google Scholar to obtain publications and articles investigating the anticancer activity of FZ in cell lines, animal tumor models, and clinical trials. For a report to be included in this review, it had to have contained FZ or BZ carbamate in either the heading or abstract. Keywords 'BZ carbamate' and 'FZ' combined with 'MT polymerization,' 'cancer,' and 'antiparasitic' were used to generate the list. Review articles were not included in our survey. All the relevant articles were initially recognized from the heading and abstract, followed by an additional examination to confirm whether the research conducted using FZ drugs pertained to cell lines or human or animal subjects. We ensured that the conducted search technique was not encyclopedic, as many publications are not included in PubMed, Naver, or Google Scholar. We evaluated the selected studies by assessing different characteristics such as cell lines and source, animal model, cancer type, and target pathway.

Author Contributions: Conceptualization, writing—original draft preparation, T.S. and J.I.L.; formatting and arrangement of the manuscript, T.S., U.J., H.L. (Hyunsu Lee) and H.L. (Hyejin Lee); funding acquisition, J.I.L. All authors have read and agreed to the published version of the manuscript.

Funding: This research received no external funding.

Institutional Review Board Statement: Not applicable.

Informed Consent Statement: Not applicable.

Data Availability Statement: Not applicable.

Conflicts of Interest: The authors declare that they have no conflict of interest.

References

1. Tahlan, S.; Kumar, S.; Kakkar, S.; Narasimhan, B. Benzimidazole scaffolds as promising antiproliferative agents: A review. *BMC Chem.* **2019**, *13*, 66. [CrossRef]
2. Armando, R.G.; Mengual Gomez, D.L.; Gomez, D.E. New drugs are not enoughdrug repositioning in oncology: An update. *Int. J. Oncol.* **2020**, *56*, 651–684. [CrossRef] [PubMed]
3. Kato, S.; Moulder, S.L.; Ueno, N.T.; Wheler, J.J.; Meric-Bernstam, F.; Kurzrock, R.; Janku, F. Challenges and perspective of drug repurposing strategies in early phase clinical trials. *Oncoscience* **2015**, *2*, 576–580. [CrossRef]
4. Bukowski, K.; Kciuk, M.; Kontek, R. Mechanisms of Multidrug Resistance in Cancer Chemotherapy. *Int. J. Mol. Sci.* **2020**, *21*, 3233. [CrossRef] [PubMed]
5. Scannell, J.W.; Blanckley, A.; Boldon, H.; Warrington, B. Diagnosing the decline in pharmaceutical R&D efficiency. *Nat. Rev. Drug Discov.* **2012**, *11*, 191–200. [CrossRef] [PubMed]
6. Bertolini, F.; Sukhatme, V.P.; Bouche, G. Drug repurposing in oncology—Patient and health systems opportunities. *Nat. Rev. Clin. Oncol.* **2015**, *12*, 732–742. [CrossRef]
7. Castro, L.S.; Kviecinski, M.R.; Ourique, F.; Parisotto, E.B.; Grinevicius, V.M.; Correia, J.F.; Wilhelm Filho, D.; Pedrosa, R.C. Albendazole as a promising molecule for tumor control. *Redox Biol.* **2016**, *10*, 90–99. [CrossRef]
8. Pantziarka, P.; Bouche, G.; Meheus, L.; Sukhatme, V.; Sukhatme, V.P.; Vikas, P. The Repurposing Drugs in Oncology (ReDO) Project. *Ecancermedicalscience* **2014**, *8*, 442. [CrossRef]
9. Hu, Y.; Ellis, B.L.; Yiu, Y.Y.; Miller, M.M.; Urban, J.F.; Shi, L.Z.; Aroian, R.V. An extensive comparison of the effect of anthelmintic classes on diverse nematodes. *PLoS ONE* **2013**, *8*, e70702. [CrossRef]
10. Nath, J.; Paul, R.; Ghosh, S.K.; Paul, J.; Singha, B.; Debnath, N. Drug repurposing and relabeling for cancer therapy: Emerging benzimidazole antihelminthics with potent anticancer effects. *Life Sci.* **2020**, *258*, 118189. [CrossRef]
11. Sultana, T.; Jan, U.; Lee, J.I. Double Repositioning: Veterinary Antiparasitic to Human Anticancer. *Int. J. Mol. Sci.* **2022**, *23*, 4315. [CrossRef] [PubMed]
12. Baeder, C.; Bahr, H.; Christ, O.; Duwel, D.; Kellner, H.M.; Kirsch, R.; Loewe, H.; Schultes, E.; Schutz, E.; Westen, H. Fenbendazole: A new, highly effective anthelmintic. *Experientia* **1974**, *30*, 753–754. [CrossRef]
13. Jiyoon, J.; Kwangho, L.; Byumseok, K. Investigation of benzimidazole anthelmintics as oral anticancer agents. *Bull. Korean Chem. Soc.* **2022**, *43*, 750–756. [CrossRef]
14. Heo, D.S. Anthelmintics as Potential Anti-Cancer Drugs? *J. Korean Med. Sci.* **2020**, *35*, e75. [CrossRef] [PubMed]
15. Crotch-Harvey, L.; Thomas, L.A.; Worgan, H.J.; Douglas, J.L.; Gilby, D.E.; McEwan, N.R. The effect of administration of fenbendazole on the microbial hindgut population of the horse. *J. Equine Sci.* **2018**, *29*, 47–51. [CrossRef] [PubMed]
16. Hayes, R.H.; Oehme, F.W.; Leipold, H. Toxicity investigation of fenbendazole, an anthelmintic of swine. *Am. J. Vet. Res.* **1983**, *44*, 1108–1111. [PubMed]
17. Schwartz, R.D.; Donoghue, A.R.; Baggs, R.B.; Clark, T.; Partington, C. Evaluation of the safety of fenbendazole in cats. *Am. J. Vet. Res.* **2000**, *61*, 330–332. [CrossRef] [PubMed]
18. Villar, D.; Cray, C.; Zaias, J.; Altman, N.H. Biologic effects of fenbendazole in rats and mice: A review. *J. Am. Assoc. Lab. Anim. Sci.* **2007**, *46*, 8–15.
19. Muser, R.K.; Paul, J.W. Safety of fenbendazole use in cattle. *Mod. Vet. Pract.* **1984**, *65*, 371–374.
20. Junquera, P. FENBENDAZOLE, Anthelmintic for Veterinary Use in CATTLE, SHEEP, GOATS, PIG, POULTRY, HORSES, DOGS and CATS against Roundworms and Tapeworms. 2015. Available online: https://parasitipedia.net/ (accessed on 1 February 2022).
21. Gull, K.; Dawson, P.J.; Davis, C.; Byard, E.H. Microtubules as target organelles for benzimidazole anthelmintic chemotherapy. *Biochem. Soc. Trans.* **1987**, *15*, 59–60. [CrossRef] [PubMed]
22. Lacey, E.; Watson, T.R. Structure-activity relationships of benzimidazole carbamates as inhibitors of mammalian tubulin, in vitro. *Biochem. Pharmacol.* **1985**, *34*, 1073–1077. [CrossRef]
23. Jordan, M.A.; Wilson, L. Microtubules as a target for anticancer drugs. *Nat. Rev. Cancer* **2004**, *4*, 253–265. [CrossRef] [PubMed]
24. Duan, Q.; Liu, Y.; Rockwell, S. Fenbendazole as a potential anticancer drug. *Anticancer Res.* **2013**, *33*, 355–362. [PubMed]
25. Spagnuolo, P.A.; Hu, J.; Hurren, R.; Wang, X.; Gronda, M.; Sukhai, M.A.; Di Meo, A.; Boss, J.; Ashali, I.; Beheshti Zavareh, R.; et al. The antihelmintic flubendazole inhibits microtubule function through a mechanism distinct from Vinca alkaloids and displays preclinical activity in leukemia and myeloma. *Blood* **2010**, *115*, 4824–4833. [CrossRef]
26. Jordan, M.A. Mechanism of action of antitumor drugs that interact with microtubules and tubulin. *Curr. Med. Chem. Anticancer Agents* **2002**, *2*, 1–17. [CrossRef]
27. Dogra, N.; Kumar, A.; Mukhopadhyay, T. Fenbendazole acts as a moderate microtubule destabilizing agent and causes cancer cell death by modulating multiple cellular pathways. *Sci. Rep.* **2018**, *8*, 11926. [CrossRef]
28. Gao, P.; Dang, C.V.; Watson, J. Unexpected antitumorigenic effect of fenbendazole when combined with supplementary vitamins. *J. Am. Assoc. Lab. Anim. Sci.* **2008**, *47*, 37–40.
29. Duan, Q.; Liu, Y.; Booth, C.J.; Rockwell, S. Use of fenbendazole-containing therapeutic diets for mice in experimental cancer therapy studies. *J. Am. Assoc. Lab. Anim. Sci.* **2012**, *51*, 224–230.
30. KalantarMotamedi, Y.; Ejeian, F.; Sabouhi, F.; Bahmani, L.; Nejati, A.S.; Bhagwat, A.M.; Ahadi, A.M.; Tafreshi, A.P.; Nasr-Esfahani, M.H.; Bender, A. Transcriptional drug repositioning and cheminformatics approach for differentiation therapy of leukaemia cells. *Sci. Rep.* **2021**, *11*, 12537. [CrossRef]

31. Yamaguchi, T.; Shimizu, J.; Oya, Y.; Horio, Y.; Hida, T. Drug-Induced Liver Injury in a Patient with Nonsmall Cell Lung Cancer after the Self-Administration of Fenbendazole Based on Social Media Information. *Case Rep. Oncol.* **2021**, *14*, 886–891. [CrossRef]
32. Park, D. Fenbendazole Suppresses Growth and Induces Apoptosis of Actively Growing H4IIE Hepatocellular Carcinoma Cells via p21-Mediated Cell-Cycle Arrest. *Biol. Pharm Bull.* **2022**, *45*, 184–193. [CrossRef] [PubMed]
33. Sharma, Y. Veterinary drug may be repurposed for human cancers: Study. *The Hindu Business Line*, 27 August 2018.
34. Dogra, N.; Mukhopadhyay, T. Impairment of the ubiquitin-proteasome pathway by methyl N-(6-phenylsulfanyl-1H-benzimidazol-2-yl)carbamate leads to a potent cytotoxic effect in tumor cells: A novel antiproliferative agent with a potential therapeutic implication. *J. Biol. Chem.* **2012**, *287*, 30625–30640. [CrossRef]
35. Yong, H.; Joo, H.-G. Involvement of reactive oxygen species in the anti-cancer activity of fenbendazole, a benzimidazole anthelmintic. *Korean J. Vet. Res.* **2020**, *60*, 79–83.
36. Sung, J.Y.; Joo, H.-G. Anti-cancer effects of Fenbendazole and Paclitaxel combination on HL-60 cells. *Prev. Vet. Med.* **2021**, *45*, 13–17. [CrossRef]
37. Peng, Y.; Pan, J.; Ou, F.; Wang, W.; Hu, H.; Chen, L.; Zeng, S.; Zeng, K.; Yu, L. Fenbendazole and its synthetic analog interfere with HeLa cells' proliferation and energy metabolism via inducing oxidative stress and modulating MEK3/6-p38-MAPK pathway. *Chem. Biol. Interact* **2022**, *361*, 109983. [CrossRef]
38. Onstot, E. Edmond man says cheap drug for dogs cured his cancer. *KOCO News 5 abc*, 26 April 2019.
39. Tippens, J. My Cancer Story Rocks. Available online: www.facebook.com (accessed on 5 December 2021).
40. Tippens, J. Get Busy Living [Internet]. Ohio. 2017. Available online: https://www.mycancerstory.rocks (accessed on 5 December 2021).
41. Lim, D.; Ha, M.; Song, I. Trends in the leading causes of death in Korea, 1983–2012. *J. Korean Med. Sci.* **2014**, *29*, 1597–1603. [CrossRef]
42. Eun-young, K. Dog dewormer goes out of stock amid rumor of efficacy for cancer. *Korea Biomedical Review*, 26 December 2019.
43. Han-soo, L. Comedian keeps dispute alive over dog vermicide's efficacy as cancer treatment. *Korea Biomedical Review*, 29 October 2019.
44. Han-sol, P. Comedian Kim Chul-min dies at 54. *The KoreaTimes*, 17 December 2021.
45. Eun-young, K. Fenbendazole has ideal structure as anticancer drug. *Korea Biomedical Review*, 11 November 2019.
46. Chan-kyong, P. Dog medicine goes viral in South Korea over claims it cures human cancer. *South China Morning Post*, 16 November 2019.
47. Hye-seon, L. Doctor fuels controversy with YouTube advice on how to take dog dewormer. *Korea Biomedical Review*, 18 November 2019.
48. Tippens, J. Mycancerstory.rocksforpets 27 August 2019. Available online: https://www.facebook.com (accessed on 7 July 2022).
49. Pan, J.X.; Ding, K.; Wang, C.Y. Niclosamide, an old antihelminthic agent, demonstrates antitumor activity by blocking multiple signaling pathways of cancer stem cells. *Chin. J. Cancer* **2012**, *31*, 178–184. [CrossRef]
50. Momtazi-Borojeni, A.A.; Abdollahi, E.; Ghasemi, F.; Caraglia, M.; Sahebkar, A. The novel role of pyrvinium in cancer therapy. *J. Cell Physiol.* **2018**, *233*, 2871–2881. [CrossRef]
51. Juarez, M.; Schcolnik-Cabrera, A.; Duenas-Gonzalez, A. The multitargeted drug ivermectin: From an antiparasitic agent to a repositioned cancer drug. *Am. J. Cancer Res.* **2018**, *8*, 317–331.
52. Mahmud, R.; Rahman, M.M.; Alam, I.; Ahmed, K.G.U.; Kabir, A.; Sayeed, S.; Rassel, M.A.; Monayem, F.B.; Islam, M.S.; Islam, M.M.; et al. Ivermectin in combination with doxycycline for treating COVID-19 symptoms: A randomized trial. *J. Int. Med. Res.* **2021**, *49*, 3000605211013550. [CrossRef]
53. Lv, J.; Shim, J.S. Existing drugs and their application in drug discovery targeting cancer stem cells. *Arch. Pharm. Res.* **2015**, *38*, 1617–1626. [CrossRef]

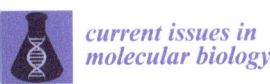

current issues in molecular biology

MDPI

Article

Deuterium Content of the Organic Compounds in Food Has an Impact on Tumor Growth in Mice

Gábor Somlyai [1,*], Lajos I. Nagy [2], László G. Puskás [2], András Papp [3], Beáta Z. Kovács [1], István Fórizs [4,5], György Czuppon [4,5] and Ildikó Somlyai [1]

1 HYD LLC for Cancer Research and Drug Development, H-1118 Budapest, Hungary
2 AVIDIN Ltd., H-6726 Szeged, Hungary
3 Department of Public Health, Albert Szent-Györgyi Medical School, University of Szeged, H-6725 Szeged, Hungary
4 Institute for Geological and Geochemical Research, Research Centre for Astronomy and Earth Sciences, H-1112 Budapest, Hungary
5 Research Centre for Astronomy and Earth Sciences, MTA Centre of Excellence, H-1121 Budapest, Hungary
* Correspondence: gsomlyai@hyd.hu

Abstract: Research with deuterium-depleted water (DDW) in the last two decades proved that the deuterium/hydrogen ratio has a key role in cell cycle regulation and cellular metabolism. The present study aimed to investigate the possible effect of deuterium-depleted yolk (DDyolk) alone and in combination with DDW on cancer growth in two in vivo mouse models. To produce DDyolk, the drinking water of laying hens was replaced with DDW (25 ppm) for 6 weeks, resulting in a 60 ppm D level in dried egg yolk that was used as a deuterium-depleted food additive. In one model, 4T1, a cell line with a high metastatic capacity to the lung was inoculated in the mice's mammary pad. After three weeks of treatment with DDW and/or DDyolk, the tumor volume in the lungs was smaller in all treated groups vs. controls with natural D levels. Tumor growth and survival in mice transplanted with an MCF-7 breast cancer cell line showed that the anticancer effect of DDW was enhanced by food containing the deuterium-depleted yolk. The study confirmed the importance of the D/H ratio in consumed water and in metabolic water produced by the mitochondria while oxidizing nutrient molecules. This is in line with the concept that the initiation of cell growth requires the cells to generate a higher D/H ratio, but DDW, DDyolk, or the naturally low-D lipids in a ketogenic diet, have a significant effect on tumor growth by preventing the cells from raising the D/H ratio to the threshold.

Keywords: deuterium-depleted water (DDW); deuterium-depleted yolk (DDyolk); anticancer drug development; D/H ratio; production of metabolic water; ketogenic diet

Citation: Somlyai, G.; Nagy, L.I.; Puskás, L.G.; Papp, A.; Kovács, B.Z.; Fórizs, I.; Czuppon, G.; Somlyai, I. Deuterium Content of the Organic Compounds in Food Has an Impact on Tumor Growth in Mice. *Curr. Issues Mol. Biol.* **2023**, *45*, 66–77. https://doi.org/10.3390/cimb45010005

Academic Editor: Dumitru A. Iacobas

Received: 20 October 2022
Revised: 6 December 2022
Accepted: 19 December 2022
Published: 21 December 2022

1. Introduction

After the discovery of the heavy isotope of hydrogen, deuterium (D), in the early 1930s, its possible role in living organisms was not investigated for 60 years [1]. The presence of D was ignored; nonetheless, its concentration in the human body is about 12–14 mmol/L (equivalent to 150 ppm). Data gathered in the meantime, however, suggested that D and its ratio to hydrogen (H) has a major impact on several cell processes [2,3]. The first research results, published in 1993, confirmed the key role of D in tumor cell growth and cancer development. An increased D/H ratio in the intracellular space was found to be a key factor in initiating cell growth [4]. Later, results showed the complex nature of the effects of D in living organisms, such as a correlation between the drinking water D level and the susceptibility of humans to depression [5]; stimulation of long-term memory in rats by deuterium-depleted water (DDW) [6] or the reversal of Mn-induced life span decrease in *Caenorhabditis elegans* [7]. Regarding the role of D in cell metabolism, research results confirmed that deuterium depletion enhanced the effect of insulin on Glucose Transporter type 4 (GLUT4) translocation in a dose-dependent manner, and potentiated glucose uptake in diabetic rats, resulting in lower serum glucose, fructose amine, and HbA1c concentrations [8].

Further studies confirmed the role of D in cancer development, in prevention [9], and proved the anticancer effect of deuterium depletion [10–14].

In the experiments revealing the role of deuterium in cell growth, cell metabolism, and physiological changes, the D level was manipulated by the application of DDW.

In the meantime, research confirmed that the D content of organic molecules, such as carbohydrates or lipids, showed significant differences, suggesting that the synthesis pathways of these molecules strongly influence D content. In organic compounds of plants, the D/H ratio may also substantially differ from that of the environment. The explanation for that lies in the metabolic processes of various plants. In plants using the so-called C3 or C4 photosynthetic pathways to fix carbon from the atmosphere, D concentration in the glucose molecules will decrease (vs. environmental water) to different extents. In plants using the C3 carbon fixation pathway (e.g., wheat, rice, barley, or spinach), the glucose D concentration is 135–137 ppm, and in plants using the C4 pathway (maize, sugar cane, millet, and sorghum) it is 141–143 ppm [15,16]. In contrast, plants using Crassulacean Acid Metabolism (CAM) photosynthesis may, under certain circumstances, raise the concentration of deuterium in the photosynthesis products. This means that the deuterium concentration in the human body is substantially affected by the plants in our diet. Investigating the D content of animal lipids led to similar conclusions. Studies revealed significantly reduced D content (down to 90 ppm) in fatty acids and that this effect can be site specific [17,18].

It is obvious that, in the case of a normal diet, the water content of the food will represent the D concentration of the local environment. But that is not the only factor influencing the final D concentration of the body, since the organic molecules of the nutrients also contain both hydrogen and deuterium which appear in the metabolic water after oxidation by the mitochondria.

So, the final D concentration of the cells depends, on the one hand, on the D concentration of the fluid intake, including the water content of the food, but, on the other hand, the D concentration of nutrient molecules also has an effect via the D level of the metabolic water produced in the mitochondria in the oxidative utilization of fats, carbohydrates, and proteins.

Recently, a ketogenic diet has been used as complementary cancer therapy with high efficacy [19]. Based on the proven anticancer effect of DDW, we postulate that the beneficial impact of a ketogenic diet in cancer results from its deuterium-depleting effect, since the mitochondria, when oxidizing fats instead of carbohydrates, produce metabolic water with as low as 118 ppm D concentration, due to the above-mentioned dissimilar D content of the various nutrients [20].

The study presented here aimed to investigate the effect of the D concentration of organic nutrient compounds on tumor development. For this purpose, foods with artificially modified depleted D content were produced, and their effect was tested in two in vivo mouse model systems to evaluate the role of altered D/H ratio of organic molecules and their effect on tumor growth.

2. Materials and Methods

2.1. Production of Deuterium-Depleted Nutrients, Measurements of D Concentration

Mouse and rat studies indicated that replacing normal drinking water with heavy water increased the D content of organic molecules within a short time [21]. Based on that, the production of deuterium-depleted yolk (DDyolk) was attempted by using DDW as drinking water for laying hens, supposing that during egg formation the D content of proteins and lipids will decrease. DDW was produced by HYD LLC for Cancer Research and Drug Development, Budapest, Hungary, from ordinary tap water by fractional distillation using a Good Manufacturing Practice conform technology. The D concentration was verified by a Liquid-Water Isotope Analyser-24d (manufactured by Los Gatos Research Inc., San Jose, CA, USA) with ±1 ppm precision. Seventeen-week-old Tetra SL hens (n = 90) were kept in a lay house exposing them to an increased day length with artificial light for 14 h per day and were fed with a mixture of grains (corn, wheat, soybean). The D concentration of the drinking water was 25 ppm. The eggs were collected from day zero and the D concentration in both the water content and the dry substance of albumen and yolk fraction was determined. The D concentration of dry substances was determined by mass spectrometry (Finnigan delta plus XP (Thermo Fisher Scientific, Waltham, MA, USA), IAEA-CH-7 and NBS-22 laboratory standards, Institute for Geological and Geochemical Research, CSFK, Budapest, Hungary) with ±1 ppm precision and reported in reference to the Vienna standard mean ocean water, VSMOW, distributed by the International Atomic Energy Agency (<150 ppm) [22,23].

The D concentration reached equilibrium 6−7 weeks after the DDW consumption had started. After 7 weeks of DDW treatment of hens, the eggs were collected and used for the experiments. The albumen and yolk were separated, freeze-dried (to 4.6−5.6% residual moisture), and the yolk was used as a food component for mice.

2.2. *Preparations Made for per os Treatment of the Mice*

Two different in vivo studies using mice were performed (see Section 2.4. for details). For the first one, lasting four weeks, dried yolk (deuterium-depleted and control) was dissolved in distilled water to 0.25 g/mL concentration at 50 °C. The mice received *per os* by gavage 400 µL of the solution daily in the first two weeks of the four-week-long experiment. The mice in the groups receiving the yolk-containing food lost vitality, attributable to the avidin content of the yolk causing biotin deficiency. So, in the second two weeks, the yolk solution was treated at 80 °C for 10 min and the adverse effects disappeared.

In the second experiment, the stress possibly caused by the gavage application was avoided by mixing the dried yolk with powdered semi-synthetic food (VRF1, Akronom Kft., Budapest, Hungary) at 7%. The mix was re-tableted and sterilized for feeding the mice.

2.3. *Cell Lines*

A metastasis-specific mouse mammary carcinoma cell line 4T1 with high metastatic capacity to the lung (purchased from ATCC, Manassas, VA, USA) was grown at 37 °C under 5% of CO_2 and 100% humidity in RPMI-1640 Medium (Gibco BRL, Carlsbad, CA, USA) containing penicillin (50 IU/mL) (Gibco BRL, Carlsbad, CA, USA), streptomycin (50 mg/mL) (Gibco BRL, Carlsbad, CA, USA), and 10% fetal bovine serum (Gibco BRL, Carlsbad, CA, USA).

An MCF-7 human breast cell line (ATCC) was grown at 37 °C under 5% of CO_2 and 100% humidity in Eagle's Minimum Essential Medium (EMEM) (Gibco BRL, Carlsbad, CA, USA) containing penicillin (50 IU/mL) (Gibco BRL, Carlsbad, CA, USA), streptomycin (50 mg/mL) (Gibco BRL, Carlsbad, CA, USA), and 10% fetal bovine serum (Gibco BRL, Carlsbad, CA, USA).

Of both cell lines, the third and fourth passages were used for inoculation. Before inoculation, the cells were trypsinized, washed, and resuspended in sterile PBS. A total of 10^5 (4T1) or 5×10^6 (MCF7) cells were counted and resuspend in appropriate serum-free medium and injected into the mammary pad in 50 µL volume.

2.4. *Description of the In Vivo Studies*

In the first, 4-week study, 8-week-old female and male BALB/cJ mice (Charles River, Innovo Kft., Hungary) were used. The mice (20 ± 4 g weight) were group-housed (5/cage), fed VRF1 commercial diet ad libitum, and housed in an animal facility under a 12 h light/dark cycle at constant temperature (22 °C) and humidity. The mice were randomly divided into six groups (n = 10, 5 males and 5 females) (Table 1). Water was provided ad libitum, with 150 ppm deuterium content (natural level) for the control animals, and with 25 ppm deuterium content for the DDW-treated mice. In this study, 4T1 cells (metastasis-specific mouse mammary carcinoma model cell line; [24]) were used.

Table 1. Groups and treatments in the first mouse study using the 4T1 cell line.

Group	Type of Treatment		
	Drinking Fluid	**Food**	**Drug**
Untreated control	Normal water	VRF1	
Treated	DDW	VRF1	
Treated	DDW	VRF1 + deuterium-depleted yolk	
Treated	Normal water	VRF1 + deuterium-depleted yolk	
Untreated control	Normal water	VRF1 + normal yolk	
Positive control	Normal water	VRF1	DU283

On day 0, the cells (100,000/animal in 50 μL) were inoculated into the mice's mammary pad, and DDW treatment was started. The administration of the yolk solution started on day 1.

Since feeding by gavage with the 400 μL solution of the dried yolk was stressful for the mice, an additional control group was set up (treated with 400 μL normal yolk) to compare the data of the other two treated groups receiving DDW and/or yolk. DU283, a compound with documented anticancer effect, was administered (3 mg/kg bw in 150 μL volume, intraperitoneal, once a day) as a positive control [25]. The growth of the primary tumor was followed on the sixth, eighth, and tenth day of treatment. At the end of the experiment, the lungs were prepared for weight measurement after incubation in formalin. The presence and size of the metastatic tumors were inferred by comparing the weight of treated mice's lungs to the average lung weight of five healthy, untreated mice; based on the correlation between the lung weight and the number of metastases, reported earlier in a paper by Ying et al. [26].

In the second mice study (Table 2), eight-week-old NSG immunodeficient mice (Charles River, Innovo Kft., 2117 Isaszeg, Hungary) were used. The mice (20 ± 4 g weight) were fed a sterile VRF1 commercial diet and sterile water ad libitum and were housed in individual ventilated cages under sterile circumstances in an animal facility under a 12 h light/dark cycle at constant temperature (22 °C) and humidity. The mice were randomized and three groups (n = 10, 5 males and 5 females) were created. The control group consumed normal water ad libitum and dried VRF1-based food supplemented with 7% normal dried yolk (see Section 2.2.). One of the treated groups consumed DDW with 25 ppm D concentration and received the same food as the control group. The second treated group consumed DDW with 25 ppm D concentration and VRF1-based food supplemented with 7% deuterium-depleted dried yolk. Five million MCF-7 cells were inoculated at the start, and the tumor volume and survival were followed up for three months. Tumor volume became measurable around day 14 and was measured every 2–3 days. Each time one mouse perished, the average tumor volume in the three groups was calculated using the previous day's volume data, divided by the number of mice alive.

Table 2. Groups and treatments in the second mouse study using the MCF-7 cell line.

Groups	Drinking Fluid	Food
Untreated control	Normal water	VRF1 supplemented with normal yolk
Treated	DDW	VRF1 supplemented with normal yolk
Treated	DDW	VRF1 supplemented with deuterium-depleted yolk

2.5. Ethical Considerations

The study was performed according to the Institutional and National Animal Experimentation and Ethics Guidelines in possession of an ethical clearance (XXIX./128/2013.; Title: Investigation of effects of macromolecules).

All animals were treated moribund and were euthanized at the observation of the first sign of torment. All operative procedures and animal care conformed strictly to the Hungarian Council on Animal Care guidelines. After treatment, the animals were randomly divided into different groups and kept under standard conditions conforming to ARRIVE (Animal Research: Reporting of in Vivo Experiments) guidelines [27] and the Guide for the Care and Use of Laboratory Animals [28].

3. Results

3.1. Production of Deuterium-Depleted Food

It was supposed that the replacement of the laying hens' normal drinking water with DDW of 25 ppm D concentration would influence the D content of the molecules synthesized in the developing eggs. The time course of D levels in water and dry substance of egg white and yolk is shown in Figure 1.

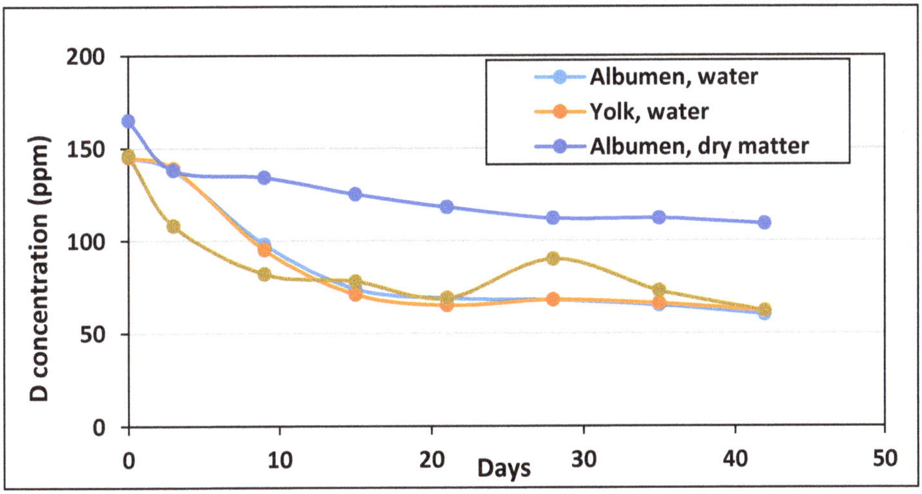

Figure 1. Changes of the D concentration in the water and the organic compounds of eggs after replacement of normal drinking water with 25 ppm DDW (the data represent the average of measurements on two independent samples.).

The D concentration of the water extracted from the yolk and egg white was equal, suggesting an equilibrium of D within the liquid phase. There was a sharp 50 ppm decrease in the first 9 days, and a further 30 ppm decrease during the next 12 days, but only a few ppm decrease in the subsequent 3 weeks. The time course of the change of D concentration in egg white and yolk was quite different. The D level of the albumen was higher already before the hens started drinking DDW, and it decreased from 160 ppm to 110 ppm by the forty-second day of the study. The D concentration of the yolk started to decrease rapidly right after the hens started consuming DDW and decreased much faster than the D concentration of the water content of the yolk, reaching equilibrium after only two weeks. By the time of collecting the eggs for the experiments, there was a 50 ppm difference between the D concentration of the egg white (110 ppm) and the yolk (60 ppm).

3.2. *Tumor Growth in Mice Inoculated with 4T1 Breast Carcinoma Cells*

The 4T1 model system was chosen as a cell line with high metastatic capacity to the lung. Therefore, the size of the primary tumor and the weight of the lung metastases were followed.

In the group consuming DDW and DDyolk, the average primary tumor size was smaller vs. control on days 6, 8, and 10 (9.9 mm^3, 30.6 mm^3, 47.0 mm^3 vs. 15.7 mm^3, 33.4 mm^3 50.3 mm^3, respectively), but the difference was not significant. Figure 2 shows the data on tumor volumes.

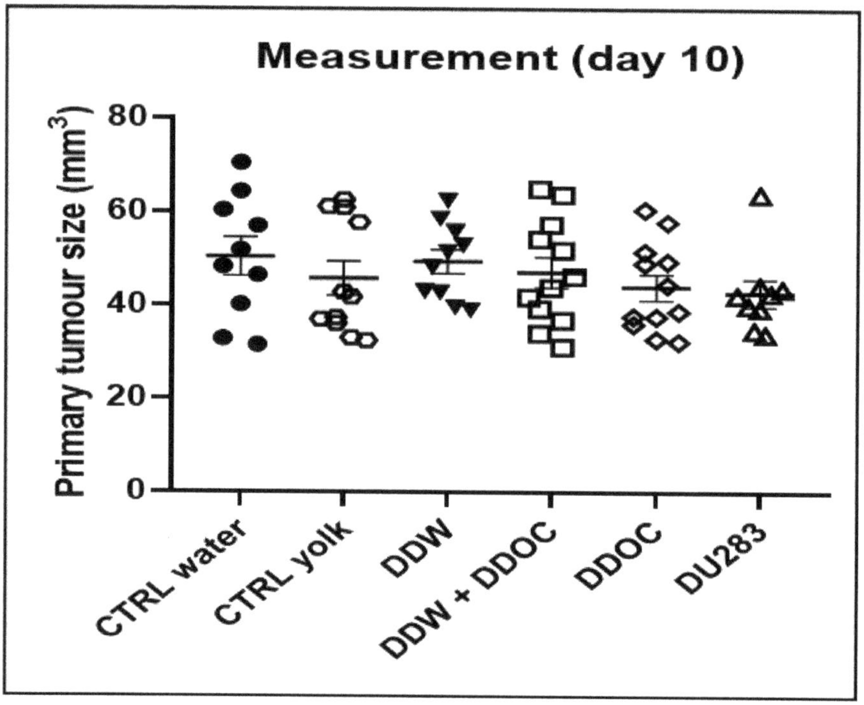

Figure 2. Tumor volume 10 days after inoculation of 4T1 cells (100,000/animal in 50 μL) into the mice's mammary pad. Dots represent individual data while the horizontal line with error bar shows mean and SD.

The average weight of the metastases in the lungs and the primary data are shown in Table 3 and Figure 3. In all four treated groups, the tumor weight was smaller compared to the two control groups, but the difference was significant in the group treated with deuterium-depleted yolk (Sidak's multiple comparisons test, $p = 0.0354$). These differences may be explained with earlier findings showing that DDW inhibits the migration of tumor cells [14].

3.3. *Tumor Size and Survival in Mice Inoculated with Human MCF-7 Breast Cancer Cell Line*

The average tumor volume data and survival data are shown in Figures 4 and 5. The tumor volume in the treated groups was somewhat smaller but the difference was not significant. According to the survival data, deuterium depletion delayed the perishment of mice as 30% of them died within 20 days in the control group but within 43 and 49 days in the DDyolk-treated and DDyolk- plus DDW-treated groups, respectively. However, the differences in median survival time were not significant (Logrank test: p = 0.0623, Gehan-Breslow-Wilcoxon test: p = 0.0689).

Table 3. Average weight of the metastases in the formalin fixed lungs in the control and treated groups.

Treatment	Weight of the Metastasis (Milligrams; mean ± SEM, n = 10))
Normal water (CTRL)	45.20 ± 4.02
DDW	32.13 ± 6.81
DDW + deuterium-depleted yolk (DDyolk)	37.86 ± 5.47
Normal water + DDyolk *	32.19 ± 4.80 *
Normal water + normal yolk (Ctrl yolk)	46.96 ± 5.30
Normal water + DU283	33.78 ± 6.71

* The tumor weight was significantly decreased compared to CTRL (p = 0.0354, Sidak's multiple comparisons test).

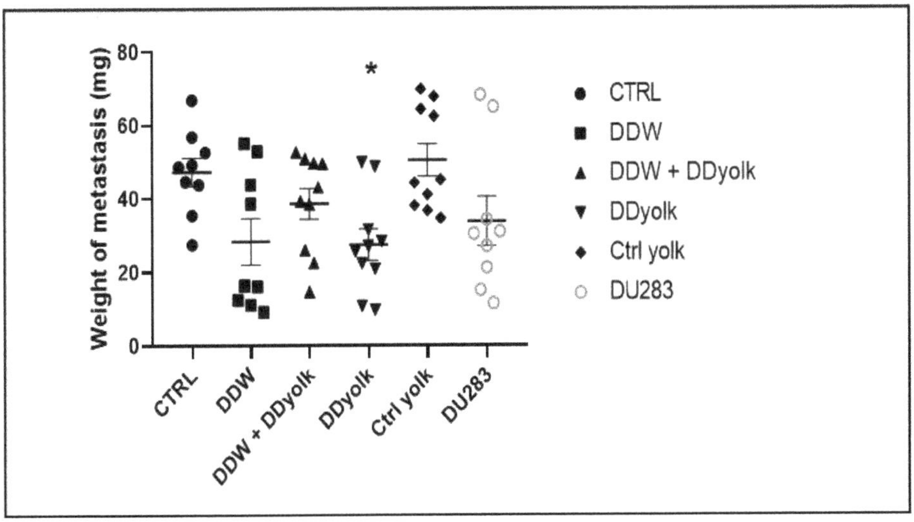

Figure 3. Weight of the metastases in the four treated and two control groups. The average weight of the metastases was lower in all four treated groups but only the group receiving deuterium-depleted yolk (*) showed significant differences (p = 0.0354) compared to the group consuming normal water and normal food. Dots represent individual data while the horizontal line with error bar shows mean and SD.

These results confirm the anticancer effect of DDW and our assumption that by administration of nutrients as deuterium-depleted organic compounds (DDyolk) to normal food, the anticancer effect of deuterium depletion can be boosted. Increasing the amount of DDyolk in the food may enhance antitumor efficacy.

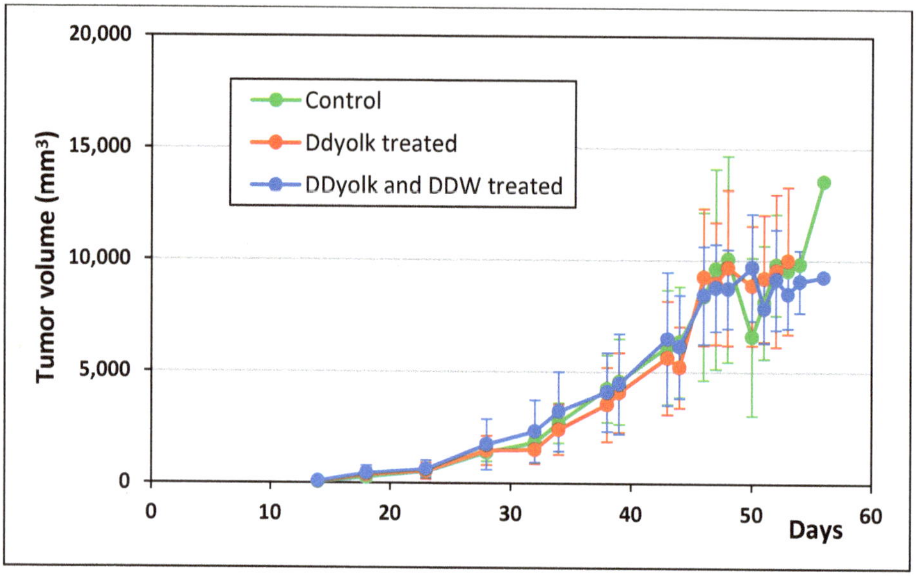

Figure 4. Changes of primary tumor volume in the control and the two treated groups inoculated with the MCF-7 cell line. The primary tumor in the mammary pad was first palpable on day 11.

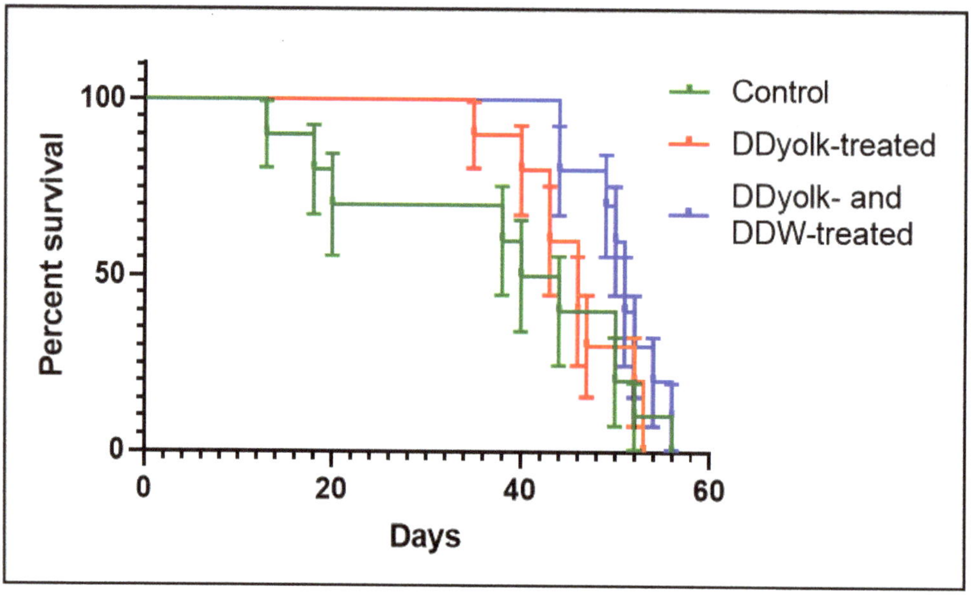

Figure 5. Survival curves of the control and the two treated groups inoculated with the MCF-7 cell line.

4. Discussion

The consumed food has an undoubtedly significant impact on cell metabolism and physiological processes in living organisms. The effects are typically attributed to the composition of the macronutrients (carbohydrates, lipids, proteins) in foods, whereas the possible role of heavy isotope of hydrogen, deuterium, has not been investigated. To preserve health, the most common dietary approach is to reduce fats and increase carbohydrates within the total caloric intake, arguing that the burden of cardiovascular diseases can be reduced this way.

Our study aimed to investigate the impact of the varying D/H ratio in food on tumor growth. To obtain deuterium-depleted nutrients, the drinking water of laying hens was replaced with DDW and the decrease of D concentration in the eggs was followed. Interestingly, the kinetics of the D level change in yolk and egg white was different which suggested that the distribution of D is strongly determined by biochemical pathways.

The mice studies presented here clearly showed the distinct antitumor effect of DDyolk. The significant weight decreases of lung metastases generated by the 4T1 cancer cells in the first study indicated that deuterium depletion may have inhibited the migration. Inhibition of lung cancer cell migration by DDW in vitro has been described earlier [14]. Tumor growth and survival data of mice inoculated with the MCF-7 breast cancer cell line confirmed our hypothesis that alterations in D concentration in organic compounds of food have an impact on tumor growth. This is related to the role of D in cell growth, namely that well-known molecular metabolic processes lead to an increasing D/H ratio which is responsible for the entering of the cells from the G1 to the S phase [4]. A recent study proved that a higher D/H ratio increased the expression of hundreds of genes with a key role in cell cycle regulation. It was concluded that by keeping the D concentration at a low level using DDW, the expression of these genes, and therefore cell growth, can be kept under control [9].

The application of a ketogenic (very low in carbs and high in fats) diet was tested in cancer patients with convincing clinical evidence of the anticancer effect of this type of nutrition. The changes in the metabolic parameters in patients during such a diet, and the beneficial antitumor effects, are well-documented [29–31]. Our results up to now suggest a common link between the antitumor effect of DDW and of ketogenic diet; the deuterium-depleting effect of both.

A carbohydrate-rich diet will result in the production of metabolic water with a D concentration close to the Standard Mean Ocean Water (SMOW) value of 155.75 ppm. However, the more complex the molecules are, such as lipids, the lower their D content, because the biochemical processes in the synthesis of these molecules show a preference to the lighter isotope of hydrogen due to the isotopic effect [2,3,18]. Consequently, increasing the ratio of fats in food intake will result in a lower D concentration of the metabolic water produced by the mitochondria.

The data in Table 4 on the D level in the dry matter of certain common foodstuffs provide evidence of the effect of nutrition on the D concentration within the human body, which in turn affects tumor growth [18,32].

Table 4. D concentration of the dry matter of different foods.

Type of Foodstuff	D concentration of Dry Matter (ppm)
Wheat flour	150 ± 1
Table sugar	146 ± 1
Cottage cheese	136 ± 1
Olive oil	130 ± 1
Butter	124 ± 1
Pork fat	118 ± 1

A high-carbohydrate and low-fat diet alone results in a higher average D concentration of the body. Consumed carbohydrates are converted into glucose for immediate energy, and into glycogen and fat as stored energy. Oxidation of glucose, immediately or after storage as glycogen, yields metabolic water with higher D level, whereas lipids become deuterium-depleted during synthesis and yield lower D in metabolic water. The D atoms not included into the lipids will be enriched in other molecules, preferably molecules with hydrogen in the exchangeable position, such as amino- and carboxyl groups of amino acids. This may explain the dissimilar D level in egg yolk and white of the DDW-treated hens (Figure 1).

Malignancies are not the only chronic disease positively influenced by reduced D levels. A recent animal study proved that the optimal D concentration of blood for reducing blood sugar levels in rats was between 125 and 140 ppm, stimulating the translocation of GLUT4 from the cytosol to the cell membrane [8]. The beneficial effect of D depletion was also confirmed in a human phase two clinical trial. DDW (105 ppm D) significantly reduced the fasting glucose level and decreased insulin resistance [33]. Another in vitro study confirmed the impact of the light isotopes of not only hydrogen, but also carbon, oxygen, and nitrogen on enzyme activity [34].

To reduce the incidence of chronic diseases, including malignant tumors, in the human population is a paramount aim. The data presented here prove that the isotopic composition of foods has a major impact on cancer cell growth. These observations raise the need to carry out further preclinical studies and human clinical trials to reveal the mechanism of deuterium depletion and to optimize the D concentration of nutrients to reduce the average D level in the blood from 145–150 ppm to 125–140 ppm.

Author Contributions: Conceptualization, G.S., L.G.P., L.I.N. and I.S.; methodology, L.G.P. and G.S.; software, A.P.; validation, L.I.N., L.G.P. and G.S.; formal analysis, L.I.N. and A.P.; investigation, L.I.N., L.G.P., I.F. and G.C.; resources, G.S. and I.S.; data curation, A.P. and L.G.P.; writing—original draft preparation, G.S., B.Z.K., I.S., A.P., I.S. and L.I.N.; writing—review and editing, A.P., G.S., I.S. and B.Z.K.; visualization, A.P., L.I.N. and B.Z.K.; supervision, G.S.; funding acquisition, G.S. All authors have read and agreed to the published version of the manuscript.

Funding: The work was supported by the European Regional Development Fund, Central Hungary Operative Program, New Széchenyi Plan (KMOP-1.1.4-11/A-2011-01-05), and HYD LLC for Cancer Research and Drug Development.

Institutional Review Board Statement: Not applicable.

Informed Consent Statement: Not applicable.

Data Availability Statement: No data supporting report.

Conflicts of Interest: The authors declare that there are no conflicts of interest that could be perceived as prejudicing the impartiality of the research reported.

Abbreviations

D	deuterium
DDW	deuterium-depleted water
DDyolk	deuterium-depleted yolk
SMOW	Standard Mean Ocean Water

References

1. Urey, H.C. Deuterium. *J. Chem. Educ.* **1962**, *39*, 583. [CrossRef]
2. Rundel, P.W.; Ehleringer, J.R.; Nagy, K.A. *Stable Isotopes in Ecological Research*; Springer: New York, NY, USA, 1988.
3. Katz, J.J.; Crespi, H.L. Isotope Effects in Biologycal Systems. In *Isotope Effects in Chemical Reactions*; Collins, C.J., Bowman, N.S., Eds.; Van Nostrand Rein-hold: New York, NY, USA, 1971; pp. 286–363.
4. Somlyai, G.; Jancsó, G.; Jákli, G.; Vass, K.; Barna, B.; Lakics, V.; Gaál, T. Naturally occurring deuterium is essential for the normal growth rate of cells. *FEBS Lett.* **1993**, *317*, 1–4. [CrossRef]
5. Strekalova, T.; Evans, M.; Chernopiatko, A.; Couch, Y.; Costa-Nunes, J.; Cespuglio, R.; Chesson, L.; Vignisse, J.; Steinbusch, H.W.; Anthony, D.C.; et al. Deuterium content of water increases depression susceptibility: The potential role of a serotonin-related mechanism. *Behav. Brain Res.* **2015**, *277*, 37–244. [CrossRef] [PubMed]
6. Mladin, C.; Ciobica, A.; Lefter, R.; Popescu, A.; Bild, W. Deuterium Depleted Water has stimulating effects on long-term memory in rats. *Neurosci. Lett.* **2014**, *583*, 154–158. [CrossRef]

7. Ávila, D.S.; Somlyai, G.; Somlyai, I.; Aschner, M. Anti-aging effects of deuterium depletion on Mn-induced toxicity in a C. elegans model. *Toxicol. Lett.* **2012**, *211*, 319–324. [CrossRef]
8. Molnár, M.; Horváth, K.; Dankó, T.; Somlyai, I.; Kovács, B.Z.; Somlyai, G. Deuterium-depleted water stimulates GLUT4 translocation in the presence of insulin, which leads to decreased blood glucose concentration. *Mol. Cell. Biochem.* **2021**, *476*, 4507–4516. [CrossRef]
9. Kovács, B.Z.; Puskás, L.G.; Nagy, L.I.; Papp, A.; Gyöngyi, Z.; Fórizs, I.; Czuppon Gy Somlyai, I.; Somlyai, G. Blocking the Increase of Intracellular Deuterium Concentration Prevents the Expression of Cancer-Related Genes, Tumor Development, and Tumor Recurrence in Cancer Patients. *Cancer Control* **2022**, *29*, 1–11. [CrossRef] [PubMed]
10. Gyöngyi, Z.; Somlyai, G. Deuterium depletion can decrease the expression of c-myc, Ha-ras and p53 gene in carcinogen-treated mice. *Vivo* **2000**, *14*, 437–440.
11. Krempels, K.; Somlyai, I.; Somlyai, G. A retrospective evaluation of the effects of deuterium depleted water consumption on four patients with brain metastases from lung cancer. *Integr. Cancer Ther.* **2008**, *7*, 172–181. [CrossRef]
12. Cong, F.S.; Zhang, Y.R.; Sheng, H.C.; Ao, Z.H.; Zhang, S.Y. Deuterium-depleted water inhibits human lung carcinoma cell growth by apoptosis. *Exp. Ther. Med.* **2010**, *1*, 277–283. [CrossRef] [PubMed]
13. Kovács, A.; Guller, I.; Krempels, K.; Somlyai, I.; Jánosi, I.; Gyöngyi, Z.; Szabó, I.; Ember, I.; Somlyai, G. Deuterium Depletion May Delay the Progression of Prostate Cancer. *J. Cancer Ther.* **2011**, *2*, 548–556. [CrossRef]
14. Somlyai, G.; Kovács BZs Somlyai, I.; Papp, A.; Nagy, L.I.; Puskás, L.G. Deuterium depletion inhibits lung cancer cell growth and migration in vitro and results in severalfold increase of median survival time of non-small cell lung cancer patients receiving conventional therapy. *J. Cancer Res. Ther.* **2021**, *9*, 12–19. [CrossRef]
15. Ziegler, H.; Osmond, C.B.; Stichler, W.; Trimborn, P. Hydrogen isotope discrimination in higher plants: Correlations with photosynthetic pathway and environment. *Planta* **1976**, *128*, 85–92. [CrossRef]
16. Sternberg, L.; Deniro, J.M.; Johnson, B.H. Isotope Ratios of Cellulose from Plants Having Different Photosynthetic Pathways. *Plant Physiol.* **1984**, *74*, 557–561. [CrossRef] [PubMed]
17. Richard, J.R.; Robins, J.; Isabelle, B.; Jia-Rong, D.; S'ebastien, G.G.; S'ebastien, P.P.; Ben-Li, Z. Measurement of 2H distribution in natural products by quantitative 2H NMR: An approach to understanding metabolism and enzyme mechanism? *Phytochem. Rev.* **2003**, *2*, 87–102. [CrossRef]
18. Robins, R.J.; Remaud, G.S.; Billault, I. Natural mechanisms by which deuterium depletion occurs in specific positions in metabolites. *Eur. Chem. Bull.* **2012**, *1*, 39–40.
19. Seyfried, T.N.; Flores, R.E.; Poff, A.M.; D'Agostino, D.P. Cancer as a metabolic disease: Implications for novel therapeutics. *Carcinogenesis* **2014**, *35*, 515–527. [CrossRef] [PubMed]
20. Somlyai, G. Deuterium Depletion Results in Several-fold Increases in the Median Survival Time of Cancer Patients durin Oncotherapy. In Proceedings of the 3rd International Congress on Deuterium Depletion, Budapest, Hungary, 7–8 May 2015.
21. Hobson, K.A.; Atwell, L.; Wassernaar, L.I. Influence of drinking water and diet on the stable-hydrogen isotope ratios of animal tissues. *Proc. Natl. Acad. Sci. USA* **1999**, *96*, 8003–8006. [CrossRef]
22. Prosser, S.J.; Scrimgeour, C.M. High-precision determination of $^2H/^1H$ in H_2 and H_2O by continuous-flow isotope ratio mass spectrometry. *Anal. Chem.* **1995**, *67*, 1992–1997. [CrossRef]
23. Scrimgeour, C.M.; Rollo, M.M.; Mudambo, S.M.; Handley, L.L.; Prosser, S.J. A simplified method for deuterium/hydrogen isotope ratio measurements on water samples of biological origin. *Biol. Mass Spectrom.* **1993**, *22*, 383–387. [CrossRef]
24. Pulaski, B.A.; Ostrad-Rosenberg, S. Mouse T1 Breast Tumor Model. *Curr. Protoc. Immunol.* **2001**. [CrossRef]
25. Szebeni, G.J.; Balog, J.A.; Demjén, A.; Alföldi, R.; Végi, L.V.; Fehér, L.Z.; Mán, I.; Kotogány, E.; Gubán, B.; Batár, P.; et al. Imidazo[1,2-*b*]pyrazole-7-carboxamides Induce Apoptosis in Human Leukemia Cells at Nanomolar Concentrations. *Molecules* **2018**, *23*, 2845. [CrossRef] [PubMed]
26. Ying, X.; Tinghong, Y.; Mengyao, W.; Yong, X.; Ningyu, W.; Xuejiao, S.; Fengtian, W.; Li, L.; Yongxia, Z.; Fangfang, Y.; et al. A Novel Cinnamide YLT26 Induces Breast Cancer Cells Apoptosis via ROS -Mitochondrial Apoptotic Pathway in Vitro and Inhibits Lung Metastasis in Vivo. *Cell. Physiol. Biochem.* **2014**, *34*, 1863–1876. [CrossRef]
27. Percie du Sert, N.; Hurst, V.; Ahluwalia, A.; Alam, S.; Avey, M.T.; Baker, M.; Browne, W.J.; Clark, A.; Cuthill, I.C.; Dirnagl, U.; et al. The ARRIVE guidelines 2.0: Updated guidelines for reporting animal research. *Br. J. Pharmacol.* **2020**, *177*, 3617–3624. [CrossRef]
28. *Guide for the Care and Use of Laboratory Animals*, 8th ed.; The National Academies Press: Washington, DC, USA, 2011. Available online: https://journals.sagepub.com/doi/10.1258/la.2012.150312 (accessed on 2 December 2022).
29. Zhou, W.; Mukherjee, P.; Kiebish, M.A.; Markis, W.T.; Mantis, J.G.; Seyfried, T.N. The calorically restricted ketogenic diet, an effective alternative therapy for malignant brain cancer. *Nutr. Metab.* **2007**, *4*, 5. [CrossRef]
30. Seyfried, T.N.; Flores, R.; Poff, A.M.; D'Agostino, D.P.; Murkherjee, P. Metabolic Therapy: A new paradigm for managing malignant brain cancer. *Cancer Lett.* **2014**, *356*, 289–300. [CrossRef]
31. Khodadadi, S.; Sobhani, N.; Mirshekar, S.; Ghiasvand, R.; Pourmasoumi, M.; Miraghajani, M.; Dehsoukhteh, S.S. Tumor Cells Growth and Survival Time with the Ketogenic Diet in Animal Models: A Systematic Review. *Int. J. Prev. Med.* **2017**, *8*, 35. [CrossRef]

32. Somlyai, G. *Deuterium Depletion—A New Way in Curing Cancer and Preserving Health*; Publish Drive: Redwood City, CA, USA, 2022; pp. 36–37.
33. Somlyai, G.; Somlyai, I.; Fórizs, I.; Czuppon Gy Papp, A.; Molnár, M. Effect of Systemic Subnormal Deuterium Level on Metabolic Syndrome Related and other Blood Parameters in Humans: A Preliminary Study. *Molecules* **2020**, *25*, 1376. [CrossRef]
34. Zhang, Z.; Meng, Z.; Beusch, C.; Gharibi, H.; Cheng, Q.; Stefano, L.; Wang, J.; Saei, A.; Vegvari, A.; Gaetani, M.; et al. Ultralight ultrafast enzymes. *Chem. Proteom.* 2021. *online 2021-12-15.* [CrossRef]

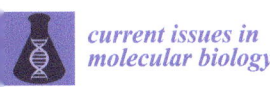

current issues in molecular biology

MDPI

Article

Loss of Nf1 and Ink4a/Arf Are Associated with Sex-Dependent Growth Differences in a Mouse Model of Embryonal Rhabdomyosarcoma

Wade R. Gutierrez [1,2,3,4], Jeffrey D. Rytlewski [4], Amanda Scherer [3,4], Grace A. Roughton [4], Nina C. Carnevale [4], Krisha Y. Vyas [4], Gavin R. McGivney [1,3,4,5], Qierra R. Brockman [3,4,6], Vickie Knepper-Adrian [4] and Rebecca D. Dodd [1,2,3,4,6,*]

1 Cancer Biology Graduate Program, University of Iowa, Iowa City, IA 52242, USA
2 Medical Scientist Training Program, University of Iowa, Iowa City, IA 52242, USA
3 Holden Comprehensive Cancer Center, University of Iowa, Iowa City, IA 52242, USA
4 Department of Internal Medicine, University of Iowa, Iowa City, IA 52242, USA
5 Department of Molecular Physiology and Biophysics, University of Iowa, Iowa City, IA 52242, USA
6 Molecular Medicine Graduate Program, University of Iowa, Iowa City, IA 52242, USA
* Correspondence: rebecca-dodd@uiowa.edu; Tel.: +1-319-335-4962

Abstract: Rhabdomyosarcoma (RMS) is an aggressive form of cancer that accounts for half of all pediatric soft tissue sarcomas. Little progress has been made in improving survival outcomes over the past three decades. Mouse models of rhabdomyosarcoma are a critical component of translational research aimed at understanding tumor biology and developing new, improved therapies. Though several models exist, many common mutations found in human rhabdomyosarcoma tumors remain unmodeled and understudied. This study describes a new model of embryonal rhabdomyosarcoma driven by the loss of Nf1 and Ink4a/Arf, two mutations commonly found in patient tumors. We find that this new model is histologically similar to other previously-published rhabdomyosarcoma models, although it substantially differs in the time required for tumor onset and in tumor growth kinetics. We also observe unique sex-dependent phenotypes in both primary and newly-developed orthotopic syngeneic allograft tumors that are not present in previous models. Using in vitro and in vivo studies, we examined the response to vincristine, a component of the standard-of-care chemotherapy for RMS. The findings from this study provide valuable insight into a new mouse model of rhabdomyosarcoma that addresses an ongoing need for patient-relevant animal models to further translational research.

Keywords: rhabdomyosarcoma; sarcoma; cancer; mouse model; Nf1; Ink4a/Arf; vincristine

Citation: Gutierrez, W.R.; Rytlewski, J.D.; Scherer, A.; Roughton, G.A.; Carnevale, N.C.; Vyas, K.Y.; McGivney, G.R.; Brockman, Q.R.; Knepper-Adrian, V.; Dodd, R.D. Loss of Nf1 and Ink4a/Arf Are Associated with Sex-Dependent Growth Differences in a Mouse Model of Embryonal Rhabdomyosarcoma. *Curr. Issues Mol. Biol.* **2023**, *45*, 1218–1232. https://doi.org/10.3390/cimb45020080

Academic Editor: Dumitru A. Iacobas

Received: 5 January 2023
Revised: 28 January 2023
Accepted: 30 January 2023
Published: 2 February 2023

1. Introduction

Rhabdomyosarcoma (RMS) is an aggressive and deadly malignancy that accounts for half of all pediatric soft tissue sarcoma diagnoses annually [1]. The majority of these tumors develop in children younger than 10 years old, with an incidence of 4.71 per million in individuals less than 20 years old [2,3]. Rhabdomyosarcomas develop from skeletal muscle stem cells called satellite cells and can arise throughout the body, including in the limbs, trunk, head and neck, and pelvic regions. Tumors present as an expanding lump or swelling and may be painful, depending on the area of development. The exact cause of RMS development is unknown. However, genetic risk factors include the presence of cancer predisposition syndromes such as Li-Fraumeni syndrome, Neurofibromatosis type 1, Costello syndrome, Noonan syndrome, Beckwith-Wiedman syndrome, and DICER1 syndrome [4]. Unfortunately, the prognosis for childhood RMS has not substantially improved since the mid-1990s [5]. The prognosis for metastatic disease, which accounts for 15% of all new diagnoses, remains grim, with a five-year survival rate of only 31% [6]. Furthermore, chemotherapies such as vincristine (VCR) that are used to treat local and metastatic disease are fraught with side effects including persistent and debilitating neuropathy.

There are two main classes of pediatric RMS: alveolar RMS (ARMS), which comprises approximately 25% of cases; and embryonal RMS (ERMS), which comprises 60% of cases [2]. Though both arise from skeletal muscle tissue, ARMS and ERMS have substantially different genetic alterations, incidences, and outcomes, and they are characterized by distinct transcriptomic, epigenetic, and proteomic signatures [7]. ARMS frequently contains characteristic PAX3-FOXO1 or PAX7-FOXO1 fusion proteins and is associated with a worse prognosis than ERMS. Mouse models of ARMS similarly contain fusion genes such as Pax3-FKHR, with or without accompanying loss of tumor suppressors such as Ink4a/Arf or Trp53 [8,9]. In contrast, ERMS is fusion-negative and contains a wide variety of mutations. Over 50% of ERMS contain alterations in RAS pathway members, including NRAS, KRAS, HRAS, NF1, and FGFR4, as well as mutations in tumor suppressor genes such as TP53 and CDKN2A (INK4A/ARF) [10]. Mouse models of ERMS contain combinations of activating mutations in proto oncogenes such as HGF/SF, HER2-neu, and Kras and inactivating mutations in tumor suppressors such as Ink4a/Arf or Trp53 [11–17]. Though much work has been done developing RMS mouse models, many common mutations and unique combinations of mutations found in human ERMS tumors remain unmodeled and understudied [9,13,18–22]. The heterogeneity of genetic alterations in ERMS necessitates the development of new ERMS mouse models containing patient-relevant mutations to better understand tumor biology and rigorously test promising therapeutic regimens.

In this study, we developed and tested a new genetically engineered mouse model (GEMM) of ERMS driven by the loss of Nf1 and Ink4a/Arf. Tumors are generated by CreER technology, with CreER expression under the control of the satellite cell specific Pax7 promoter (Pax7$^{CreER/+}$; Nf1$^{fl/fl}$; Ink4a/Arf$^{fl/fl}$, hereafter referred to as P7NI mice). Mutations in Nf1 and Ink4a/Arf are commonly found in ERMS tumors (15% and 4%, respectively), and this genetic combination follows the familiar pattern of Ras-pathway mutation coupled with the loss of a tumor suppressor that is found in other commonly-used RMS models [10,23,24]. To the best of our knowledge, this is the first report of an ERMS GEMM containing these alterations. We compared this new model to a well-established ERMS GEMM that relies on satellite cell-specific activation of oncogenic Kras^{G12D} and loss of the tumor suppressor Trp53 to drive tumor formation (Pax7$^{CreER/+}$; Kras$^{LSL\text{-}G12D/+}$; Trp53$^{fl/fl}$) [13–17]. This preclinical GEMM, here referred to as P7KP mice robustly models histologic and molecular features of human ERMS tumors. We also describe two new murine ERMS cell lines and their corresponding immune-competent allografts which can be used for future translational studies of this aggressive cancer. Furthermore, we evaluate the efficacy of vincristine, a mainstay of pediatric RMS chemotherapy regimens, on cell lines and orthotopic allografts in multiple ERMS models. Taken together, this study provides new insight into the impact of initiating mutations on ERMS tumor biology.

2. Materials and Methods

2.1. Mice

All animal experiments were performed in accordance with protocols approved by the Institutional Animal Care and Use Committee at the University of Iowa. Mouse strains were maintained in the Dodd Lab colony in the University of Iowa Office of Animal Care barrier facilities. Same sex littermates were housed in standard shoebox housing (maximum of five animals per cage) with paper bedding and fed the facility's standard irradiated mouse diet. Food and filtered drinking water were provided ad libitum. All cages and bedding were autoclaved before use. Animal health and welfare were assessed daily by veterinary staff throughout all phases of the experiment (breeding, tumor development, tumor growth). Animals were maintained on 12 h light/dark cycles and temperature and humidity-controlled rooms. Mice were maintained behind a barrier and monitored for contaminants by monthly sentinel harvest. The Pax7$^{CreER/+}$; Kras$^{LSL\text{-}G12D/+}$; Trp53$^{fl/fl}$ (P7KP) sarcoma model has been previously described and is on a 129Sv/Jae background [13,15,17]. The Pax7$^{CreER/+}$; Nf1$^{fl/fl}$; Ink4a/Arf$^{fl/fl}$ (P7NI) sarcoma model was generated by crossing previously-described Nf1$^{fl/fl}$; Ink4a/Arf$^{fl/fl}$ mice (C57Bl/6 background) [25] with Pax7$^{CreER/+}$ mice (129Sv/Jae background) [26]. Following F1 generation, Pax7$^{CreER/+}$; Nf1$^{fl/+}$; Ink4a/Arf$^{fl/+}$ mice were crossed with Nf1$^{fl/fl}$; Ink4a/Arf$^{fl/fl}$ mice to obtain Pax7$^{CreER/+}$; Nf1$^{fl/fl}$; Ink4a/Arf$^{fl/fl}$ mice. These F2 mice were bred together for at least 3 generations before obtaining experimental mice used for this study. Experimental litters usually contain 4–6 pups, with approximately half of the mice containing the Pax7$^{CreER/+}$ allele. To induce tumors in P7KP and P7NI mice, animals were injected with 50 μL of 4-hydroxytamoxifen (10 mM in DMSO, Sigma-Aldrich, St. Louis, MO, USA, H7904) in the gastrocnemius muscle. The gastrocnemius muscle is a commonly used site for tumor initiation and allows for easy and accurate assessment of tumor volume with digital calipers [13,15,17,27]. Mice were monitored a minimum of three times weekly to assess for general welfare and tumor initiation until all mice had developed tumors (final tumor development at 26 weeks post-tamoxifen injection). Following tumor initiation, tumors were measured three times weekly by a digital caliper. Tumor volume was calculated using the formula $V = (\pi \times L \times W \times H)/6$, with L, W, and H representing the length, width, and height of the tumor in mm, respectively. Tumors were considered to have initiated when they reached 200–285 mm^3 (approximately 7.5 mm × 7.5 mm × 7.5 mm). After tumors reached a volume of 1500 mm^3 (approximately 14.2 mm × 14.2 mm × 14.2 mm), the mice were euthanized. In accordance with our approved animal protocol, mice were euthanized before tumors reached the maximum approved tumor volume of 2000 mm^3. Mice were euthanized using both CO_2 asphyxiation (primary method) and cervical dislocation (secondary method). A regional autopsy of the tumor and surrounding tissue was performed and samples of tumor tissue were collected. Male and female mice between 7 and 52 weeks old were used in all studies. P7KP: 12 male mice, 12 female mice. P7NI: 16 male mice, 12 female mice.

2.2. Histological Analysis

As previously described [17], terminal tumor tissue was stored in a 10% neutral buffered formalin for fixation and subsequent paraffin embedment. Formalin-fixed paraffin-embedded tumors were sectioned and stained with hematoxylin (Vector Laboratories, Newark, CA, USA, H-3401) and eosin (Harleco, Darmstadt, Germany, 200-12) to evaluate tissue morphology. Images were taken using an inverted microscope (Fisherbrand, Waltham, MA, USA, 03-000-013) with a digital camera attachment (Amscope, Irvine, CA, USA, MU503) at 40× magnification. Scale bars: 50 μm.

2.3. Derivation of Cell Lines

As previously described [17], terminal tumor tissue was dissected with forceps and surgical scissors, washed in 5 mL PBS in a 6-well plate, then finely minced with surgical scissors. Five mL of Dissociation Buffer [Collagenase Type IV (700 units/mL, Gibco, Waltham, MA, USA, 17104-019) and dispase (65 mg/mL, Gibco, Waltham, MA, USA, 17105-041)] in PBS was added to each well. Plates were incubated for one hour at 37 °C on an orbital shaker and transferred to a tissue culture hood. Dissociated tissue was passed through a sterile 70 μM cell strainer (Fisherbrand, Waltham, MA, USA, 22363548) into a 50 mL conical vial using a 10 mL serological pipette and the plunger from a 1 mL syringe (Becton Dickinson, Franklin Lakes, NJ, USA, 309628). Cell strainers were washed with 25 mL sterile PBS into corresponding conical vials. Cell suspensions were centrifuged, and cell pellets were resuspended and plated in DMEM (Gibco, Waltham, MA, USA, 11965-092). Cells were grown in 10 cm dishes maintained in DMEM media containing 10% FBS, 1% penicillin–streptomycin (Gibco, Waltham, MA, USA, 15140-122), and 1% sodium pyruvate (Gibco, Waltham, MA, USA, 11360-070). When 90% confluency was reached, 15–35% of cells were passaged into a new dish. Cell line morphology was monitored by microscope and lines were passaged a minimum of 10 passages (range 39–46 passages) until a consistent growth rate was achieved and no stromal cells were observed in the culture. Established cell lines were frozen for use in subsequent analysis. KRIMS-3 cells were derived from an untreated P7KP primary tumor, while NIMS-1 and NIMS-2 cells were derived from untreated P7NI primary tumors.

2.4. Generation of Orthotopic Syngeneic Allografts

As previously described [17], cells were ~90% confluent on the day of injection. Cells were trypsinized, washed, and resuspended in sterile PBS containing calcium chloride and magnesium chloride. To develop orthotopic allografts, mice were injected with 50 µL of cell suspension (4×10^5 cells total) in the left gastrocnemius muscle using a 31G needle. For P7KP allografts, KRIMS-3 cells were injected into mice maintained on a KP ($Kras^{LSL\text{-}G12D/+}$; $Trp53^{fl/fl}$) background. For P7NI allografts, NIMS-1 and NIMS-2 cells were injected into mice maintained on a P7NI background. Same sex littermates were maintained in the Dodd Lab colony in the University of Iowa Office of Animal Care barrier facilities. Animals were housed in standard shoebox housing (maximum of five animals per cage) with paper bedding and fed the facility's standard irradiated mouse diet. Food and filtered drinking water were provided ad libitum. All cages and bedding were autoclaved before use. Animal health and welfare were assessed daily by veterinary staff throughout all phases of the experiment (breeding, tumor development, and tumor growth). Animals were maintained on 12 h light/dark cycles and temperature and humidity-controlled rooms. Mice were monitored a minimum of three times weekly to assess tumor initiation. Mice that had not developed tumors by 35 days after the injection of cells were euthanized. Following tumor initiation, tumors were measured three times weekly by a digital caliper. Tumors were considered to have initiated when they reached 200–285 mm^3 (approximately 7.5 mm × 7.5 mm × 7.5 mm). On the day of tumor initiation, mice were randomized to receive one dose of vincristine (Selleckchem, Houston, TX, USA, S1241) diluted in water or saline (PBS) via intraperitoneal injection. Randomization occurred by an alternating enrollment scheme. Animals were all housed on the same rack, and cage location was not changed during the study. Subsequent tumor growth measurements were collected in a blinded manner. After tumors reached a volume of 1500 mm^3 (approximately 14.2 mm × 14.2 mm × 14.2 mm), mice were euthanized. In accordance with our approved animal protocol, mice were euthanized before tumors reached the maximum approved tumor volume of 2000 mm^3. In cases of tumor regression, mice were euthanized 35 days after tumor initiation. Mice were euthanized using both CO_2 asphyxiation (primary method) and cervical dislocation (secondary method). A regional autopsy of the tumor and surrounding tissue was performed and samples of tumor tissue were collected. Male and female mice between 7 and 52 weeks old were used in all studies. KRIMS-3 allografts: 4 male mice, 8 female mice. NIMS-1 allografts: 12 male mice, 9 female mice. NIMS-2 allografts: 13 male mice, 10 female mice.

2.5. Cell Growth and Vincristine Sensitivity Assays

As previously described [17,27], KRIMS-3, NIMS-1, and NIMS-2 cells were grown in 10 cm dishes maintained in DMEM media containing 10% FBS, 1% penicillin-streptomycin (Pen-Strep, Gibco, Waltham, MA, USA, 15140-122), and 1 mM sodium pyruvate (Gibco, Waltham, MA, USA, 11360-070). For cell growth assays, cells were plated on Day 0 in a 12-well plate (2.5×10^4 cells). On Day 1, resazurin (Sigma-Aldrich, St. Louis, MO, USA, R7017) dissolved in PBS (1.5 mg/mL) was added to wells (200 µL for 12-well) and cells were returned to the tissue culture incubator for 1–2 h before being read on a microplate reader (BioTek, Winooski, VT, USA). This process was repeated 24 h later on Day 2. Fluorescence measurements from Day 2 were normalized to measurements from Day 1 for each cell line to calculate the 24 h fold growth. For vincristine dose response assays, cells were plated on Day 0 in a 96-well plate (KRIMS-3: 8.8×10^3 cells, KRIMS-4: 9.2×10^3 cells, and NIMS-1: 8×10^3 cells, NIMS-2: 5×10^3 cells). On Day 1, vincristine (Selleckchem, Houston, TX, USA, S1241) diluted in water was added at varying concentrations to wells. On Day 2, viability was assessed using a resazurin assay (20 µL per well) as described above. Fluorescence values were normalized to wells containing 0 μM vincristine. Dose-response curves were fitted using nonlinear regression, option "[inhibitor] vs. normalized response-variable slope" in GraphPad Prism (Version 9, Boston, MA, USA).

2.6. Statistics

Statistical analysis was performed using GraphPad Prism (Version 9, Boston, MA, USA) 8. In vivo, time-to-tumor and survival curves were analyzed using a Log-rank (Mantel-Cox) test with Bonferroni correction (if comparing more than two groups). In vitro data were analyzed using Welch's ANOVA and Dunnett's T3 multiple comparison tests.

3. Results

3.1. Tumor-Initiating Mutations Impact Rms Development and Growth

Both RMS models used in this study develop tumors from muscle satellite cells, which functioned as the stem cells of skeletal muscle. CreER expression was under the control of the satellite cell-specific Pax7 promoter. Under normal conditions, CreER remains in the cytoplasm. However, in the presence of tamoxifen (TMX), CreER translocates to the nucleus, allowing for Cre recombinase activity at LoxP sites. In P7KP mice ($Pax7^{CreER/+}$; $Kras^{LSL\text{-}G12D/+}$; $Trp53^{fl/fl}$), intramuscular injection of TMX caused expression of oncogeneic $Kras^{G12D}$ and a loss of Trp53 and satellite cells, resulting in RMS formation [13–17]. Similarly, in P7NI mice ($Pax7^{CreER/+}$; $Nf1^{fl/fl}$; $Ink4a/Arf^{fl/fl}$), intramuscular injection of tamoxifen led to the biallelic loss of Nf1 and Ink4a/Arf in satellite cells at the site of injection (Figure 1A). P7KP and P7NI tumors are histologically similar, with both models displaying the histopathologic hallmarks of RMS, including high cytologic variability and small round cells with eosinophilic cytoplasms and hypochromic nuclei (Figure 1D). However, though they share the same cell of origin, the difference in genetic mutations has a profound impact on tumor initiation and growth. While all animals developed ERMS tumors following TMX injection (Figure 1B), tumor latency was significantly different between the two models. Similar to previous reports [17], P7KP tumors (n = 24) initiated between 4.1 and 7.1 weeks (median 5.7 weeks) after TMX injection (Figure 1B). P7NI tumor onset (n = 28) was significantly slower, requiring 11.3 to 26.0 weeks (median 18.0 weeks) to initiate following TMX injection. Similarly, the growth kinetics of the tumors were also genotype-specific. While P7KP tumors tripled in volume within six days (range of four to seven days) after tumor initiation (Figure 1C), P7NI tumors displayed a more varied range of growth patterns, tripling in volume in nine days (range 3 to 17 days). Indeed, the majority of P7NI tumors (six of eight) grew slower than the slowest P7KP tumor which tripled in volume in seven days.

3.2. RMS Cell Lines Are Sensitive to Vincristine Treatment In Vitro

To further assess the impact of initiating mutations on tumor biology, we developed cell lines from primary RMS tumors (Figure 2A). Terminal tumor tissue was enzymatically dissociated and cultured in vitro. Cell lines were passaged a minimum of 10 times to facilitate the removal of nontumor stromal cells. We have previously published findings with two P7KP cell lines that we developed using this method, known as Kras Induced Murine Sarcoma 3 and 4 (KRIMS-3 and KRIMS-4 cells) [17]. In the current study, we derived two new P7NI cell lines called Nf1 Induced Murine Sarcoma 1 and 2 (NIMS-1 and NIMS-2 cells).

In vitro growth assays demonstrated that both KRIMS and NIMS cells grow at very similar rates (Figure 2B). As previously reported, KRIMS-4 cells grow slightly faster than KRIMS-3 cells (3.2-fold versus 2.3-fold increase after 24 h). In contrast, there is no difference in growth between NIMS-1 and NIMS-2 cells. In addition to evaluating growth rates, we assessed the sensitivity of each cell line to vincristine (VCR), a microtubule inhibitor commonly included in the multidrug chemotherapy regimen used to treat RMS (Figure 2C–G). Dose-response studies determined that IC_{50} values for P7KP cells and P7NI cells were both in the midnanomolar range, including an average of 82.6 nM for KRIMS cells (KRIMS-3: 100.5 nM and KRIMS-4: 71.2 nM) and 58.9 nM for NIMS cells (NIMS-1: 81.6 nM and

NIMS-2: 42.9 nM) (Figure 2D–G). Taken together, these data demonstrate that cells from the P7NI model closely resemble cells from previously-published P7KP models in vitro.

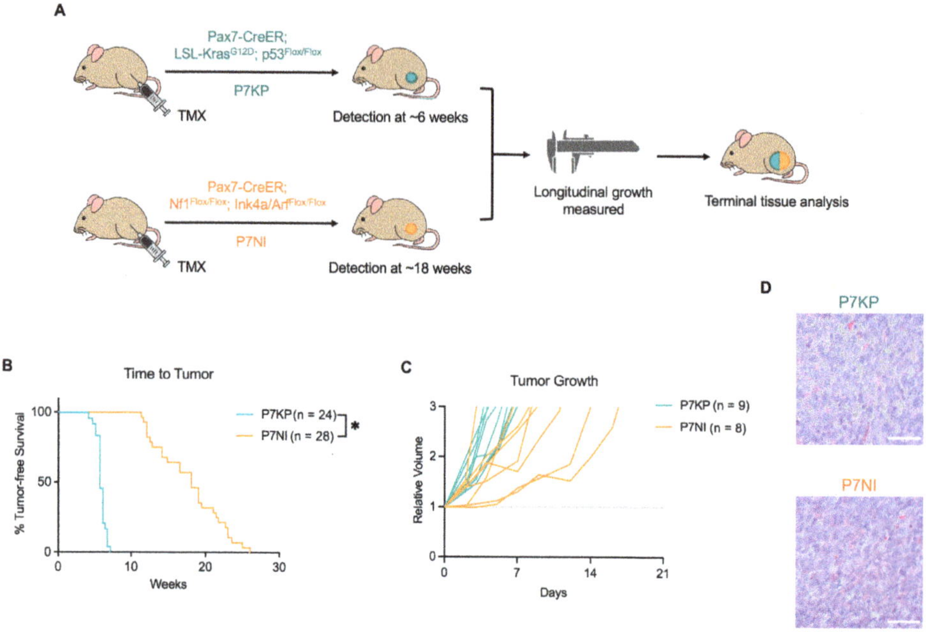

Figure 1. P7KP and P7NI tumor initiation, growth, and histology. (**A**) P7KP and P7NI tumors were induced using intramuscular injection of tamoxifen (TMX) to locally delete Trp53 and activate oncogenic $Kras^{G12D}$ (P7KP) or to locally delete Nf1 and Ink4a/Arf (P7NI). Following tumor initiation, tumor dimensions were measured by caliper three times weekly and terminal tumor tissue was collected for analysis. (**B**) P7NI tumors initiated significantly slower than P7KP tumors. (**C**) P7NI tumors also displayed a broader range of growth rates than P7KP tumors. (**D**) P7KP and P7NI tumors share similar histologic features including cells with high cytologic variability and eosinophilic cytoplasms. Scale bar: 50 μM. Log-rank (Mantel-Cox) tests used to analyze B. * $p < 0.05$.

3.3. P7NI Tumor Growth Phenotypes Are Sex Dependent

One major goal of our study was to test the sensitivity of RMS mouse models to vincristine (VCR), an integral member of the multidrug standard-of-care chemotherapy for RMS patients. To facilitate translational studies of Nf1-deleted RMS in vivo, we developed orthotopic syngeneic orthotopic allografts with the NIMS cell lines (Figure 3A). Allografts offer several advantages to primary genetically engineered tumors, including more rapid and consistent tumor initiation. Upon tumor initiation, mice were randomized to receive either vincristine (VCR) or saline control (PBS). We compared the two new NIMS allografts to the well-established KRIMS-3 allografts [17] (Figures 3B and S1A, Table 1). As previously reported, KRIMS-3 allografts arise nine to eleven days following cell implantation and show similar tumor initiation trends in male and female mice. Given the similarities between KRIMS and NIMS cells in vitro, we hypothesized that NIMS allografts would follow a similar pattern of initiation and growth. To our surprise, we observed significant sex-dependent differences in tumor growth of the NIMS allografts. In males, all mice implanted with NIMS-1 and NIMS-2 cells developed tumors within seven to nine days following cell implantation. In contrast, tumor development was substantially lower in female mice, with only six of ten female mice developing NIMS-2 tumors and seven of eight female mice developing NIMS-1 tumors. Of note, the differences between the NIMS-1 and NIMS-2 allograft initiation rates in female mice may be due to the sex of the mice from which the cell lines were derived, as NIMS-1 cells were developed from a female mouse while NIMS-2 cells were developed from a male mouse. However, this would be a NIMS-specific characteristic, as both of the previously published KRIMS-3 and KRIMS-4 cell lines derived from female mice develop allografts equally in male and female recipient mice (Figure S1B) [17].

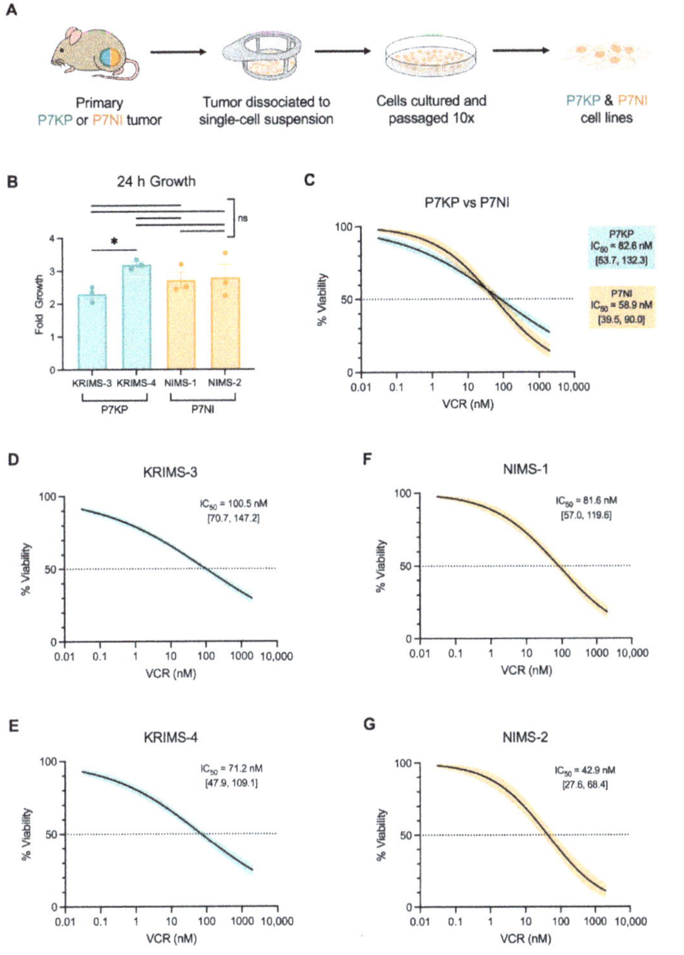

Figure 2. Development and characterization of P7KP and P7NI cell lines. (**A**) Terminal P7KP and P7NI tumors were enzymatically dissociated and cultured to develop cell lines. KRIMS (Kras Induced Murine Sarcoma) cells were derived from P7KP tumors and NIMS (Nf1 Induced Murine Sarcoma) cells were derived from P7NI tumors. (**B**) KRIMS and NIMS growth in vitro. Cells were seeded at equal densities and growth was assessed after 24 h. Data points represent independent experiments (n = 3). (**C–G**) Vincristine (VCR) dose-response curves. Cells were treated for 24 h. (**C**) Average dose-response curves by genotype. (**D–G**) VCR dose-response curves for individual cell lines. Data from 3–4 independent experiments included in each plot. Curves fitted using nonlinear regression, option "[inhibitor] vs. normalized response-variable slope" in GraphPad Prism (Version 9, Boston, MA, USA). Shaded areas represent 95% confidence intervals. Absolute IC_{50} 95% confidence intervals displayed in brackets. Welch's ANOVA and Dunnett's T3 multiple comparison test used to analyze B. * $p < 0.05$; ns = not significant.

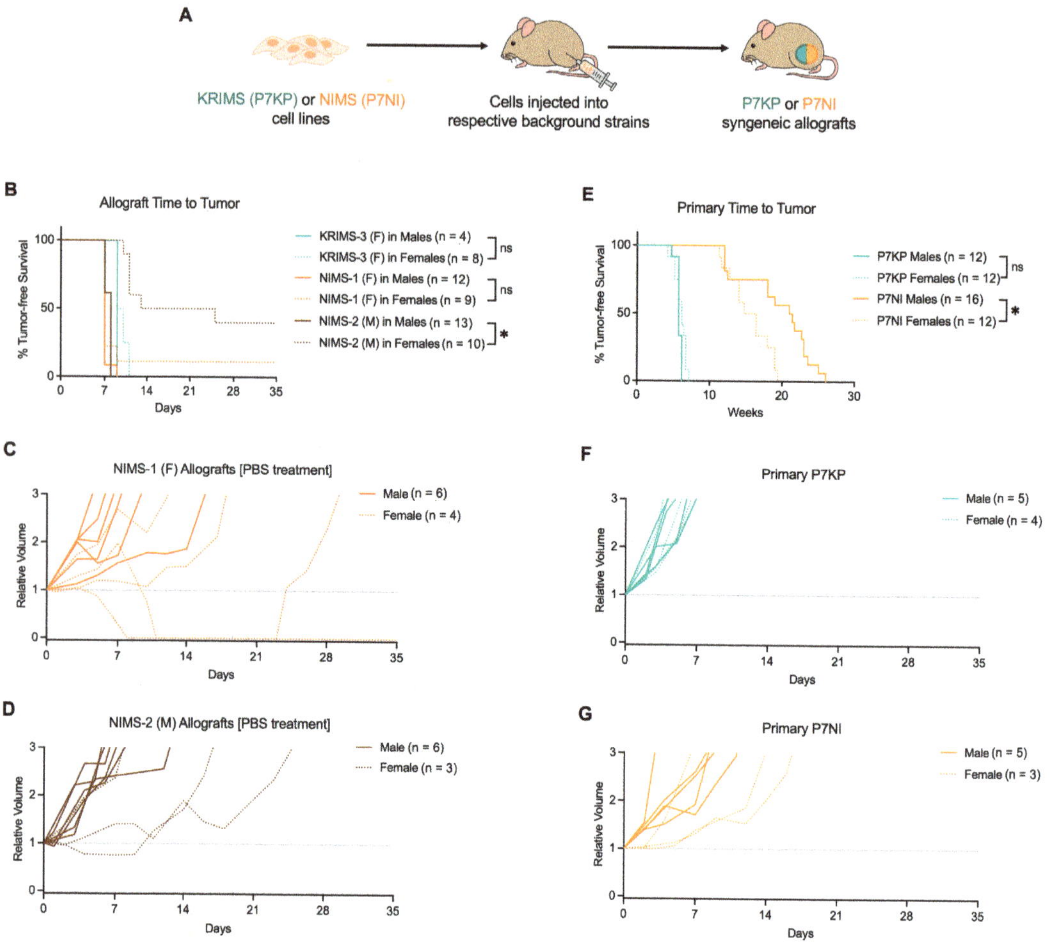

Figure 3. KRIMS and NIMS allograft development and growth. (**A**) KRIMS (P7KP) and NIMS (P7NI) cells were orthotopically implanted into mice of the same respective background strain to develop syngeneic allografts. KRIMS-3 and NIMS-1 cells were derived from tumors in female (**F**) mice, while NIMS-2 cells were derived from a tumor in a male (M) mouse. (**B**) NIMS-2 cells displayed significant sex-dependent differences in tumor initiation. (**C**,**D**) Individual NIMS-1 and NIMS-2 tumor growth curves. Tumors in females developed later than those in male mice and displayed highly variable growth patterns. (**E**) Reanalysis of data presented in Figure 1B. Tumors in female P7NI mice initiated significantly faster than in male P7NI mice. (**F**,**G**) Reanalysis of data from Figure 1C. Individual primary P7KP and P7NI tumor growth curves. Log-rank (Mantel-Cox) test with Bonferroni correction used to analyze B (adjusted $\alpha = 0.003125$) and E (adjusted $\alpha = 0.008333$). * $p < 0.003125$ in B; * $p < 0.008333$ in E. Complete statistical analysis available in Figure S1. * $p < 0.05$; ns = not significant.

In accordance with our original experimental design, mice were randomized to treatment with saline (PBS) or VCR at the time of tumor initiation. In PBS-treated Nf1-deleted tumors, we observed sex-dependent tumor growth patterns (Figure 3C–D, Table 1). In male mice, NIMS-1 and NIMS-2 allograft growth patterns were similar to primary P7NI tumors. However, in female mice, NIMS-1 allografts displayed inconsistent growth trajectories. Surprisingly, two NIMS-1 tumors in female mice regressed to unmeasurable levels. One of these tumors became detectable again after 14 days (24 days after initial cell implantation), while the other remained stably regressed at five weeks post cell implantation. Though none of the PBS-treated NIMS-2 allografts in female mice regressed to unmeasurable levels, two of the three tumors displayed substantially slower growth than the NIMS-2 allografts in male mice. Based on these findings, we reanalyzed tumor initiation data from the primary tumor models presented in Figure 1B,C. As previously reported [17], onset of P7KP primary tumors did not differ based on sex (n = 12 males, n = 12 females, Figures 3E and S1C). In contrast, primary P7NI tumors in female mice initiated significantly faster than in male mice (median 15.6 weeks for the 12 female mice versus 21.2 weeks for the 16 male mice, respectively) (Figures 3E and S1C). Similarly, longitudinal growth curves of primary P7NI tumors resembled those of PBS-treated NIMS allografts, with tumors in females trending toward slower growth than those in males (Figure 3G). P7KP primary tumors, in contrast, showed no growth differences between male and female mice (Figure 3F).

Table 1. KRIMS and NIMS allograft development, growth, and treatment.

		KRIMS-3 (F)		
Sex	Initiated	Regressed	Reached 3×	Treatment
M	Yes	No	Yes	PBS
M	Yes	No	Yes	PBS
M	Yes	No	Yes	VCR
M	Yes	No	Yes	VCR
M Total	4/4	0/4	4/4	2:2
F	Yes	No	Yes	PBS
F	Yes	No	Yes	PBS
F	Yes	No	Yes	PBS
F	Yes	No	Yes	PBS
F	Yes	No	Yes	VCR
F	Yes	No	Yes	VCR
F	Yes	No	Yes	VCR
F	Yes	No	Yes	VCR
F Total	8/8	0/8	8/8	4:4
		NIMS-1 (F)		
Sex	Initiated	Regressed	Reached 3×	Treatment
M	Yes	No	Yes	PBS
M	Yes	No	Yes	PBS
M	Yes	No	Yes	PBS
M	Yes	No	Yes	PBS
M	Yes	No	Yes	PBS
M	Yes	No	Yes	PBS
M	Yes	No	Yes	VCR
M	Yes	No	Yes	VCR
M	Yes	No	Yes	VCR
M	Yes	No	Yes	VCR
M	Yes	No	Yes	VCR
M	Yes	No	Yes	VCR
M Total	12/12	0/12	12/12	6:6
F	Yes	No	Yes	PBS
F	Yes	No	Yes	PBS
F	Yes	Yes	Yes	PBS
F	Yes	Yes	No	PBS
F	Yes	No	Yes	VCR
F	Yes	No	Yes	VCR
F	Yes	No	Yes	VCR
F	Yes	Yes	No	VCR
F	No	-	-	-
F Total	8/9	3/8	6/8	4:4

Table 1. *Cont.*

		NIMS-2 (M)		
Sex	Initiated	Regressed	Reached 3×	Treatment
M	Yes	No	Yes	PBS
M	Yes	No	Yes	PBS
M	Yes	No	Yes	PBS
M	Yes	No	Yes	PBS
M	Yes	No	Yes	PBS
M	Yes	No	Yes	PBS
M	Yes	No	Yes	VCR
M	Yes	No	Yes	VCR
M	Yes	No	Yes	VCR
M	Yes	No	Yes	VCR
M	Yes	No	Yes	VCR
M	Yes	No	Yes	VCR
M	Yes	No	Yes	VCR
M Total	13/13	0/13	13/13	6:7
F	Yes	No	Yes	PBS
F	Yes	No	Yes	PBS
F	Yes	No	Yes	PBS
F	Yes	No	Yes	VCR
F	Yes	No	Yes	VCR
F	Yes	Yes	No	VCR
F	No	-	-	-
F	No	-	-	-
F	No	-	-	-
F	No	-	-	-
F Total	6/10	1/6	5/6	3:3

3.4. Both P7KP and P7NI Allografts Are Resistant to Vincristine

For KRIMS allografts, the impact of VCR treatment was assessed using combined groups of males and females since sex affects neither the P7KP primary tumor nor KRIMS-3 allograft initiation or growth (Figures 3E,F, 4A,B and S1B). Given the substantial sex-dependent differences in NIMS allograft tumor initiation and growth, we assessed the impact of VCR treatment on males and females separately (Figure 4C–H). A single dose of VCR (1.0 mg/kg) was administered at the time of tumor initiation.

This dose was chosen since it is close to the maximum-tolerated dose in a previous study reporting that doses of VCR greater than 1.0 mg/kg in mice are associated with total body weight loss greater than 10% and death (LD_{50} = 3.8 mg/kg) [28]. We found that at this dose, VCR treatment did not impact KRIMS-3 allograft growth (Figure 4B) or survival (Figure 4E). Likewise, VCR did not affect the growth or survival of NIMS-1 and NIMS-2 allografts in male mice (Figure 4C–E). In female mice with NIMS-1 allografts, VCR had no appreciable effect, although one mouse treated with VCR experienced complete and sustained tumor regression, similar to one of the PBS-treated female NIMS-1 allograft mice described above (Figure 4F, Table 1). Similarly, in VCR-treated female mice with NIMS-2 allografts, one mouse experienced complete and sustained regression. However, this effect is not generalizable, as two VCR-treated NIMS-2 tumors in female mice showed rapid growth. Importantly, VCR did not extend survival in female mice with either NIMS-1 or NIMS-2 allografts (Figure 4H). Taken together, these data indicate that VCR has a minimal impact on tumor growth and survival in both P7KP and P7NI orthotopic syngeneic allografts, suggesting these tumors are models of chemotherapy-resistant rhabdomyosarcoma.

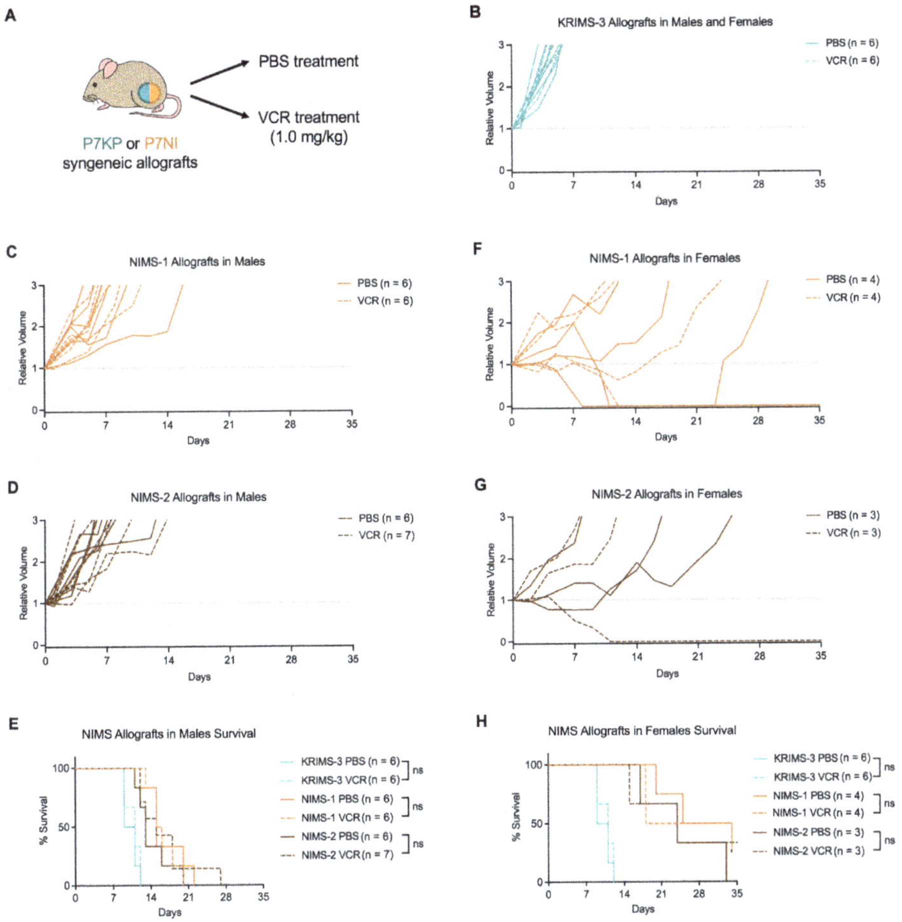

Figure 4. Impact of vincristine on allograft growth. (**A**) Mice were treated with one dose of saline (PBS) or vincristine (VCR) at the time of tumor initiation. (**B**) Individual growth curves for KRIMS-3 allografts in male and female mice. (**C**,**D**) Individual growth curves for NIMS-1 and NIMS-2 allografts in male mice. (**E**) Survival of NIMS-1 and NIMS-2 allografts in male mice and KRIMS-3 allografts in male and female mice. (**F**,**G**) Individual growth curves for NIMS-1 and NIMS-2 allografts in female mice. (**H**) Survival of NIMS-1 and NIMS-2 allografts in female mice and KRIMS-3 allografts (same KRIMS-3 allografts shown in **D**). Log-rank (Mantel-Cox) test used to analyze sets of PBS versus VCR treated in mice E and H. ns = not significant.

4. Discussion

The heterogeneity of genetic alterations in RMS necessitates the development of new mouse models to better understand RMS biology and to test promising therapeutic regimens more rigorously. In this study, we have described a new mouse model of ERMS driven by the loss of Nf1 and Ink4a/Arf in muscle satellite cells (P7NI mice). To assess the characteristics of the model, we compared it to a previously validated murine ERMS model that relies on activation of oncogenic $Kras^{G12D}$ and a loss of Trp53 (P7KP mice). We found that the primary P7NI tumors had delayed onset and grew more slowly than P7KP tumors. Cell lines derived from P7KP and P7NI tumors (KRIMS and NIMS, respectively) behaved similarly in vitro, but orthotopic syngeneic allografts generated from these cells did not follow this pattern. Unexpectedly, NIMS allografts displayed sex-dependent differences in tumor initiation and growth that were not present in KRIMS allografts. These findings prompted us to reanalyze data from the primary genetically engineered tumors. We found that the onset of P7NI tumors was five weeks earlier in female mice than in male mice. To better understand the utility of the P7NI models in preclinical studies, we assessed the sensitivity of cell lines and orthotopic syngeneic allografts to the microtubule inhibitor VCR. We found that in vitro P7KP and P7NI cell lines were equally sensitive to VCR. However, in vivo studies in orthotopic syngeneic allografts demonstrated that neither genotype was responsive to 1.0 mg/kg VCR treatment, a previously-reported maximum tolerated dose [28]. These immune-competent preclinical models of drug-resistant RMS will be a valuable tool to better understand the mechanisms underlying resistance and to identify new means of overcoming resistance.

As previously stated, to the best of our knowledge, this is the first animal model of NF1-mutated RMS, which accounts for 15% of all ERMS tumors [10]. In human ERMS, the impact of individual mutations on tumorigenesis and the age of diagnosis is not well-documented. However, mutations in TP53 have been associated with worse outcomes (shortened event-free survival times) [10]. The preservation of Trp53 in P7NI mice may partially explain the slowed tumorigenesis and tumor growth in this model compared to P7KP tumors. Other ERMS genetic signatures associated with decreased overall survival and failure-free survival include increased expression of EPHA2 (ephrin type-A receptor 2), NSMF (NMDA receptor synaptonuclear signaling and neuronal migration factor), and EPB41L4B (erythrocyte membrane protein band 4.1 like 4b) [29]. The impact of these genetic alterations on the response to specific treatments such as vincristine is not well understood due to the variable rates of genetic testing and regional differences in treatment regimens. Mouse models of ERMS, such as those described in this study, are critical for furthering our understanding of the impact of genetic changes on ERMS biology and response to treatment.

We also used this model to explore RMS response to vincristine in vitro and in vivo. Importantly, RMS patients are commonly treated with a multidrug chemotherapy regimen comprised of vincristine, actinomycin D, and cyclophosphamide (termed VAC). We chose to focus our studies on vincristine monotherapy for several reasons. First, we wanted to directly compare responses between in vitro and in vivo models. Since cyclophosphamide is a prodrug and requires exogenous activation to work in tissue culture, it is difficult to properly control its activity experimentally. Second, we were concerned about the cumulative toxicities of the VAC regimen and instead chose to use vincristine alone at the published maximum tolerated dose. Our data determined that both the P7KP and P7NI models are resistant to vincristine in vivo. Both intrinsic and acquired resistance to chemotherapy is common occurrences in human RMS. Recently, the upregulation of the Zinc finger protein GLI1 was identified in vincristine-resistant RMS and Ewings Sarcoma cell lines [30]. Other groups developed a series of 34 RMS xenografts to evaluate intrinsic and acquired chemotherapy resistance to the VAC regimen [31]. Studies such as ours describing intrinsically resistant immune-competent RMS models complement this previous work on acquired resistance using in vitro systems and immune-deficient xenografts.

The P7NI model uniquely models sex-dependent differences in tumor initiation and growth. Of note, ERMS incidence is slightly higher in males compared to females [2], although this male predominance is also seen in the majority of other childhood cancers [1,32]. However, in cohorts of pediatric patients, rates of NF1-mutated cancers are significantly higher in females than in males [33,34]. Though most of these differences were in cancers originating in the central nervous system such as optic pathway gliomas, it was possible that a similar Nf1-associated mechanism was promoting the earlier tumorigenesis in female P7NI mice. Of note, very few studies have examined sex-dependent differences in NF1-mutated tumor initiation and growth. Though studies of NF1-mutated tumors in human patients typically contain both males and females, data from the two sexes are often combined, likely due to limited cohort sizes. Additional studies are needed to better understand the impact of sex on NF1-mutated tumor biology in human and in animal models.

The newly described P7NI model has both strengths and limitations. The strengths discussed above include the unique genetics of the model, the ability to examine chemotherapy resistance, and the sexual dimorphism of tumor growth. A limitation of the primary GEMM P7NI model is slow tumor onset (median time 18 weeks) and a wide range of tumor initiation times. These factors make treatment studies in the primary model cumbersome. Though their derivative allografts had much shorter times to tumor initiation, they also displayed inconsistent initiation rates in female mice. These strengths and limitations should be taken into account when considering the use of P7NI primary or allograft tumors in experimental designs. Future studies with this model could examine the in vivo response to the VAC chemotherapy regimen and compare these outcomes to vincristine monotherapy data. Similarly, this model could be used to test targeted therapies in tumors with different initiating mutations.

In conclusion, we have demonstrated that a new mouse model of ERMS driven by the loss of Nf1 and Ink4a/Arf histologically resembles previously-published ERMS models while also displaying unique features, including delayed tumor initiation and sex-dependent growth patterns. The new P7NI primary tumor model, cell lines, and orthotopic syngeneic allografts will be useful tools in furthering our understanding of the impact of genetic heterogeneity on RMS biology. Additionally, this model may be of use in evaluating mechanisms underlying intrinsic resistance to VCR. Further investigation is needed to better understand the mechanisms driving observed sex-dependent differences in tumor growth.

Supplementary Materials: The following supporting information can be downloaded at: https://www.mdpi.com/article/10.3390/cimb45020080/s1, Figure S1: Statistical analyses and KRIMS-3 tumor growth.

Author Contributions: Conceptualization, R.D.D. and W.R.G.; methodology, W.R.G., J.D.R., A.S., G.A.R., N.C.C., K.Y.V., G.R.M., Q.R.B. and V.K.-A.; data curation, W.R.G.; writing—original draft preparation, W.R.G. and R.D.D.; writing—review and editing, W.R.G. and R.D.D.; supervision, R.D.D.; funding acquisition, R.D.D. All authors have read and agreed to the published version of the manuscript.

Funding: This research was funded by the University of Iowa Sarcoma Multidisciplinary Oncology Group, T32 GM067795 [W.R.G.], T32 GM007337 [W.R.G.], University of Iowa Dance Marathon Funding [R.D.D.], American Cancer Society Research Scholar Award [R.D.D.], and an NCI Core Grant P30 CA086862 [University of Iowa Holden Comprehensive Cancer Center].

Institutional Review Board Statement: The animal study protocol was approved by the Institutional Animal Care and Use Committee at the University of Iowa, Iowa City, IA, USA (protocol # 9101838 approved 15 October 2019) and was carried out in accordance with ARRIVE guidelines.

Data Availability Statement: The data presented in this study are available on request from the corresponding author.

Conflicts of Interest: The authors declare no conflict of interest.

References

1. Martin-Giacalone, B.A.; Weinstein, P.A.; Plon, S.E.; Lupo, P.J. Pediatric Rhabdomyosarcoma: Epidemiology and Genetic Susceptibility. *J. Clin. Med.* **2021**, *10*, 2028. [CrossRef]
2. Ognjanovic, S.; Linabery, A.M.; Charbonneau, B.; Ross, J.A. Trends in Childhood Rhabdomyosarcoma Incidence and Survival in the United States (1975–2005). *Cancer* **2009**, *115*, 4218–4226. [CrossRef]
3. Siegel, D.A.; King, J.; Tai, E.; Buchanan, N.; Ajani, U.A.; Li, J. Cancer Incidence Rates and Trends among Children and Adolescents in the United States, 2001–2009. *Pediatrics* **2014**, *134*, e945–e955. [CrossRef] [PubMed]
4. Skapek, S.X.; Ferrari, A.; Gupta, A.; Lupo, P.J.; Butler, E.; Shipley, J.; Barr, F.G.; Hawkins, D.S. Rhabdomyosarcoma. *Nat. Rev. Dis. Prim.* **2019**, *5*, 1. [CrossRef] [PubMed]
5. Smith, M.A.; Altekruse, S.F.; Adamson, P.C.; Reaman, G.H.; Seibel, N.L. Declining Childhood and Adolescent Cancer Mortality. *Cancer* **2014**, *120*, 2497–2506. [CrossRef]
6. Punyko, J.A.; Mertens, A.C.; Baker, K.S.; Ness, K.K.; Robison, L.L.; Gurney, J.G. Long-Term Survival Probabilities for Childhood Rhabdomyosarcoma. A Population-Based Evaluation. *Cancer* **2005**, *103*, 1475–1483. [CrossRef] [PubMed]

7. Stewart, E.; McEvoy, J.; Wang, H.; Chen, X.; Honnell, V.; Ocarz, M.; Gordon, B.; Dapper, J.; Blankenship, K.; Yang, Y.; et al. Identification of Therapeutic Targets in Rhabdomyosarcoma through Integrated Genomic, Epigenomic, and Proteomic Analyses. *Cancer Cell* **2018**, *34*, 411–426.e19. [CrossRef]
8. Lagutina, I.; Conway, S.J.; Sublett, J.; Grosveld, G.C. Pax3-FKHR Knock-in Mice Show Developmental Aberrations but Do Not Develop Tumors. *Mol. Cell. Biol.* **2002**, *22*, 7204–7216. [CrossRef]
9. Keller, C.; Arenkiel, B.R.; Coffin, C.M.; El-Bardeesy, N.; DePinho, R.A.; Capecchi, M.R. Alveolar Rhabdomyosarcomas in Conditional Pax3:Fkhr Mice: Cooperativity of Ink4a/ARF and Trp53 Loss of Function. *Genes Dev.* **2004**, *18*, 2614–2626. [CrossRef]
10. Shern, J.F.; Selfe, J.; Izquierdo, E.; Patidar, R.; Chou, H.-C.; Song, Y.K.; Yohe, M.E.; Sindiri, S.; Wei, J.; Wen, X.; et al. Genomic Classification and Clinical Outcome in Rhabdomyosarcoma: A Report From an International Consortium. *JCO* **2021**, *39*, 2859–2871. [CrossRef]
11. Sharp, R.; Recio, J.A.; Jhappan, C.; Otsuka, T.; Liu, S.; Yu, Y.; Liu, W.; Anver, M.; Navid, F.; Helman, L.J.; et al. Synergism between INK4a/ARF Inactivation and Aberrant HGF/SF Signaling in Rhabdomyosarcomagenesis. *Nat. Med.* **2002**, *8*, 1276–1280. [CrossRef] [PubMed]
12. Nanni, P.; Nicoletti, G.; De Giovanni, C.; Croci, S.; Astolfi, A.; Landuzzi, L.; Di Carlo, E.; Iezzi, M.; Musiani, P.; Lollini, P.-L. Development of Rhabdomyosarcoma in HER-2/Neu Transgenic P53 Mutant Mice. *Cancer Res.* **2003**, *63*, 2728–2732. [PubMed]
13. Blum, J.M.; Añó, L.; Li, Z.; Van Mater, D.; Bennett, B.D.; Sachdeva, M.; Lagutina, I.; Zhang, M.; Mito, J.K.; Dodd, L.G.; et al. Distinct and Overlapping Sarcoma Subtypes Initiated from Muscle Stem and Progenitor Cells. *Cell Rep.* **2013**, *5*, 933–940. [CrossRef] [PubMed]
14. Van Mater, D.; Añó, L.; Blum, J.M.; Webster, M.T.; Huang, W.; Williams, N.; Ma, Y.; Cardona, D.M.; Fan, C.-M.; Kirsch, D.G. Acute Tissue Injury Activates Satellite Cells and Promotes Sarcoma Formation via the HGF/c-MET Signaling Pathway. *Cancer Res.* **2015**, *75*, 605–614. [CrossRef]
15. Dodd, R.D.; Añó, L.; Blum, J.M.; Li, Z.; Van Mater, D.; Kirsch, D.G. Methods to Generate Genetically Engineered Mouse Models of Soft Tissue Sarcoma. *Methods Mol. Biol.* **2015**, *1267*, 283–295. [CrossRef]
16. Mater, D.V.; Xu, E.; Reddy, A.; Añó, L.; Sachdeva, M.; Huang, W.; Williams, N.; Ma, Y.; Love, C.; Happ, L.; et al. Injury Promotes Sarcoma Development in a Genetically and Temporally Restricted Manner. *JCI Insight* **2018**, *3*, e123687. [CrossRef] [PubMed]
17. Gutierrez, W.R.; Scherer, A.; McGivney, G.R.; Brockman, Q.R.; Knepper-Adrian, V.; Laverty, E.A.; Roughton, G.A.; Dodd, R.D. Divergent Immune Landscapes of Primary and Syngeneic Kras-Driven Mouse Tumor Models. *Sci. Rep.* **2021**, *11*, 1098. [CrossRef]
18. Fleischmann, A.; Jochum, W.; Eferl, R.; Witowsky, J.; Wagner, E.F. Rhabdomyosarcoma Development in Mice Lacking Trp53 and Fos: Tumor Suppression by the Fos Protooncogene. *Cancer Cell* **2003**, *4*, 477–482. [CrossRef]
19. Hatley, M.E.; Tang, W.; Garcia, M.R.; Finkelstein, D.; Millay, D.P.; Liu, N.; Graff, J.; Galindo, R.L.; Olson, E.N. A Mouse Model of Rhabdomyosarcoma Originating from the Adipocyte Lineage. *Cancer Cell* **2012**, *22*, 536–546. [CrossRef]
20. Comiskey Jr, D.F.; Jacob, A.G.; Sanford, B.L.; Montes, M.; Goodwin, A.K.; Steiner, H.; Matsa, E.; Tapia-Santos, A.S.; Bebee, T.W.; Grieves, J.; et al. A Novel Mouse Model of Rhabdomyosarcoma Underscores the Dichotomy of MDM2-ALT1 Function in Vivo. *Oncogene* **2018**, *37*, 95–106. [CrossRef]
21. Ragab, N.; Bauer, J.; Botermann, D.S.; Uhmann, A.; Hahn, H. Oncogenic NRAS Accelerates Rhabdomyosarcoma Formation When Occurring within a Specific Time Frame during Tumor Development in Mice. *Int. J. Mol. Sci.* **2021**, *22*, 13377. [CrossRef]
22. Nakahata, K.; Simons, B.W.; Pozzo, E.; Shuck, R.; Kurenbekova, L.; Prudowsky, Z.; Dholakia, K.; Coarfa, C.; Patel, T.D.; Donehower, L.A.; et al. K-Ras and P53 Mouse Model with Molecular Characteristics of Human Rhabdomyosarcoma and Translational Applications. *Dis. Model. Mech.* **2022**, *15*, dmm049004. [CrossRef] [PubMed]
23. Shern, J.F.; Chen, L.; Chmielecki, J.; Wei, J.S.; Patidar, R.; Rosenberg, M.; Ambrogio, L.; Auclair, D.; Wang, J.; Song, Y.K.; et al. Comprehensive Genomic Analysis of Rhabdomyosarcoma Reveals a Landscape of Alterations Affecting a Common Genetic Axis in Fusion-Positive and Fusion-Negative Tumors. *Cancer Discov.* **2014**, *4*, 216–231. [CrossRef]
24. Li, H.; Sisoudiya, S.D.; Martin-Giacalone, B.A.; Khayat, M.M.; Dugan-Perez, S.; Marquez-Do, D.A.; Scheurer, M.E.; Muzny, D.; Boerwinkle, E.; Gibbs, R.A.; et al. Germline Cancer Predisposition Variants in Pediatric Rhabdomyosarcoma: A Report From the Children's Oncology Group. *J. Natl. Cancer Inst.* **2021**, *113*, 875–883. [CrossRef] [PubMed]
25. Dodd, R.D.; Mito, J.K.; Eward, W.C.; Chitalia, R.; Sachdeva, M.; Ma, Y.; Barretina, J.; Dodd, L.; Kirsch, D.G. NF1 Deletion Generates Multiple Subtypes of Soft-Tissue Sarcoma That Respond to MEK Inhibition. *Mol. Cancer Ther.* **2013**, *12*, 1906–1917. [CrossRef] [PubMed]
26. Lepper, C.; Conway, S.J.; Fan, C.-M. Adult Satellite Cells and Embryonic Muscle Progenitors Have Distinct Genetic Requirements. *Nature* **2009**, *460*, 627–631. [CrossRef] [PubMed]
27. Gutierrez, W.R.; Scherer, A.; Rytlewski, J.D.; Laverty, E.A.; Sheehan, A.P.; McGivney, G.R.; Brockman, Q.R.; Knepper-Adrian, V.; Roughton, G.A.; Quelle, D.E.; et al. Augmenting Chemotherapy with Low-Dose Decitabine through an Immune-Independent Mechanism. *JCI Insight* **2022**, *7*, e159419. [CrossRef]
28. Harrison, S.D. An Investigation of the Mouse as a Model for Vincristine Toxicity. *Cancer Chemother. Pharmacol.* **1983**, *11*, 62–65. [CrossRef] [PubMed]
29. Hingorani, P.; Missiaglia, E.; Shipley, J.; Anderson, J.R.; Triche, T.J.; Delorenzi, M.; Gastier-Foster, J.; Wing, M.; Hawkins, D.S.; Skapek, S.X. Clinical Application of Prognostic Gene Expression Signature in Fusion Gene-Negative Rhabdomyosarcoma: A Report from the Children's Oncology Group. *Clin. Cancer Res.* **2015**, *21*, 4733–4739. [CrossRef]

30. Yoon, J.W.; Lamm, M.; Chandler, C.; Iannaccone, P.; Walterhouse, D. Up-Regulation of GLI1 in Vincristine-Resistant Rhabdomyosarcoma and Ewing Sarcoma. *BMC Cancer* **2020**, *20*, 511. [CrossRef]
31. Ghilu, S.; Morton, C.L.; Vaseva, A.V.; Zheng, S.; Kurmasheva, R.T.; Houghton, P.J. Approaches to Identifying Drug Resistance Mechanisms to Clinically Relevant Treatments in Childhood Rhabdomyosarcoma. *Cancer Drug Resist.* **2022**, *5*, 80–89. [CrossRef] [PubMed]
32. Williams, L.A.; Richardson, M.; Kehm, R.D.; McLaughlin, C.C.; Mueller, B.A.; Chow, E.J.; Spector, L.G. The Association between Sex and Most Childhood Cancers Is Not Mediated by Birthweight. *Cancer Epidemiol.* **2018**, *57*, 7–12. [CrossRef] [PubMed]
33. Uusitalo, E.; Rantanen, M.; Kallionpää, R.A.; Pöyhönen, M.; Leppävirta, J.; Ylä-Outinen, H.; Riccardi, V.M.; Pukkala, E.; Pitkäniemi, J.; Peltonen, S.; et al. Distinctive Cancer Associations in Patients With Neurofibromatosis Type 1. *JCO* **2016**, *34*, 1978–1986. [CrossRef] [PubMed]
34. Peltonen, S.; Kallionpää, R.A.; Rantanen, M.; Uusitalo, E.; Lähteenmäki, P.M.; Pöyhönen, M.; Pitkäniemi, J.; Peltonen, J. Pediatric Malignancies in Neurofibromatosis Type 1: A Population-Based Cohort Study. *Int. J. Cancer* **2019**, *145*, 2926–2932. [CrossRef]

current issues in
molecular biology

MDPI

Article

Zinc Finger E-Box Binding Homeobox 2 as a Prognostic Biomarker in Various Cancers and Its Correlation with Infiltrating Immune Cells in Ovarian Cancer

Hye-Ran Kim [1], Choong Won Seo [1], Sang Jun Han [2], Jae-Ho Lee [3,*] and Jongwan Kim [1,*]

1 Department of Biomedical Laboratory Science, Dong-Eui Institute of Technology, 54 Yangji-ro, Busanjin-gu, Busan 47230, Korea; hrkim@dit.ac.kr (H.-R.K.); seo3711@dit.ac.kr (C.W.S.)
2 Department of Biotechnology, College of Fisheries Sciences, Pukyong National University, 45 Yongso-ro, Nam-gu, Busan 48513, Korea; sjhan@pknu.ac.kr
3 Department of Anatomy, Keimyung University School of Medicine, 1095 Dalgubeol-daero, Dalseo-gu, Daegu 42601, Korea
* Correspondence: anato82@dsmc.or.kr (J.-H.L.); dahyun@dit.ac.kr (J.K.); Tel.: +82-53-580-3833 (J.-H.L.); +82-51-860-3525 (J.K.)

Abstract: This study investigated the expression of zinc finger E-box binding homeobox 2 (ZEB2), its prognostic significance in various cancers, and the correlation between ZEB2 and infiltrating immune cells and ZEB2-related proteins in ovarian cancer (OV). The Gene Expression Profiling Interactive Analysis tool was used to analyze RNA sequencing data and cancer survival rates, based on normal and tumor tissue data available in The Cancer Genome Atlas (TCGA) database. The Kaplan–Meier plotter and PrognoScan databases were used to analyze the prognostic value of ZEB2 in OV (n = 1144). The Tumor Immune Estimation Resource was used to investigate the correlation between ZEB2 and infiltrating immune cells in various cancers, including OV. High ZEB2 expression was associated with a poorer prognosis in OV. In OV, ZEB2 is positively correlated with CD8+T cells, neutrophils, macrophages, and dendritic cell invasion; and ZEB2 is negatively correlated with tumor-infiltrating B cells. The STRING database was used to investigate the correlations with ZEB2-related proteins. The results reveal that ZEB2 was positively correlated with SMAD1 and SMAD2 in OV. Our findings may serve as a potential prognostic biomarker, and provide novel insights into the tumor immunology in OV. Thus, ZEB2 may be a potential diagnostic and therapeutic target in OV.

Keywords: ovarian cancer; ZEB2; prognosis; biomarker; immune cells

Citation: Kim, H.-R.; Seo, C.W.; Han, S.J.; Lee, J.-H.; Kim, J. Zinc Finger E-Box Binding Homeobox 2 as a Prognostic Biomarker in Various Cancers and Its Correlation with Infiltrating Immune Cells in Ovarian Cancer. *Curr. Issues Mol. Biol.* **2022**, *44*, 1203–1214. https://doi.org/10.3390/cimb44030079

Academic Editors: Peter C. Hart and Dumitru A. Iacobas

Received: 30 January 2022
Accepted: 27 February 2022
Published: 1 March 2022

1. Introduction

Ovarian cancer (OV) accounts for only 3% of cancers in women. However, it is the fifth most common cause of mortality in women after lung, breast, colorectal, and pancreatic cancers [1]. According to the World Health Organization (WHO), OV is reported as one of the main causes of cancer-related deaths, and is considered a significant source of disease and death in women worldwide [2]. Moreover, OV, a major public health concern, is the most fatal gynecological malignancy, and, despite surgery and aggressive frontline treatments, most cancers ultimately recur, leading to chemotherapy-resistant diseases [3]. Recent studies have shown that this cancer recurs in approximately 50% of patients within 16 months, with a five-year overall survival rate of less than 50% [4]. Although surgery and chemotherapy have advanced significantly, the reported overall survival rates are very low. Therefore, many studies have been conducted to identify the prognostic factors, such as tumor biomarkers, to improve the prognosis of ovarian cancer. Recently, efforts have been made to improve new biomarkers, such as antibodies, cyclic tumor DNA, and micro-RNAs [5].

Zinc finger E-box binding homeobox 2 (ZEB2) gene, which is a member of the ZEB family, encodes the ZFHX1B gene and is located on chromosome 2q22 [6]. ZEB2 is a DNA-binding transcription factor that is mainly involved in epithelial-to-mesenchymal transition (EMT). EMT is a conserved process that transforms a mature epithelial-like state into a mobile mesenchymal state. It is a cellular program that is important for embryonic development, wound healing, and malignant progression of cancer. EMT plays a crucial role in the metastatic dissemination events of cancer cells: EMT confers cancer cells with a motile and invasive phenotype [7–9]. In addition to the transforming growth factor-β signaling pathway (TGF-β), which is a key inducer of EMT, the activation of several transcription factors, including ZEB1, ZEB2, SNAI1, SNAI2, TWIST1, and TWIST2, either directly or indirectly, regulate the E-cadherin promoter [10]. ZEB2 is a major transcription factor that promotes EMT under both normal and pathological conditions [11]. Emerging evidence suggests that ZEB2 plays a crucial role in EMT-induced processes, including development, differentiation, and malignant mechanisms, such as drug resistance, cancer stem cell-like characteristics, apoptosis, survival, cell cycle arrest, tumor recurrence, and metastasis [12]. ZEB2 is associated with the progression and development of malignancies such as liver cancer, cervical cancer, gastric cancer, and breast cancer [13–17]. Recent studies have shown that ZEB2 promotes peritoneal metastasis, through the process of regulating invasive cells and tumorigenesis of cancer cells in high-grade serous ovarian cancers (HGSOCs) [18]. In addition, ZEB2 had a significant correlation with poor prognosis and could improve EMT conversion in OV [19]. Thus, we performed a meta-analysis to evaluate the prognostic biomarker value of ZEB2, which is involved in the progression and development of various types of cancer, including OV.

The immune system plays a significant role in the onset and progression of cancer. Recent research has revealed that ZEB2 is expressed by the lymphatic system cells, such as T lymphocytes, B lymphocytes, and NK cells, as well as by the immune cells in myeloid systems, such as dendritic cells and macrophages, and that ZEB2 has the ability to control the transcriptional networks needed for cell differentiation and maintenance [20]. The interaction between ZEB2, which is associated with immune cells, and the immune cells that infiltrate OV is unclear. Thus, we investigated whether there was a correlation between ZEB2 and the infiltrating immune cells in various cancers, including OV.

In the present study, we investigated ZEB2 in various types of cancer, based on The Cancer Genome Atlas (TCGA) data from public databases. We investigated the correlation of ZEB2 with prognosis and infiltrating immune cells in various types of cancer, including OV. Furthermore, we focused the correlation between ZEB2-related proteins in OV.

2. Materials and Methods

2.1. Gene Expression Profiling Interactive Analysis (GEPIA)

The GEPIA database, which is a web server tool consisting of normal and tumor tissue samples from the TCGA and GTEx projects [21–23], was used to analyze the differences in ZEB2 expression between normal and tumor tissues, based on RNA sequencing.

2.2. Kaplan–Meier Plotter Database Analysis

The Kaplan–Meier plotter is based on an online database [24] and is capable of assessing the association of genes with survival in various types of cancer, including ovarian cancer ($n = 1144$). It includes survival rates, such as progression-free survival (PFS), overall survival (OS), and post-progression survival (PPS), and clinical data (including stage, grade, and TP53 mutation). The correlations between ZEB2 expression and survival in ovarian cancer were examined and presented with the hazard ratio, 95% confidence intervals, and log rank p-value computed.

2.3. PrognoScan Database Analysis

The PrognoScan database is widely used to evaluate biological relationships between gene expression and survival rates, such as disease-free survival, overall survival, and relapse-free survival [25]. It includes a large-scale collection of publicly available cancer microarray datasets with clinical information. We used this database to identify the correlation between ZEB2 mRNA expression and survival in ovarian cancer with a Cox p-value < 0.05.

2.4. Tumor Immune Estimation Resource (TIMER) Analysis

TIMER is used to analyze tumor-infiltrating immune cells in ovarian cancer [26]. TIMER assesses the tumor-infiltrating immune cells based on the statistical analysis of gene expression profiles [27]. We analyzed the correlation between ZEB2 and infiltrating immune cells, such as CD4+T cells, CD8+T cells, neutrophils, macrophages, dendritic cells, and B cells in different cancer types, including ovarian cancer.

2.5. STRING Analysis

The STRING protein network database can aid in the prediction of potential protein interactions. It is a precomputed global resource for the exploration and analysis of protein interactions. We used the STRING database to investigate whether ZEB2 has any protein-protein interactions.

2.6. Statistical Analysis

All statistical analyses were performed using the Statistical Package for the Social Sciences (SPSS), version 25.0, for Windows (IBM, Armonk, NY, USA). Survival results are presented with p-values from a log-rank test, and Kaplan–Meier survival curves were generated using PrognoScan online tools. The correlation of gene expression was evaluated in the TIMER database using Spearman's correlation analysis. A two-tailed p-value of <0.05 was considered to indicate statistical significance.

3. Results

3.1. mRNA Levels of ZEB2 in Various Types of Cancer

To determine differences in ZEB2 mRNA expression between tumor and normal tissues, ZEB2 expression in normal samples and multiple cancer types, including ovarian cancer, was analyzed using the GEPIA database. The results revealed that ZEB2 expression was lower in adrenocortical carcinoma (ACC), bladder urothelial carcinoma (BLCA), breast invasive carcinoma (BRCA), cervical squamous cell carcinoma (CESC), colon adenocarcinoma (COAD), kidney chromophobe (KICH), lung adenocarcinoma (LUAD), lung squamous cell carcinoma (LUSC), ovarian serous cystadenocarcinoma (OV), pheochromocytoma and paraganglioma (PCPG), prostate adenocarcinoma (PRAD), rectum adenocarcinoma (READ), thyroid carcinoma (THCA), thymoma (THYM), uterine corpus endometrial carcinoma (UCEC), and uterine carcinosarcoma (UCS) than in normal tissue. However, ZEB2 expression was higher in acute myeloid leukemia (LAML), brain lower grade glioma (LGG), pancreatic adenocarcinoma (PAAD), and skin cutaneous melanoma (SKCM) than in normal tissue (Figure 1). These results suggest that the expression of ZEB2 was lower in OV compared to normal tissue. In addition, it was confirmed that the mRNA level of ZEB2 was different in various cancers.

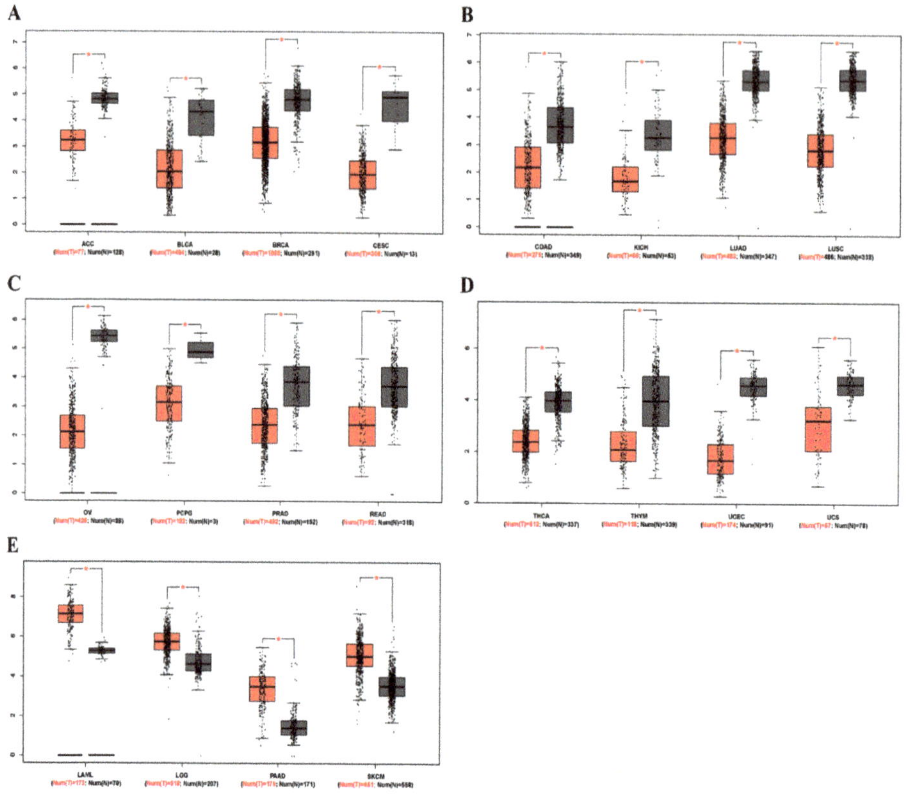

Figure 1. The mRNA levels of ZEB2 in different types of cancer. The expression levels of ZEB2 were analyzed using the GEPIA database. (**A–D**) Low expression of ZEB2 in various cancer tissues compared with normal tissue. (**E**) High expression of ZEB1 in various cancer tissues compared with normal tissue. Adrenocortical carcinoma, ACC; bladder urothelial carcinoma, BLCA; breast invasive carcinoma, BRCA; cervical squamous cell carcinoma, CESC; colon adenocarcinoma, COAD; kidney chromophobe, KICH; lung adenocarcinoma, LUAD; lung squamous cell carcinoma, LUSC; ovarian serous cystadenocarcinoma, OV; pheochromocytoma and paraganglioma, PCPG; prostate adenocarcinoma, PRAD; rectum adenocarcinoma, READ; thyroid carcinoma, THCA; thymoma, THYM; uterine corpus endometrial carcinoma, UCEC; uterine carcinosarcoma, UCS; acute myeloid leukemia, LAML; brain lower grade glioma, LGG; pancreatic adenocarcinoma, PAAD; skin cutaneous melanoma, SKCM (* $p < 0.05$).

3.2. *The Prognostic Value of ZEB2 Expression in Various Types of Cancer*

We investigated whether ZEB2 expression correlated with the prognosis of OV. Therefore, the effect of ZEB2 expression on survival rates was evaluated using the Kaplan–Meier plotter and PrognoScan databases. Survival rates, such as PFS, OS, and PPS of ZEB2 in OV, were analyzed. The findings revealed that high ZEB2 expression had significantly shorter survival times than those with low expression (Figure 2). High ZEB2 expression was associated with poorer prognosis in OV (PFS, HR = 1.35, $p = 0.0015$; OS, HR = 1.35, $p = 0.0036$; PPS, HR = 1.33, $p = 0.0018$; Figure 2A–C). These findings demonstrate the prognostic significance of ZEB2 in OV. Next, we investigated the relationship between ZEB2 and the clinicopathological characteristics of ovarian cancer using the Kaplan–Meier Plotter database, and the results are shown in Table 1. High ZEB2 expression correlated with poorer PFS in stages II (HR = 2.5, $p = 0.043$), II + III (HR = 1.34, $p = 0.0056$), II + III + IV (HR = 1.33, $p = 0.0034$), III (HR = 1.29, $p = 0.017$), III + IV (HR = 1.37, $p = 0.0017$), and IV (HR = 1.37, $p = 0.0017$). In addition, high ZEB2 expression correlated with poorer OS in stages II + III + IV (HR = 1.29, $p = 0.025$) and III + IV (HR = 1.26, $p = 0.048$). High ZEB2 expression correlated with poorer PPS in stages II + III (HR = 1.31, $p = 0.041$) and II + III + IV (HR = 1.29, $p = 0.034$). High ZEB2 expression correlated with poorer PFS, OS, and PPS in Grade II + III (PFS, HR = 1.33, $p = 0.0061$; OS, HR = 1.31, $p = 0.0016$; PPS, HR = 1.31, $p = 0.033$) and III (PFS, HR = 1.38, $p = 0.012$; OS, HR = 1.29, $p = 0.044$; PPS, HR = 1.37, $p = 0.035$). These findings revealed the prognostic significance of ZEB2 expression based on clinicopathological characteristics, especially in Grade II + III and Grade III OV. Moreover, the tumor suppressor gene p53 (TP53) mutation was associated with poorer OS in the wild type (HR = 2.83, $p = 0.046$). To further examine the prognostic potential of ZEB2 in different cancer types, we analyzed the PrognoScan database. Poor prognosis was identified in cancers of the blood, brain, breast, colorectal, and lung cancers, including ovarian cancer (Supplementary Table S1). Taken together, ZEB2 expression was associated with poorer prognosis in OV and other cancers.

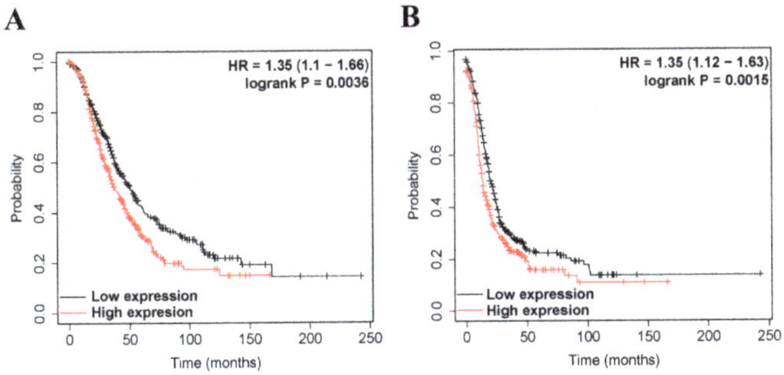

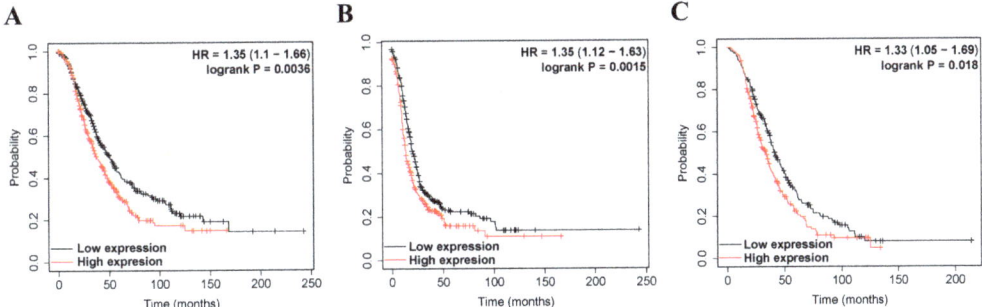

Figure 2. The prognostic significance of high expression of ZEB2 in cancers. The prognostic value of ZEB2 was analyzed using the Kaplan–Meier plotter. Survival curves OS (**A**), PFS (**B**), and PPS (**C**) of ZEB2 in OV. Overall survival, OS; progression-free survival, PFS; post-progression survival, PPS; ovarian serous cystadenocarcinoma, OV.

Table 1. Association between ZEB2 and clinicopathological characteristics in ovarian cancer. The clinicopathological characteristics of ZEB2 were analyzed using the Kaplan–Meier plotter.

Clinicopathological Characteristics	Progression-Free Survival (n = 614)			Overall Survival (n = 1144)			Post-progression Survival (n = 138)		
	n	Hazard Ratio	p-Value	n	Hazard Ratio	p-Value	n	Hazard Ratio	p-Value
STAGE									
I	74	1.3 (0.36–4.7)	0.68	51	1.57 (0.39–6.32)	0.52	7	-	-
I+ II	115	2.13 (0.94–4.8)	0.064	83	1.2 (0.43–3.4)	0.73	20	1.85 (0.52–6.7)	0.33
II	14	2.5 (1.0–6.4)	0.043	32	2.7 (0.52–14)	0.22	13	5.26 (0.59–48)	0.093
II + III	465	1.34 (1.09–1.64)	**0.0056**	458	1.21 (0.95–1.54)	0.13	325	1.31 (1.01–1.7)	**0.041**
II + III + IV	535	1.33 (1.1–1.61)	**0.0034**	519	1.29 (1.03–1.61)	**0.025**	374	1.29 (1.02–1.64)	**0.034**
III	424	1.29 (1.05–1.6)	**0.017**	426	1.26 (0.99–1.62)	0.063	312	1.25 (0.96–1.62)	0.097
III + IV	494	1.37 (1.12–1.66)	**0.0017**	487	1.26 (1.0–1.58)	**0.048**	361	1.22 (0.96–1.56)	0.099
IV	70	1.65 (0.99–2.75)	**0.05**	61	1.13 (0.63–2.01)	0.69	49	1.1 (0.59–2.1)	0.76
GRADE									
I	54	4.16 (0.86–20)	0.054	41	1.12 (0.39–3.22)	0.84	9	-	-
I+ II	189	1.38 (0.97–1.96)	0.074	203	1.17 (0.78–1.75)	0.44	118	1.12 (0.72–1.74)	0.62
II	161	1.27 (0.88–1.83)	0.19	162	1.18 (0.76–1.83)	0.46	109	1.09 (0.68–1.74)	0.71
II + III	476	1.33 (1.08–1.64)	**0.0061**	554	1.31 (1.05–1.62)	**0.016**	349	1.31 (1.02–1.68)	**0.033**
III	315	1.38 (1.07–1.77)	**0.012**	392	1.29 (1.01–1.66)	**0.044**	240	1.37 (1.02–1.84)	**0.035**
IV	18	-	-	18	1.06 (0.39–2.9)	0.91	18	-	-

Table 1. *Cont.*

Clinicopathological Characteristics	Progression-Free Survival (n = 614)			Overall Survival (n = 1144)			Post-progression Survival (n = 138)		
	n	Hazard Ratio	p-Value	n	Hazard Ratio	p-Value	n	Hazard Ratio	p-Value
TP53 mutation									
Mutated	124	1.26 (0.87–1.84)	0.22	124	1.18 (0.81–1.73)	0.38	116	1.02 (0.7–1.5)	0.92
Wild type	19	2.2 (0.77–6.33)	0.13	19	2.83 (0.98–8.16)	**0.046**	17	2.02 (0.7–5.8)	0.18

Bold values indicate $p < 0.05$.

3.3. Correlation between ZEB2 and Infiltrating Immune Cells in Various Types of Cancer

The survival times of patients with several cancers are determined by the quantity and activity status of tumor-infiltrating lymphocytes [28,29]. We explored the correlation between ZEB2 and infiltrating immune cells in various cancers, including OV, using the TIMER database. We analyzed this correlation in various types of cancer, including OV. The results revealed that ZEB2 was significantly positively correlated with the infiltration levels of CD8+T cells (R = 0.24, $p = 132 \times 10^{-6}$), neutrophils (R = 0.576, $p = 224 \times 10^{-25}$), macrophages (R = 0.475, $p = 197 \times 10^{-17}$), and dendritic cells (R = 0.232, $p = 222 \times 10^{-6}$). However, ZEB2 was significantly negatively correlated with the infiltration levels of B cells (R = −0.194, $p = 214 \times 10^{-5}$) in OV (Figure 3). Moreover, ZEB2 correlates with the infiltration levels of CD4+T cells in 20 cancer types, CD8+T cells in 24 cancer types, B cells in 18 cancer types, neutrophils in 29 cancer types, macrophages in 26 cancer types, and dendritic cells in 28 cancer types (Supplementary Table S2). These findings suggest that ZEB2 expression correlates with the infiltration of immune cells in different cancer types, including OV.

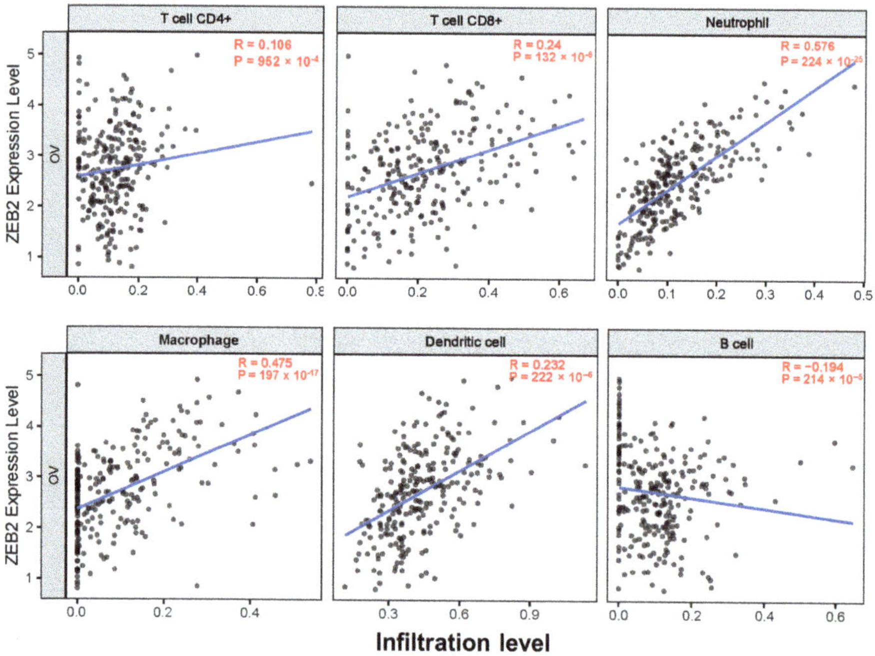

Figure 3. Correlation between ZEB2 expression and infiltrating immune cells in ovarian cancer. The correlation between ZEB2 and infiltrating immune cells (CD4+T cells, CD8+T cells, neutrophils, macrophages, dendritic cells, and B cells) was analyzed using the TIMER database.

3.4. Correlation between ZEB1 and ZEB2 Expression in Various Types of Cancer

To identify the correlation between ZEB1 and ZEB2 expression, we analyzed this correlation in 32 cancer types using the TIMER database (Table 2). Interestingly, these findings revealed that ZEB1 and ZEB2 were positively correlated in all cancer types.

Table 2. Correlation between ZEB1 and ZEB2 expression in various types of cancer.

Cancer Type	R	*p*
Adrenocortical carcinoma	0.30	**0.007**
Bladder urothelial carcinoma	0.82	**0.000**
Breast invasive carcinoma	0.80	**0.000**
Cervical squamous cell carcinoma andendocervical adenocarcinoma	0.71	**0.000**
Cholangiocarcinoma	0.70	**0.000**
Colon adenocarcinoma	0.85	**0.000**
Lymphoid neoplasm diffuse large B-cellLymphoma	0.79	**0.000**
Esophageal carcinoma	0.87	**0.000**
Glioblastoma multiforme	0.49	**0.000**
Head and neck squamous cell carcinoma	0.87	**0.000**
Kidney chromophobe	0.71	**0.000**
Kidney renal clear cell carcinoma	0.75	**0.000**
Kidney renal papillary cell carcinoma	0.47	**0.000**
Brain lower grade glioma	0.54	**0.000**
Liver hepatocellular carcinoma	0.57	**0.000**
Lung adenocarcinoma	0.78	**0.000**
Lung squamous cell carcinoma	0.76	**0.000**
Mesothelioma	0.51	**0.000**
Ovarian serous cystadenocarcinoma	0.78	**0.000**
Pancreatic adenocarcinoma	0.88	**0.000**
Pheochromocytoma and paraganglioma	0.12	**0.097**
Prostate adenocarcinoma	0.86	**0.000**
Rectum adenocarcinoma	0.89	**0.000**
Sarcoma	0.14	**0.029**
Skin cutaneous melanoma	0.20	**0.000**
Stomach adenocarcinoma	0.82	**0.000**
Testicular germ cell tumors	0.76	**0.000**
Thyroid carcinoma	0.69	**0.000**
Thymoma	0.22	**0.017**
Uterine corpus endometrial carcinoma	0.71	**0.000**
Uterine carcinosarcoma	0.52	**0.000**
Uveal melanoma	0.61	**0.000**

Bold values indicate $p < 0.05$.

3.5. Correlations with ZEB2-Related Proteins in Various Types of Cancer

We investigated the correlations with ZEB2-related proteins using the TIMER and STRING databases. We analyzed this correlation in various types of cancer, including OV. These findings reveal that ZEB2 protein was positively correlated with ZEB1 ($R = 0.776$, $p = 345 \times 10^{-64}$), SMAD1 ($R = 0.271$, $p = 171 \times 10^{-8}$), and SMAD2 ($R = 0.282$, $p = 602 \times 10^{-9}$) in OV. However, ZEB2 did not significantly affect other SMADs proteins in OV. In addition, ZEB2 did not significantly affect CTBPs in the OV (Figure 4). Moreover, ZEB2 correlates with SMAD1 in 28 cancer types, SMAD2 in 27 cancer types, SMAD3 in 22 cancer types, SMAD5 in 28 cancer types, CTBP1 in 14 cancer types, and CTBP2 in 22 cancer types (Supplementary Table S3). These findings indicate that ZEB2 correlates with SMADs and CTBPs in various types of cancer.

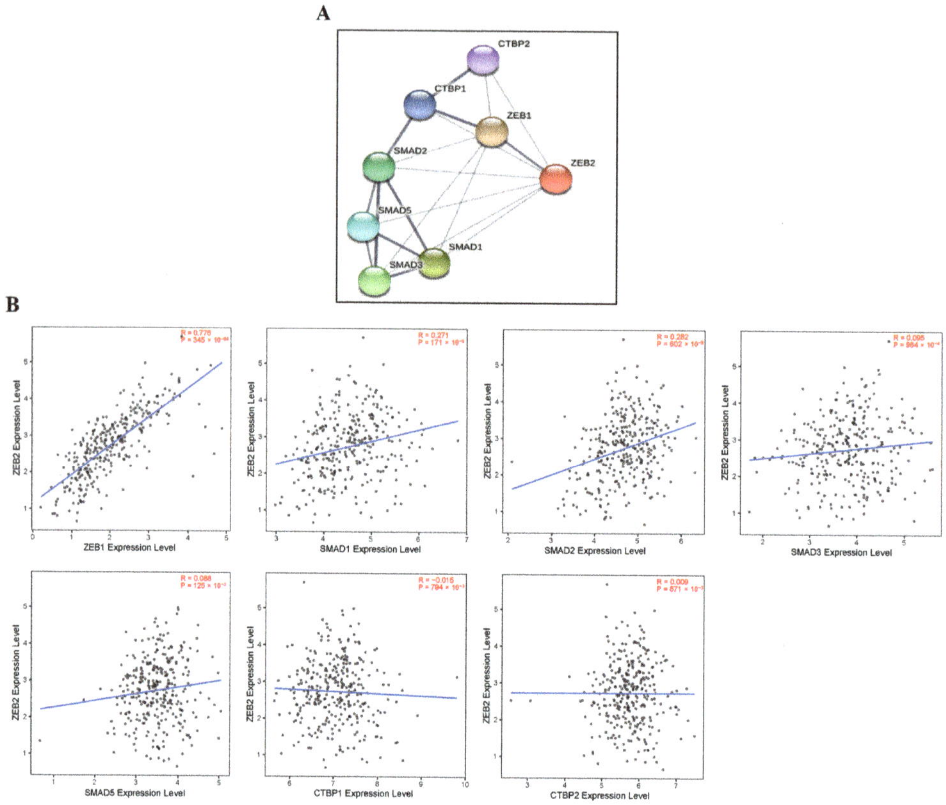

Figure 4. Correlations with ZEB2-related proteins in ovarian cancer. ZEB2 network database in OV was analyzed using the STRING online database (**A**). The correlation between ZEB2 and ZEB2-related proteins (ZEB1, SMAD1, SMAD2, SMAD3, SMAD5, CTBP1 and CTBP2) was analyzed using the TIMER database (**B**).

4. Discussion

OV, the most common gynecological malignancy, is characterized by a relatively high incidence and poor prognosis. In this study, the correlation with ZEB2 was confirmed by factor investigation to improve the prognosis of OV. ZEB2 is a transcription factor belonging to the human ZEB family. ZEB2 regulates gene expression by interacting with specific activators or repressors in various cancers. It is related to the development of cancer stem cell characteristics and treatment resistance, and is recognized as a reliable prognostic marker for cancer patient outcomes [30]. ZEB2 plays a significant role in EMT during tumor invasion and metastasis in a variety of human malignancies. EMT is a biological process characterized by the transformation of an epithelial cell phenotype to a mesenchymal phenotype, which is associated with increased cell motility and invasion. ZEB2 overexpression is known to have an aggressive correlation in a variety of cancers that may be involved in malignant transformation. Our results show that ZEB2 expression was lower in OV and many cancers compared to normal tissue. However, ZEB2 is relatively highly expressed in tumors such as leukemia, brain glioma, pancreatic adenocarcinoma, and melanoma. Previous studies suggest that ZEB2 is associated with a bad prognosis in numerous kinds of cancer [18,19,31]. Nevertheless, the prognostic significance of ZEB2 in OV remains unclear. Thus, we performed a comprehensive meta-analysis to evaluate the prognostic value of ZEB2 in different types of cancer, including OV. Our results show that high ZEB2 expression correlates with poorer PFS, OS, and PPS in OV. Moreover, it correlates with worse PFS, OS, and PPS in grades II + III and III. This result is in agreement with previous studies in OV patients, suggesting its metastatic potential [18,19]. Although its expression was decreased in OV compared to normal tissue, higher EZH2 expression induced peritoneal metastasis, such as cancer stem cells. The tumor-suppressing gene TP53 is involved in cell cycle control and apoptosis after DNA damage. However, when mutations occur, DNA-damaged cells can escape apoptosis and turn into cancer cells [32]. We showed that TP53 mutations were associated with unfavorable OS in the wild type. These findings suggest that ZEB2 is a potential prognostic marker for OV.

Cancer is caused by a variety of causes over a long period of time and is originally caused by cancer cells avoiding the immune system with the advantage of being one's own cells. Moreover, these cancer cells are known to affect the onset, growth, and metastasis of tumors by interacting with immune cells, such as T and B lymphocytes, macrophages, neutrophils, and dendritic cells in the body [33]. Cancer cells, by engaging in a dynamic crosstalk with immune cells, exhibit EMT plasticity to adapt to the changing microenvironment they encounter in the primary tumor, during metastasis, and at distant sites. Tumorigenicity and invasiveness are important acquired characteristics for the development and progression of cancer, and could be regulated by transcription factors associated with EMT, such as ZEB1, ZEB2, SNAI1, SLUG, and STAT3 [34]. Elevated ZEB2 is known to correlate with the acquisition and function of CSCs. Moreover, ZEB2 plays an essential role in NK cell maturation, CD8+T cell differentiation, and dendritic cell development [35–37]. To the best of our knowledge, our study is the first to explore the correlation between ZEB2 and the immune infiltrate in OV. Our findings reveal that ZEB2 is positively correlated with the infiltration levels of CD8+T cells, neutrophils, macrophages, and dendritic cells in OV. However, ZEB2 is negatively correlated with B cell infiltration levels. These findings suggest that ZEB2 may be a potential diagnostic and therapeutic target in patients with OV. Further research is needed to confirm the correlation between ZEB2 expression and infiltrating immune cells.

ZEB1 and ZEB2 have many similarities in transcriptional regulation; they have different expression patterns and molecular and biological roles, such as cell differentiation and disease progression regulation [38,39]. Both ZEB1 and ZEB2 induce EMT and enhance cancer progression. They transform into mesenchymal cells during the EMT process; epithelial cells lose their adhesiveness and become migratory and invasive [40,41]. TGF-β is a crucial cytokine that promotes EMT [42]. TGF-β promotes their expression in some types of normal and malignant cells, as well as certain other EMT-related transcription factors such as Snail and Slug [43,44]. TGF-β reduces E-cadherin expression in mammary epithelial cells via inducing ZEB1 and ZEB2. In recent years, ZEB1 and ZEB2 have been found in a variety of malignancies [45,46]. However, the correlation between ZEB1 and ZEB2 in cancer is yet to be elucidated. Herein, we identified a correlation between ZEB1 and ZEB2 in various cancers. Interestingly, our findings reveal that ZEB2 has a positive correlation in all cancers. ZEB1 and ZEB2 may have multiple functions that will be elucidated by analyses of specific cancer types in the future. The detailed mechanisms of ZEB1 and ZEB2 in cancer should be elucidated.

According to published studies, EMT is an essential mechanism for tumor progression, intrusion, and metastasis, and ZEB2 has been reported as a key transcription factor for EMT [41,47]. ZEB2 has been identified as a protein that interacts with receptor-active SMAD in the pathway related to signals from other members of the TGF-β superfamily [48]. In addition, the TGF-β pathway has been found in breast cancer, non-small cell lung cancer, and other tumors that promote EMT progression [49]. TGF-β stimulates the expression of other EMT-related transcription factors. TGF-β binds to type I and type II receptors, and transduces signals via SMAD and non-SMAD signaling pathways. The TGF-β type I receptor is activated by ligand stimulation and phosphorylates the receptor-regulated SMADs, SMAD2, and SMAD3, which form trimeric complexes with the common partner SMAD and SMAD4. Activation of TGF-β receptors causes the phosphorylation and nuclear translocation of SMAD proteins, which then participate in the regulation of target gene expression. ZEB2 can also mediate transcriptional repression via cooperation with activated SMADs, or through recruitment of the corepressor CTBP, as well as histone deacetylase complexes [50]. Our results show that ZEB2 was positively correlated with SMAD1 and SMAD2 proteins in OV.

5. Conclusions

In conclusion, our results demonstrate a lower expression of ZEB2 in OV; however, patients with high ZEB2 expression may induce poor prognosis. To understand its clinical significance, the criteria of high ZEB2 expression should be investigated in OV, since the majority of OV did not have ZEB2 expression. Moreover, we suggest that increased ZEB2 expression is correlated with infiltrating immune cells and SMAD1 and SMAD2 in OV. Taken together, our findings suggest that ZEB2 could be a potential prognostic biomarker and may provide novel insights into tumor immunology. However, depending on intratumor heterogeneity, the expression level and prognosis of numerous genes may differ even within tumors. Therefore, transcriptome analysis of single cancer cells in tumor heterogeneity will be required to demonstrate the possibility as a potential prognostic biomarker of ZEB2 in various cancers including OV. In addition, the detailed mechanisms by which ZEB2 contributes to the correlation in cancer should be elucidated further.

Supplementary Materials: The following supporting information can be downloaded at: https://www.mdpi.com/article/10.3390/cimb44030079/s1, Table S1: Survival analysis of ZEB2 in various cancers using prognoscan database. Table S2: Correlation between ZEB2 and infiltrating immune cells in different types of cancer using TIMER database. Table S3: Correlations with ZEB2-related proteins in various types of cancer using STRING network database.

Author Contributions: Conceptualization, J.K. and J.-H.L.; methodology, J.K. and J.-H.L.; software, H.-R.K., C.W.S. and S.J.H.; validation, H.-R.K. and S.J.H.; formal analysis, C.W.S.; investigation, J.K. and H.-R.K.; resources, J.K. and H.-R.K.; data curation, S.J.H. and C.W.S.; writing—original draft preparation, J.K., H.-R.K. and J.-H.L.; writing—review and editing, J.K., H.-R.K., C.W.S., S.J.H. and J.-H.L.; visualization, H.-R.K.; supervision, J.K. and J.-H.L.; project administration, J.K., H.-R.K., S.J.H. and J.-H.L.; funding acquisition, S.J.H. All authors have read and agreed to the published version of the manuscript.

Funding: This research was supported by the National Research Foundation Grant funded by the Korean Government (NRF-2021R1C1C1003333 to S.J.H.).

Institutional Review Board Statement: Not applicable.

Informed Consent Statement: Not applicable.

Data Availability Statement: Not applicable.

Conflicts of Interest: The authors declare no conflict of interest.

References

1. Roett, M.A.; Evans, P. Ovarian cancer: An overview. *Am. Fam. Physician* **2009**, *80*, 609–616. [PubMed]
2. Lisio, M.A.; Fu, L.; Goyeneche, A.; Gao, Z.H.; Telleria, C. High-Grade Serous Ovarian Cancer: Basic Sciences, Clinical and Therapeutic Standpoints. *Int. J. Mol. Sci.* **2019**, *20*, 952. [CrossRef]
3. Odunsi, K. Immunotherapy in ovarian cancer. *Ann. Oncol.* **2017**, *28*, viii1–viii7. [CrossRef] [PubMed]
4. Berek, J.S.; Kehoe, S.T.; Kumar, L.; Friedlander, M. Cancer of the ovary, fallopian tube, and peritoneum. *Int. J. Gynaecol. Obstet.* **2018**, *143*, 59–78. [CrossRef]
5. Elias, K.M.; Guo, J.; Bast, R.C., Jr. Early Detection of Ovarian Cancer. *Hematol. Oncol. Clin. N. Am.* **2018**, *32*, 903–914. [CrossRef] [PubMed]
6. Yang, L.P.; Lin, Q.; Mu, X.L. MicroRNA-155 and FOXP3 jointly inhibit the migration and invasion of colorectal cancer cells by regulating ZEB2 expression. *Eur. Rev. Med. Pharmacol. Sci.* **2019**, *23*, 6131–6138.
7. Ma, F.; Li, W.; Liu, C.; Li, W.; Yu, H.; Lei, B.; Ren, Y.; Li, Z.; Pang, D.; Qian, C. MiR-23a promotes TGF-beta1-induced EMT and tumor metastasis in breast cancer cells by directly targeting CDH1 and activating Wnt/beta-catenin signaling. *Oncotarget* **2017**, *8*, 69538–69550. [CrossRef]

8. Zhao, Z.; Zhou, W.; Han, Y.; Peng, F.; Wang, R.; Yu, R.; Wang, C.; Liang, H.; Guo, Z.; Gu, Y. EMT-Regulome: A database for EMT-related regulatory interactions, motifs and network. *Cell Death Dis.* **2017**, *8*, e2872. [CrossRef]
9. Chaffer, C.L.; San Juan, B.P.; Lim, E.; Weinberg, R.A. EMT, cell plasticity and metastasis. *Cancer Metastasis Rev.* **2016**, *35*, 645–654. [CrossRef]
10. Liang, H.; Yu, T.; Han, Y.; Jiang, H.; Wang, C.; You, T.; Zhao, X.; Shan, H.; Yang, R.; Yang, L.; et al. LncRNA PTAR promotes EMT and invasion-metastasis in serous ovarian cancer by competitively binding miR-101-3p to regulate ZEB1 expression. *Mol. Cancer* **2018**, *17*, 119. [CrossRef]
11. Wang, P.; Liu, X.; Han, G.; Dai, S.; Ni, Q.; Xiao, S.; Huang, J. Downregulated lncRNA UCA1 acts as ceRNA to adsorb microRNA-498 to repress proliferation, invasion and epithelial mesenchymal transition of esophageal cancer cells by decreasing ZEB2 expression. *Cell Cycle* **2019**, *8*, 2359–2376. [CrossRef] [PubMed]
12. Fardi, M.; Alivand, M.; Baradaran, B.; Farshdousti Hagh, M.; Solali, S. The crucial role of ZEB2: From development to epithelial-to-mesenchymal transition and cancer complexity. *J. Cell Physiol.* **2019**. Online ahead of print. [CrossRef]
13. Zhang, X.; Xu, X.; Ge, G.; Zang, X.; Shao, M.; Zou, S.; Zhang, Y.; Mao, Z.; Zhang, J.; Mao, F.; et al. miR-498 inhibits the growth and metastasis of liver cancer by targeting ZEB2. *Oncol. Rep.* **2019**, *41*, 1638–1648. [CrossRef] [PubMed]
14. Yu, X.; Wang, W.; Lin, X.; Zheng, X.; Yang, A. Roles of ZEB2 and RBM38 in liver cancer stem cell proliferation. *J. BUON* **2020**, *25*, 1390–1394.
15. Feng, S.; Liu, W.; Bai, X.; Pan, W.; Jia, Z.; Zhang, S.; Zhu, Y.; Tan, W. LncRNA-CTS promotes metastasis and epithelial-to-mesenchymal transition through regulating miR-505/ZEB2 axis in cervical cancer. *Cancer Lett.* **2019**, *465*, 105–117. [CrossRef] [PubMed]
16. Wang, F.; Zhu, W.; Yang, R.; Xie, W.; Wang, D. LncRNA ZEB2-AS1 contributes to the tumorigenesis of gastric cancer via activating the Wnt/beta-catenin pathway. *Mol. Cell. Biochem.* **2019**, *456*, 73–83. [CrossRef]
17. Meng, L.; Yue, X.; Zhou, D.; Li, H. Long non coding RNA OIP5-AS1 promotes metastasis of breast cancer via miR-340-5p/ZEB2 axis. *Oncol. Rep.* **2020**, *44*, 1662–1670. [CrossRef]
18. Li, Y.; Fei, H.; Lin, Q.; Liang, F.; You, Y.; Li, M.; Wu, M.; Qu, Y.; Li, P.; Yuan, Y.; et al. ZEB2 facilitates peritoneal metastasis by regulating the invasiveness and tumorigenesis of cancer stem-like cells in high-grade serous ovarian cancers. *Oncogene* **2021**, *40*, 5131–5141. [CrossRef]
19. Prislei, S.; Martinelli, E.; Zannoni, G.F.; Petrillo, M.; Filippetti, F.; Mariani, M.; Mozzetti, S.; Raspaglio, G.; Scambia, G.; Ferlini, C. Role and prognostic significance of the epithelial-mesenchymal transition factor ZEB2 in ovarian cancer. *Oncotarget* **2015**, *6*, 18966–18979. [CrossRef]
20. Scott, C.L.; Omilusik, K.D. ZEBs: Novel Players in Immune Cell Development and Function. *Trends Immunol.* **2019**, *40*, 431–446. [CrossRef]
21. Chen, B.; Lai, J.; Dai, D.; Chen, R.; Li, X.; Liao, N. JAK1 as a prognostic marker and its correlation with immune infiltrates in breast cancer. *Aging* **2019**, *11*, 11124–11135. [CrossRef]
22. Gu, Y.; Li, X.; Bi, Y.; Zheng, Y.; Wang, J.; Li, X.; Huang, Z.; Chen, L.; Huang, Y.; Huang, Y. CCL14 is a prognostic biomarker and correlates with immune infiltrates in hepatocellular carcinoma. *Aging* **2020**, *12*, 784–807. [CrossRef] [PubMed]
23. Tang, Z.; Li, C.; Kang, B.; Gao, G.; Li, C.; Zhang, Z. GEPIA: A web server for cancer and normal gene expression profiling and interactive analyses. *Nucleic Acids Res.* **2017**, *45*, W98–W102. [CrossRef] [PubMed]
24. Györffy, B.; Lanczky, A.; Eklund, A.C.; Denkert, C.; Budczies, J.; Li, Q.; Szallasi, Z. An online survival analysis tool to rapidly assess the effect of 22,277 genes on breast cancer prognosis using microarray data of 1809 patients. *Breast Cancer Res. Treat.* **2010**, *123*, 725–731. [CrossRef] [PubMed]
25. Mizuno, H.; Kitada, K.; Nakai, K.; Sarai, A. PrognoScan: A new database for meta-analysis of the prognostic value of genes. *BMC Med. Genom.* **2009**, *2*, 18. [CrossRef]
26. Li, T.; Fan, J.; Wang, B.; Traugh, N.; Chen, Q.; Liu, J.S.; Li, B.; Liu, X.S. TIMER: A Web Server for Comprehensive Analysis of Tumor-Infiltrating Immune Cells. *Cancer Res.* **2017**, *77*, e108–e110. [CrossRef]
27. Reddel, R.R. Telomere maintenance mechanisms in cancer: Clinical implications. *Curr. Pharm. Des.* **2014**, *20*, 6361–6374. [CrossRef]
28. Japanese Gastric Cancer Association. Japanese gastric cancer treatment guidelines 2014 (ver. 4). *Gastric Cancer* **2017**, *20*, 1–19. [CrossRef]
29. Ohtani, H. Focus on TILs: Prognostic significance of tumor infiltrating lymphocytes in human colorectal cancer. *Cancer Immun.* **2007**, *7*, 4.
30. Soen, B.; Vandamme, N.; Berx, G.; Schwaller, J.; Van Vlierberghe, P.; Goossens, S. ZEB Proteins in Leukemia: Friends, Foes, or Friendly Foes? *Hemasphere* **2018**, *2*, e43. [CrossRef]
31. Katsura, A.; Tamura, Y.; Hokari, S.; Harada, M.; Morikawa, M.; Sakurai, T.; Takahashi, K.; Mizutani, A.; Nishida, J.; Yokoyama, Y.; et al. ZEB1-regulated inflammatory phenotype in breast cancer cells. *Mol. Oncol.* **2017**, *11*, 1241–1262. [CrossRef] [PubMed]
32. Liu, J.; Ma, Q.; Zhang, M.; Wang, X.; Zhang, D.; Li, W.; Wang, F.; Wu, E. Alterations of TP53 are associated with a poor outcome for patients with hepatocellular carcinoma: Evidence from a systematic review and meta-analysis. *Eur. J. Cancer* **2012**, *48*, 2328–2338. [CrossRef] [PubMed]
33. Teng, F.; Tian, W.Y.; Wang, Y.M.; Zhang, Y.F.; Guo, F.; Zhao, J.; Gao, C.; Xue, F.X. Cancer-associated fibroblasts promote the progression of endometrial cancer via the SDF-1/CXCR4 axis. *J. Hematol. Oncol.* **2016**, *9*, 8–23. [CrossRef] [PubMed]

34. Xavier, P.L.P.; Cordeiro, Y.G.; Rochetti, A.L.; Sangalli, J.R.; Zuccari, D.A.P.C.; Silveira, J.C.; Bressan, F.F.; Fukumasu, H. ZEB1 and ZEB2 transcription factors are potential therapeutic targets of canine mammary cancer cells. *Vet. Comp. Oncol.* **2018**, *16*, 596–605. [CrossRef] [PubMed]
35. van Helden, M.J.; Goossens, S.; Daussy, C.; Mathieu, A.L.; Faure, F.; Marçais, A.; Vandamme, N.; Farla, N.; Mayol, K.; Viel, S.; et al. Terminal NK cell maturation is controlled by concerted actions of T-bet and Zeb2 and is essential for melanoma rejection. *J. Exp. Med.* **2015**, *212*, 2015–2025. [CrossRef] [PubMed]
36. Dominguez, C.X.; Amezquita, R.A.; Guan, T.; Marshall, H.D.; Joshi, N.S.; Kleinstein, S.H.; Kaech, S.M. The transcription factors ZEB2 and T-bet cooperate to program cytotoxic T cell terminal differentiation in response to LCMV viral infection. *J. Exp. Med.* **2015**, *212*, 2041–2056. [CrossRef]
37. Omilusik, K.D.; Best, J.A.; Yu, B.; Goossens, S.; Weidemann, A.; Nguyen, J.V.; Seuntjens, E.; Stryjewska, A.; Zweier, C.; Roychoudhuri, R.; et al. The transcriptional repressor ZEB2 promotes the terminal differentiation of CD8+ effector and memory T cell populations during infection. *J. Exp. Med.* **2015**, *212*, 2027–2039. [CrossRef]
38. Zhang, P.; Sun, Y.; Ma, L. ZEB1: At the crossroads of epithelial-mesenchymal transition, metastasis and therapy resistance. *Cell Cycle* **2015**, *14*, 481–487. [CrossRef]
39. Vandewalle, C.; Comijn, J.; De Craene, B.; Vermassen, P.; Bruyneel, E.; Andersen, H.; Tulchinsky, E.; Van Roy, F.; Berx, G. SIP1/ZEB2 induces EMT by repressing genes of different epithelial cell-cell junctions. *Nucleic Acids Res.* **2005**, *33*, 6566–6578. [CrossRef]
40. Nieto, M.A.; Huang, R.Y.; Jackson, R.A.; Thiery, J.P. EMT: 2016. *Cell* **2016**, *166*, 21–45. [CrossRef]
41. Thiery, J.P.; Acloque, H.; Huang, R.Y.J.; Nieto, M.A. Epithelial-mesenchymal transitions in development and disease. *Cell* **2009**, *139*, 871–890. [CrossRef] [PubMed]
42. Lamouille, S.; Xu, J.; Derynck, R. Molecular mechanisms of epithelial-mesenchymal transition. *Nat. Rev. Mol. Cell Biol.* **2014**, *15*, 178–196. [CrossRef] [PubMed]
43. Heldin, C.H.; Vanlandewijck, M.; Moustakas, A. Regulation of EMT by TGFbeta in cancer. *FEBS Lett.* **2012**, *586*, 1959–1970. [CrossRef]
44. Miyazono, K.; Ehata, S.; Koinuma, D. Tumor-promoting functions of transforming growth factor-beta in progression of cancer. *J. Med. Sci.* **2012**, *117*, 143–152.
45. Diaferia, G.R.; Balestrieri, C.; Prosperini, E.; Nicoli, P.; Spaggiari, P.; Zerbi, A.; Natoli, G. Dissection of transcriptional and cis-regulatory control of differentiation in human pancreatic cancer. *EMBO J.* **2016**, *35*, 595–617. [CrossRef]
46. Krebs, A.M.; Mitschke, J.; Lasierra Losada, M.; Schmalhofer, O.; Boerries, M.; Busch, H.; Boettcher, M.; Mougiakakos, D.; Reichardt, W.; Bronsert, P.; et al. The EMT-activator Zeb1 is a key factor for cell plasticity and promotes metastasis in pancreatic cancer. *Nat. Cell Biol.* **2017**, *19*, 518–529. [CrossRef] [PubMed]
47. Kalluri, R.; Weinberg, R.A. The basics of epithelial-mesenchymal transition. *J. Clin. Investig.* **2009**, *119*, 1420–1428. [CrossRef] [PubMed]
48. Verschueren, K.; Remacle, J.E.; Collart, C.; Kraft, H.; Baker, B.S.; Tylzanowski, P.; Nelles, L.; Wuytens, G.; Su, M.T.; Bodmer, R.; et al. SIP1, a novel zinc finger/homeodomain repressor, interacts with Smad proteins and binds to 5′-CACCT sequences in candidate target genes. *Biol. Chem.* **1999**, *274*, 20489–20498. [CrossRef]
49. Wen, H.; Qian, M.; He, J.; Li, M.; Yu, Q.; Leng, Z. Inhibiting of self-renewal, migration and invasion of ovarian cancer stem cells by blocking TGF-beta pathway. *PLoS ONE* **2020**, *15*, e0230230.
50. Postigo, A.A.; Dean, D.C. ZEB represses transcription through interaction with the corepressor CtBP. *Proc. Natl. Acad. Sci. USA* **1999**, *96*, 6683–6688. [CrossRef]

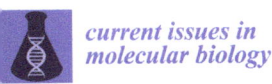

current issues in
molecular biology

Article

Circulating Exosomal miR-1290 for Diagnosis of Epithelial Ovarian Cancer

Hyeji Jeon [1,2,†], Su Min Seo [3,*,†], Tae Wan Kim [2,4], Jaesung Ryu [2,4], Hyejeong Kong [2,4], Si Hyeong Jang [5], Yong Soo Jang [1], Kwang Seock Kim [2], Jae Hoon Kim [6], Seongho Ryu [3,*] and Seob Jeon [1,2,*]

1 Department of Obstetrics and Gynecology, College of Medicine, Soonchunhyang University, Cheonan Hospital, Cheonan 31151, Korea; hjjeon@schmc.ac.kr (H.J.); marqu2moon@naver.com (Y.S.J.)
2 Soonchunhyang Innovative Convergence Research Center, Soonchunhyang University, Cheonan Hospital, Cheonan 31151, Korea; ktwdreem@naver.com (T.W.K.); rjs652@naver.com (J.R.); angelkonghj@gmail.com (H.K.); kimks5005gt@gmail.com (K.S.K.)
3 Soonchunhyang Institute of Med-Bio Sciences (SIMS), Soonchunhyang University, Cheonan 31151, Korea
4 Department of Medical Life Science, Soonchunhyang University, Asan 31538, Korea
5 Department of Pathology, College of Medicine, Soonchunhyang University, Cheonan Hospital, Cheonan 31151, Korea; 82632@schmc.ac.kr
6 Department of Obstetrics and Gynecology, Gangnam Severance Hospital, Yonsei University College of Medicine, Seoul 05029, Korea; jaehoonkim@yuhs.ac.kr
* Correspondence: niceolvia@naver.com (S.M.S.); seonghoryu@gmail.com (S.R.); sjeon4595@gmail.com (S.J.); Tel.: +82-41-570-2150 (S.J.)
† These authors contributed equally to this work.

Abstract: The aim of the study was to develop a new diagnostic biomarker for identifying serum exosomal miRNAs specific to epithelial ovarian cancer (EOC) and to find out target gene of the miRNA for exploring the molecular mechanisms in EOC. A total of 84 cases of ovarian masses and sera were enrolled, comprising EOC ($n = 71$), benign ovarian neoplasms ($n = 13$). We detected expression of candidate miRNAs in the serum and tissue of both benign ovarian neoplasm group and EOC group using real-time polymerase chain reaction. Immunohistochemistry were constructed using formalin fixed paraffin embedded (FFPE) tissue to detect expression level of suppressor of cytokine signaling 4 (SOCS4). In the EOC group, miRNA-1290 was significantly overexpressed in serum exosomes and tissues as compared to benign ovarian neoplasm group (fold change ≥ 2, $p < 0.05$). We observed area under the receiver operating characteristic curve (AUC) for miR-1290, using a cut-off of 0.73, the exosomal miR-1290 from serum had AUC, sensitivity, and specificity values of 0.794, 69.2 and 87.3, respectively. In immunohistochemical study, expression of SOCS4 in EOC was lower than that in benign ovarian neoplasm. Serum exosomal miR-1290 could be considered as a biomarker for differential diagnosis of EOC from benign ovarian neoplasm and SOCS4 might be potential target gene of miR-1290 in EOC.

Keywords: ovarian cancer; exosomal microRNA; biomarker; early diagnosis

Citation: Jeon, H.; Seo, S.M.; Kim, T.W.; Ryu, J.; Kong, H.; Jang, S.H.; Jang, Y.S.; Kim, K.S.; Kim, J.H.; Ryu, S.; et al. Circulating Exosomal miR-1290 for Diagnosis of Epithelial Ovarian Cancer. *Curr. Issues Mol. Biol.* **2022**, *44*, 288–300. https://doi.org/10.3390/cimb44010021

Academic Editor: Peter C. Hart

Received: 9 December 2021
Accepted: 7 January 2022
Published: 9 January 2022

1. Introduction

Ovarian cancer is the second most common cancer among in women worldwide, accounting for about 2.5% of all cancers in women, but has the highest mortality rate of about 5 percent of all cancers; the 5-year survival rate of ovarian cancer is 90% in stage I and 75% in stage II, but less than 30% in stages III and IV [1]. Ovarian cancer, unlike other gynecologic cancers, is difficult to diagnose in early stages because it has few specific symptoms, and at the time of diagnosis more than 75% of the cases are found in advanced stages III or IV of the International Federation of Gynecology and Obstetrics (FIGO) stages [2]. As the death rate has not reduced by more than the incidence rate, this suggests that the improvements and advancements in ovarian cancer screening and treatment have only a modest impact in lowering ovarian cancer death rate.

A representative biomarker in the early screening of epithelial ovarian cancer is serum levels of cancer 125 (CA125). Since it was first described in 1983 by Bast et al. that CA125 was expressed increasingly in epithelial ovarian cancer, CA125 has been widely used as an early screening test for ovarian cancer [3]. However, it is elevated in more than 80% of patients with advanced stage ovarian cancer but only elevated in 50~60% of patients with stage I ovarian cancer, which is somewhat less sensitive in the early stage [4]. Although reproductive-age women with elevated levels of serum CA125 are more likely to develop malignant ovarian tumors, the usefulness of elevated levels of serum CA125 decrease in distinguishing malignant and benign conditions such as endometriosis, uterine fibroids, and pelvic inflammatory disease, pregnancy and menstruation [5]. To compensate for this, in 2009, Moore et al. described the risk of malignant ovarian cancer can be predicted by using ROMA (risk of malignancy) scores calculated from CA125 and human epididymis secretory protein 4(HE4) but Van Gorp et al. reported that the ROMA score is not superior for detecting ovarian cancer when compared to CA125 alone [6,7]. OVA1 is the first test cleared by the U.S. Food and Drug Administration (FDA) for aiding in the pre-surgical evaluation of a woman's ovarian mass for cancer. According to a recent prospective clinical trial, OVA1 test was more sensitive in detecting ovarian cancer than clinical impression and CA125 [8]. Thus, there is a need for useful biomarkers to detect ovarian cancer in early stages. Exosomes are membrane-bounded extracellular vehicles, 30–100 nm in size, which are produced in endosomes of almost all eukaryotic cells [9,10]. Exosomes and extracellular vesicles are found in body fluids including blood (plasma or serum), urine, feces, ascites, etc. Cancer cells produce more exosomes than normal cells, and it is reported that cancer-derived exosomes can promote invasion and proliferation by intercellular communication in tumor microenvironment [10,11]. Recently, many biomolecules such as mRNAs, proteins, miRNAs have been identified in exosomes; therefore, researchers have had a great deal of interest in exosomes mainly because exosomes may play a role in cell to cell signaling through the transport of miRNAs, growth factors, and other small molecules [12]. miRNA-carrying exosomes secreted by tumor cells are likely to be non-invasive biomarkers and potential targetable factors [13]. Recent studies have shown that detection of specific exosomes could be a novel diagnostic tool. In fact, research is underway to apply exosome contents as biomarkers and capsules for therapeutic delivery [14,15]. The aim of the study was to identify serum exosomal miRNAs as a biomarker for early and differential diagnosis of EOC and we compared its expression both in tissues and exosome in serum of patients with EOC and benign ovarian neoplasm, and also tried to find out the expression of the target gene of the miRNA in tissues of EOC and benign ovarian neoplasm.

2. Materials and Methods

2.1. Serum and Tissues Specimens from Patients

All serum and tissue samples were obtained from patients who underwent primary cyto-reductive surgery or tumor resection for EOC and benign ovarian neoplasm, respectively, in Soonchunhyang University Cheonan Hospital and Gangnam Severance Hospital at Yonsei University between 2000 and 2019. Written informed consent was obtained from all patients. The present study was approved by the Institutional Review Board (IRB) of Soonchunhyang University Cheonan Hospital (IRB Number: 2019-10-013-008).

2.2. RNA Isolation and Assessment

Total RNA, including miRNA, was isolated from tissue samples using miRNeasy Mini kit (Qiagen, Hilden, Germany). Tissues were homogenized (IKA Works, Staufen, Germany) with 700 μL QIAzol lysis buffer (Qiagen). Homogenates were processed according to the manufacturer's instructions. RNA was eluted with RNase-free water. The integrity of the RNA was confirmed using Agilent RNA 6000 Pico Kit and Small RNA Kit on Agilent 2100 Bioanalyzer (Agilent Technologies, Santa Clara, CA, USA).

2.3. Serum Exosomal RNA Isolation and Assessment

For exosomal RNA sequencing, serum samples from the patients with 3 benign ovarian neoplasm and 5 EOC were used. Exosomes were isolated from serum using ExoQuick isolation agent (System Bioscience, Palo Alto, CA, USA) in accordance with the manufacturer's instructions. Serum samples (1000 μL) were centrifuged at 3000× *g* for 15 min to remove cells and cell debris. The supernatants were mixed with ExoQuick reagent (System Biosciences, Palo Alto, CA, USA) and incubated for 30 min at 4 °C. After incubation, the samples were centrifuged at 1500× *g* for 30 min to generate an exosome pellet that was resuspended in 100 μL of sterile phosphate-buffered saline (PBS). Total RNA was extracted from the exosomes using the miRNeasy Mini Kit (Qiagen, Hilden, Germany). Exosome suspensions were mixed with 700 μL QIAzol lysis buffer (Qiagen), and the mixtures were processed according to the manufacturer protocol. The RNA was eluted in 20 μL RNase-free water (Qiagen). The size distribution of purified RNA was assessed using an Agilent 2100 Bioanalyzer with an RNA Pico Chip and Small RNA Chip (Agilent Technologies, Santa Clara, CA, USA).

2.4. Small RNA Library Preparation and Sequencing

The samples were processed to produce exosomal RNA (10 ng) as the input for each library. Small RNA libraries were constructed using the SMARTer smRNA-Seq Kit for Illumina (Takara Bio, Kusatsu, Japan) following the manufacturer's directions. Validation of libraries was performed using Agilent Technologies 2100 Bioanalyzer and DNA High Sensitivity Chips. We assessed the quantity of libraries using qPCR according to qPCR Quantification Protocol Guide (KAPA Library Quantification kits for Illumina Sequencing platforms). These libraries were qualified using TapeStation D1000 ScreenTape (Agilent Technologies, Waldbronn, Germany). Pooled libraries were sequenced on an Illumina HiSeq 2500 (Illumina, San Diego, CA, USA) for generating 101 bp single-end reads. Image decomposition and quality values calculation were performed using modules of the Illumina pipeline. Macrogen (Seoul, Korea) processed all steps for next-generation sequencing analysis.

2.5. Identification of Known and Novel miRNAs

Sequence alignment and detection of known and novel miRNAs were performed using miRDeep2 algorithm (Berlin Institute for Medical Systems Biology at the Max-Delbruck-Center for Molecular Medicine, Berlin-Buch, Germany). Prior to performing sequence alignment, *Homo sapiens* hg19 reference genome was retrieved from UCSC genome browser and indexed using Bowtie (1.2.3-07/05/2019; http://bowtie-bio.sourceforge.net/, accessed on 20 November 2021) to align sequencing reads to reference sequences. Sequence alignment was then performed for *Homo sapiens* matured and precursor miRNAs obtained from miRBase v21 (http://www.mirbase.org/, accessed on 20 November 2021).

2.6. Proportions of miRNAs and Other RNAs

Uniquely clustered reads were aligned to the reference genome, miRBase v21, and non-coding RNA database RNA central release 10.0 (https://rnacentral.org/, accessed on 20 November 2021) to classify known miRNAs and other types of RNAs, respectively.

2.7. Statistical Analysis of Differential miRNA Expression

Reads for miRNAs were subjected to Relative Log Expression (RLE) normalization with DESeq2 R library (Genome Biology Unit, European Molecular Biology Laboratory, Heidelberg, Germany). For preprocessing, mature miRNAs with zero counts across more than 60% of all samples were excluded. We added 1 with normalized read count of filtered miRNAs to facilitate log2 transformation to make the correlation plot. For each miRNA, logCPM (Counts Per Million) and log fold change were calculated between the test and control. A statistical hypothesis test for the comparison of the two groups was conducted using binomial Wald Test in DESeq2. $|\text{Fold change}| \geq 2$ and $p < 0.05$ were used to identify differentially expressed miRNAs between two groups. Hierarchical clustering analysis using complete linkage and Euclidean distance was performed to display expression patterns of differentially expressed miRNAs that satisfied |fold change| ≥ 2 and $p < 0.05$. Differentially expressed genes were analyzed and visualized using R 3.6.1 (The R Foundation for Statistical Computing, Vienna, Austria).

2.8. miRNA Preparation and Validation by qRT-PCR

Two independent validations using qRT-PCR in FFPE and serum samples from the same patients were performed. In FFPE validation, 15 benign and 67 malignant FFPE samples were used. Total RNA, including miRNA, was extracted from Formalin-fixed, paraffin-embedded (FFPE) samples using miRNeasy FFPE kit (Qiagen, Hilden, Germany) according to the manufacturer's instructions. In serum validation, 13 sera from benign patients and 71 sera from malignant patients were utilized for qRT-PCR. The RNA concentrations were quantified using a NanoDrop™2000 (Thermo, Waltham, MA, USA). Subsequently, RNA was reverse-transcribed using the TaqMan MicroRNA Reverse Transcription Kit (Applied Biosystems) for two upregulated (miR-1246 and miR-1290) miRNAs. The quantitative RT-PCR was performed on StepOnePlus® Real-Time PCR System (Applied Biosystems) following the manufacturer's recommendation with a standard relative quantification thermal cycling program. RNU48 (Applied Biosystems, Waltham, MA, USA) was served as an endogenous control. The relative expression level of each miRNA between two groups was determined by the $-\Delta\Delta$Ct method.

2.9. Cell Transfection & Inhibition of miR-1290

We used SKOV3-seeded cell density 5×10^5 per well for 6-well cell culture plate and cultured in 37 °C CO_2 incubator for 24 h miR-1290 inhibitor (mirVana miRNA inhibitor, Life Technologies, Carlsbad, CA, USA) and Lipofectamine RNAiMAX Transfection Reagent (Life Technologies, Carlsbad, CA, USA) were diluted in serum-free media of the 1:1 ratio. Prepared Transfection reagent-miR inhibitor solution was treated 250 μL per well. The cells were transfected by 25 pM of inhibitor in triplicate on 6-well cell culture plate and incubated for 72 h. Additionally, to exclude the influence of transfection reagents on gene expression, we used the untreated cell line SKOV3 as a control. Thereafter, transfected cells were treated with trypsin to harvest cell pellets.

2.10. miRNA Preparation and Validation by qRT-PCR

Total RNA was isolated from cell lines using the Hybrid-R™ RNA extraction kit (GeneAll, Seoul, Korea). Reverse transcription was performed using the ReverTra Ace® qPCR RT kit (Toyobo, Osaka, Japan). Real time PCR was performed using the SYBR® Green Real-time PCR Master Mix kit (Toyobo) and the following primer pairs: SOCS4 forward: 5′-ACC AAG AAA GGA AGC ACA GC-3′ and reverse 5′-TGA TCG AGG TGG GAA AGG AC-3′; and GAPDH forward: 5′-TGT TCG TCA TGG GTG TGA AC-3′ and reverse: 5′-GCA GGG ATG ATG TTC TGG AG-3′. The PCR cycle included one cycle at 95 °C for 3 min, followed by 40 cycles at 95 °C for 15 s, 60 °C at 15 s, and 72 °C for 25 s.

2.11. Western Blotting

Cells were washed in phosphate-buffered saline (PBS) and lysed in Pro-Prep™ protein extraction solution (INtRON, Seongnam, Korea). The lysate was centrifuged and the supernatant was denatured by boiling. The protein concentration was determined by the bicinchoninic acid (BCA) assay. Equal quantities of protein (30 μg/lane) were resolved by 10% sodium dodecyl sulfate-polyacrylamide gel electrophoresis (SDS-PAGE) and transferred onto an Immobilon polyvinylidene difluoride membrane (Millipore, Billerica, MA, USA). The membrane was then blocked for 1 h in 5% skim milk. Membranes were incubated overnight at 4 °C with anti-human SOCS4 antibody (Mybiosource, San Diego, CA, USA, MBS7043907) diluted 1:1000 and incubated with the diluted 1:1000 secondary antibody for 2 h at room temperature. The signal was detected using ECL Western detection reagents

(Advansta, San Jose, CA, USA, K-12049-D50) and a Molecular Images were captured by CheBi (Cellgentek, Deajeon, Chemi-luminescence Bioimaging Instrument (Thermo Fischer, Waltham, MA, USA)).

2.12. Immunohistochemistry

Paraffin-embedded patient tissue blocks sectioned at 4 μm thickness. The slides were allowed to dry for a day and were left to warm at 60 °C degrees for an hour. For antigen retrieval, 3% H_2O_2 and 95 °C antigen retriever buffer were used. For permeability, 0.2% triton solution was treated and for blocking, 5% BSA in PBS was treated for 15 min. The 1st antibody was treated with Rabbit Polyclonal antibody SOCS4 (1:100, Mybiosource, MBS7043907) and incubated at 4 °C overnight. Additionally, Goat anti-Rabbit IgG (H + L), HRP (1:100, Thermofisher, A11008) was treated as a secondary antibody for 1 h at room temperature. Stained with DAB Substrate Kit (3,3′-diaminobenzidine, VECTORLABS, SK-4100) and counterstaining with 50% Hematoxylin for 30 s. Then, the slide was dried at 37 °C for 1 h (Mounted with Eukitt® Quick-hardening mounting medium (Sigma Aldrich, St. Louis, MO, USA, 03989-100)). Slides were interpreted twice by two pathologists.

2.13. Statistical Analysis

Statistical evaluation was performed with IBM® SPSS Statistics 21 (Chicago, IL, USA) and GraphPad Prism Software version 6.0 (San Diego, CA, USA). Statistical relevance of the relative expression between benign and malignant samples was analyzed by the unpaired *t* test. The values were presented as the median, the interquartile range and the standard deviation. A p-value ≤ 0.05 was considered to be statistically significant. Receiver operating characteristic (ROC) curve analysis and the area under the curve (AUC) were used to assess diagnostic performance for each biomarker individually (estimate the feasibility of using the serum exosomal miRNA expression levels as diagnostic markers for discriminating OC patients from benign patients). The results were presented as odds ratios with 95% confidence intervals and p-values.

3. Results

3.1. Baseline Clinical Characteristics

First, we have confirmed the baseline clinical characteristics of the patients with EOC group and benign ovarian neoplasm group. Most of the histology in malignancy group and benign ovarian neoplasm group is high-grade serous carcinoma and serous cystadenoma, respectively. The median age was 51 years (range 24–81). 33 and 30 (44.26 and 42.25%) patients were FIGO stage I and II, 34 and 41 (50.75 and 57.75%) patients were FIGO stage ≥ III at diagnosis from FFPE and serum samples (Tables 1 and 2). Tumor marker CA125 was above normal range around 80% in EOC group and over 25% in benign ovarian neoplasm group ($p < 0.001$).

Table 1. Characteristics of the FFPE sample from patients.

FFPE	Total (n = 82)	Type		p-Value
		Benign (n = 15)	EOC (n = 67)	
Age (years, mean ± SD)	51.80 ± 11.74	46.27 ± 15.72	53.04 ± 10.4	0.1297
FIGO stage (%)	n (%)	n (%)	n (%)	
Stage1	26 (38.81)	-	26 (38.81)	-
Stage2	7 (10.44)	-	7 (10.44)	
Stage3	26 (38.81)	-	26 (38.81)	
Stage4	8 (11.94)	-	8 (11.94)	

Table 1. *Cont.*

FFPE	Total (n = 82)	Type		p-Value
		Benign (n = 15)	EOC (n = 67)	
CA 125 (U/mL)				
≥35	60 (73.17)	4 (26.67)	56 (83.58)	<0.001
<35	22 (26.83)	11 (73.33)	11 (16.42)	
miR-1290				
≥1.71	66 (80.49)	65 (97.01)	1 (6.67)	<0.001
<1.71	16 (19.51)	2 (2.99)	14 (93.33)	

Table 2. Characteristics of the serum sample from patients.

Serum	Total (n = 84)	Type		p-Value
		Benign (n = 13)	EOC (n = 71)	
Age (years, mean ± SD)	51.81 ± 11.66	45.46 ± 14.36	52.97 ± 10.82	0.0933
FIGO stage (%)	***n* (%)**	***n* (%)**	***n* (%)**	
Stage1	25 (35.21)	-	25 (35.21)	>0.99
Stage2	5 (7.04)	-	5 (7.04)	
Stage3	34 (47.89)	-	34 (47.89)	
Stage4	7 (9.86)	-	7 (9.86)	
CA 125 (U/mL)				
≥35	64 (76.19)	6 (46.15)	58 (81.69)	0.011
<35	20 (23.81)	7 (53.85)	13 (18.31)	
miR-1290				
≥0.73	66 (78.57)	4 (30.77)	62 (87.32)	<0.001
<0.73	18 (21.43)	9 (69.23)	9 (12.68)	

3.2. Differentially Expressed miRNAs in Malignant Ovarian Cancer Patients Based on RNA Sequencing: Identification of Candidate miRNAs

To investigate the tissue origin of miRNAs, we compared the miRNA expression between matched serum and tissue samples. Based on this, we identified 81 miRNAs (44 upregulated and 37 downregulated miRNAs) in the tissues of six patients with EOC compared with patients with benign ovarian neoplasm. Furthermore, a total of 26 exosomal miRNAs (15 upregulated and 11 downregulated miRNAs) were identified as differentially expressed between sera of EOC group and benign ovarian neoplasm group ($|\text{Fold change}| \geq 2$ and $p < 0.05$; Figure 1).

Results are also shown in a volcano plot from serum and tissue samples (Figure 2). Only seven miRNAs (hsa-miR-1246, hsa-miR-1290, has miR-21-5p, hsa-miR-7-5p, hsa-miR-93-5p, hsa-miR-16-5p and hsa-miR-29c-3p) were identified as differentially expressed in both matched tissue and serum samples (Table 3).

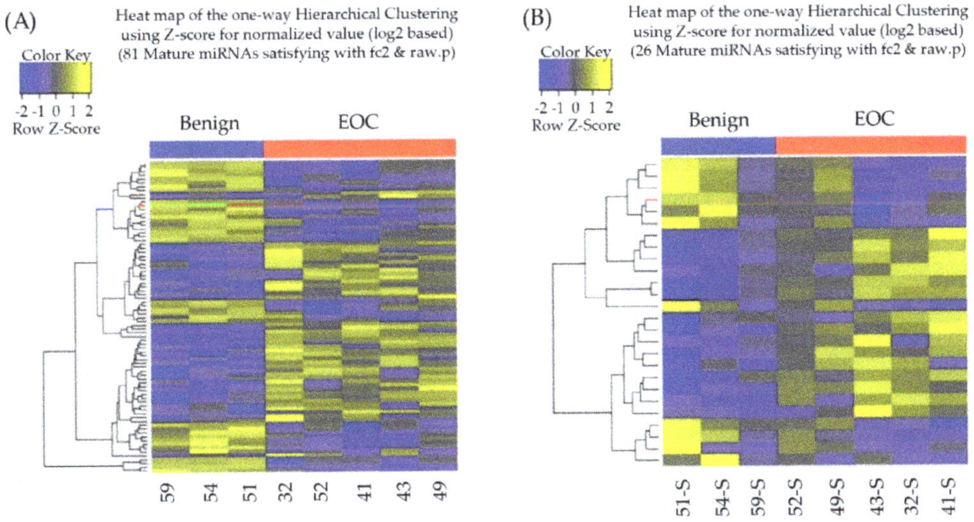

Figure 1. (**A**) Heatmap showing z score of miRNAs from benign tumors ($n = 3$) and malignant tumors ($n = 5$) with 44 upregulated (yellow) and 37 downregulated (blue) miRNAs. (**B**) Heat map showing z score of exosomal miRNAs from serum of benign tumors ($n = 3$) and malignant tumors ($n = 5$) with 15 upregulated (yellow) and 11 downregulated (blue) miRNAs.

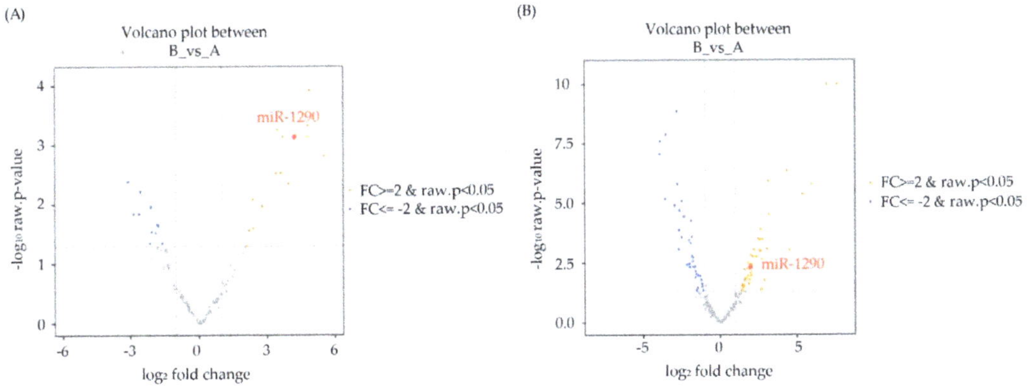

Figure 2. (**A**) Serum. (**B**) FFPE: Volcano plot showing differences in microRNAs (miRNAs) expression between benign and malignant ovarian cancer patients. Yellow dots represent upregulated miRNAs and blue dots represent downregulated miRNAs.

Table 3. Fold change of miRNAs expressions in EOC group compared to benign ovarian neoplasm group.

Mature miRNA	Fold Change in Tissue	Fold Change in Serum
has-miR-1246	6.78	27.61
has-miR-1290	3.66	27.04
has-miR-21-5p	4.1	−3.43
has-miR-7-5p	41.33	−4.27
has-miR-93-5p	3.76	−4.39
has-miR-16-5p	2.63	−8.58
has-miR-29c-3p	−2.4	−3.05

3.3. Validation of Candidate miRNAs by qRT-PCR

From the differentially expressed miRNAs, we chose to focus on miR-1246 and miR-1290 which were the topmost upregulated miRNAs among common miRNAs between tissue and serum exosome. To demonstrate the expression levels of candidate miRNAs to ovarian cancer tissues and sera, their levels were examined by qRT-PCR in FFPE and sera of patients with EOC. Both miR-1246 and miR-1290 showed consistent upregulation in EOC FFPE samples (n = 67) compared with benign ovarian neoplasm FFPE samples (n = 15) ($p < 0.01$ for miR-1246 and miR-1290). In serum, the expression level of miR1246 and miR-1290 was higher in EOC patients (n = 71) compared to benign ovarian neoplasm patients (n = 13). The p-value of miR-1290 was 0.0005 but miR-1246 was not significant (Figure 3).

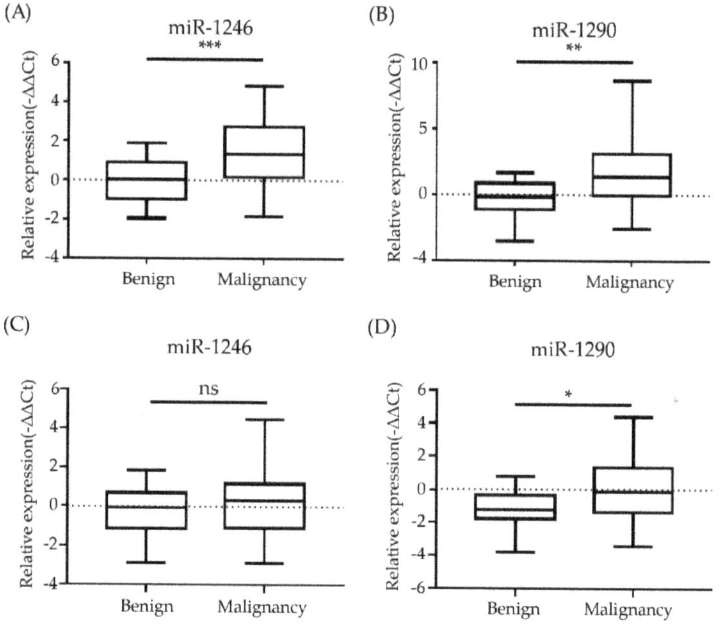

Figure 3. Evaluation of the two selected miRNA expression values in FFPE and serum samples by quantitative RT-PCR (**A**,**B**) FFPE: A total of 45 (15 benign ovarian neoplasm, 67 EOC) (**C**,**D**) Serum: A total of 84 (13 benign ovarian neoplasm, 71 EOC) were used for qRT-PCR. * $p < 0.05$, ** $p < 0.01$, *** $p < 0.001$. ns: not significant.

3.4. Diagnostic Value of Exosomal miRNA 1290 for Epithelial Ovarian Cancer

The diagnostic performance value of serum miR-1290 was calculated by ROC curve analysis to discriminate EOC group from benign ovarian neoplasm group. Using a cut-off of 1.71, the miR-1290 from tissue had AUC, sensitivity, specificity positive predictive (PP) and negative predictive (NP) values of 0.988, 93.3, 97.0, 87.5 (0.616–0.985) and 98.5% (0.918–1.000), respectively. Using a cut-off of 0.73, the exosomal miR-1290 from serum had AUC, sensitivity, specificity, PP and NP values of 0.794, 69.2, 87.3, 50 (0.26–0.74) and 94% (0.852–0.983), respectively (Figure 4 and Supplementary Figure S3). Although the diagnostic performance of serum exosomal miRNA-1290 was not better than CA125, the combination of serum exosomal miR-1290 and CA125 significantly improved AUC value from 0.775 to 0.856 compared to serum exosomal miR-1290 alone ($p < 0.001$; Figure 4); however comparing to CA125 alone, neither serum exosomal miR-1290 alone nor the combination of serum exosomal miR-1290 and CA125 improved AUC value (Supplementary Figure S1).

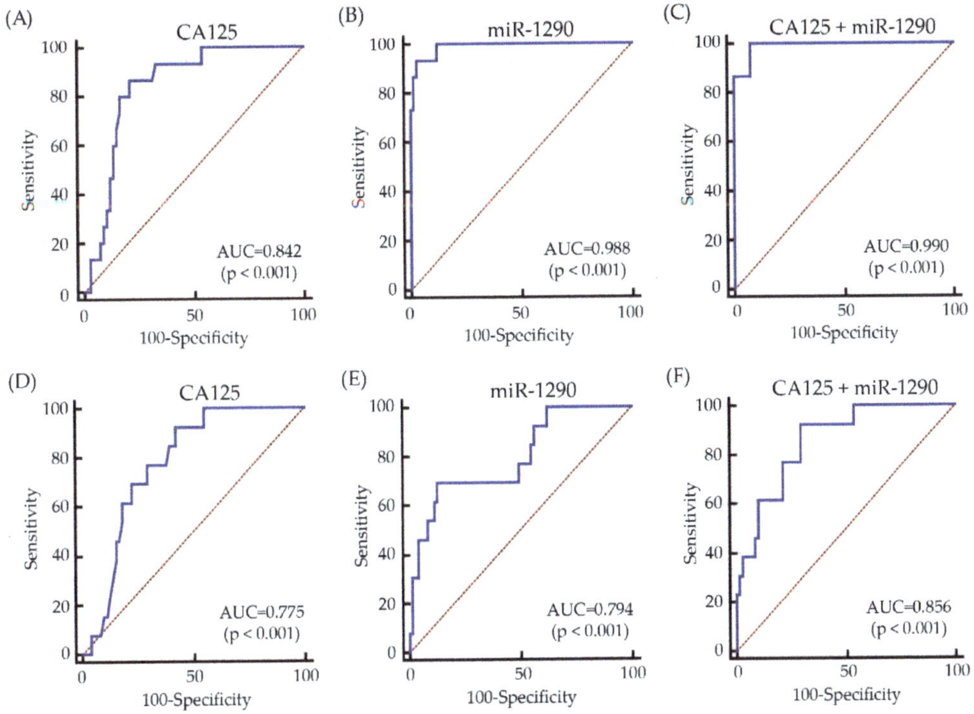

Figure 4. ROC curves for the identification of patients with EOC vs. benign ovarian neoplasm controls based on the expression of CA125, miR-1290, and the combination of both. The AUC values are shown on the graphs. (**A–C**) FFPE, (**D–F**) Serum.

3.5. Decreased Expression of SOCS4 in Malignancy Group That Was Negatively Regulated by miRNA 1290

To determine the targets of miR-1290 were used web tools include Targetscan and miRanda, SOCS4 was selected as a potential target of miR-1290. After treatment with miR-1290 inhibitor, miR-1290 expression level in SKOV3 cells was decreased (Figure 5A) whereas mRNA and protein expression of SOCS4 was significantly increased in SKOV cells ($p < 0.001$; Figure 5B,C and Supplementary Figure S2). In the present study, immunohistochemistry revealed that expression of SOCS4 decreased significantly in EOC group ($n = 38$) compared to benign ovarian neoplasm group ($n = 12$) ($p < 0.001$; Figure 5D,F). In benign ovarian neoplasm group, there were no patients with grade 0 or grade 1 SOCS4 expression, while 63% of EOC has grade 0 or grade 1 expression.

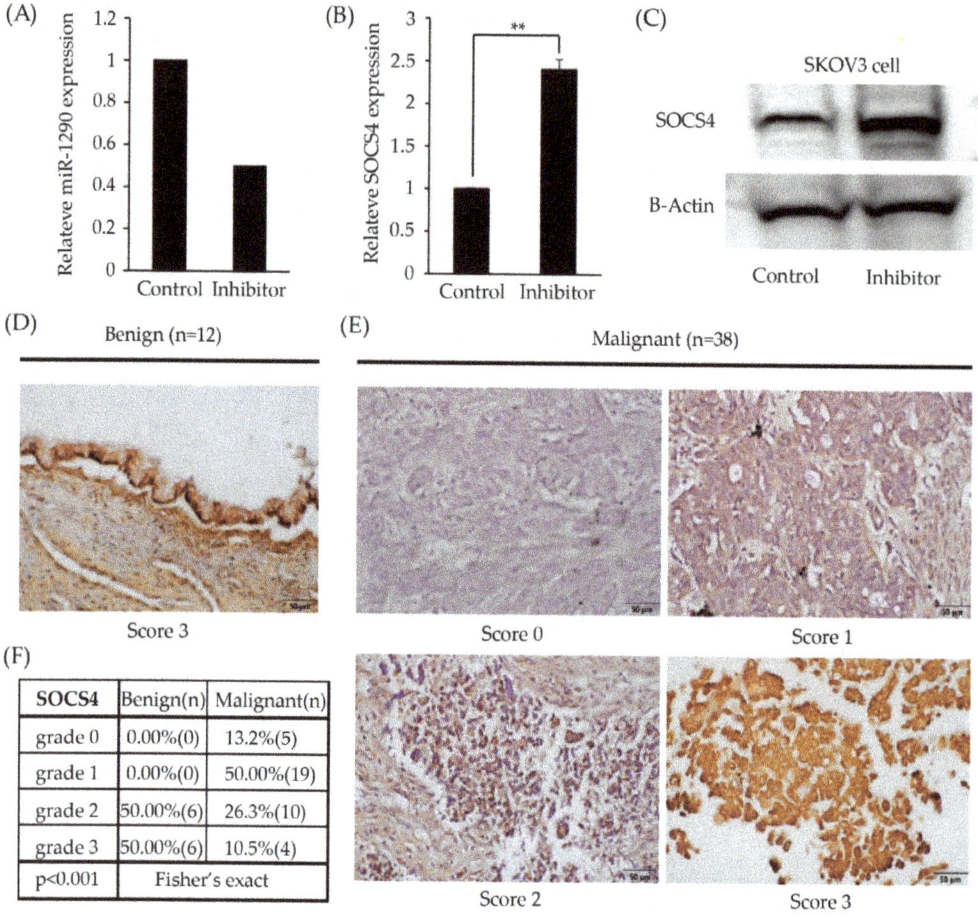

SOCS4	Benign(n)	Malignant(n)
grade 0	0.00%(0)	13.2%(5)
grade 1	0.00%(0)	50.00%(19)
grade 2	50.00%(6)	26.3%(10)
grade 3	50.00%(6)	10.5%(4)
p<0.001	Fisher's exact	

Figure 5. SOCS4 was a target of miR-1290. (**A**) The miR-1290 expression level was decreased after treatment of miR-1290 inhibitor in SKOV3 cell lines by RT-qPCR. (**B**,**C**) SOCS4 expression was increased in miR-1290 inhibitor treated SKOV3 cells by RT-qPCR and Western blot. B-Actin was used as the loading control. Representative images of nuclear SOCS4 immunohistochemistry staining in EOC and benign ovarian neoplasm (scale bar = 50 μm). (**D**) Benign ovarian neoplasm and (**E**) Ovarian cancer, detection of SOCS4 protein expressions in patients' tissues via immunohistochemistry method. (**F**) The expression level of SOCS4 in ovarian cancer tissues is significantly decreased compared with that of benign ovarian tumor. SOCS, suppressor of cytokine signaling. ** $p < 0.0001$.

4. Discussion

In this study, based on RNA sequencing, analysis of serum exosomal miRNA expression profiles revealed that three miRNAs including hsa-miR-1246, hsa-miR-1290 and hsa-miR-29c-3p exhibited the same regulation direction in both tumor tissues and sera. We think it is extremely important that if certain exosomal miRNA increases in the serum, same miRNA should increase in the cancer tissue. Therefore, we performed qRT-PCR to detect expression levels of hsa-miR-1246, hsa-miR-1290 and hsa-miR-29c-3p both in the serum and the tissue of patients with epithelial ovarian cancer (EOC) and benign ovarian neoplasm, and we confirmed that the expression of tissue miR1246 and miR-1290 were statistically higher in patients with EOC than in patients with benign ovarian neoplasm, whereas the expression of serum exosomal miR-1290 was statistically higher in patients with EOC than in patients with benign ovarian neoplasm. Serum and tissue miR-1290 was significantly elevated in patients with EOC compared to patients with benign ovarian neoplasm and could discriminate malignancy versus benign ovarian neoplasm with an area under the curve (AUC) of 0.932 and 0.812, respectively, suggesting its potential as both a useful biomarker for differential diagnosis and prognostic factor. The combination of miR-1290 and CA125 significantly improved the AUC value from 0.812 to 0.95.4. The OVA1 test seems to be more sensitive than CA125 or miR-1290 in the preoperative clinical setting; however, a higher false positive rate is still an unresolved issue.

In 2008, Taylor et al. reported eight exosomal miRNAs (miR-21, miR-141, miR-200a, miR-200b, miR-200c, miR-203, miR-205, miR-214) extracted from serum of ovarian cancer patients. It was reported for the first time that there was an increase compared to benign ovarian tumors. In that study, circulating tumor exosomes were isolated using an anti-EpCAM-modified MACS procedure, and the microRNA profile of ovarian tumors was compared with that of tumor exosomes isolated from the same patient. These results mean that circulating miRNA profiles accurately reflects the tumor profiles [16]. In a cohort of 56 high-grade serous ovarian cancer (HGSOC) patients, Shah et al. reported that Serum miR-375, along with Ca125, is a biomarker that discriminates between normal women and patients with ovarian cancer [17].

Kobayashi et al. also reported that miR-1290 is a specific discriminator for HGSOC suggesting the potential of this miRNA as a biomarker for HGSOC. However, they did not use next generation RNA sequencing but used miRNA microarray to detect candidate miRNAs, and they validated these miRNAs not in tumor tissues from patients but only in cell lines [18].

miR-1290 was first reported in human embryonic stem cells [19]. It was reported that upregulation of miR-1290 was associated with progression of various cancers, including pancreatic cancer [20], esophageal squamous cell carcinoma [21] and colon cancer [22]. Zhang et al. indicated that miR-1290 was a tumor-initiating, cell-specific RNA, which were crucial drivers of tumor initiation and progression in non-small cell lung cancer (NSCLC) [23]. A previous study reported that miR-1290 was upregulated in tissues and serum samples from patients with lung adenocarcinoma and correlated with poor prognosis, and suppressor of cytokine signaling 4 (SOCS4) was target of miR-1290, by targeting SOCS4, miR-1290 facilitated the JAK/STAT3 and PI3K/AKT pathways [24].

SOCS family is a group of cytokine-inducible negative regulators by inhibiting multiple signaling pathways, especially the JAK/STAT signaling pathway [25,26]. SOCS4 was previously shown to be associated with earlier tumor stage and better clinical prognosis in breast cancer [27].

In this study, the expression levels of SOCS4 were examined in EOC and benign ovarian neoplasm. Decreased expression of SOCS4 was shown in EOC compared to benign ovarian neoplasm. Although we did not undertake clinic pathological analysis due to limited patients' medical information and small sample size, to our knowledge this is first report that showed difference of SOCS4 expression in EOC and benign ovarian neoplasm. Several limitations should be acknowledged. The sample size is too small to reach a solid conclusion. Large and prospective registry-embedded trials would be needed to strengthen our hypothesis that serum miR-1290 can serve as a biomarker of EOC. Additionally, the mechanisms and potential pathways between miR-1290 and other target genes, including SOCS4 in EOC, need to be explored. Despite the small sample size, our results show that miR-1290 might be a useful biomarker which can discriminate epithelial ovarian cancer from benign ovarian neoplasm. Further studies would be required to explore other target genes of miR1290 and to identify the true target genes of miR-1290 in EOC, in order to clarify its role.

Supplementary Materials: The following supporting information can be downloaded at: https://www.mdpi.com/article/10.3390/cimb44010021/s1, Figure S1: The comparison diagnostic performance of miR-1290 with CA125 alone (A), CA125 with the combination (B), and miR-1290 with the combination (C). Figure S2: SOCS4 Western blotting full image. Figure S3: The dot plot data for the identification of patients with EOC vs. benign ovarian neoplasm controls based on the expression of CA125 and miR-1290. (A,B) FFPE, (C,D) Serum.

Author Contributions: Conceptualization, S.J. and S.R.; methodology, and software, T.W.K., J.R., H.K., S.H.J., J.H.K. and Y.S.J.; validation, and formal analysis, H.J. and S.M.S.; investigation, T.W.K., J.R. and H.K.; resources, S.J. and S.R.; all authors contributed to the writing, review, and/or revision of the manuscript. All authors have read and agreed to the published version of the manuscript.

Funding: This research was supported by the Bio and Medical Technology Development Program of the National Research Foundation (NRF) funded by the Korean Government (MSIT) (No. NRF-2019M3E5D1A02069068) and was supported by the Soonchunhyang University Research Fund 202200007.

Institutional Review Board Statement: The present study was approved by the Institutional Review Board (IRB) of Soonchunhyang University Cheonan Hospital (IRB Number: 2019-10-013-008) and adhered to the principles in the Declaration of Helsinki. Informed consent was obtained from each patient.

Informed Consent Statement: Not applicable.

Data Availability Statement: Not applicable.

Acknowledgments: Some paraffin blocks were provided by the Korea Gynecologic Cancer Bank through the Bio & Medical Technology Development Program of Ministry of Education, Science and Technology, Korea (NRF-2017M3A9B8069610).

Conflicts of Interest: The authors declare no conflict of interest.

References

1. Siegel, R.L.; Miller, K.D.; Jemal, A. Cancer statistics, 2019. *CA Cancer J. Clin.* **2019**, *69*, 7–34. [CrossRef] [PubMed]
2. Goff, B.; Mandel, L.; Muntz, H.G.; Melancon, C.H. Ovarian carcinoma diagnosis. *Cancer* **2000**, *89*, 2068–2075. [CrossRef]
3. Bast, R.; Feeney, M.; Lazarus, H.; Nadler, L.M.; Colvin, R.B.; Knapp, R.C. Reactivity of a monoclonal antibody with human ovarian carcinoma. *J. Clin. Investig.* **1981**, *68*, 1331–1337. [CrossRef]
4. Sölétormos, G.; Duffy, M.J.; Abu Hassan, S.O.; Verheijen, R.H.; Tholander, B.; Bast, R.C.; Gaarenstroom, K.N.; Sturgeon, C.M.; Bonfrer, J.M.; Petersen, P.H.; et al. Clinical Use of Cancer Biomarkers in Epithelial Ovarian Cancer: Updated Guidelines from the European Group on Tumor Markers. *Int. J. Gynecol. Cancer* **2016**, *26*, 43–51. [CrossRef]
5. Van Calster, B.; Timmerman, D.; Bourne, T.; Testa, A.C.; Van Holsbeke, C.; Domali, E.; Jurkovic, D.; Neven, P.; Van Huffel, S.; Valentin, L. Discrimination between benign and malignant adnexal masses by specialist ultrasound examination versus serum CA-125. *J. Natl. Cancer Inst.* **2007**, *99*, 1706–1714. [CrossRef]
6. Moore, R.G.; McMeekin, D.S.; Brown, A.K.; DiSilvestro, P.; Miller, M.C.; Allard, W.J.; Gajewski, W.; Kurman, R.; Bast, R.C., Jr.; Skates, S.J. A novel multiple marker bioassay utilizing HE4 and CA125 for the prediction of ovarian cancer in patients with a pelvic mass. *Gynecol. Oncol.* **2009**, *112*, 40–46. [CrossRef]
7. Van Gorp, T.; Cadron, I.; Despierre, E.; Daemen, A.; Leunen, K.; Amant, F.; Timmerman, D.; De Moor, B.; Vergote, I. HE4 and CA125 as a diagnostic test in ovarian cancer: Prospective validation of the Risk of Ovarian Malignancy Algorithm. *Br. J. Cancer* **2011**, *104*, 863–870. [CrossRef]
8. Miller, R.W.; Smith, A.; DeSimone, C.P.; Seamon, L.; Goodrich, S.; Podzielinski, I.; Sokoll, L.; van Nagell, J.R., Jr.; Zhang, Z.; Ueland, F.R. Performance of the American College of Obstetricians and Gynecologists' ovarian tumor referral guidelines with a multivariate index assay. *Obstet Gynecol.* **2011**, *117*, 1298–1306. [CrossRef]
9. Green, T.M.; Alpaugh, M.L.; Barsky, S.H.; Rappa, G.; Lorico, A. Breast Cancer-Derived Extracellular Vesicles: Characterization and Contribution to the Metastatic Phenotype. *BioMed Res. Int.* **2015**, *2015*, 634865. [CrossRef]
10. Wang, M.; Yu, F.; Ding, H.; Wang, Y.; Li, P.; Wang, K. Emerging Function and Clinical Values of Exosomal MicroRNAs in Cancer. *Mol. Ther. Nucleic Acids* **2019**, *16*, 791–804. [CrossRef]
11. Sempere, L.F.; Keto, J.; Fabbri, M. Exosomal MicroRNAs in Breast Cancer towards Diagnostic and Therapeutic Applications. *Cancers* **2017**, *9*, 71. [CrossRef] [PubMed]
12. Meng, Y.; Sun, J.; Wang, X.; Hu, T.; Ma, Y.; Kong, C.; Piao, H.; Yu, T.; Zhang, G. Exosomes: A Promising Avenue for the Diagnosis of Breast Cancer. *Technol. Cancer Res. Treat.* **2019**, *18*, 1533033818821421. [CrossRef]

13. Bahrami, A.; Aledavood, A.; Anvari, K.; Hassanian, S.M.; Maftouh, M.; Yaghobzade, A.; Salarzaee, O.; ShahidSales, S.; Avan, A. The prognostic and therapeutic application of microRNAs in breast cancer: Tissue and circulating microRNAs. *J. Cell. Physiol.* **2018**, *233*, 774–786. [CrossRef]
14. Van der Meel, R.; Fens, M.H.; Vader, P.; van Solinge, W.W.; Eniola-Adefeso, O.; Schiffelers, R.M. Extracellular vesicles as drug delivery systems: Lessons from the liposome field. *J. Control. Release* **2014**, *195*, 72–85. [CrossRef]
15. Liang, Y.; Duan, L.; Lu, J.; Xia, J. Engineering exosomes for targeted drug delivery. *Theranostics* **2021**, *11*, 3183–3195. [CrossRef]
16. Taylor, D.D.; Gercel-Taylor, C. MicroRNA signatures of tumor-derived exosomes as diagnostic biomarkers of ovarian cancer. *Gynecol. Oncol.* **2008**, *110*, 13–21. [CrossRef]
17. Shah, J.S.; Gard, G.B.; Yang, J.; Maidens, J.; Valmadre, S.; Soon, P.S.; Marsh, D.J. Combining serum microRNA and CA-125 as prognostic indicators of preoperative surgical outcome in women with high-grade serous ovarian cancer. *Gynecol. Oncol.* **2018**, *148*, 181–188. [CrossRef]
18. Kobayashi, M.; Sawada, K.; Nakamura, K.; Yoshimura, A.; Miyamoto, M.; Shimizu, A.; Ishida, K.; Nakatsuka, E.; Kodama, M.; Hashimoto, K.; et al. Exosomal miR-1290 is a potential biomarker of high-grade serous ovarian carcinoma and can discriminate patients from those with malignancies of other histological types. *J. Ovarian Res.* **2018**, *11*, 81. [CrossRef]
19. Morin, R.D.; O'Connor, M.D.; Griffith, M.; Kuchenbauer, F.; Delaney, A.; Prabhu, A.-L.; Zhao, Y.; McDonald, H.; Zeng, T.; Hirst, M.; et al. Application of massively parallel sequencing to microRNA profiling and discovery in human embryonic stem cells. *Genome Res.* **2008**, *18*, 610–621. [CrossRef]
20. Li, A.; Yu, J.; Kim, H.; Wolfgang, C.L.; Canto, M.I.; Hruban, R.H.; Goggins, M. MicroRNA Array Analysis Finds Elevated Serum miR-1290 Accurately Distinguishes Patients with Low-Stage Pancreatic Cancer from Healthy and Disease Controls. *Clin. Cancer Res.* **2013**, *19*, 3600–3610. [CrossRef]
21. Li, M.; He, X.-Y.; Zhang, Z.-M.; Li, S.; Ren, L.-H.; Cao, R.-S.; Feng, Y.-D.; Ji, Y.-L.; Zhao, Y.; Shi, R.-H. MicroRNA-1290 promotes esophageal squamous cell carcinoma cell proliferation and metastasis. *World J. Gastroenterol.* **2015**, *21*, 3245–3255. [CrossRef]
22. Wu, J.; Ji, X.; Zhu, L.; Jiang, Q.; Wen, Z.; Xu, S.; Shao, W.; Cai, J.; Du, Q.; Zhu, Y.; et al. Up-regulation of microRNA-1290 impairs cytokinesis and affects the reprogramming of colon cancer cells. *Cancer Lett.* **2013**, *329*, 155–163. [CrossRef]
23. Zhang, W.C.; Chin, T.M.; Yang, H.; Nga, M.E.; Lunny, D.P.; Lim, E.K.; Sun, L.L.; Pang, Y.H.; Leow, Y.N.; Malusay, S.R.; et al. Tumour-initiating cell-specific miR-1246 and miR-1290 expression converge to promote non-small cell lung cancer progression. *Nat. Commun.* **2016**, *7*, 11702. [CrossRef]
24. Xiao, X.; Yang, D.; Gong, X.; Mo, D.; Pan, S.; Xu, J. miR-1290 promotes lung adenocarcinoma cell proliferation and invasion by targeting SOCS4. *Oncotarget* **2018**, *9*, 11977–11988. [CrossRef]
25. Starr, R.; Willson, T.A.; Viney, E.M.; Murray, L.J.; Rayner, J.R.; Jenkins, B.J.; Gonda, T.J.; Alexander, W.S.; Metcalf, D.; Nicola, N.A.; et al. A family of cytokine-inducible inhibitors of signalling. *Nature* **1997**, *387*, 917–921. [CrossRef] [PubMed]
26. McCormick, S.; Heller, N.M. Regulation of Macrophage, Dendritic Cell, and Microglial Phenotype and Function by the SOCS Proteins. *Front. Immunol.* **2015**, *6*, 549. [CrossRef]
27. Sasi, W.; Jiang, W.G.; Sharma, A.; Mokbel, K. Higher expression levels of SOCS 1,3,4,7 are associated with earlier tumour stage and better clinical outcome in human breast cancer. *BMC Cancer* **2010**, *10*, 178. [CrossRef]

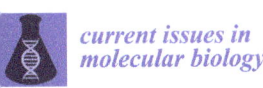

current issues in molecular biology

MDPI

Article

Personalized 3-Gene Panel for Prostate Cancer Target Therapy

Sanda Iacobas [1] and Dumitru Andrei Iacobas [2,*]

1 Department of Pathology, New York Medical College, Valhalla, NY 10595, USA; sandaiacobas@gmail.com
2 Personalized Genomics Laboratory, Center for Computational Systems Biology, Roy G. Perry College of Engineering, Prairie View A&M University, Prairie View, TX 77446, USA
* Correspondence: daiacobas@pvamu.edu

Abstract: Many years and billions spent for research did not yet produce an effective answer to prostate cancer (PCa). Not only each human, but even each cancer nodule in the same tumor, has unique transcriptome topology. The differences go beyond the expression level to the expression control and networking of individual genes. The unrepeatable heterogeneous transcriptomic organization among men makes the quest for universal biomarkers and "fit-for-all" treatments unrealistic. We present a bioinformatics procedure to identify each patient's unique triplet of PCa Gene Master Regulators (GMRs) and predict consequences of their experimental manipulation. The procedure is based on the Genomic Fabric Paradigm (GFP), which characterizes each individual gene by the independent expression level, expression variability and expression coordination with each other gene. GFP can identify the GMRs whose controlled alteration would selectively kill the cancer cells with little consequence on the normal tissue. The method was applied to microarray data on surgically removed prostates from two men with metastatic PCas (each with three distinct cancer nodules), and DU145 and LNCaP PCa cell lines. The applications verified that each PCa case is unique and predicted the consequences of the GMRs' manipulation. The predictions are theoretical and need further experimental validation.

Keywords: AP5M1; BAIAP2L1; CRISPR; ENTPD2; master regulator; LOC145474; MTOR; PRRG1; VIM; WFDC3

Citation: Iacobas, S.; Iacobas, D.A. Personalized 3-Gene Panel for Prostate Cancer Target Therapy. *Curr. Issues Mol. Biol.* **2022**, *44*, 360–382. https://doi.org/10.3390/cimb44010027

Academic Editor: Arumugam R. Jayakumar

Received: 18 November 2021
Accepted: 11 January 2022
Published: 15 January 2022

1. Introduction

For decades, cancer genomists have struggled to identify gene biomarkers whose altered sequence (e.g., Reference [1]) or/and expression (e.g., Reference [2]) is/are indicative for the prostate cancer (PCa) and could serve in active surveillance [3] of PCa development. The skillful handling of biomarkers was hoped to increase the survival rate (e.g., Reference [4]), destroy the cancer cells (e.g., Reference [5]) and reduce their proliferation (e.g., Reference [6]) and spreading (e.g., Reference [7]). Biomarkers' "smart" manipulation was also thought to block the PCa recurrence after various types of treatments (e.g., References [8,9]).

However, increasing evidence indicates that most cancerous prostates harbor genetically distinct independently developing malign clones [10]. This tumor heterogeneity [11,12], both at the histopathological and transcriptomic [13,14] levels, within the prostate of one patient, as well as among patients, complicates significantly the diagnostic and treatment options [15–17]. Moreover, together with the biomarker(s) whose altered sequence or expression level is thought indicative for that cancer type, hundreds of other genes are mutated or/and regulated in each cancer nodule with respect to the surrounding cancer-free tissue [18]. The unique combination of affected genes in each human is the direct result of the never-repeatable association of favoring factors affecting the entire body: race, age, medical history, habits, diet, stress, climate, etc. The set of the affected genes depends also on the specific local conditions (microbiome and cellular environment). This explains the observed wide diversity of PCa forms and the large spectrum of treatment outcomes. Therefore, it is imperious to go beyond precision medicine [12,19] to treatments tailored to the unrepeatable characteristics of every single patient at each moment of his or her life.

This report presents the Gene Master Regulator (GMR) method [20,21] to identify the most legitimate targets for a gene therapy that would selectively kill the cancer cells from the prostate [22]. Although we present three-gene panels to erase three distinct cancer clones in the profiled prostate, the method could be used for as many as relevant cancer nodules are found in the tissue. The GMR of a cell phenotype in a tissue is the gene whose strictly controlled expression level regulates the major functional pathways by coordinating the expressions of most of their genes. Owing to the uniqueness of the transcriptome, each cell phenotype of the tumor has a distinct gene hierarchy. Therefore, the GMR approach personalizes the gene treatment for each patient to destroy as many as possible cancer nodules of his affected prostate.

The GMR method is based on the Genomic Fabric Paradigm (GFP) [23] that takes advantage of profiling thousands of genes at a time on multiple biological replicates. GFP assigns to each quantified gene the independent variables: average expression level (AVE), relative expression variability (REV) and correlation (COR) with expression of each other gene [24]. Regardless of (microarray or RNA-sequencing) platform, adding REVs and CORs values increases by four orders of magnitude the amount of useful information provided by the transcriptomic study.

AVE is used by all oncogenomists to determine whether that gene was up/downregulated or turned on/off by cancer with respect to the normal tissue. In almost all publications, AVE is the single variable considered for individual genes.

Although profiling additional biological replicas was required initially only for statistical relevance of the results, it also gives us very important clues about the cell priorities in controlling the random fluctuations of the gene expression. The biological replicas can be formally considered as instances of the same system subjected to (non-significantly regulating) different environmental conditions. Thus, REV indicates how sensitive that gene is to slight environmental changes beyond the inherent stochastic nature of the chemical reactions involved in the transcription. In all transcriptomic studies, we found genes that are very stably expressed (low REV) and genes with high expression variability (high REV) across biological replicas. Low REV indicates strong control of the expression level by cellular the homeostatic mechanisms, most likely because the right expression of that gene is critical for the cell phenotypic expression, survival, proliferation and integration in the multicellular tissue. By contrast, expressions of other genes are left to fluctuate (high REV) to ensure cell adaptation to the environmental continuous changes [25].

The profiling expressions of thousands of genes at a time on biological replicas allows us to quantify how many fluctuations in the expression of one gene are correlated/coordinated with fluctuations of each other gene across biological replicas. COR analysis responds to the "Principle of Transcriptomic Stoichiometry" [25], a generalization of the Law of Multiple Proportions from chemistry [26]. The principle requires coordinated expression of genes whose encoded products are linked in functional pathways.

2. Materials and Methods

2.1. Prostate Tissues and Cell Lines

This report uses transcriptomic profiles generated in the NYMC IacobasLab by profiling the surgically removed prostates of two men, hereafter denoted as patients "ABCN" and "PQMZ". For comparison, we added the expression data from two human prostate cancer cell lines: the androgen-sensitive LNCaP [27] and the not-hormone-sensitive DU145 [28]. Expression data obtained in our lab from the LNCaP cells (hereafter denoted by "L") were deposited at Reference [29], and those from the DU145 cells (hereafter denoted by "D") were deposited at Reference [30].

From the "ABCN" patient, we profiled the primary cancer nodule "A" (Gleason Score GS = 4 + 5 = 9); the secondary cancer nodules "B" and "C", each with GS = (4 + 4 = 8);

and the surrounding normal tissue "N". Gene-expression data of "ABCN" were deposited at Reference [31] for the primary nodule "A" and the cancer-free margins "N", and at Reference [32] for the secondary nodules "B" and "C", and partially analyzed in a recent paper [33]. Patient "PQMZ" had prostatic adenocarcinoma involving 75% of bilateral lobes, with extensive perineural invasion, multifoci of extraprostatic extension that affected also the bilateral seminal vesicles. Gene-expression data from this patient are available at Reference [34] for the nodule "M" (GS = 4 + 5 = 9) and the cancer-free resection margins "Z", and at Reference [35] for the cancer nodules "P" and "Q", each with GS = 4 + 5 = 9. From each nodule, we collected a ~2 mm area from the center and then split it into 4 parts to limit the possibility that the collected quarters contain cells from different clones.

The study, conducted according to the guidelines of the Declaration of Helsinki, was part of Dr. Iacobas's project approved by the Institutional Review Boards (IRBs) of the New York Medical College (NYMC) and Westchester Medical Center (WMC) Committees for Protection of Human Subjects. The approved IRB (L11,376 from 2 October 2015) granted access to frozen cancer specimens from the WMC Pathology Archives and depersonalized pathology reports, waiving the patients' informed consent.

The experimental protocol (RNA extraction, fluorescent labeling, hybridization with the microarray and washing and scanning the chip), as well as the primary analysis of the fluorescent values (filtering, background subtraction and normalization to the median of valid spots in all profiled samples) are detailed on the Gene Expression Omnibus website hosting the deposited datasets [29–32,34,35].

2.2. Transcriptomic Characteristics of Individual Genes

AVE, REV and COR values were computed to account for the non-uniform numbers of spots probing redundantly numerous genes in Agilent (Agilent, Santa Clara, CA, U.S.A.) 4 × 44 k human dual-color microarrays (configuration G2519F, platform GPL13497 [36]).

$$AVE_i^{(sample)} = \frac{1}{R_i}\sum_{k=1}^{R_i} \mu_{i,k}^{(sample)} = \frac{1}{R_i}\sum_{k=1}^{R_i}\left(\frac{1}{4}\sum_{j=1}^{4} a_{i,k,j}^{(sample)}\right), \quad where:$$
$$sample = \text{"N","A","B","C","Z","P","Q","M","L","D"} \qquad (1)$$
$$R_i = \text{ number of spots probing redundantly gene "i"},$$
$$a_{i,k,j}^{(sample)} = \text{expression of gene "}i\text{" probed by spot "}k\text{" on biological replica "}j\text{" in "sample"}$$

$$REV_i^{(sample)} = \underbrace{\frac{1}{2}\left(\sqrt{\frac{r_i}{\chi^2(r_i;0.975)}} + \sqrt{\frac{r_i}{\chi^2(r_i;0.025)}}\right)}_{\text{correction coefficient}} \sqrt{\underbrace{\frac{1}{R_i}\sum_{k=1}^{R_i}\left(\frac{s_{ik}^{(sample)}}{\mu_{ik}^{(sample)}}\right)^2}_{\text{pooled CV}}} \times 100\% \qquad (2)$$

$\chi^2(r_i;\alpha)$= chi-square for $r_i(= 4R_i - 1 =$ number of degrees of freedom) and probability α
μ_{ik}= average expression of gene i probed by spot k $(= 1, \ldots, R_i)$ in the 4 biological replicas
s_{ik} = standard deviation of the expression level of gene i probed by spot k

The correction coefficient is the mid chi-square interval estimate of the unit standard deviation and takes values from 2.15 for genes probed by one spot each to 1.05 for genes probed by 11 spots (e.g., *TP53*).

$$COR_{ig}^{(sample)} = \frac{\sum_{k_i=1}^{R_i}\sum_{k_g=1}^{Rg}\left(\sum_{j=1}^{4}\left(a_{i,k,j}^{(sample)} - AVE_i^{(sample)}\right)\left(a_{g,k,j}^{(sample)} - AVE_g^{(sample)}\right)\right)}{\sqrt{\sum_{k_i=1}^{R_i}\left(\sum_{j=1}^{4}\left(a_{i,k,j}^{(sample)} - AVE_i^{(sample)}\right)^2\right)\sum_{k_g=1}^{R_g}\left(\sum_{j=1}^{4}\left(a_{g,k,j}^{(sample)} - AVE_g^{(sample)}\right)^2\right)}} \qquad (3)$$

In Equation (3), *COR* ($-1 \leq$ COR ≤ 1) is the Pearson product-momentum correlation coefficient between the ($\log_2$) expression levels of genes "*i*" and "*g*". For genes probed by 1 spot each, $p < 0.05$ significant synergistic/antagonistic expression was assigned if

$|COR| \geq 0.95$. If the genes are probed by 2 spots each, then significant coordination occurs for $|COR| \geq 0.71$, and so on with cutoff diminishing for larger numbers of spots probing each gene [25].

REV and *COR* were used to determine the Gene Commanding Height (*GCH*) [33], which establishes the gene hierarchy in the profiled phenotype:

$$GCH_i^{(sample)} = \underbrace{\frac{\langle REV\rangle^{(sample)}}{REV_i^{(sample)}}}_{\text{estimate of the transcription control}} \times \underbrace{\exp\left(4\left.\overline{\left(COR_{ig}^{(sample)}\right)^2}\right|_{\forall g\neq i}\right)}_{\text{measure of expression coordination}}, \quad where: \tag{4}$$
$$\langle\,\rangle = \text{median}, \ \overline{(\,)^2} = \text{average of the square values}$$

The top gene of the hierarchy (highest *GCH*) is the Gene Master Regulators (*GMR*s) of that phenotype, whose altered expression should have the largest consequences. The hierarchies of the four groups of samples were used to identify the top 3 genes whose *GCH* scores in the three cancer nodules are far above the corresponding scores in the cancer-free tissue.

2.3. Transcriptomic Alterations in Cancer

A gene was considered as statistically ($p < 0.05$) significantly regulated in a cancer nodule ("cancer") with respect to the normal tissue from the same tumor if the absolute fold-change x and the p-value ($p_i^{(normal\rightarrow cancer)}$) of the heteroscedastic t-test of the means equality in the two regions satisfied the composite criterion Equation (5). Our cutoff of the absolute fold-change considers the combined effects of the biological variability and technical noise, providing a much accurate criterion for expression regulation than any other arbitrary ($1.5\times$, 2.0) fold-change requirement.

$$\begin{cases} \left|x_i^{(normal\rightarrow cancer)}\right| > CUT_i^{(normal\rightarrow cancer)} = 1 + \frac{1}{100}\sqrt{2\left(\left(REV_i^{(normal)}\right)^2 + \left(REV_i^{(cancer)}\right)^2\right)} \\ p_i^{(normal\rightarrow cancer)} < 0.05 \end{cases}$$
$$where: cancer = \text{"A","B","C","M","P","Q"} \quad normal = \text{"N","Z"} \tag{5}$$
$$x_i^{(normal\rightarrow cancer)} \equiv \begin{cases} \frac{\mu_i^{(cancer)}}{\mu_i^{(normal)}}, & \text{if } \mu_i^{(cancer)} > \mu_i^{(normal)} \\ -\frac{\mu_i^{(normal)}}{\mu_i^{(cancer)}}, & \text{if } \mu_i^{(cancer)} < \mu_i^{(normal)} \end{cases}$$

The p-value was computed with Bonferroni correction for multiple testing [37] in the case of several spots probing redundantly the same gene.

Instead of the uniform ± 1 contribution to the altered transcriptome used in the traditional percentage measure of the significantly up/downregulated genes, we considered the Weighted Individual Gene Regulation (*WIR*). *WIR* analysis is not limited to the significantly regulated genes; it is applied to any gene. *WIR* was used to compute the Weighted Pathway Regulation (*WPR*) to average the contributions of all genes assigned to that functional pathway:

$$WIR_i^{(normal\rightarrow cancer)} = AVE_i^{(normal)} \frac{x_i^{(normal\rightarrow cancer)}}{\left|x_i^{(normal\rightarrow cancer)}\right|}\left(\left|x_i^{(normal\rightarrow cancer)}\right| - 1\right)\left(1 - p_i^{(normal\rightarrow cancer)}\right)$$
$$WPR_\Gamma^{(normal\rightarrow cancer)} = \left.\overline{WIR_i^{(normal\rightarrow cancer)}}\right|_{i\in\Gamma}, \quad \Gamma = \text{functional pathaway} \tag{6}$$

3. Results

3.1. Overview

In total, we quantified the expression of 14,203 distinct unigenes in all three cancer nodules and the surrounding cancer-free tissue from the surgically removed prostate of the "PQMZ" man. The 403,620,854 (total number of AVE, REV and COR) values resulted from this experiment were used to illustrate the analyses below. In order to show the uniqueness of the three-gene target panel, we used also the expression values in the three cancer nodules and cancer-free surroundings of the "ABCN" man (14,908 genes), in the "L" cells (15,278 genes) and in the "D" cells (16,126 genes).

Table 1 presents the three genes with the largest expression levels in each of the four profiled regions of patient "PQMZ" and in the four regions of the patient "ABCN". Remarkably, *RPL13* was among the three genes with the largest expressions in all regions, and in both patients, it was downregulated in all cancer nodules with respect to the corresponding normal tissues. The robust downregulation of *RPL13* in all six cancer nodules (by −1.70×/−1.38×/−1.36× in "P"/"Q"/"M" vs. "Z" and by −2.15×/−1.30×/−1.50× in "A"/"B"/"C" vs. "N") explains the reduced immune response in cancer. Our results add new evidence about the extra-ribosomal roles of *RPL13*, particularly in activating the innate response [38].

Table 1. Three genes with the largest average expression levels (AVE) in each of the four profiled regions of the "PQMZ" patient and in the four regions of the "ABCN" patient. The largest 3 AVE values in each phenotype have a gray background.

Gene	Description	AVE-P	AVE-Q	AVE-M	AVE-Z
CYTB	mitochondrially encoded cytochrome b	231	237	186	266
RPL13	**ribosomal protein L13**	215	2365	269	366
ACTG2	actin, gamma 2, smooth muscle, enteric	204	126	119	325
NPY	neuropeptide Y	76	471	10	123
RPL13	**ribosomal protein L13**	215	265	269	366
RPL7A	ribosomal protein L7a	106	310	141	241
RPL13	**ribosomal protein L13**	215	265	269	366
ZNF865	zinc finger protein 865	182	267	203	292
PQLC2	PQ loop repeat containing 2	172	179	199	203
RPL13	**ribosomal protein L13**	215	265	269	366
ACTG2	actin, gamma 2, smooth muscle, enteric	204	126	119	325
MYH11	myosin, heavy chain 11, smooth muscle	137	101	139	307

Gene	Description	AVE-A	AVE-B	AVE-C	AVE-N
RPL13	**ribosomal protein L13**	288	477	415	621
PQLC2	PQ loop repeat containing 2	227	266	338	490
CYTB	mitochondrially encoded cytochrome b	215	173	146	382
RPL13	**ribosomal protein L13**	288	477	415	621
RPS27	ribosomal protein S27	158	365	294	212
RPL32	ribosomal protein L32	164	315	299	359
RPL13	**ribosomal protein L13**	288	477	415	621
ZNF865	zinc finger protein 865	132	215	395	600
RPS8	ribosomal protein S8	153	233	348	504
RPL13	**ribosomal protein L13**	288	477	415	621
ZNF865	zinc finger protein 865	132	215	395	600
RPS2	ribosomal protein S2	164	260	337	517

Table 2 lists the three genes with the most controlled expression (low REV) in each of the four profiled regions of the patient "PQMZ" and in the four regions of the patient "ABCN". Supplementary Table S1 presents the three genes with the most variable expression (high REV) across biological replicas in all profiled regions of the two patients.

Table 2. Three most stably (low REV, darker gray background) expressed genes in each of the four profiled regions of patients "PQMZ" and "ABCN".

Gene	Description	REV-P	REV-Q	REV-M	REV-Z
FKBP9	FK506 binding protein 9	0.32	8.80	10.87	9.27
ZBTB2	zinc finger and BTB domain containing 2	0.50	12.39	5.03	16.51
NUBPL	nucleotide binding protein-like	0.69	12.19	5.48	10.73
TBRG4	transforming growth factor beta regulator 4	11.07	0.59	1.56	8.06
DNAJC24	DnaJ (Hsp40) homolog, subfamily C, member 24	11.62	0.85	7.19	4.16
UBE3B	ubiquitin protein ligase E3B	5.32	1.00	6.51	9.46
TMEM186	transmembrane protein 186	9.32	20.45	0.28	12.49
NDUFA6-AS1	NDUFA6 antisense RNA 1 (head to head)	28.15	8.46	0.50	15.82
LMAN2L	lectin, mannose-binding 2-like	8.05	8.20	0.52	8.00
COPS5	COP9 signalosome subunit 5	51.46	27.83	7.37	0.12
ARPC5L	actin related protein 2/3 complex, subunit 5-like	19.90	22.18	7.13	0.14
DAZAP1	DAZ associated protein 1	7.14	18.87	3.84	0.22
Gene	**Description**	**REV-A**	**REV-B**	**REV-C**	**REV-N**
ENTPD2	ectonucleoside triphosphate diphosphohydrolase 2	0.29	33.36	15.97	8.66
COMMD9	COMM domain containing 9	1.04	23.01	4.49	3.82
MIEN1	migration and invasion enhancer 1	1.14	13.78	9.01	8.67
SSX3	synovial sarcoma, X breakpoint 3	39.80	0.94	17.16	12.18
FCRL5	Fc receptor-like 5	121.40	1.41	36.36	38.96
RANBP2	RAN binding protein 2	48.44	1.63	11.21	10.48
BAIAP2L1	BAI1-associated protein 2-like 1	64.94	11.35	0.40	8.03
FAM71E1	family with sequence similarity 71, member E1	41.61	48.01	0.81	11.93
LILRB3	leukocyte immunoglobulin-like receptor, subfamily B, member 3	22.77	32.51	0.91	27.55
MRPS12	mitochondrial ribosomal protein S12	34.96	42.93	11.21	0.32
TOR1A	torsin family 1, member A	11.09	17.64	6.28	0.42
DENND1B	DENN/MADD domain containing 1B	8.18	45.86	15.79	0.47

Table 2 has some very interesting results. First, each of the four regions from each of the two patients appears to have different priorities in controlling the transcripts' abundances. Our results indicate that the most controlled genes are critical for preserving the phenotype. Thus, *FKBP9*, the most controlled gene in nodule "P", is known for promoting malignant behavior of glioblastoma cells [39] and poor prognosis of PCa patients [40]. *TBRG4*, the most stably expressed gene in "Q", was reported as being actively involved in myeloma [41], squamous carcinoma [42], osteosarcoma [43], glioblastoma [44], leukemia [45] and lung cancer [46]. The list of stably expressed genes also includes a long noncoding RNA, *NDUFA6-AS1*, identified recently as a biomarker for the prognostic of thyroid cancer [47]. We believe that the strict control of the expression of *COPS5* is related to its role in controlling the progression of PCa [48].

Supplementary Table S1, which presents the most variably expressed genes across biological replicas, is also interesting by indicating that the adaptation to the environmental fluctuations is carried out by distinct sets of genes in each profiled region. Thus, excepting *UBE2I*, common for nodule "M" and normal tissue "Z", there is no overlap of the most variably expressed three genes in profiled regions.

3.2. Independent Variables

Figure 1 serves as an example of the independency of AVE, REV and COR for the first 50 alphabetically ordered genes involved in the mTOR signaling [49] in the cancer nodules "P", "Q", and "M" and the normal surrounding tissue "Z" of the "PQMZ" patient prostate. Figure 1c presents the expression coordination with *MTOR* (mechanistic target of rapamycin (serine/threonine kinase)).

Although the *MTOR* gene and its partners in the mTOR signaling pathway were selected for their roles in the development, proliferation and migration of cancer cells [50], any other subset of genes would confirm the independence of the AVE, REV and COR characteristics.

Within this selection, genes such as ATPase, H+ transporting, lysosomal 13 kDa and V1 subunit G2 (*ATP6V1G2*) have very low expression (AVE = 0.14 in "Q"), and genes such as ATPase, H+ transporting, lysosomal 14 kDa and V1 subunit F (*ATP6V1F*) have much higher expression (34.5 in "Q"). Likewise, there are very stably (e.g., DEP domain containing five (*DEPDC5*), REV = 0.25% in "Z") and very unstably (e.g., frizzled class receptor 10 (*FZD10*), REV = 87.48% in "Q") expressed genes.

In addition to the clear independence of the three variables, of note are also the differences among the three equally histopathologically ranked cancer nodules from the same prostate. These findings prove both the transcriptomic heterogeneity of the PCa and the uniqueness of each nodule. The transcriptomes of the three nodules differ not only by expression levels of individual genes but also in their expression variability (indicating different strengths of controlling mechanisms) and expression coordination (distinct gene networking in pathways). For instance, *FZD10*, known for its role in breast cancer [51] was very unstably expressed in "Q" and "P" (REV = 59.96%), but it was stably expressed enough in "M" (REV = 6.16%). These values suggest that the right expression of *FZD10* was more important for the "M" cells than for the "P" and "Q" cells. Moreover, eukaryotic translation initiation factor 4E (*EIF4E*) was synergistically expressed with *MTOR* in "Q", but antagonistically expressed with *MTOR* in "P" and "M", meaning that *EIF4E* might work as an activator of *MTOR* in "Q" but as an inhibitor in "P" and "M", which indicates that targeting *EIF4E* may have opposite clinical results [52]. Warning: the "opposite roles", suggested by a pure theoretical speculation about in-phase and in-antiphase fluctuations of the expressions of *MTOR* and *EIF4E* in distinct nodules, needs rigorous experimental validation.

3.3. The Power of the Weighted Individual (Gene) Regulation (WIR) Score to Identify the Main Contributors to the Cancer Phenotype

Figure 2 presents the regulation of the first 50 alphabetically ordered mTOR signaling genes in the three cancer nodules with respect to the cancer-free surrounding tissue. The regulation of genes is presented as uniform ± 1 contribution (as in the percentage of significantly up/downregulated genes), expression ratio "x" and Weighted Individual Gene Regulation (WIR).

The power of WIR to discriminate the genes according to their contribution is outstanding. For instance, *DEPTOR* (DEP domain containing MTOR-interacting protein) and *DVL1* (disheveled segment polarity protein 1) have similar expression ratios in "Q" (2.03$\times$ and 2.09$\times$) but substantially different WIRs (1.10 and 17.13).

Remarkably, although most of the significant regulations go the same way in all three nodules (e.g., *ATP6V1AB2*, *ATP6V1C1*, *ATV6V1H*, *FLCN*, *FZD1* and *FZD3*), suggesting a shared PCa transcriptomic signature. However, none of the similarly regulated genes was listed among the PCa biomarkers in the NIH-NCI GDC Data Portal [53], and there are also opposite regulations (*AKT1*, *DVL2* and *DVL3*). The significant opposite regulations presented in Table 3 and Figure 2 indicate that, even within the same tumor, each cancer nodule has its own cancer transcriptomic topology. Of note is that almost all human genes (20,237) were found to be altered in at least one case of PCa included in the portal [53], and the top PCa biomarkers are among the top biomarkers for many other cancers. For instance, *TP53* (#1 for PCa) is also #1 for lung, breast, head and neck, ovary cancers and among the top five for almost all other cancers. Therefore, we believe that it makes no sense to continue using the transcriptomic signature in classifying the PCas and that the commercially available assays for cancer diagnostic (e.g., References [54–56]) have disputable prediction value, as discussed in a previous paper [33]).

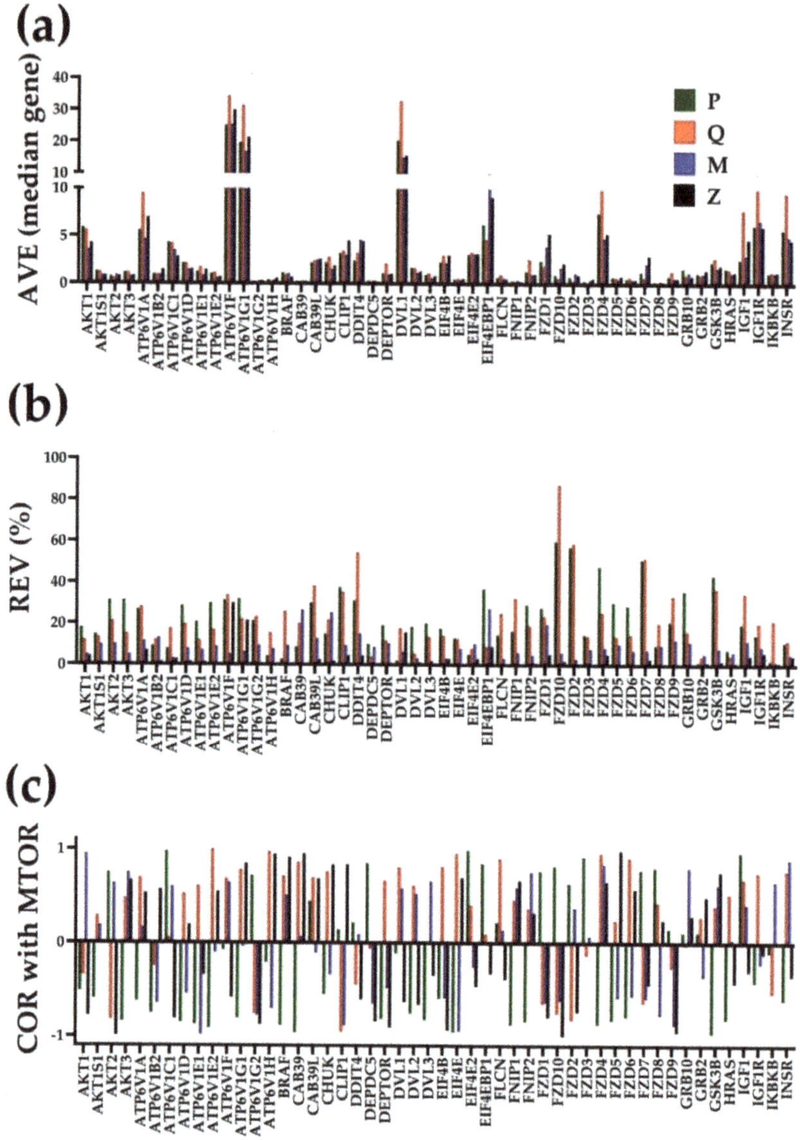

Figure 1. Independent transcriptomic characteristics of the first 50 alphabetically ordered genes of the mTOR signaling pathway in the three cancer nodules ("P", "Q" and "M") and the surrounding normal prostate tissue ("Z"). (**a**) Average expression levels (AVE) in expressions of the median gene. (**b**) Relative expression variability (REV). (**c**) Expression correlation (COR) with *MTOR*.

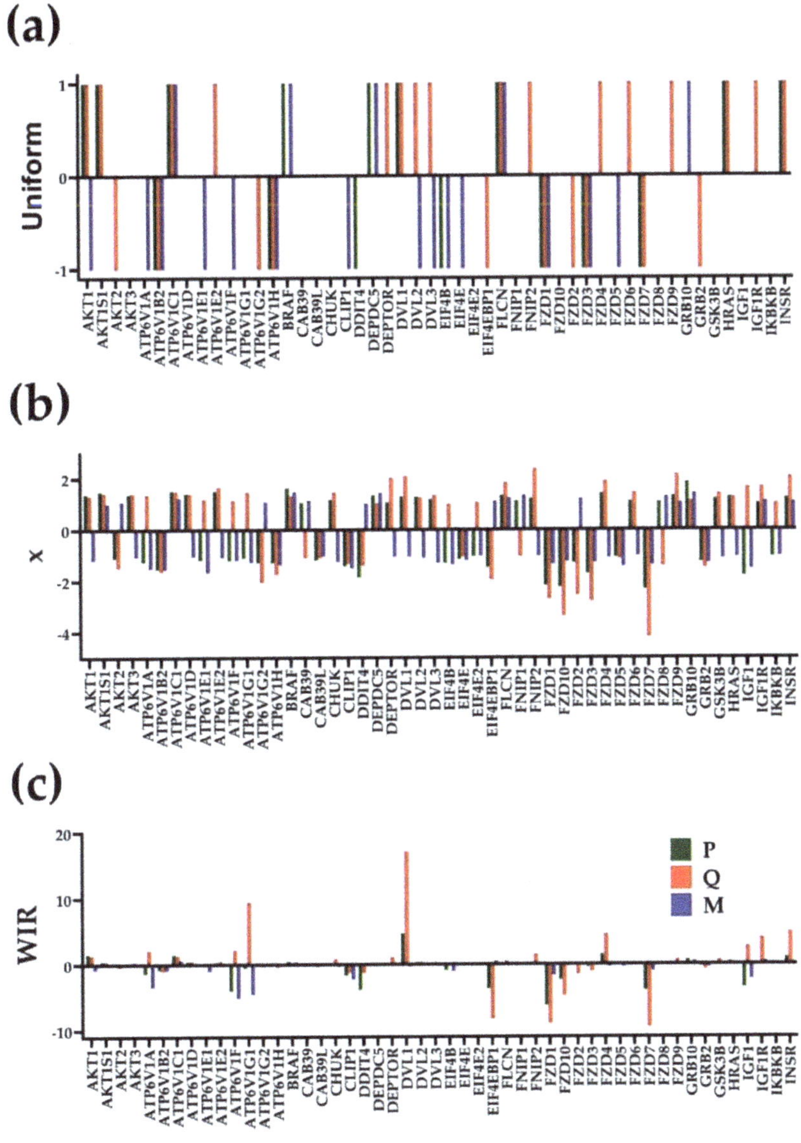

Figure 2. Regulation of the first 50 alphabetically ordered mTOR signaling genes. (**a**) As uniform ±1 contribution. (**b**) Expression ratio "x" (negative for downregulation). (**c**) Weighted Individual Gene Regulation (WIR). Note the differences among the nodules.

3.4. Each Cancer Nodule Has Its Own "Transcriptomic Signature"

Table 3 presents the three genes with the largest positive and the three genes with the largest negative contributions to the transcriptomic alterations in each of the three cancer nodules of the patients "PQMZ" and "ABCN".

It is interesting to note that, in Table 3, each nodule has different sets of the three largest contributors and that these contributors are involved in a wide diversity of functional pathways. These data indicate a large spectrum of possible molecular mechanisms responsible for the formation of cancer nodules in the prostate. Moreover, while some genes were regulated the same way in all three nodules (*CNN1, RNA28S5, RLN1,* and *ACTG2*), others (*RPS8, MARC1,* and *PSCA*) were regulated in only one or two nodules. There are also genes (e.g., *NPY, IGKC, IGHG1,* and *SNORD3B-1*) that were even oppositely regulated in one nodule than in the other two. These results indicate the unique response of each region to cancer and that restoration of the normal expression level of some genes might have opposite effects on distinct nodules.

Table 3. Three genes with the largest positive and negative contributions to the transcriptomic alterations in the cancer nodules of the "PQMZ" and "ABCN" patients.

Gene	Description	WIR-P	WIR-Q	WIR-M
PSCA	prostate stem cell antigen	**54**	138	0
KLK12	kallikrein-related peptidase 12	**40**	121	0
BASP1	brain abundant, membrane attached signal protein 1	**32**	39	0
RPS8	ribosomal protein S8	**−383**	0	−123
CNN1	calponin 1, basic, smooth muscle	**−403**	−680	−568
RNA28S5	RNA, 28S ribosomal 5	**−604**	−314	−26
MARC1	mitochondrial amidoxime reducing component 1	0	**9180**	0
NPY	neuropeptide Y	−50	**348**	−1345
PSCA	prostate stem cell antigen	54	**138**	0
CNN1	calponin 1, basic, smooth muscle	−403	**−680**	−568
LTF	lactotransferrin	−301	**−1682**	22
RLN1	relaxin 1	−86	**−2658**	−46
IGKC	immunoglobulin kappa constant	−51	−144	**60**
IGHG1	immunoglobulin heavy constant gamma 1	−29	−83	**52**
SNORD3B-1	small nucleolar RNA, C/D box 3B-1	−3	−35	**49**
ACTG2	actin, gamma 2, smooth muscle, enteric	−170	−514	**−561**
CNN1	calponin 1, basic, smooth muscle	−403	−680	**−568**
NPY	neuropeptide Y	−50	348	**−1345**
Gene	**Description**	**WIR-A**	**WIR-B**	**WIR-C**
IGLL5	immunoglobulin lambda-like polypeptide 5	**59**	141	56
TPM2	tropomyosin 2 (beta)	**50**	9	−9
ACTG2	actin, gamma 2, smooth muscle, enteric	**46**	2	−29
RPS8	ribosomal protein S8	**−1139**	−575	−190
RPL14	ribosomal protein L14	**−1182**	−734	−85
ZNF865	zinc finger protein 865	**−2073**	−1027	−239
IGLL5	immunoglobulin lambda-like polypeptide 5	59	**141**	56
MDK	midkine (neurite growth-promoting factor 2)	28	**133**	148
RPS27	ribosomal protein S27	−51	**127**	56
NPIPB5	nuclear pore complex interacting protein family, member B5	−937	**−1077**	−98
KLK3	kallikrein-related peptidase 3	−585	**−1248**	−439
SPON2	spondin 2, extracellular matrix protein	−332	**−1575**	−1288
MDK	midkine (neurite growth-promoting factor 2)	28	133	**148**
IFI27	interferon, alpha-inducible protein 27	13	55	**123**
HMGN2	high mobility group nucleosomal binding domain 2	6	107	**116**
CYTB	mitochondrially encoded cytochrome b	−238	−446	**−600**
LOC101929612	mitochondrially encoded cytochrome c oxidase III	−97	−279	**−800**
SPON2	spondin 2, extracellular matrix protein	−332	−1575	**−1288**

3.5. Tumor Heterogeneity of Gene Networks

Figure 3 presents the expression coordinations of *AKT2* (v-akt murine thymoma viral oncogene homolog 2), with its partners central to the prostate-cancer development [57] in all four profiled regions from the patients "PQMZ" and "ABCN". Figure 3 also presents the coordinations of *MTOR* with mTORC1 (*RAPTOR*, ACT1S1, DEPTOR, MLST8, TELO2 and *TTI1*) and mTORC2 (*RICTOR*, MAPKAP1, PRR5, DEPTOR, MLST8, TELO2 and *TTI1*) partners [49] in the same regions.

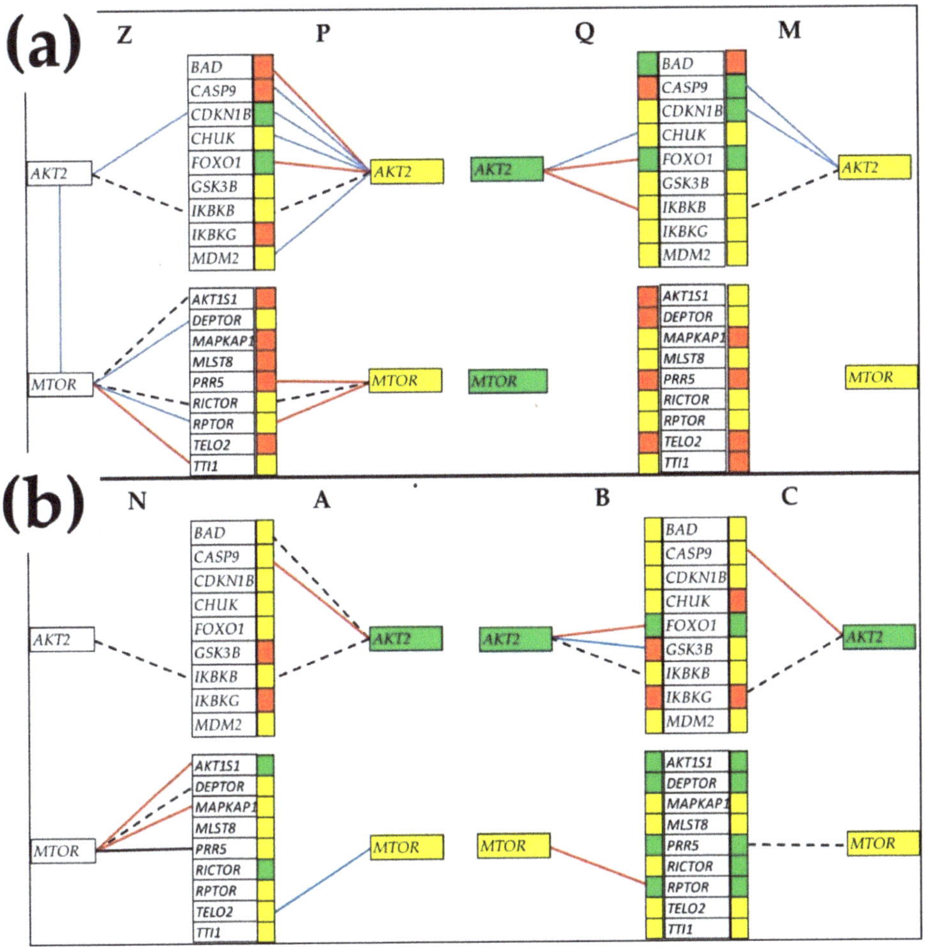

Figure 3. Expression coordinations of *AKT2* with its partners, central to the prostate cancer development, and the coordinations of *MTOR* with its partners from mTORC1 and mTORC2 in the four profiled regions of the (**a**) "PQMZ" patient and (**b**) the "ABCN" patient. A continuous red/blue line indicates a statistically ($p < 0.05$) significant synergistic/antagonistic expression of the linked gene, while a dashed black line indicates a statistically ($p < 0.05$) significant independent expression of the paired genes. Missing lines mean lack of statistical significance of the expression coordination between the two genes. Red/green background specifies significant up/downregulation of that gene in the indicated cancer nodule ("P", "Q", "M", "A", "B" and "C") with respect to the corresponding cancer-free surrounding tissue ("Z" or "N"), while yellow background means that the expression difference was not significant according to our composite criterion.

Of note, in Figure 3 are the substantial differences in both expression regulations with respect to the surrounding normal tissue and in expression coordination not only between the two patients but also among the cancer nodules of each patient. These results extend the notion of transcriptomic heterogeneity of the tumor [11–14] to the formation of gene networks that could be even more important for the cell behavior than the heterogeneity of the gene-expression levels.

Interestingly, *MTOR* is not significantly coordinately expressed with any of its MTORC1 and MTORC2 alleged partners in the nodules "Q", "M", "B" and "C", indicating major remodeling of the mTOR signaling in these cancer clones. Altogether, the differences in gene networking among the profiled groups of samples show that the pathways built by Kyoto Encyclopedia for Genes and Genomes (KEGG) [58] are not universal and can be used only as a general reference. The same conclusion is valid for the pathways built by other specialized software, such as Ingenuity Pathway Analysis [59], DAVID [60] and even the old GenMapp and MAPPFinder [61].

3.6. Gene Master Regulators

Figure 4 presents the GMRs of all the profiled regions from the two patients and the two cancer cell lines. For each GMR, the graph shows the GCH scores in all profiled sample types. Note that each group of samples has a distinct GMR and the substantial difference between the GCH score in the region the GMR commands and in the other regions from the same tumor. For instance, *FKBP9*, the GMR of the region "P" has the GCHs: 158.86 (in "P"), but 3.66 (in "Q"), 1.96 (in "M"), 2.59 (in "Z"), 8.24 (in "A"), 2.52 (in "B"), 2.36 (in "C"), 1.47 (in "N"), 7.46 (in "L") and 16.59 (in "D"). The large difference between the GCH scores in the cancer clone and the normal tissue indicates that manipulation of that gene would have major transcriptomic consequences in the cancer but practically nothing in the healthy tissue. This observation makes the GMR approach suitable to design cancer gene therapies.

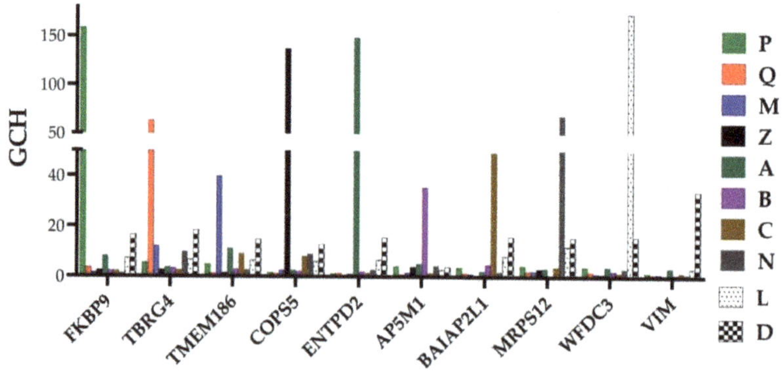

Figure 4. GMRs of the profiled regions from the prostates of the "PQMZ" and "ABCN" patients and from the cancer cell lines "L" and "D". Note that the GCH scores in the other samples are substantially lower the score in the sample commanded by the GMR.

The GMRs of the cancer nodules are not necessary regulated with respect to the normal tissue, but as evident from Table 2, they are among the most stably expressed genes in that region. The expression level of the GMR is allowed to fluctuate within a very narrow interval, because it regulates the expression of numerous other genes. For instance, *FKBP9*, the GMR of "P", was similarly downregulated in all three cancer nodules: x = −1.52 (WIR = −7.44) in "P", x = −1.59 (WIR = −8.51) in "Q" and x = −1.49 (WIR = −6.97) in "M". *TBRG4*, the GMR of "Q", was upregulated in "P" (x = 1.25, WIR = 0.46) and "Q" (x = 1.73, WIR = 1.39) but not in "M", while *TMEM186*, the GMR of "M", was not regulated in any of the profiled cancer nodules. *ENTPD2*, the GMR of "A", was not regulated in "A", but it was upregulated in "B" and "C"; *AP5M1*, the GMR of "B", was upregulated in "B" but not regulated in either "A" or "C"; and *BAIAP2L1*, the GMR of "C", was upregulated in "B" and "C", but not in "A".

3.7. What Experimental Manipulation of the GMR Would Do to the Cancer Nodule's Metabolism?

We used our software #PATHWAY# [22] to identify all KEGG pathways [58] that include genes with statistically significant synergistic/antagonistic expression correlation with the GMR in each profiled cancer nodule. In all cancer nodules from both patients, the most significantly correlated genes with the GMR were from the metabolic pathways, indicating that targeting the GMR would have most dramatic consequences on the cell metabolism.

Figure 5 presents the significantly synergistically and antagonistically expressed genes with the corresponding GMR in each of the nodules "P", "Q" and "M" of the "PQMZ" patient, and what one might expect from the significant manipulation of the GMR expression. Thus, by therapeutically increasing the expression of *FKBP9* (downregulated in all three cancer nodules with respect to "Z"), the expressions of its synergistic partner genes would be pushed up, while the antagonistically expressed ones would be pushed down. Although the proposed therapeutic overexpression will restore the normal expression of *FKBP9* in all three cancer nodules, while upregulating it in "Z", it would have significant consequences only on "P", owing to the low *FKBP9* GCH scores in the other three regions. Large metabolic disturbances on the respective commanded cancer nodules, but not in the other regions, are also expected by knocking down *TBRG4* and *TMEM186*, as illustrated in panels Figure 5b,c.

Figure 6 presents the significantly synergistically and antagonistically expressed genes with the corresponding GMR in each of the nodules "A", "B" and "C" of the "ABCN" patient, and what one might expect from the significant manipulation of the GMR expression. Of note are, again, the different regulations of the metabolic genes in the cancer nodules of the same tumor, as well as between the two tumors (compare with Figure 5).

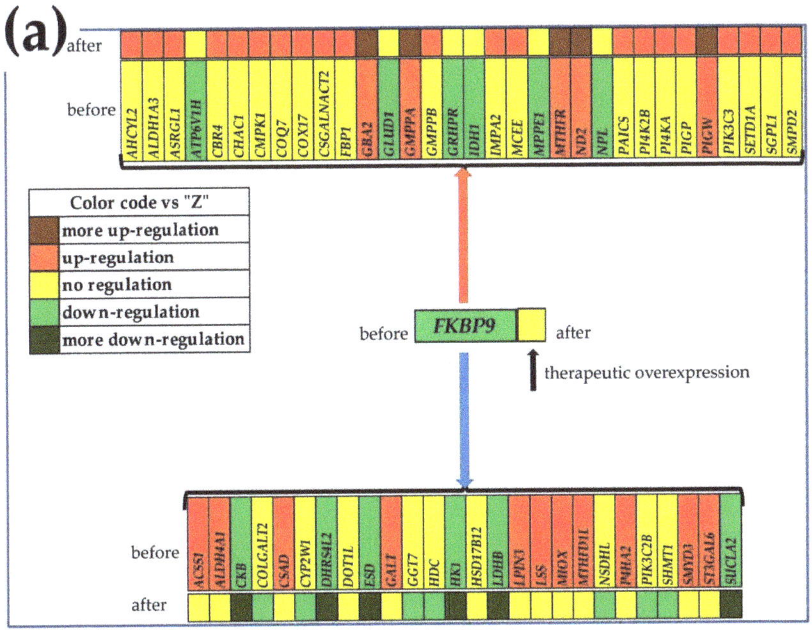

Figure 5. *Cont.*

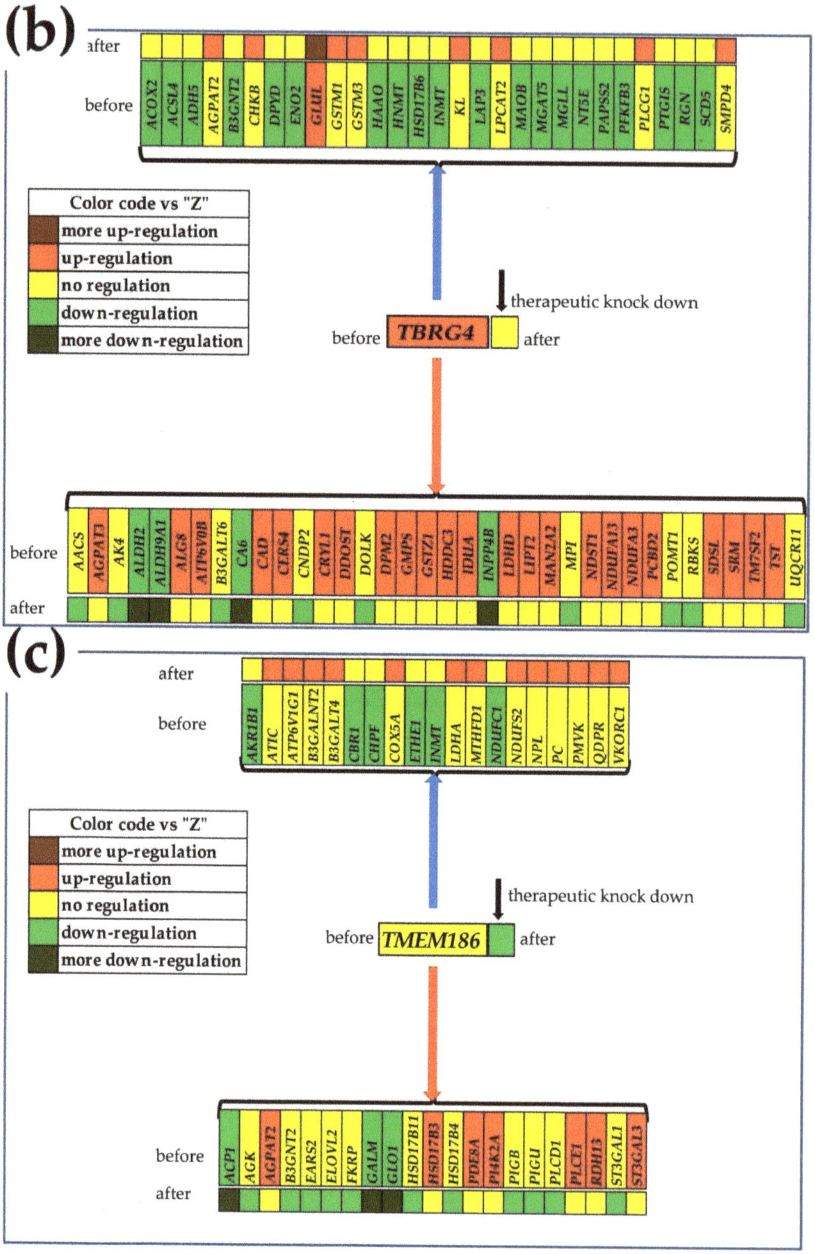

Figure 5. Significantly synergistically and antagonistically expressed metabolic genes with the corresponding GMR in the nodules (**a**) "P", (**b**) "Q" and (**c**) "M" of the "PQMZ" patient and the predicted regulations after the therapeutic alteration of the GMR. The red/blue arrow indicates the genes synergistically/antagonistically expressed with the GMR in that nodule. Gene symbol background indicates the status of that gene in the mentioned cancer nodule with respect to the surrounding normal tissue "Z" before (observed) and after the treatment (predicted). Note the different regulations of the metabolic genes in the cancer nodules.

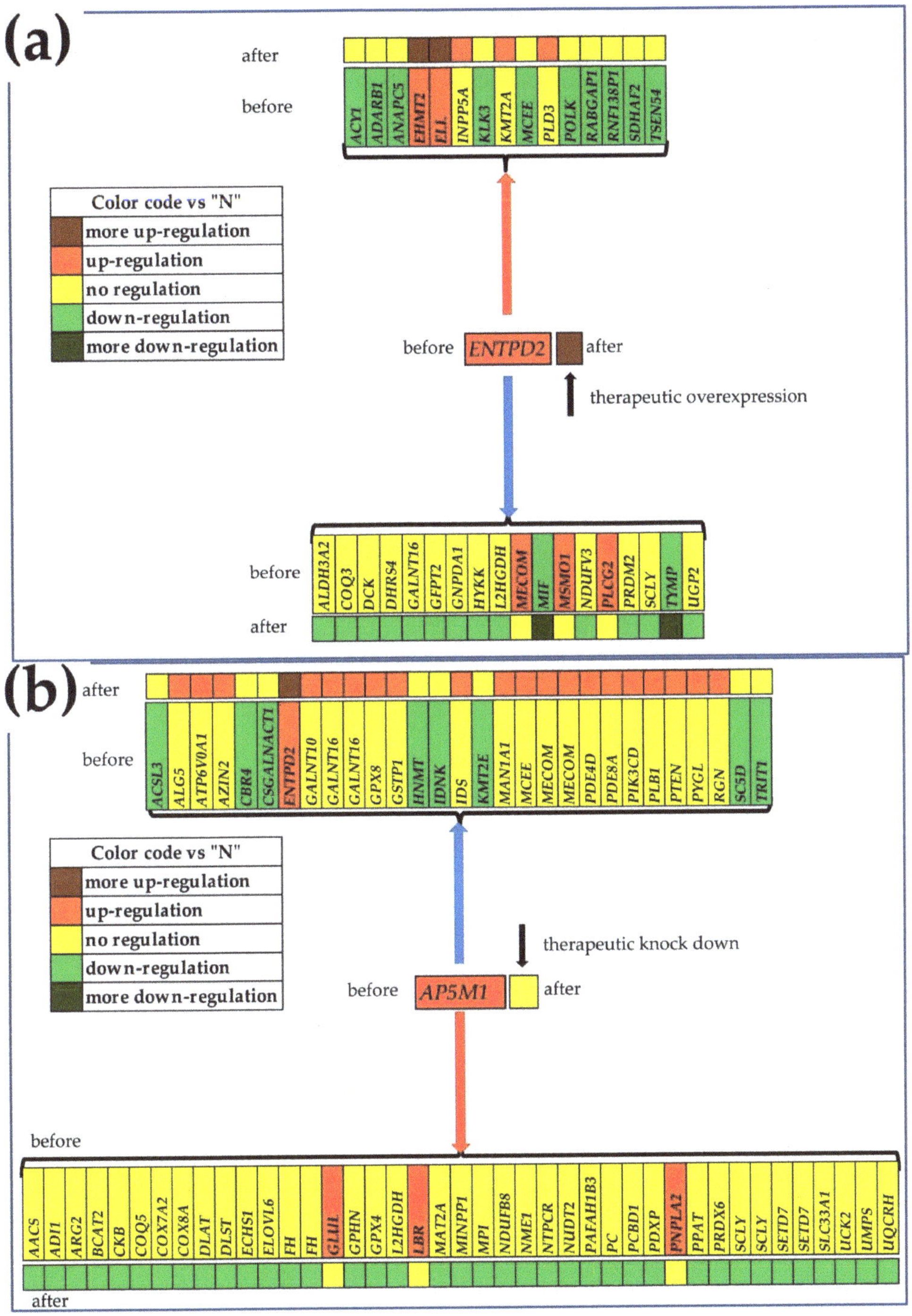

Figure 6. *Cont.*

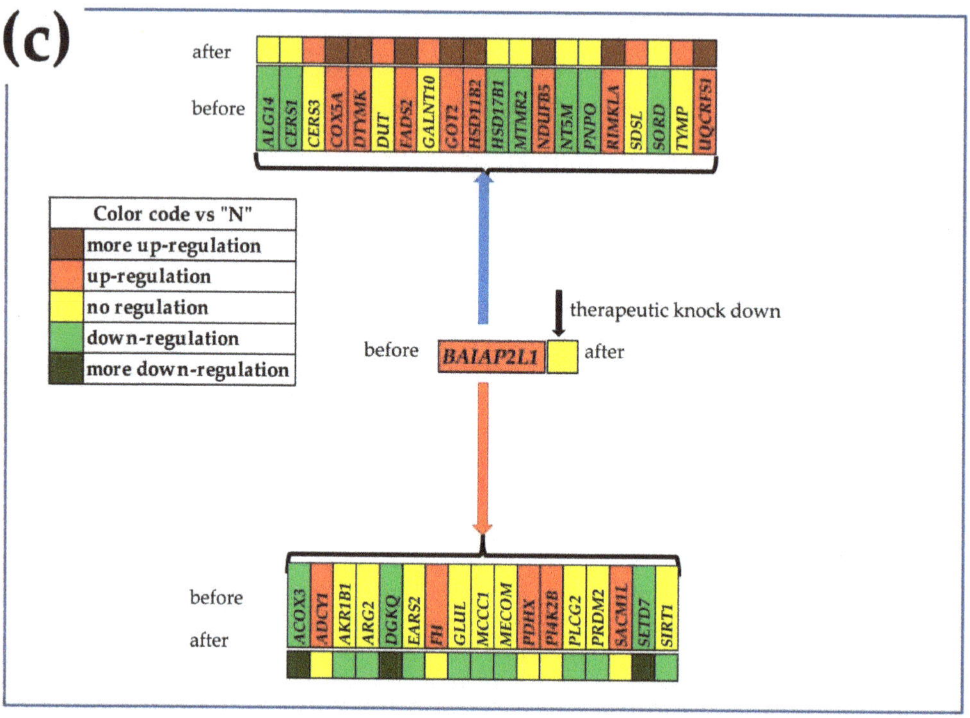

Figure 6. Significantly synergistically and antagonistically expressed metabolic genes with the corresponding GMR in the nodules (**a**) "A", (**b**) "B" and (**c**) "C" of the "ABCN" patient and the predicted regulations after the therapeutic alteration of the GMR. The red/blue arrow indicates the genes synergistically/antagonistically expressed with the GMR in that nodule. Gene symbol background indicates the status of that gene in the mentioned cancer nodule with respect to the surrounding normal tissue "N".

4. Discussion

The purpose of this study was to provide justification and a framework for the development of a personalized gene-therapy approach for prostate-cancer management. The manuscript detailed the theoretical foundations for a new and effective prostate cancer treatment. As defined by the US Food and Drug Administration (FDA), the gene therapy "seeks to modify or manipulate the expression of a gene or to alter the biological properties of living cells for therapeutic use" [62]. The FDA also issued the Guidance for Industry for "Long term follow-up after administration of human gene therapy products" [63]. According to the FDA [62], gene therapy replaces a disease-causing altered gene with a healthy copy of it or/and inactivates a disease-causing altered gene that is not functioning properly or/and introduces a new or modified gene. Gene-therapy products can be delivered as plasmid DNAs, using modified viral or bacterial vectors, or through gene editing technology. As such, gene therapy is part of the targeted therapies that, in the vision of the USA National Cancer Institute, include hormone therapies, signal transduction inhibitors, gene-expression modulators, apoptosis inducers, angiogenesis inhibitors, immunotherapies and toxin-delivery molecules [64].

The development of effective PCa therapies targeting selected genes or their downstream products was the objective of many research groups, and numerous publications detail their findings (e.g., References [65–69]. Some therapies that work against the consequences of altered genes (e.g., use of olaparib for HRR gene-mutated metastatic castration-resistant prostate cancer [70,71]) were already granted FDA approval (all FDA-approved therapies for PCa are listed in Reference [72] and partially discussed in Reference [73]). Gene therapy targets not only protein-coding genes but also long non-coding RNAs [74] and microRNAs [75]. PCa FDA-approved therapies also include inhibitors of the androgen receptor signaling, such as enzalutamide [76,77] and darolutamide [78,79]. However, the endless diversity and the strong impact of certain individual characteristics on the PCa progression and treatment outcomes raise serious doubts about the value of such "good-for-everybody" targeted therapies, refocusing the research on personalized solutions.

The report is based on transcriptomic data obtained in IacobasLab from 10 groups of samples: three distinct cancer nodules and the surrounding normal tissue from each of two PCa patients and two standard PCa cell lines. All experimental data were documented in the NCBI Gene Expression Omnibus. Because we tailor our approach of the prostate-cancer genomics and gene therapy on the uniqueness of each affected man, and the data clearly show that, even within the same tumor, each cancer nodule has a distinct transcriptome, the sample size is not important.

The power of the Genomic Fabric Paradigm [18] used here comes from extending the workable transcriptomic information by four orders of magnitude by considering for each quantified gene all the readily available independent characteristics, namely AVE, REV and COR, with each other gene. While the AVE (the average expression level across biological replicas) is used in all studies to determine whether the gene was significantly regulated in cancer with respect to the normal tissue, the ignored REV and COR bring additional fundamental information. Thus, REV (relative expression variability) tells about a cell's priorities in limiting the expression fluctuations of that gene, with genes critical for the cellular phenotypic expression constrained within very narrow intervals (low REV). On the other hand, COR (expression correlation) shows how much the expression fluctuations of one gene are correlated with the fluctuations of another gene; such information is essential for the formation and maintenance of gene networks.

The independence and complementarity of these types of characteristics was proved in Figure 1 for the first 50 alphabetically ordered genes from the mTOR signaling pathway in the three cancer nodules and the surrounding normal prostate tissue. These genes were used only to illustrate the independence of AVE, REV and COR, but any other gene subset would equally prove the independence. It was also proved in previous publications for apoptosis in human thyroid [21], chemokine signaling in human kidney [18] and mouse cortex [80], evading apoptosis in human prostate [33], ionic channels in mouse heart [81] and PI3K–AKT signaling in mouse hippocampus [82]).

Our research confirmed the numerous reports about the heterogeneity of PCa tumors at the gene-expression-profile level [10–17]. For this, we included Table 1 for the genes with the largest expression in each profiled region, and Table 3 and Figure 2 for the genes' contributions to the cancer transcriptomic departure from the normal tissue. Use of the Weighted Individual Gene Regulation (WIR) and of the Weighted Pathway Regulation (WPR) provided more accurate measures of the transcriptomic alterations than the traditional percentages of upregulated and downregulated genes. Our results with distinct regulations even at the cancer nodule level within the same tumor (see Table 3 and Figure 2 in the present report and Figures 2–4 in [18] and Figures 2–4 in Reference [33]) question the idea of "transcriptomic signature" that is believed to be common for large cancer-affected populations. We consider such cancer transcriptomic signatures as artefacts of the meta-analyses combining transcriptomic data from many individuals (not always demographically grouped as race, age and other major criteria), collected by numerous labs, using (some)times distinct transcriptomic platforms and experimental protocols.

More importantly, we proved that the heterogeneity extends to the expression control (see Table 2 for the most stably and Supplementary Table S1 for the most unstably expressed genes). It extends also to the expression coordination (see Figure 3 for the networking of *AKT2* and *MTOR* with their partners). These heterogeneities are even more important than the heterogeneity of the expression level, because manipulation of the same gene may have different consequences in the distinct regions of the tumor. Thus, experimental overexpression or knockdown of one gene may trigger different responses of the homeostatic control mechanisms (higher in regions with low REV) and may differently remodel the gene networks.

The substantial heterogeneity of the gene-expression level and control and networking other genes require us to go beyond the bulk tissue and profile each cancer clone in the tumor separately. This strategy (when served by adequate analytical tools) is a pre-requisite for designing personalized therapeutic strategies to selectively destroy the cancer clones with minimal impact on the surrounding healthy tissue.

We have introduced the GCH (Gene Commanding Height) score to determine how influential a gene is in a particular region/condition and identified the GMR (Gene Master Regulator) of each profiled phenotype as its top gene (highest GCH). The GMR enjoys the strongest protection of expression (lowest REV), while being coordinately expressed with many other genes. As such, "smart" manipulation of the GMR expression beyond critical limits is expected to selectively kill (or at least block the proliferation) of the cells it commands. Since the GMRs of the cancer clones have very low GCH in the surrounding healthy tissue, targeting the cancer GMRs would have little effect on the normal cells. However, these are theoretical predictions that need further experimental validation.

Many investigators looked before us for transcription factors (e.g., References [83,84]) or hormone receptors (e.g., Reference [85]) as master regulators that can be used in targeted cancer therapies. However, our GMR approach is fundamentally distinct from the traditional quest for master regulators both by the selection method and by the gene coverage. Thus, instead of the molecular affinity (same regardless the cell phenotype) used to identify the transcription factors (see References [86–88]), we rank the genes according to their expression control and coordination with other genes (GCH) in that particular specimen. The GCH-based gene hierarchy is not only personalized for the profiled tissue, but it is also dynamic, being reorganized during the progression of the disease, in response to a treatment and to other environmental changes. Moreover, we do not restrict the GMR quest to transcription factors and hormone receptors; any coding and non-coding RNA can be chosen by the cell itself as its GMR if its strictly controlled expression level regulates major functional pathways via expression coordination with their genes.

Figure 4 shows that the GMR in one region has practically very little role in any other regions (much lower GCH), which is exceptionally important for designing a gene therapy that would selectively affect the phenotype ruled by the GMR.

In this report for metabolism (Figures 5 and 6) and in previous publications for basal transcription, RNA polymerase and cell cycle [18], apoptosis [21] and enzymes [33], we included predictions of what might happen when targeting the GMR by a gene therapy. Unfortunately, there was no possibility to experimentally validate any of these predictions on the patients from which we have collected the prostate tissues. However, the GMR approach involving a monotonically ascendant relationship between the GCH score and the overall transcriptomic changes was validated by us on two human thyroid cancer cell lines stably transfected with four genes [20,22]. Thus, transfection of the BCPAP (papillary) and 850C (anaplastic) thyroid cancer cells with *DDX19B* (DEAD-Box Helicase 19B), *NEMP1* (nuclear envelope integral membrane protein 1), *PANK2* (pantothenate kinase 2) and *UBALD1* (UBA-like domain containing 1) induced significantly larger transcriptomic alterations in the cells where these genes had higher GCH.

The worst scenario of our GMR approach is when manipulation of a gene with top GCH scores in two cancer nodules is beneficial for one cancer nodule but detrimental for the other. Although this situation is very unlikely (never found something even close in our studies), the solution is to go to the next in line gene that has either similar effects in the two nodules or is irrelevant (low GCH) in the second nodule.

5. Conclusions

For now, the approved FDA PCa treatments [72] are considered effective for all men, regardless of race, age, medical history, habits and other risk factors whose dynamic combination makes each of them unique at each stage. However, there is enough evidence that each man responds differently to the same treatment, that the outcomes change in time for the same person and that, in most cases, there is little improvement. In contrast, we propose to identify the most legitimate gene targets that will selectively destroy the cancer clones of the prostate of the current patient, now. Although a "good-for-everybody" drug seems much more advantageous from economical point of view, in time, our approach may become economically competitive. There are already numerous FDA-approved cellular and gene therapy products [89], and, with the right stimulus, the industry will soon produce drugs based on CRISPR or other types of constructs to target almost all genes. When this will be the case, the oncologist will perform the transcriptomic analysis of tumor biopsies, identify the GMR of the cancer clones, and order and administrate the right product to his patient. The treatment will have similar costs as the actual gene therapy but would be more efficient for the PCa-affected person. Certainly, there are many more men with PCa (>3.1 mil diagnosed in the USA [90]) than protein-coding genes (~20,000), so that, on average, 15 of them share the same protein-coding GMR in one cancer nodule. However, we recommend a personalized three-gene cocktail to destroy the most aggressive three cancer clones in the prostate, and the chance of two men sharing the same gene triplet is less than 1 in one trillion.

Caution: Although very attractive, the GMR approach of the PCa therapy is still a theory that needs rigorous experimental validation.

If the tumor heterogeneity is very high (as determined by scRNA-sequencing), then one may have too many GMRs to consider. In such a case, one may use our method to establish the gene hierarchy for the entire tumor, considering the average GCH scores across the distinct nodules. The same could be performed by re-analyzing the gene expression profiles of many persons from the GFP perspective. Although not lethal for any nodule, the top genes will have enough influence on all of them. Apparently, this method will restore the concept of "one gene fits all", but with a totally different way to select the target gene(s).

Supplementary Materials: The following supporting information can be downloaded at https://www.mdpi.com/article/10.3390/cimb44010027/s1. Table S1: Three genes with the largest expression variability across biological replicas in all profiled regions of the two patients.

Author Contributions: Conceptualization, D.A.I.; methodology, S.I. and D.A.I.; software, D.A.I.; validation, S.I. and D.A.I.; formal analysis, D.A.I.; investigation, D.A.I.; resources, D.A.I.; data curation, D.A.I.; writing—original draft preparation, S.I.; writing—review and editing, D.A.I.; visualization, D.A.I.; supervision, D.A.I.; project administration, D.A.I.; funding acquisition, D.A.I. All authors have read and agreed to the published version of the manuscript.

Funding: This research received no external funding.

Institutional Review Board Statement: The study was conducted according to the guidelines of the Declaration of Helsinki. At the time of the experiment (2016), the study was part of D.A. Iacobas' project approved by the Institutional Review Boards (IRB) of the New York Medical College's (NYMC) and Westchester Medical Center (WMC) Committees for Protection of Human Subjects. The approved IRB (L11,376 from 2 October 2015) granted access to frozen cancer specimens from the WMC Pathology Archives and depersonalized pathology reports, waiving the patient's informed consent.

Informed Consent Statement: Patient consent was waived due to the use of depersonalized pathology reports.

Data Availability Statement: Raw and processed gene-expression data were deposited and are publicly accessible at https://www.ncbi.nlm.nih.gov/geo/query/acc.cgi?acc=GSE72333, https://www.ncbi.nlm.nih.gov/geo/query/acc.cgi?acc=GSE72414, accessed on 1 September 2021. https://www.ncbi.nlm.nih.gov/geo/query/acc.cgi?acc=GSE133891, accessed on 1 September 2021. https://www.ncbi.nlm.nih.gov/geo/query/acc.cgi?&acc=GSE133906, accessed on 1 September 2021. https://www.ncbi.nlm.nih.gov/geo/query/acc.cgi?acc=GSE168718, accessed on 1 September 2021. https://www.ncbi.nlm.nih.gov/geo/query/acc.cgi?acc=GSE183889. accessed on 1 September 2021.

Acknowledgments: D.A.I. was funded by the Texas A&M University System Chancellor's Research Initiative (CRI) for the Center for Computational Systems Biology at Prairie View University.

Conflicts of Interest: The authors declare no conflict of interest.

References

1. Hatano, K.; Nonomura, N. Genomic Profiling of Prostate Cancer: An Updated Review. *World J. Men's Health* **2021**, *39*. [CrossRef]
2. Spratt, D.E. Prostate Cancer Transcriptomic Subtypes. *Adv. Exp. Med. Biol.* **2019**, *1210*, 111–120. [CrossRef]
3. Manceau, C.; Fromont, G.; Beauval, J.-B.; Barret, E.; Brureau, L.; Créhange, G.; Dariane, C.; Fiard, G.; Gauthé, M.; Mathieu, R.; et al. Biomarker in Active Surveillance for Prostate Cancer: A Systematic Review. *Cancers* **2021**, *13*, 4251. [CrossRef]
4. Evans, R.; Hawkins, N.; Dequen-O'Byrne, P.; McCrea, C.; Muston, D.; Gresty, C.; Ghate, S.R.; Fan, L.; Hettle, R.; Abrams, K.R.; et al. Exploring the Impact of Treatment Switching on Overall Survival from the PROfound Study in Homologous Recombination Repair (HRR)-Mutated Metastatic Castration-Resistant Prostate Cancer (mCRPC). *Target. Oncol.* **2021**, *16*, 613–623. [CrossRef] [PubMed]
5. Rogers, O.C.; Rosen, D.M.; Antony, L.; Harper, H.M.; Das, D.; Yang, X.; Minn, I.; Mease, R.C.; Pomper, M.G.; Denmeade, S.R. Targeted delivery of cytotoxic proteins to prostate cancer via conjugation to small molecule urea-based PSMA inhibitors. *Sci. Rep.* **2021**, *11*, 14925. [CrossRef]
6. Qureshi, Z.; Ahmad, M.; Yang, W.-X.; Tan, F.-Q. Kinesin 12 (KIF15) contributes to the development and tumorigenicity of prostate cancer. *Biochem. Biophys. Res. Commun.* **2021**, *576*, 7–14. [CrossRef] [PubMed]
7. Stachurska, A.; Elbanowski, J.; Kowalczyńska, H.M. Role of α5β1 and αvβ3 integrins in relation to adhesion and spreading dynamics of prostate cancer cells interacting with fibronectin underin vitroconditions. *Cell Biol. Int.* **2012**, *36*, 883–892. [CrossRef]
8. Moghaddam, S.; Jalali, A.; O'Neill, A.; Murphy, L.; Gorman, L.; Reilly, A.-M.; Heffernan, Á.; Lynch, T.; Power, R.; O'Malley, K.J.; et al. Integrating Serum Biomarkers into Prediction Models for Biochemical Recurrence Following Radical Prostatectomy. *Cancers* **2021**, *13*, 4162. [CrossRef]
9. Jillson, L.K.; Rider, L.C.; Rodrigues, L.U.; Romero, L.; Karimpour-Fard, A.; Nieto, C.; Gillette, C.; Torkko, K.; Danis, E.; Smith, E.E.; et al. MAP3K7 Loss Drives Enhanced Androgen Signaling and Independently Confers Risk of Recurrence in Prostate Cancer with Joint Loss of CHD1. *Mol. Cancer Res.* **2021**, *19*, 1123–1136. [CrossRef] [PubMed]
10. Kulac, I.; Roudier, M.P.; Haffner, M.C. Molecular Pathology of Prostate Cancer. *Surg. Pathol. Clin.* **2021**, *14*, 387–401. [CrossRef] [PubMed]
11. Tolkach, Y.; Kristiansen, G. The Heterogeneity of Prostate Cancer: A Practical Approach. *Pathobiology* **2018**, *85*, 108–116. [CrossRef]
12. Tu, S.-M.; Zhang, M.; Wood, C.G.; Pisters, L.L. Stem Cell Theory of Cancer: Origin of Tumor Heterogeneity and Plasticity. *Cancers* **2021**, *13*, 4006. [CrossRef]
13. Berglund, E.; Maaskola, J.; Schultz, N.; Friedrich, S.; Marklund, M.; Bergenstråhle, J.; Tarish, F.; Tanoglidi, A.; Vickovic, S.; Larsson, L.; et al. Spatial maps of prostate cancer transcriptomes reveal an unexplored landscape of heterogeneity. *Nat. Commun.* **2018**, *9*, 2419. [CrossRef]
14. Brady, L.; Kriner, M.; Coleman, I.; Morrissey, C.; Roudier, M.; True, L.D.; Gulati, R.; Plymate, S.R.; Zhou, Z.; Birditt, B.; et al. Inter- and intra-tumor heterogeneity of metastatic prostate cancer determined by digital spatial gene expression profiling. *Nat. Commun.* **2021**, *12*, 1426. [CrossRef]
15. Melo, C.M.; Vidotto, T.; Chaves, L.P.; Lautert-Dutra, W.; dos Reis, R.B.; Squire, J.A. The Role of Somatic Mutations on the Immune Response of the Tumor Microenvironment in Prostate Cancer. *Int. J. Mol. Sci.* **2021**, *22*, 9550. [CrossRef] [PubMed]
16. Makino, T.; Izumi, K.; Iwamoto, H.; Mizokami, A. Treatment Strategies for High-Risk Localized and Locally Advanced and Oligometastatic Prostate Cancer. *Cancers* **2021**, *13*, 4470. [CrossRef] [PubMed]
17. Burgess, L.; Roy, S.; Morgan, S.; Malone, S. A Review on the Current Treatment Paradigm in High-Risk Prostate Cancer. *Cancers* **2021**, *13*, 4257. [CrossRef]
18. Iacobas, D.A.; Mgbemena, V.E.; Iacobas, S.; Menezes, K.M.; Wang, H.; Saganti, P.B. Genomic Fabric Remodeling in Metastatic Clear Cell Renal Cell Carcinoma (ccRCC): A New Paradigm and Proposal for a Personalized Gene Therapy Approach. *Cancers* **2020**, *12*, 3678. [CrossRef]
19. Mateo, J.; McKay, R.; Abida, W.; Aggarwal, R.; Alumkal, J.; Alva, A.; Feng, F.; Gao, X.; Graff, J.; Hussain, M.; et al. Accelerating precision medicine in metastatic prostate cancer. *Nat. Rev. Cancer* **2020**, *1*, 1041–1053. [CrossRef] [PubMed]
20. Iacobas, D.A.; Tuli, N.; Iacobas, S.; Rasamny, J.K.; Moscatello, A.; Geliebter, J.; Tiwari, R.M. Gene master regulators of papillary and anaplastic thyroid cancer phenotypes. *Oncotarget* **2018**, *9*, 2410–2424. [CrossRef]
21. Iacobas, D.A. Biomarkers, Master Regulators and Genomic Fabric Remodeling in a Case of Papillary Thyroid Carcinoma. *Genes* **2020**, *11*, 1030. [CrossRef] [PubMed]
22. Iacobas, S.; Ede, N.; Iacobas, D.A. The Gene Master Regulators (GMR) Approach Provides Legitimate Targets for Personalized, Time-Sensitive Cancer Gene Therapy. *Genes* **2019**, *10*, 560. [CrossRef]
23. Iacobas, D.A. The Genomic Fabric Perspective on the Transcriptome Between Universal Quantifiers and Personalized Genomic Medicine. *Biol. Theory* **2016**, *11*, 123–137. [CrossRef]
24. Iacobas, D.A. Powerful quantifiers for cancer transcriptomics. *World J. Clin. Oncol.* **2020**, *11*, 679–704. [CrossRef] [PubMed]

25. Iacobas, D.A.; Iacobas, S.; Lee, P.R.; Cohen, J.E.; Fields, R.D. Coordinated Activity of Transcriptional Networks Responding to the Pattern of Action Potential Firing in Neurons. *Genes* **2019**, *10*, 754. [CrossRef] [PubMed]
26. Petrucci, R.H.; Harwood, W.S.; Herring, F.G. *General Chemistry: Principles and Modern Applications*, 8th ed.; Prentice Hall: Hoboken, NJ, USA, 2002; p. 37, ISBN 978-0-13-014329-7.
27. Horoszewicz, J.S.; Leong, S.S.; Kawinski, E.; Karr, J.P.; Rosenthal, H.; Chu, T.M.; Mirand, E.A.; Murphy, G.P. LNCaP model of human prostatic carcinoma. *Cancer Res.* **1983**, *43*, 1809–1818. [PubMed]
28. Alimirah, F.; Chen, J.; Basrawala, Z.; Xin, H.; Choubey, D. DU-145 and PC-3 human prostate cancer cell lines express androgen receptor: Implications for the androgen receptor functions and regulation. *FEBS Lett.* **2006**, *580*, 2294–2300. [CrossRef]
29. Remodeling of DNA Transcription Genomic Fabric in Capridine-Treated LNCaP Human Prostate Cancer Cell Line. Available online: https://www.ncbi.nlm.nih.gov/geo/query/acc.cgi?acc=GSE72414 (accessed on 1 September 2021).
30. Remodeling of Major Genomic Fabrics and Their Interplay in Capridine-Treated DU145 Classic Human Prostate Cancer. Available online: https://www.ncbi.nlm.nih.gov/geo/query/acc.cgi?acc=GSE72333 (accessed on 1 September 2021).
31. Gene Commanding Height (GCH) Hierarchy in the Cancer Nucleus and Cancer-Free Resection Margins from a Surgically Removed Prostatic Adenocarcinoma of a 65y Old Black Man. Available online: https://www.ncbi.nlm.nih.gov/geo/query/acc.cgi?&acc=GSE133906 (accessed on 1 September 2021).
32. Genomic Fabric Remodeling in Prostate Cancer. Available online: https://www.ncbi.nlm.nih.gov/geo/query/acc.cgi?acc=GSE168718 (accessed on 1 September 2021).
33. Iacobas, S.; Iacobas, D. A Personalized Genomics Approach of the Prostate Cancer. *Cells* **2021**, *10*, 1644. [CrossRef]
34. Gene Commanding Height (GCH) Hierarchy in the Cancer Nucleus and Cancer-Free Resection Margins from a Surgically Removed Prostatic Adenocarcinoma of a 47y Old White Man. Available online: https://www.ncbi.nlm.nih.gov/geo/query/acc.cgi?acc=GSE13389 (accessed on 1 September 2021).
35. Transcriptomic Heterogeneity of the Prostate Cancer. Available online: https://www.ncbi.nlm.nih.gov/geo/query/acc.cgi?acc=GSE183889 (accessed on 9 December 2021).
36. Agilent-026652 Whole Human Genome Microarray 4 × 44K v2. Available online: https://www.ncbi.nlm.nih.gov/geo/query/acc.cgi?acc=GPL13497 (accessed on 1 September 2021).
37. Stranger, B.E.; Forrest, M.S.; Clark, A.G.; Minichiello, M.J.; Deutsch, S.; Lyle, R.; Hunt, S.; Kahl, B.; Antonarakis, S.E.; Tavaré, S.; et al. Genome-Wide Associations of Gene Expression Variation in Humans. *PLoS Genet.* **2005**, *1*, e78. [CrossRef]
38. Guan, J.; Han, S.; Wu, J.; Zhang, Y.; Bai, M.; Abdullah, S.W.; Sun, S.; Guo, H. Ribosomal Protein L13 Participates in Innate Immune Response Induced by Foot-and-Mouth Disease Virus. *Front. Immunol.* **2021**, *12*. [CrossRef]
39. Xu, H.; Liu, P.; Yan, Y.; Fang, K.; Liang, D.; Hou, X.; Zhang, X.; Wu, S.; Ma, J.; Wang, R.; et al. FKBP9 promotes the malignant behavior of glioblastoma cells and confers resistance to endoplasmic reticulum stress inducers. *J. Exp. Clin. Cancer Res.* **2020**, *39*, 44. [CrossRef] [PubMed]
40. Jiang, F.-N.; Dai, L.-J.; Yang, S.-B.; Wu, Y.-D.; Liang, Y.-X.; Yin, X.-L.; Zou, C.-Y.; Zhong, W.-D. Increasing of FKBP9 can predict poor prognosis in patients with prostate cancer. *Pathol.—Res. Pract.* **2020**, *216*, 152732. [CrossRef]
41. Macauda, A.; Piredda, C.; Clay-Gilmour, A.I.; Sainz, J.; Buda, G.; Markiewicz, M.; Barington, T.; Ziv, E.; Hildebrandt, M.A.T.; Belachew, A.A.; et al. Expression quantitative trait loci of genes predicting outcome are associated with survival of multiple myeloma patients. *Int. J. Cancer* **2021**, *149*, 327–336. [CrossRef] [PubMed]
42. Wang, J.; Luo, Q.; Liu, M.; Zhang, C.; Jia, Y.; Tong, R.; Yang, L.; Fu, X. TBRG4 silencing promotes progression of squamous cell carcinoma via regulation of CAV-1 expression and ROS formation. *Cell. Mol. Biol.* **2020**, *66*, 157–164. [CrossRef]
43. Huang, F.; Zhou, P.; Wang, Z.; Zhang, X.-L.; Liao, F.-X.; Hu, Y.; Chang, J. Knockdown of TBRG4 suppresses proliferation, invasion and promotes apoptosis of osteosarcoma cells by downregulating TGF-β1 expression and PI3K/AKT signaling pathway. *Arch. Biochem. Biophys.* **2020**, *686*, 108351. [CrossRef]
44. Varghese, R.T.; Liang, Y.; Guan, T.; Franck, C.T.; Kelly, D.F.; Sheng, Z. Survival kinase genes present prognostic significance in glioblastoma. *Oncotarget* **2016**, *7*, 20140–20151. [CrossRef] [PubMed]
45. Bashanfer, S.A.A.; Saleem, M.; Heidenreich, O.; Moses, E.J.; Yusoff, N.M. Disruption of MAPK1 expression in the ERK signalling pathway and the RUNX1-RUNX1T1 fusion gene attenuate the differentiation and proliferation and induces the growth arrest in t(8;21) leukaemia cells. *Oncol. Rep.* **2019**, *41*, 2027–2040. [CrossRef] [PubMed]
46. Wang, A.; Zhao, C.; Liu, X.; Su, W.; Duan, G.; Xie, Z.; Chu, S.; Gao, Y. Knockdown of TBRG4 affects tumorigenesis in human H1299 lung cancer cells by regulating DDIT3, CAV1 and RRM2. *Oncol. Lett.* **2018**, *15*, 121–128. [CrossRef]
47. Rao, Y.; Liu, H.; Yan, X.; Wang, J. In Silico Analysis Identifies Differently Expressed lncRNAs as Novel Biomarkers for the Prognosis of Thyroid Cancer. *Comput. Math. Methods Med.* **2020**, *2020*, 3651051. [CrossRef]
48. Danielpour, D.; Purighalla, S.; Wang, E.; Zmina, P.M.; Sarkar, A.; Zhou, G. JAB1/COPS5 is a putative oncogene that controls critical oncoproteins deregulated in prostate cancer. *Biochem. Biophys. Res. Commun.* **2019**, *518*, 374–380. [CrossRef] [PubMed]
49. mTOR Signaling Pathway. Available online: https://www.genome.jp/pathway/hsa04150 (accessed on 1 September 2021).
50. Hua, H.; Kong, Q.; Zhang, H.; Wang, J.; Luo, T.; Jiang, Y. Targeting mTOR for cancer therapy. *J. Hematol. Oncol.* **2019**, *12*, 71. [CrossRef]
51. Gong, C.; Qu, S.; Lv, X.-B.; Liu, B.; Tan, W.; Nie, Y.; Su, F.; Liu, Q.; Yao, H.; Song, E. BRMS1L suppresses breast cancer metastasis by inducing epigenetic silence of FZD10. *Nat. Commun.* **2014**, *5*, 5406. [CrossRef] [PubMed]

52. Wendel, H.-G.; Silva, R.L.; Malina, A.; Mills, J.R.; Zhu, H.; Ueda, T.; Watanabe-Fukunaga, R.; Fukunaga, R.; Teruya-Feldstein, J.; Pelletier, J.; et al. Dissecting eIF4E action in tumorigenesis. *Genes Dev.* **2007**, *21*, 3232–3237. [CrossRef]
53. The Cancer Genome Atlas. Available online: https://portal.gdc.cancer.gov/ (accessed on 20 October 2021).
54. Oncotype IQ. Available online: https://www.oncotypeiq.com/en-US/prostate-cancer/patients-and-caregivers/early-stage-gps/why-oncotype-dx-gps (accessed on 11 February 2021).
55. Foundation Medicine. Available online: https://www.foundationmedicine.com/genomic-testing (accessed on 12 July 2020).
56. Prostate Cancer is Manageable. Available online: https://decipherbio.com/ (accessed on 8 February 2021).
57. Prostate Cancer Pathway. Available online: https://www.genome.jp/pathway/hsa05215 (accessed on 1 September 2021).
58. KEGG Pathway Database. Available online: https://www.genome.jp/kegg/pathway.html (accessed on 1 September 2021).
59. QIAGEN Ingenuity Pathway Analysis. Available online: https://digitalinsights.qiagen.com/products-overview/discovery-insights-portfolio/analysis-and-visualization/qiagen-ipa/ (accessed on 20 October 2021).
60. The Database for Annotation, Visualization and Integrated Discovery (DAVID). Available online: https://david.ncifcrf.gov/ (accessed on 20 October 2021).
61. Prickett, D.; Watson, M. Use of GenMAPP and MAPPFinder to analyse pathways involved in chickens infected with the protozoan parasite *Eimeria*. *BMC Proc.* **2009**, *3*, S7. [CrossRef] [PubMed]
62. What Is Gene Therapy? Available online: https://www.fda.gov/vaccines-blood-biologics/cellular-gene-therapy-products/what-gene-therapy (accessed on 1 September 2021).
63. Long Term Follow-Up After Administration of Human Gene Therapy Products. Guidance for Industry. Available online: https://www.fda.gov/media/113768/download (accessed on 20 October 2021).
64. U.S.A. National Cancer Institute: Targeted Cancer Therapies. Available online: https://www.cancer.gov/about-cancer/treatment/types/targeted-therapies/targeted-therapies-fact-sheet (accessed on 20 October 2021).
65. Harrington, K.J.; Spitzweg, C.; Bateman, A.R.; Morris, J.C.; Vile, R.G. Gene therapy for prostate cancer: Current status and future prospects. *J. Urol.* **2001**, *166*, 1220–1233. [CrossRef]
66. Ikegami, S.; Tadakuma, T.; Ono, T.; Suzuki, S.; Yoshimura, I.; Asano, T.; Hayakawa, M. Treatment efficiency of a suicide gene therapy using prostate-specific membrane antigen promoter/enhancer in a castrated mouse model of prostate cancer. *Cancer Sci.* **2004**, *95*, 367–370. [CrossRef]
67. Satoh, T.; Irie, A.; Egawa, S.; Baba, S. In Situ Gene Therapy for Prostate Cancer. *Curr. Gene Ther.* **2005**, *5*, 111–119. [CrossRef] [PubMed]
68. Cai, Z.; Lv, H.; Cao, W.; Zhou, C.; Liu, Q.; Li, H.; Zhou, F. Targeting strategies of adenovirus-mediated gene therapy and virotherapy for prostate cancer. *Mol. Med. Rep.* **2017**, *16*, 6443–6458. [CrossRef]
69. Talkar, S.S. Gene Therapy for Prostate Cancer: A Review. *Endocrine, Metab. Immune Disord.—Drug Targets* **2021**, *21*, 385–396. [CrossRef]
70. De Bono, J.; Mateo, J.; Fizazi, K.; Saad, F.; Shore, N.; Sandhu, S.; Chi, K.N.; Sartor, O.; Agarwal, N.; Olmos, D.; et al. Olaparib for Metastatic Castration-Resistant Prostate Cancer. *N. Engl. J. Med.* **2020**, *382*, 2091–2102. [CrossRef] [PubMed]
71. U.S. Food and Drug Administration, FDA. FDA Approves Olaparib for Hrr Gene-Mutated Metastatic Castration-Resistant Prostate Cancer. Available online: https://www.fda.gov/drugs/resources-information-approved-drugs/fda-approves-olaparib-hrr-gene-mutated-metastatic-castration-resistant-prostate-cancer (accessed on 1 September 2021).
72. U.S. Food and Drug Administration, FDA. Approved Treatments for Prostate Cancer. Available online: https://search.usa.gov/search?utf8=%E2%9C%93&affiliate=fda&sort_by=&query=prostate+cancer&commit=Search (accessed on 20 September 2021).
73. Zhang, J.; Sun, J.; Bakht, S.; Hassan, W. Recent Development and Future Prospects of Molecular Targeted Therapy in Prostate Cancer. *Curr. Mol. Pharmacol.* **2021**, *14*. [CrossRef]
74. Morgan, R.; da Silveira, W.A.; Kelly, R.C.; Overton, I.; Allott, E.H.; Hardiman, G. Long non-coding RNAs and their potential impact on diagnosis, prognosis, and therapy in prostate cancer: Racial, ethnic, and geographical considerations. *Expert Rev. Mol. Diagn.* **2021**, *21*, 1257–1271. [CrossRef] [PubMed]
75. Oh-Hohenhorst, S.J.; Lange, T. Role of Metastasis-Related microRNAs in Prostate Cancer Progression and Treatment. *Cancers* **2021**, *13*, 4492. [CrossRef]
76. FDA Approves Enzalutamide for Metastatic Castration-Sensitive Prostate Cancer. Available online: https://www.fda.gov/drugs/resources-information-approved-drugs/fda-approves-enzalutamide-metastatic-castration-sensitive-prostate-cancer (accessed on 20 October 2021).
77. Leblanc, E.; Ban, F.; Cavga, A.D.; Lawn, S.; Huang, C.-C.F.; Mohan, S.; Chang, M.E.K.; Flory, M.R.; Ghaidi, F.; Lingadahalli, S.; et al. Development of 2-(5,6,7-Trifluoro-1H-Indol-3-yl)-quinoline-5-carboxamide as a Potent, Selective, and Orally Available Inhibitor of Human Androgen Receptor Targeting Its Binding Function-3 for the Treatment of Castration-Resistant Prostate Cancer. *J. Med. Chem.* **2021**, *64*, 14968–14982. [CrossRef]
78. FDA Approves Darolutamide for Non-Metastatic Castration-Resistant Prostate Cancer. Available online: https://www.fda.gov/drugs/resources-information-approved-drugs/fda-approves-darolutamide-non-metastatic-castration-resistant-prostate-cancer (accessed on 20 October 2021).
79. Devos, G.; Devlies, W.; De Meerleer, G.; Baldewijns, M.; Gevaert, T.; Moris, L.; Milonas, D.; Van Poppel, H.; Berghen, C.; Everaerts, W.; et al. Neoadjuvant hormonal therapy before radical prostatectomy in high-risk prostate cancer. *Nat. Rev. Urol.* **2021**, *18*, 739–762. [CrossRef] [PubMed]

80. Iacobas, D.; Wen, J.; Iacobas, S.; Schwartz, N.; Putterman, C. Remodeling of Neurotransmission, Chemokine, and PI3K-AKT Signaling Genomic Fabrics in Neuropsychiatric Systemic Lupus Erythematosus. *Genes* **2021**, *12*, 251. [CrossRef]
81. Iacobas, S.; Amuzescu, B.; Iacobas, D.A. Transcriptomic uniqueness and commonality of the ion channels and transporters in the four heart chambers. *Sci. Rep.* **2021**, *11*, 2743. [CrossRef] [PubMed]
82. Iacobas, D.A.; Wen, J.; Iacobas, S.; Putterman, C.; Schwartz, N. TWEAKing the Hippocampus: The Effects of TWEAK on the Genomic Fabric of the Hippocampus in a Neuropsychiatric Lupus Mouse Model. *Genes* **2021**, *12*, 1172. [CrossRef]
83. Pratap, J.; Lian, J.B.; Javed, A.; Barnes, G.L.; Van Wijnen, A.J.; Stein, J.L.; Stein, G.S. Regulatory roles of Runx2 in metastatic tumor and cancer cell interactions with bone. *Cancer Metastasis Rev.* **2006**, *25*, 589–600. [CrossRef]
84. Orr-Urtreger, A.; Bar-Shira, A.; Matzkin, H.; Mabjeesh, N.J. The homozygous P582S mutation in the oxygen-dependent degradation domain of HIF-1α is associated with increased risk for prostate cancer. *Prostate* **2006**, *67*, 8–13. [CrossRef]
85. Kinyamu, H.K.; Collins, J.B.; Grissom, S.F.; Hebbar, P.B.; Archer, T.K. Genome wide transcriptional profiling in breast cancer cells reveals distinct changes in hormone receptor target genes and chromatin modifying enzymes after proteasome inhibition. *Mol. Carcinog.* **2008**, *47*, 845–885. [CrossRef] [PubMed]
86. Latchman, D.S. Transcription factors: An overview. *Int. J. Exp. Pathol.* **1993**, *74*, 417–422. [CrossRef]
87. Kawaji, H. Computational predictions of transcription factor binding sites. *Tanpakushitsu Kakusan Koso.* **2004**, *49*, 2877–2881.
88. Ji, H.; Wong, W.H. Computational Biology: Toward Deciphering Gene Regulatory Information in Mammalian Genomes. *Biometrics* **2006**, *62*, 645–663. [CrossRef] [PubMed]
89. U.S. Food and Drug Administration. Approved Cellular and Gene Therapy Products. Available online: https://www.fda.gov/vaccines-blood-biologics/cellular-gene-therapy-products/approved-cellular-and-gene-therapy-products (accessed on 20 September 2021).
90. American Cancer Society: Key Statistics for Prostate Cancer. Available online: https://www.cancer.org/cancer/prostate-cancer/about/key-statistics.html (accessed on 20 October 2021).

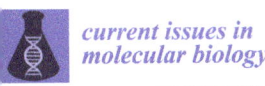

current issues in molecular biology

MDPI

Article

MiR-942-3p as a Potential Prognostic Marker of Gastric Cancer Associated with AR and MAPK/ERK Signaling Pathway

Wenjia Liu [1,2], Nanjiao Ying [2,*], Xin Rao [1] and Xiaodong Chen [1,3,*]

1 School of Electronics and Information Engineering, Hangzhou Dianzi University, Hangzhou 310018, China
2 School of Automation, Hangzhou Dianzi University, Hangzhou 310018, China
3 School of Electronic Engineering and Computer Science, Queen Mary University of London, London E1 4NS, UK
* Correspondence: yingnj@hdu.edu.cn (N.Y.); xiaodongchen@hdu.edu.cn (X.C.)

Abstract: Gastric cancer is a common tumor with high morbidity and mortality. MicroRNA (miRNA) can regulate gene expression at the translation level and various tumorigenesis processes, playing an important role in tumor occurrence and prognosis. This study aims to screen miRNA associated with gastric cancer prognosis as biomarkers and explore the regulatory genes and related signaling pathways. In this work, R language was used for the standardization and differential analysis of miRNA and mRNA expression profiles. Samples were randomly divided into a testing group and a training group. Subsequently, we built the five miRNAs (has-miR-9-3p, has-miR-135b-3p, has-miR-143-5p, has-miR-942-3p, has-miR-196-3p) prognostic modules, verified and evaluated their prediction ability by the Cox regression analysis. They can be used as an independent factor in the prognosis of gastric cancer. By predicting and analyzing potential biological functions of the miRNA target genes, this study found that the AR gene was not only a hub gene in the PPI network, but also associated with excessive survival of patients. In conclusion, this study demonstrated that hsa-miR-942-3p could be a potential prognostic marker of gastric cancer associated with the AR and MAPK/ERK signaling pathways. The results of this study provide insights into the occurrence and development of gastric cancer.

Keywords: gastric cancer; prognosis; biomarker; hsa-miR-942-3p; AR; MAPK/ERK signaling pathway

Citation: Liu, W.; Ying, N.; Rao, X.; Chen, X. MiR-942-3p as a Potential Prognostic Marker of Gastric Cancer Associated with AR and MAPK/ERK Signaling Pathway. *Curr. Issues Mol. Biol.* **2022**, *44*, 3835–3848. https://doi.org/10.3390/cimb44090263

Academic Editor: Dumitru A. Iacobas

Received: 18 July 2022
Accepted: 22 August 2022
Published: 24 August 2022

1. Introduction

Gastric cancer is one of the most common tumors and its overall survival rate is only about 10% [1]. Some treatments are developing rapidly, including surgery, radiotherapy, chemotherapy, and targeted therapy. However, the recurrence rate and poor prognosis remain a troubling issue. At present, some biomarkers related to the occurrence and prognosis of gastric cancer have been found [2] but their reliability has not been completely verified. Therefore, it is essential to screen new biomarkers or therapeutic targets for the prognosis of gastric cancer patients.

MicroRNA (miRNA) is a non-coding molecule, which can regulate gene expression at the translation level. Some studies have shown that miRNAs regulate various tumorigenesis processes (cell proliferation, cell differentiation, and cell apoptosis) by combining tumor suppressor genes or oncogenes. Yang L et al. found that miR-9-3p was a down-regulated gene of glioma cells. Its low expression resulted in increased levels of Herpud1 that could protect glioma cells from apoptosis [3]. Chen Z et al. showed that miR-143-5p could promote cadmium-induced apoptosis of LLC-PK1 cells by acting on the target gene AKT3 and inhibiting the Akt/Bad signaling pathway [4]. Ma R et al. verified that up-regulated miR-196b could induce a proliferative phenotype, leading to a poor prognosis in glioblastoma patients [5]. Chen M et al. showed that miR-135b could play the role of oncogenes by regulating the PI3K/Akt, HIF-1/FIH, Hippo, p53 signaling pathways, promote tumor cell proliferation, migration, invasion, promote tumor angiogenesis, affect the prognosis of tumor patients, and reduce the total survival and survival time. Moreover, the expression of miR-135b in serum can be used as a biomarker for the diagnosis of a tumor [6]. In addition, miRNA also plays a great role in the treatment of gastric cancer [7]. Lin A et al. concluded that miRNA-449b was associated with the occurrence of gastric cancer and lymph node metastasis [8]. Ma X et al. found that the expression level of miRNA-375 in gastric cancer was related to the degree of tumor differentiation, which could be considered a clinical monitoring target [9]. Han W and Su X found that miRNA-30c showed low expression in gastric cancer tissues and was involved in the occurrence and development of gastric cancer by changing cell proliferation, apoptosis, and cell cycle [10]. With this in mind, the studies of miRNA in gastric cancer still need to be pushed forward and further investigated.

In this study, we constructed, validated, and evaluated five miRNAs and the results showed that they could be used as independent prognostic factors in gastric cancer. More importantly, we detected the target gene AR of hsa-miR-942-3p which was the core target gene and closely related to the prognosis and survival of gastric cancer patients. In short, hsa-miR-942-3p may be a potential prognostic marker of gastric cancer related to the AR and mitogen-activated protein kinase (MAPK)/extracellular signal-regulated kinase (ERK) signaling pathways.

2. Materials and Methods

2.1. Data Downloading and Processing

The miRNA and mRNA profiles data were gained from The Cancer Genome Atlas (TCGA) database (https://www.cancer.gov, accessed on 20 August 2022 (Table 1). The miRNA expression profiles included 45 normal and 446 tumor samples, and the mRNA expression profiles included 32 normal and 375 tumor samples. Clinical information (443) for all gastric cancer samples was also downloaded (Table 2).

Table 1. The miRNA and mRNA expression profiles information.

	Variables	miRNA Expression Profiles	mRNA Expression Profiles
Case	Count	436	380
	Primary Site	Stomach	stomach
	Program	TCGA	TCGA
	Project	TCGA-STAD	TCGA-STAD
Files	Count	491	407
	Data Category	Transcriptome Profiling	Transcriptome Profiling
	Data Type	Isoform Expression Quantification	Gene Expression Quantification
	Workflow Type	BCGSC miRNA Profiling	HTSeq-Counts

Table 2. All patient information.

Variables		Case	Percentage (%)
Gender	Male	285	64.3
	Female	158	35.7
Age (years)	Range	30–90	
	Median	68	3.1
Futime (day)	Range	0–3720	
	Median	422	
Fustat	1	171	38.6
	0	272	61.3
Clinical stage	I	59	13.2
	II	130	29.2
	III	183	41.1
	IV	44	9.9
	Unknown	27	6

Table 2. *Cont.*

Variables		Case	Percentage (%)
T stage	T1	23	5
	T2	93	20.8
	T3	198	44.6
	T4	119	26.7
	TX	10	2.2
Lymph node stage	N0	132	29.7
	N1	119	26.8
	N2	85	19.1
	N3	88	19.7
	NX	17	3.8
	Unknown	2	0.4
Metastatic	M0	391	88.2
	M1	30	6.7
	MX	22	4.9

2.2. *Detection of Differentially Expressed miRNAs and mRNA Combined with Clinical Information*

Standardization and differential analysis of expression profiles were performed using R language ($p < 0.05$ and $|\log FC| > 1.0$) [11]. Thereafter, clinical information on patients was combined with the disposed of miRNAs and mRNAs.

2.3. *Construction of Sample Grouping and Prognostic Module*

Samples were divided into training group and testing group randomly by R language package. Univariate Cox regression analysis was used to detect the miRNAs with $p < 0.05$ in the training group. Multivariate Cox regression was used to build the miRNA module prognostic biomarkers with different overall survival [12]. Then, we established the risk score of a prognostic miRNA signature and detected the Proportional Hazards Assumption of the Cox module. The module was used to assess the survival prognosis of patients in three groups by the Kaplan–Meier curve. Log-rank tests were classified into a high-risk and low-risk group according to the risk score of the median value grouping. R language ("survivalROC" package) was used to evaluate miRNA predictive power by receiver operating characteristic (ROC) curve [13].

2.4. *Independent Prognostic Ability of miRNA*

The univariate Cox regression was analyzed to test the relationship between the prognostic miRNA and the overall survival of patients in the training group. Clinical factors were also analyzed by multivariate Cox regression to serve as independent prognostic elements.

2.5. *miRNA Target Genes Prediction and Functions Analysis*

The miRNA information was downloaded from three prediction databases (targetScan, miRTarBase, and miRDB). The target genes of miRNA were obtained and crosschecked in at least two databases. Using the Cytoscape and Venn software to draw the relation between miRNAs and the target genes. Differentially expressed genes (DEGs) and target genes were taken at the intersection to test whether these target genes were involved in the progression of gastric cancer. Kyoto Encyclopedia of Genes and Genomes (KEGG) enrichment pathway and Gene Ontology (GO) analysis displayed the potential function of all the intersection genes through R language ("org.Hs.eg.db" package and "clusterProfiler" package) [14].

2.6. *Screening Core Target Genes and Survival Analysis*

The protein–protein interaction (PPI) network between the target genes was obtained from STRING websites [15] while the medium confidence is 0.400. Then, the top ten hub genes were detected through Cytoscape plug-in CytoHubba. In addition, Kaplan–Meier curves were used to detect whether the intersection genes showed a relationship with overall survival.

3. Results

3.1. Detection of Differentially Expressed miRNAs and Differentially Expressed mRNAs

The miRNA expression profiles displayed 267 differentially expressed miRNAs (DEmiRNAs) (185 up-regulated and 82 down-regulated) (Figures 1 and 2). The mRNA expression profiles displayed 7531 differentially expressed mRNAs (DEmRNAs) (4395 up-regulated and 3136 down-regulated) (adjust p-value < 0.05 and |logFC| > 1.0).

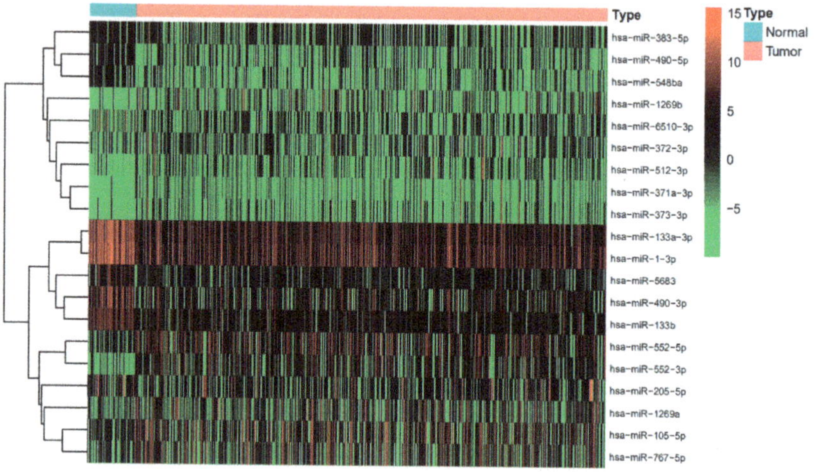

Figure 1. Clustering heatmap of differentially expressed miRNA.

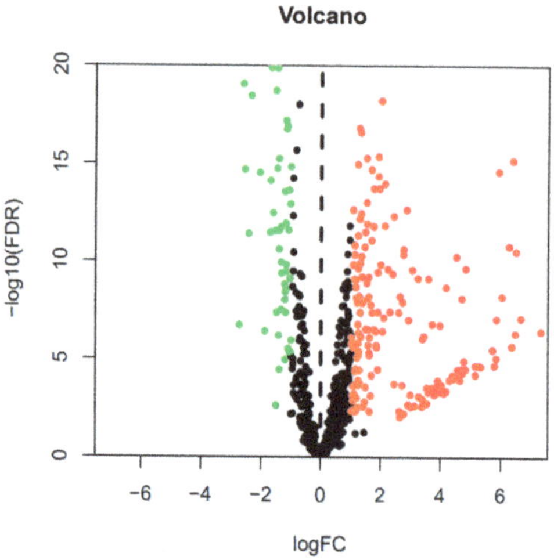

Figure 2. Volcanic maps of differentially expressed miRNAs.

3.2. Five miRNAs Associated with Overall Survival

All 389 groups (miRNA expression profiles) were divided into training group (196) and testing group (193) randomly. Univariate Cox regression analysis revealed that fifteen miRNAs were related to overall survival in the training group. Multivariate Cox regression analysis selected five miRNAs (hsa-miR-9-3p, hsa-miR-135b-3p, hsa-miR-143-5p, hsa-miR-942-3p, and hsa-miR-196b-3p) from the fifteen miRNAs finally (Table 3). Besides, the Kaplan–Meier curve also showed that the five miRNAs were related to overall survival (Figure 3).

Table 3. Univariate Cox regression and multivariate Cox regression of differentially expressed miRNAs.

ID	Univariate Cox Regression				Multivariate Cox Regression				
	HR	HR.95L	HR.95H	*p*-Value	Coef	HR	HR.95L	HR.95H	*p*-Value
hsa-miR-96-5p	0.761	0.642	0.903	0.002					
hsa-miR-7-5p	0.801	0.695	0.923	0.002					
hsa-let-7e-3p	1.379	1.112	1.71	0.003					
hsa-miR-143-5p	1.265	1.077	1.487	0.004	0.134	1.144	0.961	1.361	0.129
hsa-miR-942-3p	0.727	0.586	0.902	0.004	−0.178	0.837	0.663	1.056	0.132
hsa-miR-183-5p	0.806	0.69	0.942	0.007					
hsa-miR-196b-3p	0.648	0.468	0.897	0.009	−0.307	0.736	0.527	1.027	0.072
hsa-miR-125a-5p	1.401	1.067	1.839	0.015					
hsa-miR-135b-3p	0.799	0.665	0.96	0.017	−0.148	0.862	0.706	1.052	0.144
hsa-miR-30a-3p	1.21	1.024	1.428	0.025					
hsa-miR-652-5p	0.784	0.623	0.986	0.037					
hsa-miR-9-3p	1.17	1.008	1.359	0.039	0.147	1.159	0.989	1.358	0.069
hsa-miR-99a-3p	1.175	1.007	1.372	0.040					
hsa-miR-139-5p	1.221	1.007	1.48	0.042					
hsa-miR-137-3p	1.16	1.000	1.346	0.049					

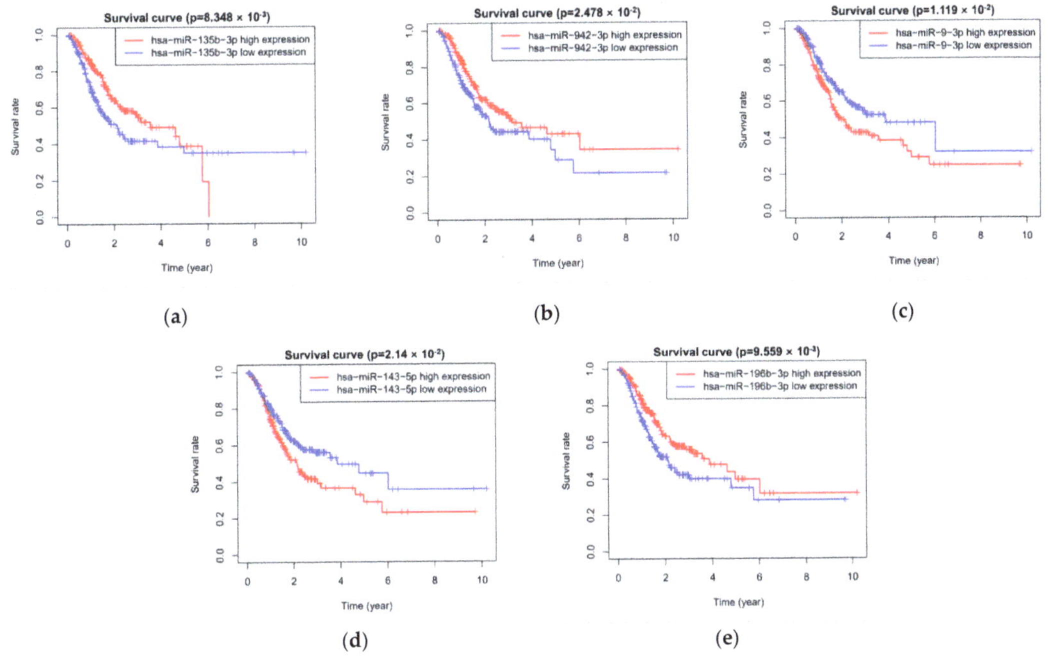

Figure 3. Five miRNAs associated with overall survival. (**a**) has-miR-135b-3p; (**b**) has-miR-942-3p; (**c**) has-miR-9-3p; (**d**) has-miR-143-5p; and (**e**) has-miR-196b-3p.

3.3. Prediction and Assessment of Five miRNAs for Overall Survival in Three Groups

According to the median value grouping of risk score, the Kaplan—Meier curve displayed that the high-risk group had worse survival than the low-risk group in the training group ($p = 1.417 \times 10^{-4}$), the testing group ($p = 2.131 \times 10^{-2}$), and the whole group ($p = 1.436 \times 10^{-5}$; Figure 4a–c). The area under curve (AUC) of ROC for the five miRNAs severally attained 0.719, 0.660, and 0.689 in the training group, the testing group, and the whole group (Figure 4d–f), which indicated that the five miRNAs perform well in predicting the overall survival of gastric cancer patients. Furthermore, patients with high-risk scores had a higher death rate than those with low-risk scores in the three groups (Figure 4g–i).

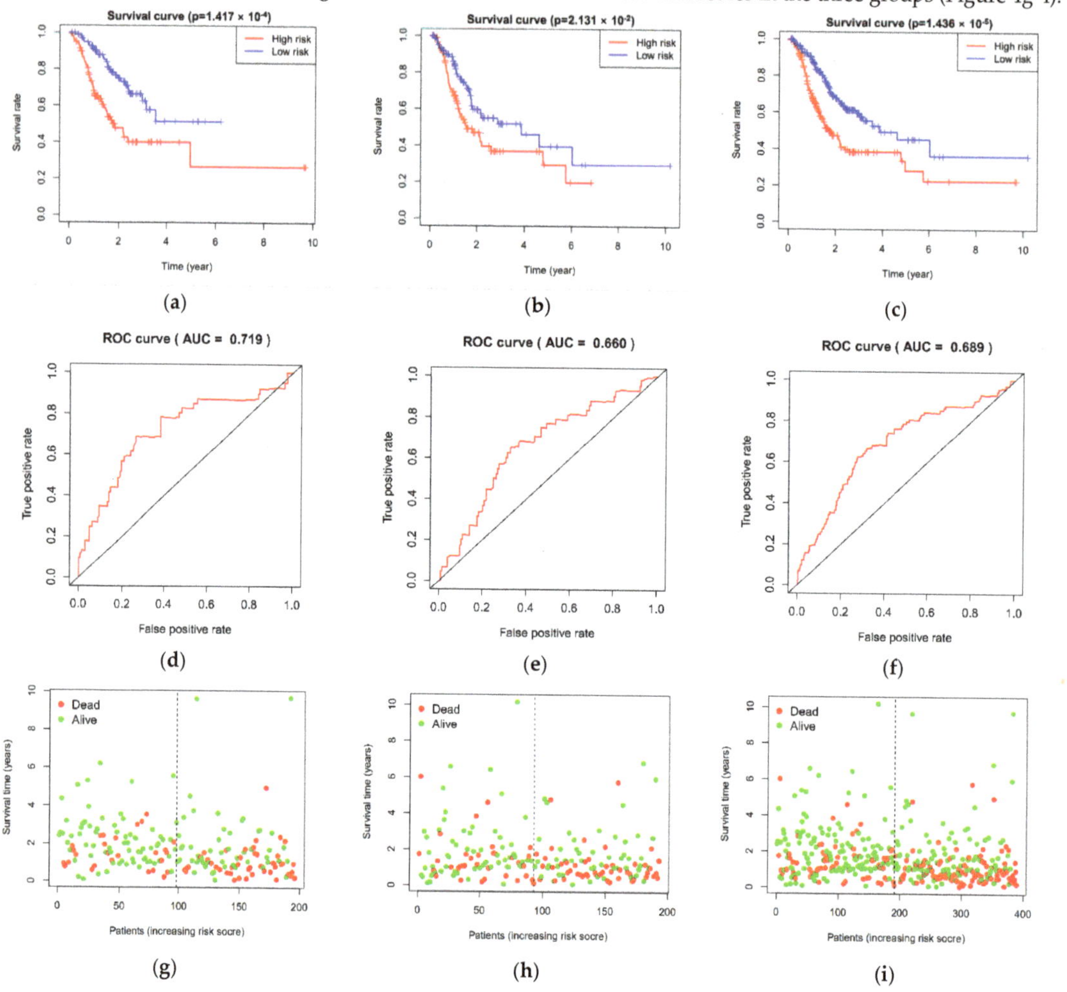

Figure 4. Verification and assessment of the five miRNAs. Kaplan–Meier curves in the (**a**) training group, (**b**) testing group, (**c**) whole group; The AUC curves in the (**d**) training group, (**e**) testing group, (**f**) whole group; Survival status of patients in high-risk and low-risk in the (**g**) training group, (**h**) testing group, (**i**) whole group.

3.4. Independence of the Five miRNAs

Based on the univariate and multivariate Cox regression analysis, the five miRNAs were related to the overall survival of patients (HR = 1.726, 95% CI = 1.396–2.136, $p < 0.001$). They were also independent in overall survival considering other clinical elements (HR = 1.971, 95% CI = 1.557–2.494, $p < 0.001$). Other clinical features include age, gender, stage, T stage, metastasis, and lymph node stage (Table 4).

Table 4. Univariate and multivariate Cox regression of clinical features.

Clinical Features	Univariate Cox Regression				Multivariate Cox Regression			
	HR	HR.95L	HR.95H	*p*-Value	HR	HR.95L	HE.95H	*p*-Value
Age	1.015	0.999	1.032	0.062	1.027	1.010	1.045	0.002
Gender	1.225	0.853	1.760	0.271	1.510	1.027	2.218	0.036
Grade	1.278	0.908	1.800	0.160	1.115	0.781	1.591	0.550
Stage	1.607	1.294	1.996	<0.001	1.210	0.807	1.815	0.357
T	1.288	1.038	1.599	0.022	1.215	0.911	1.621	0.186
M	1.880	1.013	3.489	0.045	1.818	0.844	3.917	0.127
N	1.361	1.170	1.584	<0.001	1.233	0.987	1.540	0.065
riskScore	1.726	1.395	2.136	<0.001	1.971	1.557	2.494	<0.001

3.5. Target Genes Prediction of Five miRNAs

The target genes were obtained and crosschecked in at least two databases. The predicted results showed that the five miRNAs (has-miR-9-3hashsa-miR-196hasp, hsa-miR-135b-3p, hsa-miR-942-3p, and hsa-miR-143-5p) overlapping target genes were 996, 54, 224, 457, and 767, respectively. The results were shown in Figure 5. Then, the above detected 7531 DEmRNAs (4395 up-regulated and 3136 down-regulated) were used to determine whether these target genes were involved in the development of gastric cancer.

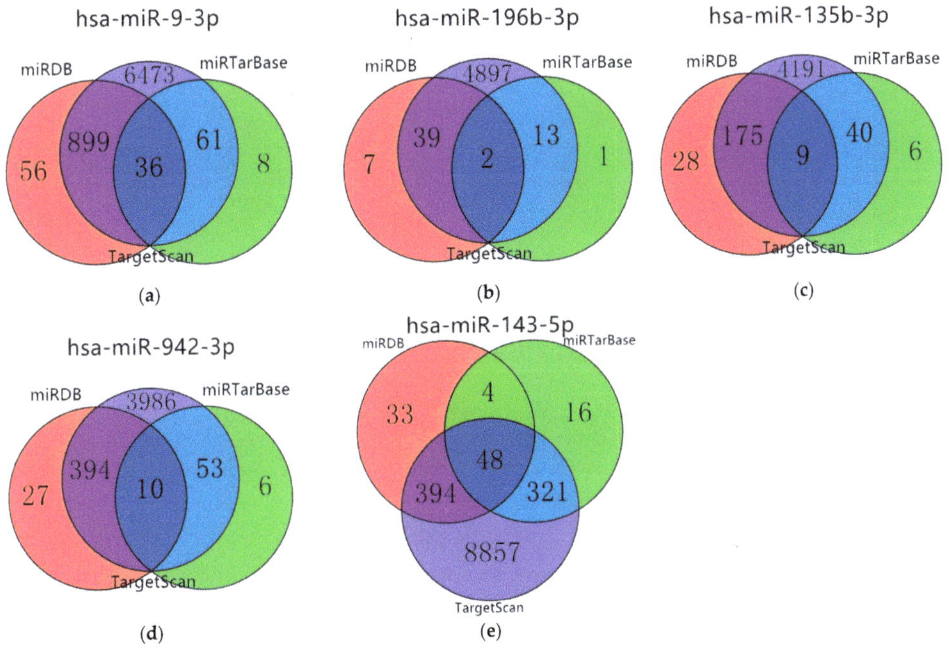

Figure 5. Venn diagram of target genes. (**a**) hsa-miR-9-3p; (**b**) hsa-miR-196-3p; (**c**) hsa-miR-135b-3p; (**d**) hsa-miR-942-3p; (**e**) hsa-miR-143-5.

Figure 6a displayed the regulatory network between five miRNAs and 196 target genes. There are 121 overlapping genes between the target genes of down-regulated miRNAs (hsa-miR-143-5p, hsa-miR-9-3p) (1661) and up-regulated mRNAs (4395). There were 75 overlapping genes between the target genes of up-regulated miRNAs (hsa-miR-135-3p, hsa-miR-196b-3p, hsa-miR-942-3p) (713) and down-regulated mRNAs (3136), as shown in Figure 6b,c.

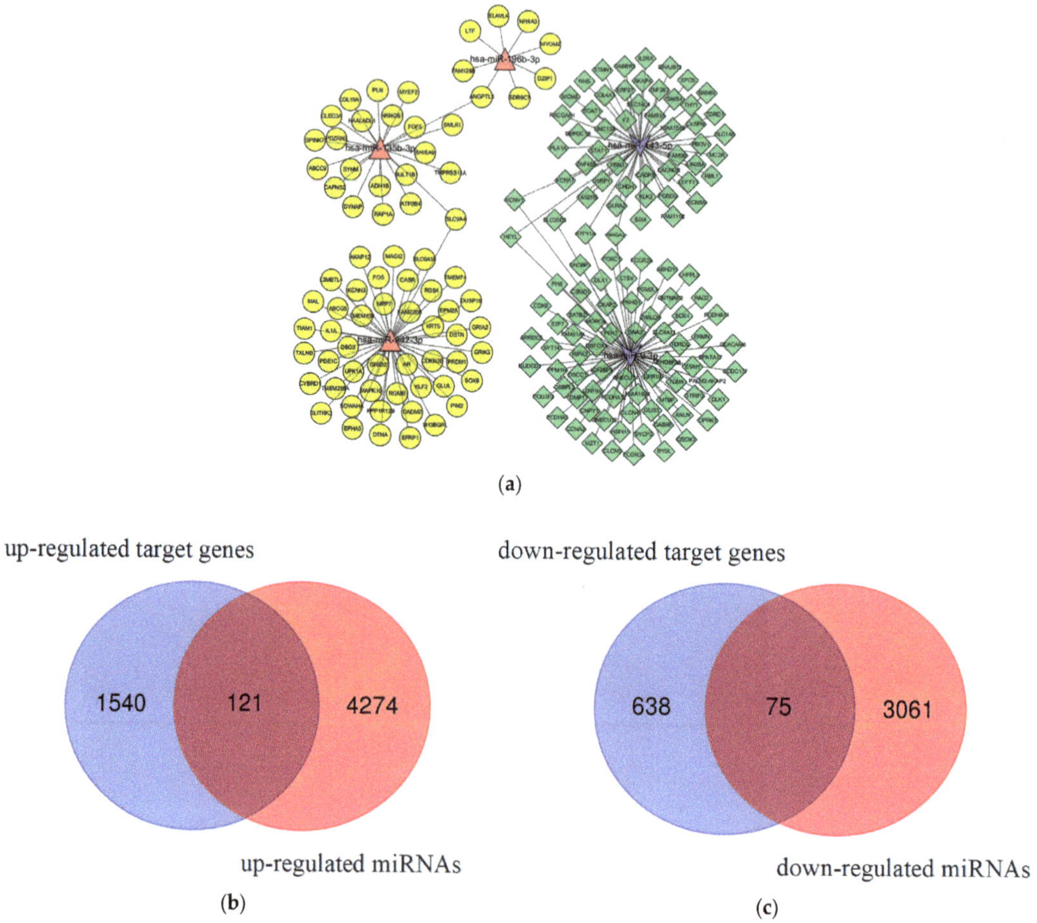

Figure 6. Network diagram of miRNAs regulating mRNAs. (**a**) Regulatory network between five miRNAs and 196 target genes. Red triangles mean 3 up-regulated miRNAs, blue arrows mean 2 down-regulated miRNAs, (**b**) green rhomboids mean 121 up-regulated mRNAs, and (**c**) yellow circles mean 75 down-regulated mRNAs.

3.6. Target Genes Functional Enrichment Analysis

The GO results in the top fifteen terms, including biological process (BP), cellular component (CC), and molecular function (MF) were displayed in dot plot (Figure 7a–c). BP mainly contained cell cycle G1/S phase transition, urogenital and renal system development; CC mainly contained transmembrane transporter complex, transporter complex, and apical part of cell; MF mainly contained ion channel and substrate-specific channel activity. KEGG pathways analysis results were mainly enriched in the neuroactive ligand–receptor interaction, cAMP signaling pathway, and the MAPK signaling pathway (Figure 7d and Table 5).

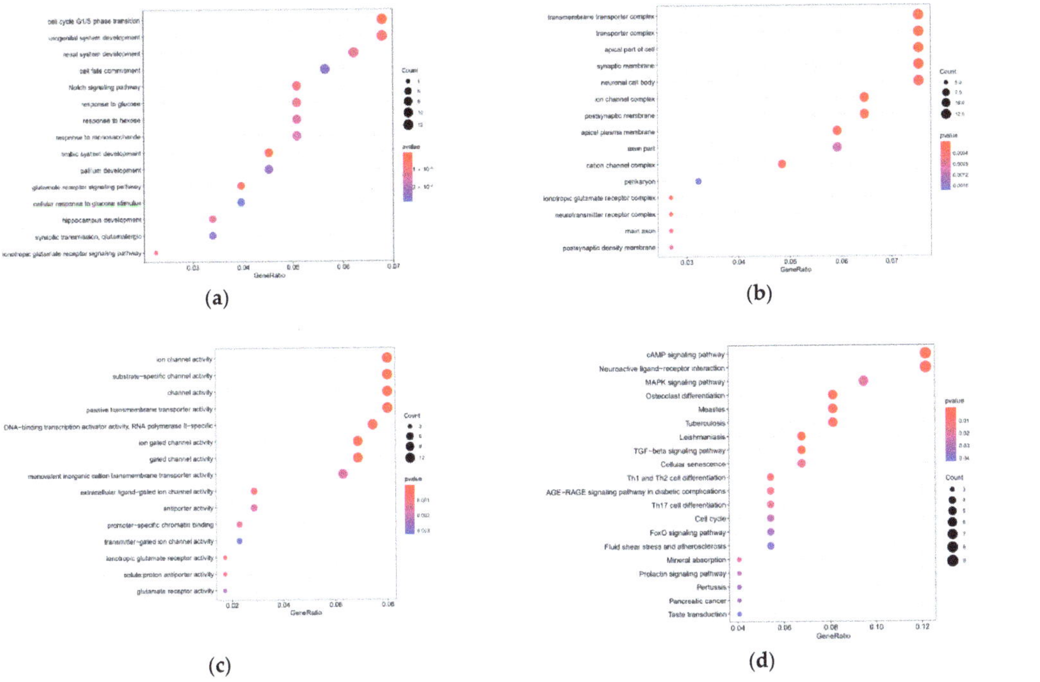

Figure 7. Functional enrichment analysis of target genes. (**a**) BP; (**b**) CC; (**c**) MF; (**d**) KEGG signaling pathways.

Table 5. KEGG signaling pathways of the target genes.

ID	Description	*p*-Value	*Q*-Value	Count	Gene
hsa04024	cAMP signaling pathway	0.0002	0.0272	9	TIAM1/FOS/GRIA2/MAPK10/PLN/MC2R/ATP2B4/GABBR2/RAP1A
hsa05140	Leishmaniasis	0.0007	0.0603	5	FCGR3A/FOS/STAT1/IL1A/FCGR2A
hsa04380	Osteoclast differentiation	0.0011	0.0603	6	FCGR3A/FOS/MAPK10/STAT1/IL1A/FCGR2A
hsa05162	Measles	0.0017	0.0603	6	CDK6/FOS/MAPK10/STAT1/IL1A/IL2RA
hsa04350	TGF-beta signaling pathway	0.0017	0.0603	5	CDKN2B/RGMB/LEFTY1/BAMBI/RBL1
hsa04080	Neuroactive ligand-receptor interaction	0.0039	0.1148	9	GRIA2/GRID2/MC2R/F2/GLRA2/GABRP/GRIK3/GABBR2/OPRK1
hsa05152	Tuberculosis	0.0062	0.1570	6	FCGR3A/MAPK10/RIPK2/STAT1/IL1A/FCGR2A
hsa04658	Th1 and Th2 cell differentiation	0.0101	0.2256	4	FOS/MAPK10/STAT1/IL2RA
hsa04933	AGE-RAGE signaling pathway	0.0135	0.2506	4	MAPK10/STAT1/IL1A/COL4A1
hsa04218	Cellular senescence	0.0159	0.2506	5	CDK6/CDKN2B/IL1A/CCNA2/RBL1
hsa04978	Mineral absorption	0.0162	0.2506	3	SLC6A19/CYBRD1/ATP2B4
hsa04659	Th17 cell differentiation	0.0169	0.2506	4	FOS/MAPK10/STAT1/IL2RA
hsa04010	MAPK/ERK signaling pathway	0.0187	0.2554	7	CACNG8/FOS/MAPK10/IL1A/STMN1/FGF5/RAP1A
hsa04917	Prolactin signaling pathway	0.0266	0.3244	3	FOS/MAPK10/STAT1
hsa04110	Cell cycle	0.0274	0.3244	4	CDK6/CDKN2B/CCNA2/RBL1
hsa04068	FoxO signaling pathway	0.0326	0.3247	4	CDKN2B/MAPK10/KLF2/RAG2
hsa05133	Pertussis	0.0329	0.3247	3	FOS/MAPK10/IL1A
hsa05212	Pancreatic cancer	0.0329	0.3247	3	CDK6/MAPK10/STAT1
hsa05418	Fluid shear stress and atherosclerosis	0.0392	0.3471	4	FOS/MAPK10/KLF2/IL1A
hsa04742	Taste transduction	0.0410	0.3471	3	PDE1C/TAS2R5/GABBR2

3.7. Hub Genes of PPI Network and Survival Analysis of Target Genes

The PPI network included a total of 196 target genes. The ten hub genes (CCNA2, GRIA2, FOS, AR, RACGAP1, RBFOX1, LIN28A, DSCC1, GRID2, OPRK1) from PPI network were screened by Cytoscape plug-in CytoHubba (Figure 8 and Table 6). Besides, the Kaplan–Meier curve indicated that the expression of eight genes (AKAP12, AR, DEIP1, PCDHA11, PCDHA12, P115, SH3BGRL, TMEM108) was correlated with survival prognosis (Figure 9).

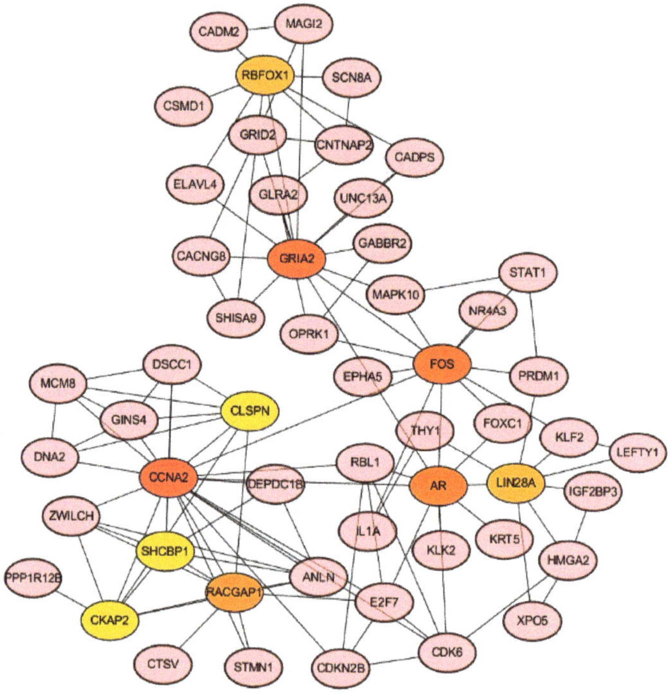

Figure 8. Core genes of PPI network.

Table 6. Identification of the top ten core genes.

Node_Name	MCC	DMNC	MNC	Degree	EPC	Bottle Neck	Ec Centricity	Closeness	Radiality	Betweenness	Stress	Clustering Coefficient
CCNA2	236	0.264	18	18	35.308	19	0.097	43.961	7.799	2843.324	5374	0.235
GRIA2	22	0.233	8	13	33.492	71	0.129	44.383	8.065	3792.232	8494	0.128
FOS	17	0.220	7	12	34.894	84	0.111	46.076	8.159	3863.375	7988	0.091
AR	33	0.329	7	10	35.226	18	0.111	43.369	7.987	2490.437	5478	0.200
RACGAP1	131	0.408	8	9	33.683	2	0.086	33.073	7.110	210.833	484	0.389
RBFOX1	12	0.238	6	8	28.267	4	0.111	35.586	7.525	953.058	2332	0.179
LIN28A	9	0.309	3	8	28.652	3	0.097	36.095	7.517	790.409	2030	0.107
DSCC1	28	0.454	5	7	30.653	3	0.086	32.073	7.094	392.500	704	0.333
GRID2	13	0.285	6	7	28.277	2	0.111	34.919	7.525	366.790	918	0.286
OPRK1	10	0.463	3	7	25.901	13	0.097	37.251	7.721	1633.949	3034	0.143

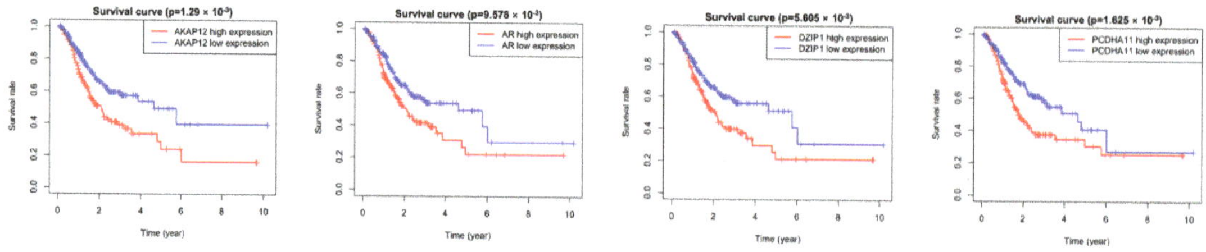

(a) (b) (c) (d)

Figure 9. *Cont.*

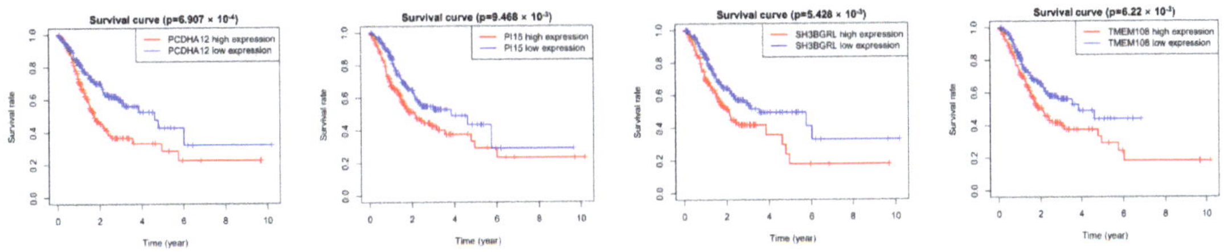

(e) (f) (g) (h)

Figure 9. Eight target genes associated with overall survival. (**a**) AKAP12; (**b**) AR; (**c**) DEIP1/DZIP1; (**d**) PCDHA11; (**e**) PCDHA12; (**f**) P115; (**g**) SH3BGRL; (**h**) TMEM108.

3.8. The Working Mechanism of AR and Its Potential Relationship with the MAPK/ERK Signaling Pathway

From the above, we can conclude that the Androgen Receptor (AR) was not only a hub gene in the PPI network but also associated with excessive survival of patients. AR can regulate the transcription of genes and express new proteins, ultimately changing the function of cells. Figure 10 shows a typical AR working mechanism. AR usually forms a complex with heat shock proteins (HSPs) in the cytoplasm. The binding of AR to androgen (such as 5α-dihydrotestosterone, DHT) alters its conformation, and HSPs are subsequently released. Under the action of coactivators, androgen–AR complexes are transferred to the nucleus and recognize androgen response elements in the form of homodimer to regulate downstream target gene expression.

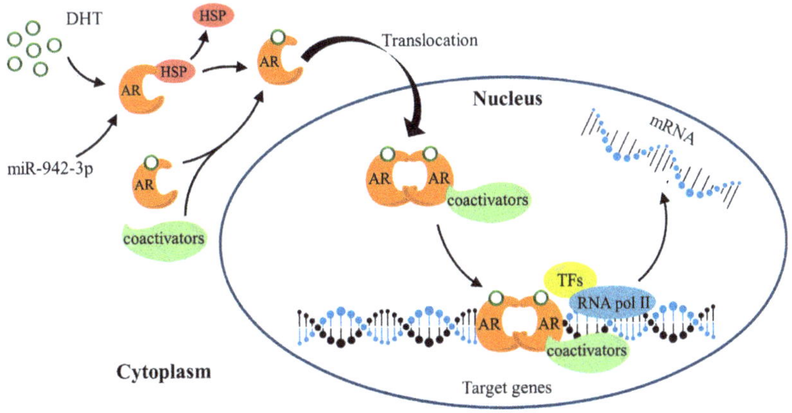

Figure 10. AR working mechanism. DHT: 5α-dihydrotestosterone; HSPs: heat shock proteins; TFs: transcription factors; RNA pol II: RNA polymerase II.

In the absence of androgen, AR may depend on the MAPK/ERK signaling pathway to play its role. Figure 11 shows the potential relationship between the AR and MAPK/ERK signaling pathways. In the cytoplasm, AR can interact with several signaling molecules, including phosphoinositide 3-kinase (PI3K), Src family kinase (Src), Ras GTPase (Ras), and protein kinase C (PKC), which in turn converge on the MAPK/ERK pathway. Then, the MAPK/ERK enters the nucleus, where it translocates and interacts with transcription factors that regulate the expression of genes associated with cell proliferation.

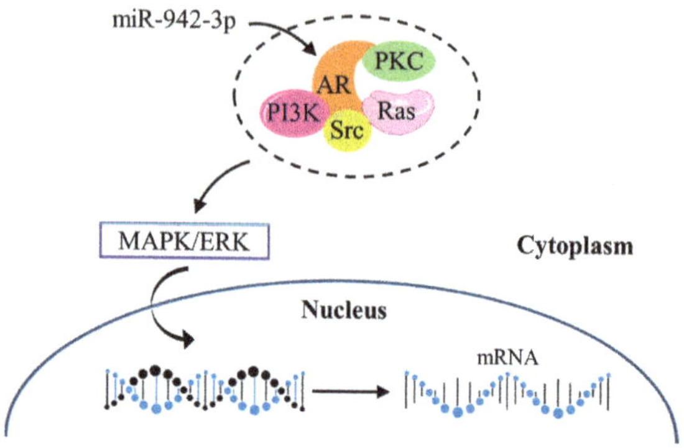

Figure 11. Potential relationship between the AR and MAPK/ERK signaling pathways. PI3K: phosphoinositide 3-kinase; Src: Src family kinase; RAS: Ras GTPase; PKC: protein kinase C.

4. Discussion

Gastric cancer is one of the most common tumors with high morbidity and mortality. Therefore, the detection of sensitive specific biomarkers for gastric cancer is urgent. Many studies indicated miRNAs could regulate expression in vivo, and it plays an essential role in the biological process of human malignancy [16]. Currently, some miRNAs have been used as potential prognostic indicators for tumors, such as miR-191 [17], miR-1908 [18], miR-217 [19], and miR-200c [20]. Previously, a variety of miRNAs were discovered in many prognostic markers for tumors [21,22], especially for gastric cancer [23].

In this study, we obtained 267 DEmiRNAs. All samples were divided into training group and testing group randomly. Then, the five miRNAs were constructed in the training group. At the same time, based on the median grouping of risk score, these five miRNAs were proved in the testing group and the whole group, respectively. Kaplan–Meier curves showed that overall survival was significantly lower in the high-risk group than in the low-risk group among the three groups. By ROC curve, the overall survival of the five miRNAs among the three groups showed better predictive ability. Subsequently, the Cox regression analysis indicated that the five miRNAs were independent of overall survival.

The target genes of five miRNAs were predicted in order to in-depth understand the regulatory mechanisms of these five miRNAs. GO analysis showed that the target genes were correlated with cell cycle G1/S phase transition, urogenital and renal system development, transmembrane transporter complex, transporter complex and apical part of the cell, ion channel, substrate-specific channel, and channel activity. The signaling pathways were enriched in the cAMP and MAPK signaling pathways and the Neuroactive ligand–receptor interaction. Park et al. pointed out that the cAMP signaling pathway inhibited the degradation of the HDAC8 and the expression of TIPRL in lung cancer cells, and also increased cisplatin-induced apoptosis [24]. Jagriti Pal et al. showed that the neuroactive ligand–receptor interaction pathway had a poor prognosis in patients with glioma [25]. The MAPK/ERK signaling pathway was essential in regulating cellular processes, such as cell differentiation, division, proliferation, and apoptosis.

The top ten hub genes (CCNA2, GRIA2, FOS, AR, RACGAP1, RBFOX1, LIN28A, DSCC1, GRID2, OPRK1) of target genes were detected by Cytoscape. Moreover, the Kaplan–Meier curve showed that eight target genes (AKAP12, AR, DEIP1, PCDHA11, PCDHA12, P115, SH3BGRL, TMEM108) were related to survival prognosis. Unexpectedly, AR was a hub gene in the PPI network, and it had a relationship with the excessive survival of patients. AR is a nuclear transcription factor, it can recognize and combine specific DNA sequences on target factors, thereby regulating the transcription of the gene and expressing new proteins, which ultimately changes the function of cells and promotes cell differentiation and the development of tissues and organs [26–28]. Salma S et al. showed that the p14ARF tumor suppressor could restrain AR activity and prevent apoptosis in prostate cancer cells [29]. Peng L et al. verified that AR could be directly combined with LAMA4, and it was related to enhanced cisplatin resistance in gastric cancer, providing a new mechanism for the treatment of drug-resistant gastric cancer [30]. In addition, AR may depend on the MAPK/ERK signaling pathway to function. Specifically, AR can interact with a variety of signaling molecules (PI3K, Src, Ras, and PKC) in the cytoplasm, which in turn converge on the MAPK/ERK pathway [31,32]. MAPK/ERK then enters the nucleus, where it translocates and interacts with transcription factors to regulate the expression of genes involved in cell proliferation [33].

In a word, this study found that the MAPK/ERK signaling pathway may help AR signal transduction and promote the interaction between AR and transcription factors, leading to cell proliferation. At the same time, AR is a target gene of the has-miR-942-3p, which well verifies the important role of the has-miR-942-3p in the occurrence and prognosis of gastric cancer.

5. Conclusions

This study built the five miRNAs (has-miR-9-3p, has-miR-135b-3p, has-miR-143-5p, has-miR-942-3p, has-miR-196-3p) prognostic modules, also verified and evaluated the prediction ability of the five miRNAs by grouping. They can be used as an independent factor in the prognosis of gastric cancer. By predicting the target genes to explore the potential biological functions, our results could provide a deeper understanding of the occurrence and development. This study identified the AR gene regulated by has-miR-942-3p which may depend on the MAPK/ERK signaling pathway to promote the proliferation of cancer cells. In future experiments, we will further explore the regulatory mechanisms of other miRNAs (has-miR-9-3p, has-miR-135b-3p, has-miR-143-5p, has-miR-196-3p) to provide effective prediction and treatment targets for gastric cancer patients.

Author Contributions: Conceptualization and supervision, W.L. and X.C.; methodology, W.L., N.Y., X.R. and X.C.; software, W.L. and N.Y.; validation, W.L., N.Y. and X.R.; writing—original draft, W.L., N.Y. and X.R.; visualization, W.L. and X.C.; Writing—review and editing, W.L. and N.Y. All authors have read and agreed to the published version of the manuscript.

Funding: This research received no external funding.

Institutional Review Board Statement: Not applicable.

Informed Consent Statement: Not applicable.

Data Availability Statement: The miRNA expression profiles, mRNA expression profiles data, and clinical information of all gastric cancer samples were gained from The Cancer Genome Atlas (TCGA) database (https://www.cancer.gov (accessed on 20 August 2022)).

Acknowledgments: We thank TCGA for providing access to miRNA and mRNA profiles, and their clinical patient information. We appreciate the professors at Hangzhou Dianzi University for providing valuable insights that helped us significantly improve the quality of the manuscript.

Conflicts of Interest: The authors declare no conflict of interest.

References

1. Orditura, M.; Galizia, G.; Sforza, V.; Gambardella, V.; Fabozzi, A.; Laterza, M.M.; Andreozzi, F.; Ventriglia, J.; Savastano, B.; Mabilia, A.; et al. Treatment of gastric cancer. *World J. Gastroenterol.* **2014**, *20*, 1635–1649. [CrossRef] [PubMed]
2. Gires, O. Lessons from common markers of tumor-initiating cells in solid cancers. *Cell. Mol. Life Sci.* **2011**, *68*, 4009–4022. [CrossRef] [PubMed]
3. Yang, L.; Mu, Y.; Cui, H.; Liang, Y.; Su, X. MiR-9-3p augments apoptosis induced by H_2O_2 through down regulation of Herpud1 in glioma. *PLoS ONE* **2017**, *12*, e0174839.
4. Chen, Z.; Gu, D.; Zhou, M.; Yan, S.; Shi, H.; Cai, Y. Regulation of miR-143-5p on cadmium-induced renal cell apoptosis and its mechanism. *J. Nanjing Med. Univ. Nat. Sci. Ed.* **2015**, *35*, 490–495.
5. Ma, R.; Yan, W.; Zhang, G.; Lv, H.; Liu, Z.; Fang, F.; Zhang, W.; Zhang, J.; Tao, T.; You, Y.; et al. Upregulation of miR-196b confers a poor prognosis in glioblastoma patients via inducing a proliferative phenotype. *PLoS ONE* **2012**, *7*, e38096.
6. Chen, M.; Ma, L.; Teng, Y. Research Progress of miRNA-135b in Tumor Development. *Med. Recapitul.* **2018**, *24*, 4836–4841.

7. Verma, H.K.; Bhaskar, L. MicroRNA a small magic bullet for gastric cancer. *Gene* **2020**, *753*, 144801. [CrossRef]
8. Lin, A.; Wang, G.; Duan, R.; Li, T. Expression of miRNA-449b in gastric cancer and lymph node metastasis. *Acta Acad. Med. Wannan* **2015**, *34*, 113–116.
9. Ma, X.; Lu, K.; Lei, J.; Ji, L.; Shi, L. Relationship between the methylation and expression of microRNA-375 and its clinicopathological characteristics in gastric cancer. *Prog. Mod. Biomed.* **2018**, *18*, 1931–1935.
10. Han, W.; Su, X. The mechanism of miRNA-30c in gastric cancer. In Proceedings of the National Symposium on Tumor Epidemiology and Etiology, Chengdu, China, 19 August 2015.
11. Robinson, M.D.; McCarthy, D.J.; Smyth, G.K. edgeR: A bioconductor package for differential expression analysis of digital gene expression data. *Bioinformatics* **2010**, *26*, 139–140. [CrossRef]
12. Yao, T.; Liu, Y.; Li, C.; Hu, L. Regression model analysis of survival data-Cox proportional hazard regression model analysis of survival data. *Sichuan Ment. Health* **2020**, *33*, 27–32.
13. Zhe, W.; Ryan, M. TModel-free posterior inference on the area under the receiver operating characteristic curve. *J. Stat. Plan. Inference* **2020**, *209*, 174–186.
14. Yu, G.; Wang, L.G.; Han, Y.; He, Q. clusterProfiler: An R package for comparing biological themes among gene clusters. *OMICS J. Integr. Biol.* **2012**, *16*, 284–287. [CrossRef] [PubMed]
15. Szklarczyk, D.; Morris, J.H.; Cook, H. The STRING database in 2017: Quality-controlled protein-protein association networks, made broadly accessible. *Nucleic Acids Res.* **2017**, *45*, D362–D368. [CrossRef] [PubMed]
16. Bertoli, G.; Cava, C.; Castiglioni, I. MicroRNAs: New Biomarkers for Diagnosis, prognosis, Therapy prediction and Therapeutic Tools for Breast Cancer. *Theranostics* **2015**, *5*, 1122–1143. [CrossRef] [PubMed]
17. Gao, X.; Xie, Z.; Wang, Z.; Cheng, K.; Liang, K.; Song, Z. Overexpression of miR-191 predicts poor prognosis and promotes proliferation and invasion in esophageal squamous cell carcinoma. *Yonsei Med. J.* **2017**, *58*, 1101–1110. [CrossRef] [PubMed]
18. Teng, C.; Zheng, H. Low expression of microRNA-1908 predicts a poor prognosis for patients with ovarian cancer. *Oncol. Lett.* **2017**, *14*, 4277–4281. [CrossRef]
19. Yang, J.; Zhang, H.F.; Qin, C.F. MicroRNA-217 functions as a prognosis predictor and inhibits pancreatic cancer cell proliferation and invasion via targeting E2F3. *Eur. Rev. Med. Pharmacol.* **2017**, *21*, 4050–4057.
20. Si, L.; Tian, H.; Yue, W.; Li, L.; Li, S.; Gao, C.; Qi, L. Potential use of microRNA-200c as a prognostic marker in non-small cell lung cancer. *Oncol. Lett.* **2017**, *14*, 4325–4330. [CrossRef]
21. Liang, B.; Zhao, J.; Wang, X. A three-microRNA signature as a diagnostic and prognostic marker in clear cell renal cancer: An in silico analysis. *PLoS ONE* **2017**, *12*, e0180660. [CrossRef]
22. Shi, X.H.; Li, X.; Zhang, H.; He, R.Z.; Zhao, Y.; Zhou, M.; Pan, S.T.; Zhao, C.L.; Feng, Y.C.; Wang, M.; et al. A five-microRNA signature for survival prognosis in pancreatic adenocarcinoma based on TCGA data. *Sci. Rep.* **2018**, *8*, 7638. [CrossRef] [PubMed]
23. Zhang, C.; Zhang, C.D.; Ma, M.H. Three-microRNA signature identified by bioinformatics analysis predicts prognosis of gastric cancer patients. *World J. Gastroenterol.* **2018**, *24*, 1206–1215. [CrossRef] [PubMed]
24. Park, J.Y.; Juhnn, Y.S. cAMP signaling increases histone deacetylase 8 expression via the Epac2-Rap1A-Akt pathway in H1299 lung cancer cells. *Exp. Mol. Med.* **2017**, *49*, e297. [CrossRef] [PubMed]
25. Jagriti, P.; Vikas, P.; Anupam, K.; Kavneet, K.; Chitra, S.; Kumaravel, S. Genetic landscape of glioma reveals defective neuroactive ligand receptor interaction pathway as a poor prognosticator in glioblastoma patients. *Cancer Res.* **2017**, *77*, 2454. [CrossRef]
26. Higa, G.M.; Fell, R.G. Sex hormone receptor repertoire in breast cancer. *Int. J. Breast Cancer* **2013**, *2013*, 284036. [CrossRef]
27. Li, D.; Zhou, W.; Pang, J.; Tang, Q.; Zhong, B.; Shen, C.; Xiao, L.; Hou, T. A magic drug target: Androgen receptor. *Med. Res. Rev.* **2019**, *39*, 1485–1514. [CrossRef]
28. Li, J.; Al-Azzawi, F. Mechanism of androgen receptor action. *Maturitas* **2009**, *63*, 142–148. [CrossRef]
29. Salma, S.; Stephen, J.L.; Christopher, A.L.; Alan, P.L.; Maria, M. The p14ARF tumor suppressor restrains androgen receptor activity and prevents apoptosis in prostate cancer cells. *Cancer Lett.* **2020**, *483*, 12–21.
30. Peng, L.; Li, Y.; Wei, S.; Li, X.; Zhang, G. LAMA4 activated by Androgen receptor induces the cisplatin resistance in gastric cancer. *Biomed. Pharmacother.* **2020**, *124*, 109667. [CrossRef]
31. McCubrey, J.A.; Steelman, L.S.; Chappell, W.H.; Abrams, S.L.; Wong, E.W.; Chang, F.; Lehmann, B.; Terrian, D.M.; Milella, M.; Tafuri, A.; et al. Roles of the Raf/MEK/ERK pathway in cell growth, malignant transformation and drug resistance. *Biochim. Biophys. Acta* **2007**, *1773*, 1263–1284. [CrossRef]
32. Roberts, P.J.; Der, C.J. Targeting the Raf-MEK-ERK mitogen-activated protein kinase cascade for the treatment of cancer. *Oncogene* **2007**, *26*, 3291–3310. [CrossRef] [PubMed]
33. Liao, R.S.; Ma, S.; Miao, L.; Li, R.; Yin, Y.; Raj, G.V. Androgen receptor-mediated non-genomic regulation of prostate cancer cell proliferation. *Transl. Androl. Urol.* **2013**, *2*, 187–196. [PubMed]

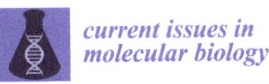

current issues in molecular biology

MDPI

Article

Cyclin D1 Serves as a Poor Prognostic Biomarker in Stage I Gastric Cancer

Se-Il Go [1,†], Gyung Hyuck Ko [2,†], Won Sup Lee [3,*], Jeong-Hee Lee [2], Sang-Ho Jeong [4], Young-Joon Lee [4], Soon Chan Hong [5] and Woo Song Ha [5]

1 Department of Internal Medicine, Gyeongsang National University Changwon Hospital, Institute of Health Sciences, Gyeongsang National University College of Medicine, Changwon 51472, Korea; gose1@gnuh.co.kr
2 Department of Pathology, Institute of Health Sciences, Gyeongsang National University College of Medicine, Jinju 52727, Korea; gyunghko@gnu.ac.kr (G.H.K.); jhlee7@gnu.ac.kr (J.-H.L.)
3 Department of Internal Medicine, Institute of Health Sciences, Gyeongsang National University College of Medicine, Jinju 52727, Korea
4 Department of Surgery, Gyeongsang National University Changwon Hospital, Institute of Health Sciences, Gyeongsang National University College of Medicine, Changwon 51472, Korea; shjeong@gnu.ac.kr (S.-H.J.); yjlee@gnu.ac.kr (Y.-J.L.)
5 Department of Surgery, Institute of Health Sciences, Gyeongsang National University College of Medicine, Jinju 52727, Korea; hongsc@gnu.ac.kr (S.C.H.); woosongha@gnu.ac.kr (W.S.H.)
* Correspondence: lwshmo@gnu.ac.kr; Tel.: +82-55-750-8733
† These authors contributed equally to this work.

Abstract: TNM stage still serves as the best prognostic marker in gastric cancer (GC). The next step is to find prognostic biomarkers that detect subgroups with different prognoses in the same TNM stage. In this study, the expression levels of epidermal growth factor receptor (EGFR) and cyclin D1 were assessed in 96 tissue samples, including non-tumorous tissue, adenoma, and carcinoma. Then, the prognostic impact of EGFR and cyclin D1 was retrospectively investigated in 316 patients who underwent R0 resection for GC. EGFR positivity increased as gastric tissue became malignant, and cyclin D1 positivity was increased in all the tumorous tissues. However, there was no survival difference caused by the EGFR positivity, while the cyclin D1-postive group had worse overall survival (OS) than the cyclin D1-negative group in stage I GC (10-year survival rate (10-YSR): 62.8% vs. 86.5%, $p = 0.010$). In subgroup analyses for the propensity score-matched (PSM) cohort, there were also significant differences in the OS according to the cyclin D1 positivity in stage I GC but not in stage II and III GC. Upon multivariate analysis, cyclin D1 positivity was an independent prognostic factor in stage I GC. In conclusion, cyclin D1 may be a useful biomarker for predicting prognosis in stage I GC.

Keywords: cyclin D1; epidermal growth factor receptor (EGFR); stage I gastric cancer; early gastric cancer; node-negative gastric cancer

Citation: Go, S.-I.; Ko, G.H.; Lee, W.S.; Lee, J.-H.; Jeong, S.-H.; Lee, Y.-J.; Hong, S.C.; Ha, W.S. Cyclin D1 Serves as a Poor Prognostic Biomarker in Stage I Gastric Cancer. *Curr. Issues Mol. Biol.* **2022**, *44*, 1395–1406. https://doi.org/10.3390/cimb44030093

Academic Editor: Peter C. Hart

Received: 1 March 2022
Accepted: 17 March 2022
Published: 20 March 2022

1. Introduction

Gastric adenocarcinoma, commonly referred to as gastric cancer (GC), is the third most common cause of cancer death worldwide [1] and the second most common cause of cancer death in Korea [2]. Tremendous efforts have been made to detect predictive biomarkers [3], but the pathologic TNM stage still serves as one of the best prognostic markers in GC. The TNM stages are determined from I to IV according to the depth of tumor invasion, regional lymph node metastasis, and distant metastasis. However, we know that heterogeneous subgroups with different prognoses exist within the same TNM stage; widespread metastasis is sometimes observed several months after the surgical resection of early gastric cancer (EGC) for which adjuvant chemotherapy is not applied, according to clinical guidelines. In addition, we previously demonstrated that heterogeneous subgroups with different prognoses exist in EGC by reporting that CD44 variant 9 and Ki-67 expression served as prognostic biomarkers in EGC [4,5]. Therefore, the next step is to find prognostic biomarkers that detect subgroups with different clinical features and prognoses within the same TNM stage. With a literature review, we also found that the poor prognostic factors in GC would exhibit rapid growth, metastasis, and drug resistance for GC [6–8]. We believe that the biomarkers that are associated with the above features are relevant to poor prognosis in GC.

Epidermal growth factor receptor (EGFR) is a member of the ErbB family of receptors, a subfamily of four closely related receptor tyrosine kinases. It is found to be overexpressed in various cancers, including colorectal cancer, pancreatic cancer, and GC [9–12]. High EGFR expression is associated with an increased risk of metastasis and drug resistance, and the inhibition of EGFR leads to a reduction in cancer migration and angiogenesis and an increase in drug sensitivity in cancers [13,14]. In addition, EGFR gene amplification is related to lymph node metastases in GC [15]. Therefore, we thought that EGFR expression deserves to be investigated to determine whether it is a biomarker in GC.

The next candidate is cyclin D1, a protooncogene that plays a positive regulation role in cancer progression [16]. Increased cyclin D1 expression is an early cell proliferation event that is stimulated by growth factors or other mitogens [17]. Cyclin D1 has previously been used as a biomarker for cell proliferation and prognosis in various types of cancer as much as the Ki-67 proliferation index has [18–21]. We considered it worthwhile to investigate the role of cyclin D1 as a prognostic biomarker in GC.

Therefore, we planned this study to determine whether EGFR and the cyclin D1 protein serve as prognostic biomarkers in GC. In addition, subgroup analysis and propensity score-matched analysis were conducted to identify a specific group in which the biomarkers could be applied usefully.

2. Materials and Methods

2.1. Patients and Specimens

First, as a pilot study to evaluate the associations between the biomarker expression and tumor development, 96 tissue samples including 23 non-tumorous tissues (6 normal mucosa, 5 *Helicobacter pylori*-related gastritis, and 12 intestinal metaplasia), 24 adenoma (12 low-grade adenoma and 12 high-grade adenoma), and 49 GC (23 EGC and 26 advanced gastric cancer (AGC)) were used. Then, we retrospectively reviewed 316 consecutive patients who underwent R0 resection for GC from 2004 to 2013 at a single institution. Patients who underwent R1/R2 resection or endoscopic mucosal resection and had metastasis to a distant organ were excluded. The demographics, clinical data, and histologic findings were obtained through an electronic medical record review. The TNM stage was classified according to the 8th edition AJCC staging system for gastric cancer. Histologic classification was performed using the WHO classification (G1 well differentiated, G2 moderately differentiated, and G3 poorly differentiated and undifferentiated) and Lauren classification (intestinal, diffuse, and mixed). EGC was defined as a tumor confined to the mucosa or submucosa regardless of the presence of lymph node metastases. AGC was defined as a tumor invading the muscularis propria or deeper layers.

The expression levels of each biomarker were determined by immunohistochemical (IHC) staining of tissue microarray (TMA) sections. A 2 mm diameter core tissue in each case was arrayed in a new recipient paraffin block. The TMA section of each slide was then deparaffinized, rehydrated, and incubated in 3% H_2O_2 to prevent non-specific background staining. After heating in a microwave oven at 700 W for 20 min with 10 mmol/L citrate buffer (pH 6.0) and incubating for 10 min with Ultra V Block (Lab Vision, Fremont, CA, USA) at room temperature, slides were incubated for 32 and 44 min with primary monoclonal antibodies specific to Cyclin D1 and EGFR, respectively. The BenchMark XT (VENTANA, Tucson, AZ, USA) was used for IHC staining. The expression level of the biomarkers was scored and interpreted by two pathologists blind to the patients' clinical data as follows: negative (<1% of tumor cells were stained) and positive (≥1% of tumor cells were stained; 1+ (1–20%), 2+ (21–50%), and 3+ (>50%)) (Figure S1, see Supplementary Materials).

2.2. Statistical Analysis

Overall survival (OS) was calculated as the time from surgery to cancer-related death or last follow-up. Missing data on the date and cause of death in electronic medical records were obtained from the National Statistical Office of Korea [21]. A Kaplan–Meier curve was plotted for survival. A log-rank test was performed to compare the survival probability. The median follow-up duration was calculated by the reverse Kaplan–Meier method. The Cox regression model was used to analyze multiple variables influencing patient survival. The final model was internally validated by bootstrap resampling (200 replications). If there was a discrepancy in the variables listed as the baseline characteristics of patients between the positive and negative groups for each biomarker, propensity score matching (PSM) was performed to reduce the probability of selection bias. The expression of the biomarkers (negative vs. positive) was regressed by a logistic regression analysis for the conventional prognostic factors as follows: depth of invasion, nodal status, TNM stage, histologic differentiation, and Lauren classification. A nearest-neighbor matching algorithm with a 1:1 ratio was applied to the PSM. A two-sided p-value < 0.05 was considered significant. The MatchIt package in R software version 4.0.5 (The R Foundation for Statistical Computing, Vienna, Austria) was used for the PSM. All other statistical analyses were performed with the Stata software version 16.1 (Stata Corp., College Station, TX, USA).

3. Results

3.1. Baseline Characteristics

The baseline characteristics in the unmatched cohort are presented in Table 1. In an unmatched cohort of 316 patients, the median age of patients was 65 years (interquartile range (IQR), 56–70). The male-to-female ratio was about 2 to 1. The most frequent type of surgery was subtotal gastrectomy (221/316, 69.9%), as the majority of the primary tumors were located in the lower third of the stomach (215/316, 68.0%). The number of patients with EGC and AGC was the same (158:158). Node-positive diseases were observed in 122 of 316 (38.6%) patients. The proportion of poorly differentiated histology in the WHO classification and intestinal type in the Lauren classification was higher than other histologic subtypes.

The cyclin D1 positivity was 19.3% (61 of 316). Fifty-seven and four patients had 1+ and 2+ cyclin D1 expressions, respectively. None of the patients had 3+ cyclin D1 expression. Patients were evenly distributed overall with respect to the variables presented in the baseline characteristics, except that the cyclin D1-negative group tended to have a more advanced stage ($p = 0.088$). EGFR positivity was 13.0% (41 of 316). Thirty-three, two, and six patients had 1+, 2+, and 3+ expressions of EGFR, respectively. There were no significant differences in the baseline characteristics between the EGFR-positive and EGFR-negative groups. These findings suggest that cyclin D1 and EGFR expression positivity were relatively low in GC and that there was an imbalance in terms of the cancer stage between the cyclin D1-positive and cyclin D1-negative groups.

3.2. The Expression Patterns of Each Biomarker According to the Progression of Carcinogenesis and Advancement of Malignancy in GC

The expression patterns of each biomarker are presented in Figure 1, which revealed that EGFR expression increased with the progression of carcinogenesis and the advancement of malignancy, whereas cyclin D1 expression was increased in all tumorous tissues (adenoma, EGC, and AGC) compared to normal tissues and did not increase with advancement from EGC to AGC. The positive expression rates of cyclin D1 were 8.7% (2/23), 33.3% (8/24), 17.4% (4/23), and 38.5% (10/26) in the control tissue, adenoma, EGC, and AGC, respectively ($p = 0.061$). In the tumor group (adenoma, EGC, and AGC), the positive expression rate of cyclin D1 was high compared to the control group (non-tumorous tissue) (30.1% (22/73) vs. 8.7% (2/23), $p = 0.038$). The positive expression rates of EGFR were 0%, 0%, 8.7% (2/23), and 26.9% (7/26) in the control tissue, adenoma, EGC, and AGC, respectively ($p = 0.003$). These findings indicate that EGFR expression positivity increased as the gastric tissue became malignant and that cyclin D1 expression positivity was increased in all of the tumorous tissues.

Table 1. Baseline characteristics of patients.

	Cyclin D1-Negative (*n* = 255)	Cyclin D1-Positive (*n* = 61)	*p*	EGFR-Negative (*n* = 275)	EGFR-Positive (*n* = 41)	*p*
Median (IQR) age, years	65 (56–70)	64 (54–70)	0.890	65 (56–70)	66 (56–69)	0.958
Sex			0.504			0.544
Male	164 (64.3)	42 (68.9)		181 (65.8)	25 (61.0)	
Female	91 (35.7)	19 (31.2)		94 (34.2)	16 (39.0)	
Location			0.117			0.997
Upper	26 (10.2)	12 (19.7)		33 (12.0)	5 (12.2)	
Middle	51 (20.0)	12 (19.7)		55 (20.0)	8 (19.5)	
Lower	178 (69.8)	37 (60.7)		187 (68.0)	28 (68.3)	
Operation			0.957			0.699
Subtotal gastrectomy	180 (70.6)	41 (67.2)		190 (69.1)	31 (75.6)	
Total gastrectomy	60 (23.5)	16 (26.2)		67 (24.4)	9 (22.0)	
Proximal gastrectomy	12 (4.7)	3 (4.9)		14 (5.1)	1 (2.4)	
Wedge resection	3 (1.2)	1 (1.6)		4 (1.5)	0	
Depth of invasion			0.669			0.616
EGC	126 (49.4)	32 (52.5)		139 (50.6)	19 (46.3)	
AGC	129 (50.6)	29 (47.5)		136 (49.5)	22 (53.7)	
Nodal status			0.650			0.455
N0	155 (60.8)	39 (63.9)		171 (62.2)	23 (56.1)	
N+	100 (39.2)	22 (36.1)		104 (37.8)	18 (43.9)	
AJCC 8th edition staging			0.088			0.577
I	141 (55.3)	35 (57.4)		156 (56.7)	20 (48.8)	
II	43 (16.9)	16 (26.2)		51 (18.6)	8 (19.5)	
III	71 (27.8)	10 (16.4)		68 (24.7)	13 (31.7)	
Tumor size			0.819			0.367
≤4 cm	117 (45.9)	27 (44.3)		128 (46.6)	16 (39.0)	
>4 cm	138 (54.1)	34 (55.7)		147 (53.5)	25 (61.0)	
WHO classification			0.597			0.389
Well-differentiated	56 (22.0)	10 (16.4)		60 (21.8)	6 (14.6)	
Moderately differentiated	77 (30.2)	21 (34.4)		82 (29.8)	16 (39.0)	
Poorly differentiated and others *	122 (47.8)	30 (49.2)		133 (48.4)	19 (46.3)	
Lauren classification			0.287			0.096
Intestinal	179 (70.2)	47 (77.1)		192 (69.8)	34 (82.9)	
Diffuse or mixed	76 (29.8)	14 (23.0)		83 (30.2)	7 (17.1)	

IQR: interquartile range; EGC: early gastric cancer; AGC: advanced gastric cancer; AJCC: American Joint Committee on Cancer. * Undifferentiated and mucinous adenocarcinoma as well as signet-ring cell carcinoma were included.

3.3. Overall Survival According to Cyclin D1 and EGFR Expression in Whole Cohort

The median follow-up duration was 73 months. In the whole cohort, the median OS was not reached, and the 5- and 10-YSRs were 70.4% and 58.7%, respectively. When survival was compared according to the expression of each biomarker, there were no differences in the OS between the negative and positive expression groups for each biomarker (Figure 2). In subgroup analyses, the cyclin D1-positive group had shorter survival than the cyclin D1-negative group in stage I GC (Figure 3). The 5- and 10-YSRs were 89.8% and 86.5% in the cyclin D1-negative group and 84.4% and 62.8% in the cyclin D1-positive group in stage I GC, respectively ($p = 0.010$, Figure 3C). In EGC, the cyclin D1-negative group had better survival than cyclin D1-positive with borderline statistical significance ($p = 0.062$, Figure 3A), but not in AGC ($p = 0.541$, Figure 3E). However, the EGFR positivity did not affect the survival in any subgroup (Figure 3B,D,F). On a forest plot, the cyclin D1-positive group showed a poor prognosis compared to the cyclin D1-negative group in stage I and in node-negative disease (Figure 4). Upon multivariate analysis in stage I GC, cyclin D1 positivity was an independent poor prognostic factor (hazard ratio (HR) 2.801, 95% confidence interval (CI) 1.221–6.426, $p = 0.015$, Table 2)).

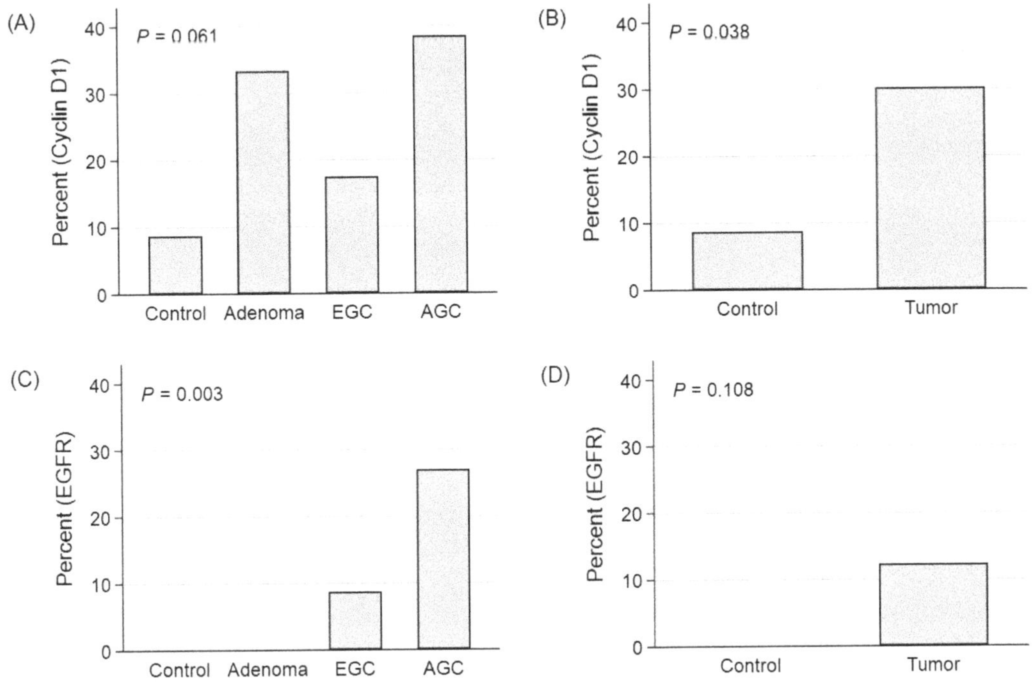

Figure 1. Expression of the biomarkers according to the progression of carcinogenesis. (**A**,**B**) Cyclin D1 and (**C**,**D**) EGFR.

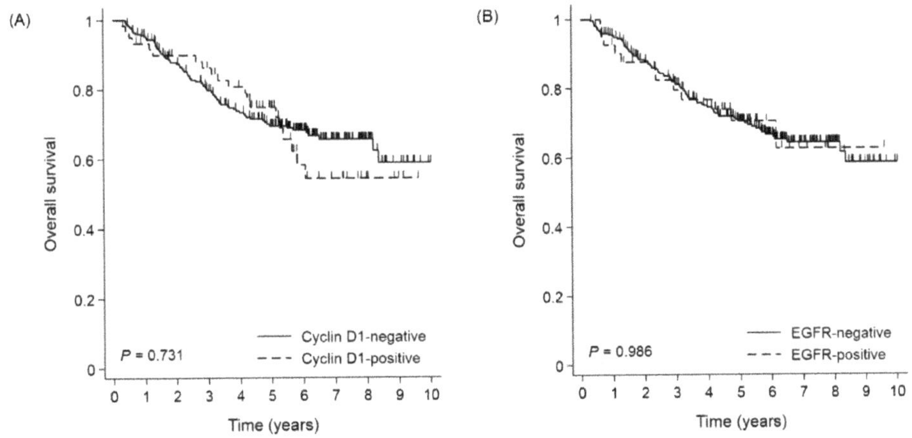

Figure 2. Overall survival by (**A**) cyclin D1 and (**B**) EGFR positivity in total cohort.

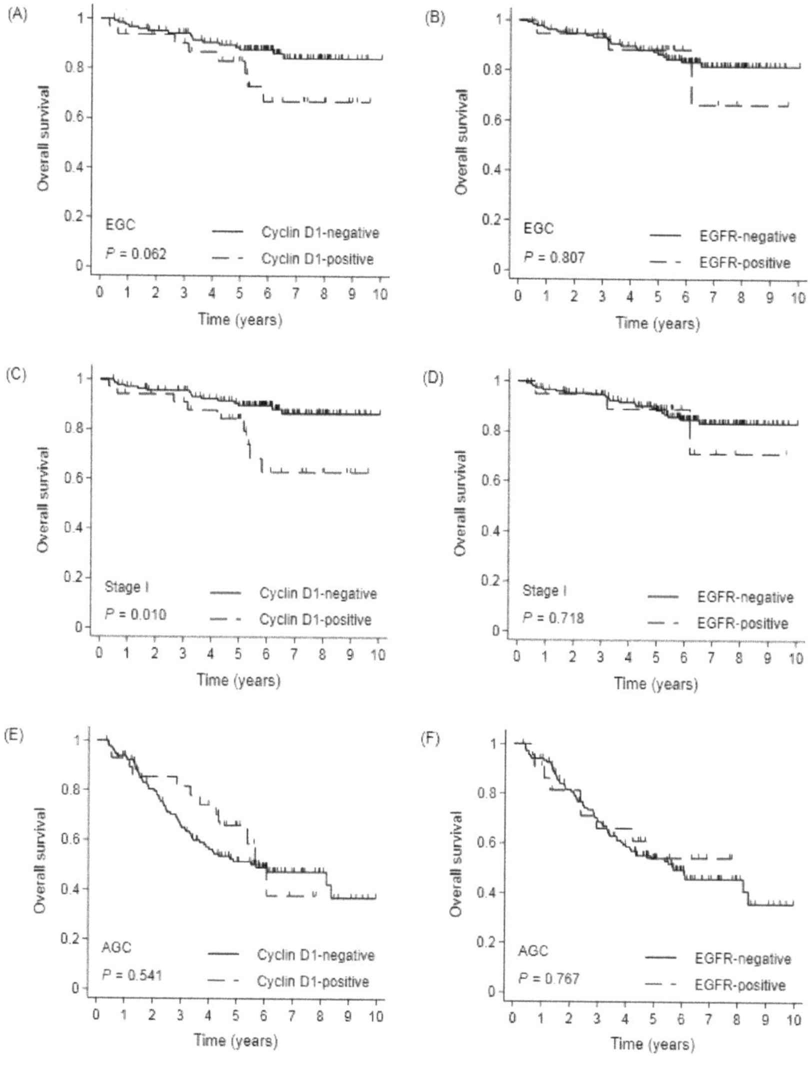

Figure 3. Overall survival by (**A**,**C**,**E**) cyclin D1 and (**B**,**D**,**F**) EGFR positivity in EGC, stage I gastric cancer, and AGC. EGC: early gastric cancer; AGC: advanced gastric cancer.

3.4. Survival According to Cyclin D1 Expression in Propensity Score-Matched Cohort

Given the imbalance in tumor stages among the basic characteristics of patients the between cyclin D1-positive and -negative groups, survival was reassessed between these two groups in the PSM cohort. After PSM, tumor stage and histologic classification were well balanced between the cyclin D1-negative and -positive groups (Table S1 and Figure S2). In 122 patients from the PSM cohort, the 5- and 10-YSRs were 88.7% and 88.7% in the cyclin D1-negative group and 74.8% and 54.4% in the cyclin D1-positive group, respectively ($p = 0.002$, Figure 5). In subgroup analyses, similarly to the result in the pre-PSM cohort (whole cohort), the cyclin D1-positive group had a poor prognosis compared to the cyclin D1-negative group in patients with EGC or stage I GC (Figure 6A,B), while there were no differences in the OS in patients with AGC or stage II-III GC (Figure 6C,D). Overall, the cyclin D1-positive group had a worse prognosis than the cyclin D1-negative group in the majority of subgroups (Figure 7). In a multivariate analysis of the PSM cohort, cyclin D1 expression was an independent prognostic factor (HR 3.630, 95% CI 1.450–9.086, $p = 0.006$). On bootstrap resampling, the statistical significance of the cyclin D1-positive group was internally validated (Table 3).

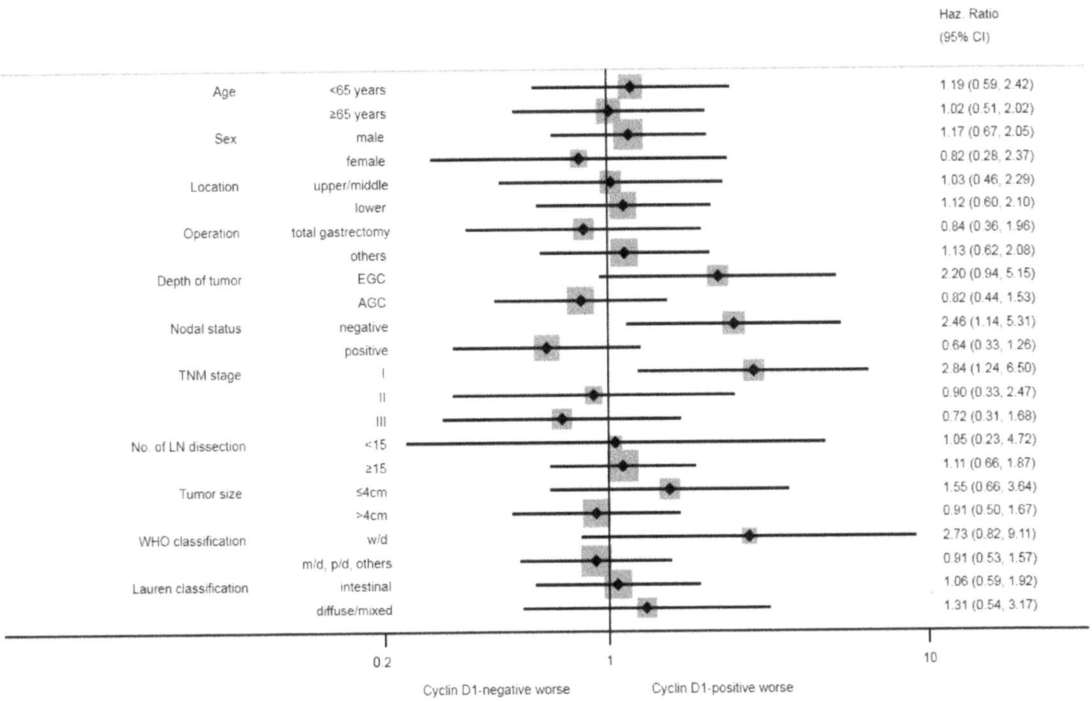

Figure 4. Forest plot for subgroup analyses of overall survival by cyclin D1 positivity.

Table 2. Cox regression for overall survival in stage I gastric cancer.

	Univariate			Multivariate			
	HR	**95% CI**	***p***	**HR**	**95% CI**	***p***	***p* (Bootstrap)**
Age (≥65 vs. <65)	2.623	1.087–6.328	0.032	2.679	1.110–6.466	0.028	0.042
Sex (male vs. female)	3.384	1.009–11.347	0.048	3.547	1.056–11.909	0.041	0.876
Location (upper/middle vs. lower)	1.086	0.465–2.538	0.849				
Operation (total gastrectomy vs. others)	1.894	0.752–4.774	0.176				
Depth of invasion (AGC vs. EGC)	0.651	0.153–2.770	0.561				
Nodal status (positive vs. negative)	2.035	0.477–8.685	0.337				
Tumor size (>4 cm vs. ≤4 cm)	0.687	0.273–1.730	0.425				
WHO classification (others vs. well-differentiated)	1.096	0.454–2.645	0.839				
Lauren classification (diffuse/mixed vs. intestinal)	1.055	0.419–2.658	0.910				
Cyclin D1 (positive vs. negative)	2.836	1.238–6.498	0.014	2.801	1.221–6.426	0.015	0.023
EGFR (positive vs. negative)	1.250	0.372–4.200	0.718				

HR: hazard ratio; CI: confidence interval; AGC: advanced gastric cancer; EGC: early gastric cancer.

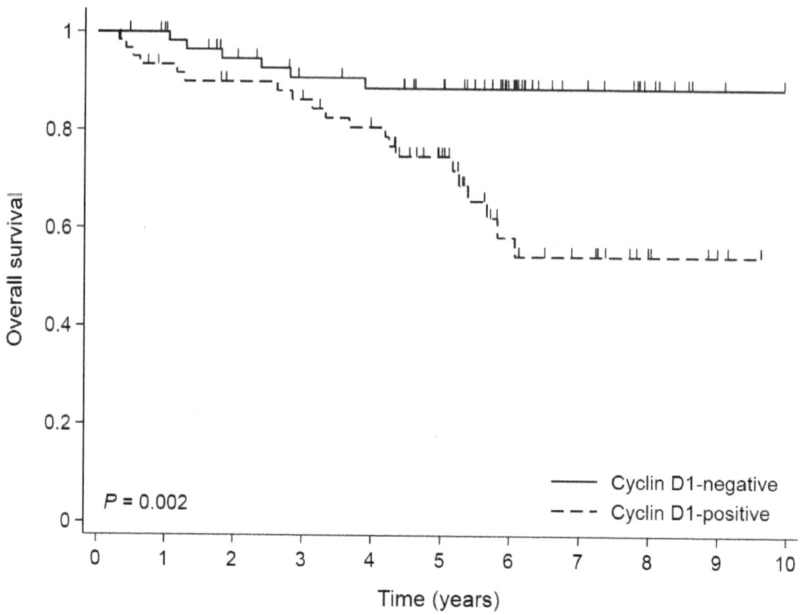

Figure 5. Overall survival by cyclin D1 positivity in propensity score-matched cohort.

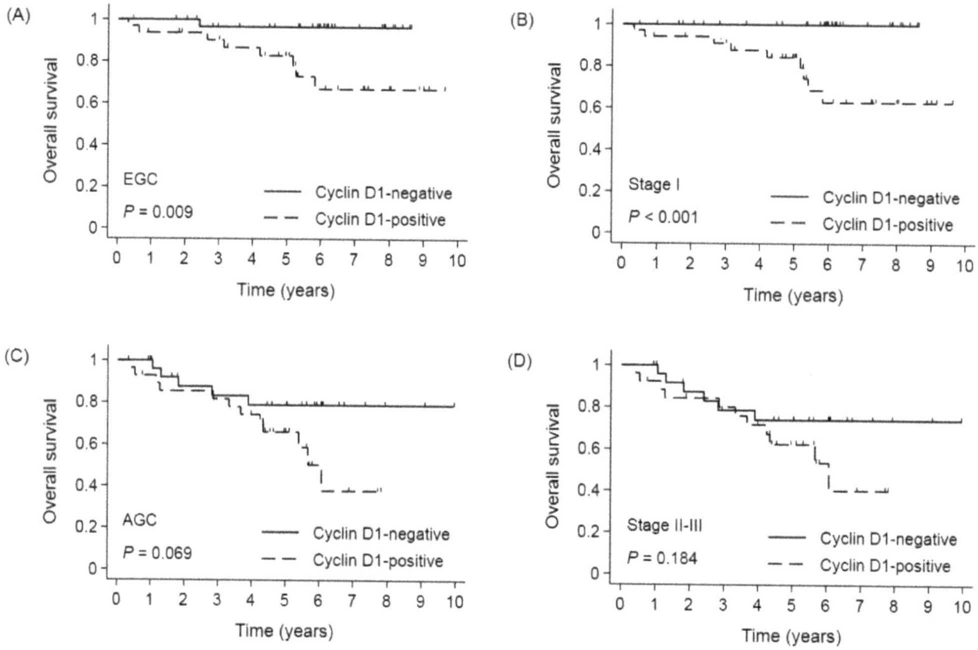

Figure 6. Overall survival by cyclin D1 in (**A**) EGC, (**B**) stage I gastric cancer, (**C**) AGC, and (**D**) stage II-III gastric cancer in propensity score-matched cohort. EGC: early gastric cancer; AGC: advanced gastric cancer.

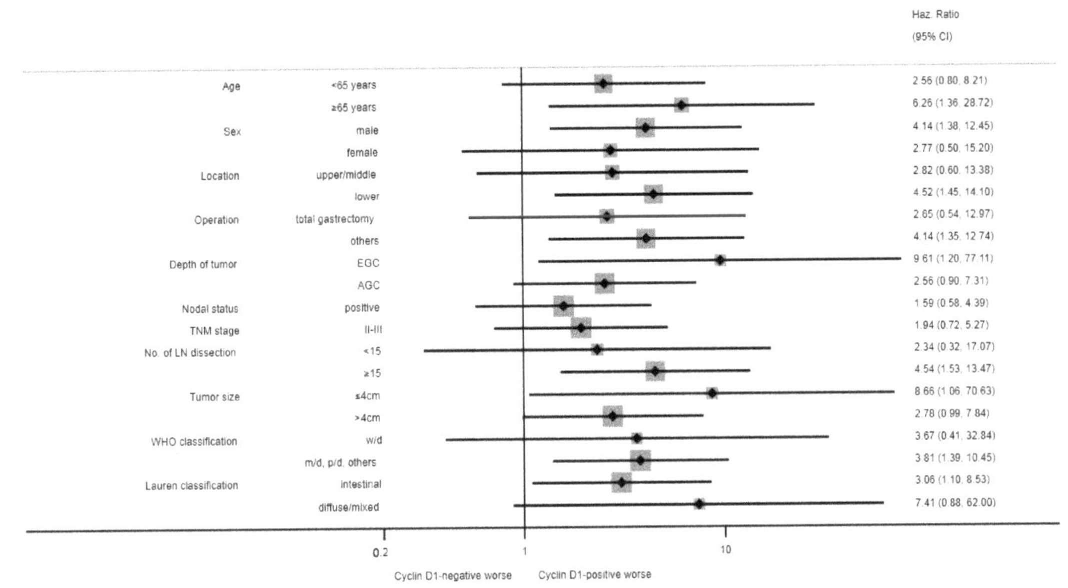

Figure 7. Forest plot for subgroup analyses of overall survival by cyclin D1 positivity in propensity score-matched cohort.

Table 3. Cox regression for overall survival after PSM.

	Univariate			Multivariate			
	HR	**95% CI**	***p***	**HR**	**95% CI**	***p***	***p* (Bootstrap)**
Age (≥65 vs. <65)	0.833	0.385–1.803	0.643				
Sex (male vs. female)	1.881	0.755–4.688	0.175				
Location (upper/middle vs. lower)	1.266	0.575–2.791	0.558				
Operation (total gastrectomy vs. others)	1.940	0.863–4.362	0.109				
Depth of invasion (AGC vs. EGC)	2.564	1.139–5.769	0.023	1.033	0.328–3.255	0.956	0.960
Nodal status (positive vs. negative)	3.847	1.736–8.526	0.001	3.169	1.014–9.906	0.047	0.077
Tumor size (>4 cm vs. ≤4 cm)	2.102	0.913–4.838	0.081	1.403	0.576–3.414	0.456	0.479
WHO classification (others vs. well-differentiated)	0.931	0.351–2.469	0.885				
Lauren classification (diffuse/mixed vs. intestinal)	1.313	0.551–3.126	0.539				
Cyclin D1 (positive vs. negative)	3.831	1.532–9.578	0.004	3.630	1.450–9.086	0.006	0.030

HR: hazard ratio; CI: confidence interval; AGC: advanced gastric cancer; EGC: early gastric cancer.

4. Discussion

This study was designed to find biomarkers that detect subgroups with different prognoses within the same stage. To accomplish this aim, we assessed the role of EGFR and cyclin D1 according to the depth of invasion and cancer stage. We also assessed each biomarker expression profile in normal cells, adenoma, EGC, and AGC to estimate the role of the biomarkers in gastric carcinogenesis. From the above results, we found that EGFR positivity was only observed in AGC (Figure 1). However, high EGFR expression failed to select a group with poor prognosis in AGC, while many previous studies reported that the overexpression of EGFR was related to tumor growth and drug resistance [13,22,23]. In addition, our result was different from previous data that suggested that EGFR amplification was associated with lymph node metastases [24–26]. This discrepancy can be explained. First, there are many signals that induce EGFR positivity in IHC, not just EGFR amplification [27–29]. Second, if the candidate biomarker is tested in the cohort including a subgroup in which the biomarker has no impact on predicting prognosis, the biomarker may come out as ineffective by reducing the statistical effect size [30]. This means that although EGFR positivity may play a role as a prognostic biomarker in a small subgroup, its role as a biomarker with statistical significance could be hidden due to the heterogeneity of the study group. Therefore, we also assessed the prognostic role of each biomarker's role in various subgroups stratified by clinically important variables such as age, sex, location, operation type, depth of tumor, nodal status, number of lymph nodes dissected, tumor size, and histologic classification. However, EGFR did not serve as a prognostic biomarker in any subgroup even though EGFR prognostic roles were tested in various subgroups.

On the other hand, cyclin D1 positivity showed a role as a prognostic factor in the node-negative GC as well as in TNM stage I GC, while cyclin D1 positivity was increased in all of the tumorous tissues without any significant increase in the positivity with the progression of carcinogenesis (adenoma to AGC). The statistical significance of this result increased when the PSM was performed despite the decrease in the number of patients, which means a significant reduction in statistical power. Cyclin D1 expression is tightly regulated in normal cells but is overexpressed in various ways in cancer. The overexpression of cyclin D1 contributes to uncontrolled cell proliferation and plays a central role in cancer carcinogenesis [16,17]. The overexpression of cyclin D1 has already been used as a cell proliferation and prognosis-related biomarker in several tumors together with the Ki-67 proliferation index [31,32]. The Ki-67 proliferation index also serves as a prognostic biomarker in EGC [4,33]. In addition, increased expression of Ki-67 and cyclin D1 has been reported to be associated with the development of precancerous lesions such as pancreatic intraepithelial neoplasia [34]. Therefore, in stage I GC, the proliferation index is considered to play an important role in the prognosis.

The strength of the study is that we assessed the EGFR and cyclin D1 expression profile from normal cells to cancer cells and then investigated its prognostic capability in GC within the same TNM stage. In previous studies suggesting the association of cyclin D1 expression with prognosis in GC, the prognostic impact of cyclin D1 by the TNM stage was not reported [18,35]. In addition, to find the hidden predictive role of the biomarkers, we also assessed the prognostic role of each biomarker in various subgroups. In contrast, there are several limitations in the study. First, this study is retrospectively designed. To overcome the potential bias caused by the retrospective study design, we additionally analyzed the PSM cohort and internally validated the multivariate Cox regression model using bootstrap resampling. Second, our criteria for the positive expression of each biomarker have not been validated in other studies. There are no standard criteria, and various criteria indicating positive expression of EGFR and cyclin D1 have been suggested [27,28,36,37]. A prospectively designed study using various criteria for positive expression of the biomarkers is warranted to confirm our results.

In conclusion, this study suggests that cyclin D1, but not EGFR, can be a useful biomarker in predicting the prognosis of stage I GC. Our results raise the question of whether adjuvant therapy is needed in those with positive cyclin D1.

Supplementary Materials: The following supporting information can be downloaded at: https://www.mdpi.com/article/10.3390/cimb44030093/s1, Figure S1: Representative findings of immunohistochemical staining for the biomarkers in gastric cancer cells (×200); Figure S2: Jitter plot to assess the distribution of propensity scores; Table S1: Baseline characteristics of patients after propensity score matching.

Author Contributions: Conceptualization and design, S.-I.G., G.H.K. and W.S.L.; investigation, S.-I.G., G.H.K., W.S.L., J.-H.L., S.-H.J. and Y.-J.L.; resources, G.H.K., S.C.H. and W.S.H.; formal analysis, S.-I.G. and W.S.L.; supervision, W.S.L., S.C.H. and W.S.H.; writing—original draft, S.-I.G., G.H.K. and W.S.L.; writing—review and editing, all authors; funding acquisition, S.-I.G. and W.S.L. All authors have read and agreed to the published version of the manuscript.

Funding: This study was funded by the biomedical research institute fund (GNUHBRIF-2018-0009) from the Gyeongsang National University Hospital, and a grant of the National R&D Program for Cancer Control, Ministry for Health, Welfare and Family Affairs, Republic of Korea (0820050).

Institutional Review Board Statement: The study was conducted according to the guidelines of the Declaration of Helsinki, and approved by the Institutional Review Board of Gyeongsang National University Changwon Hospital (protocol code: GNUCH-IRB-2022-01-027, date of approval: 27 January 2022).

Informed Consent Statement: Due to the retrospective nature of study, informed consent was waived.

Data Availability Statement: The data presented in this study are available on request from the corresponding author. The data are not publicly available due to their containing information that could compromise the privacy of research participants.

Acknowledgments: This work was supported by biomedical research institute fund (GNUHBRIF-2018-0009) from the Gyeongsang National University Hospital, and a grant of the National R&D Program for Cancer Control, Ministry for Health, Welfare and Family Affairs, Republic of Korea (0820050).

Conflicts of Interest: The authors declare no conflict of interest.

References

1. Bray, F.; Ferlay, J.; Soerjomataram, I.; Siegel, R.L.; Torre, L.A.; Jemal, A. Global cancer statistics 2018: GLOBOCAN estimates of incidence and mortality worldwide for 36 cancers in 185 countries. *CA Cancer J. Clin.* **2018**, *68*, 394–424. [CrossRef] [PubMed]
2. Hong, S.; Won, Y.J.; Lee, J.J.; Jung, K.W.; Kong, H.J.; Im, J.S.; Seo, H.G. Community of Population-Based Regional Cancer, R. Cancer Statistics in Korea: Incidence, Mortality, Survival, and Prevalence in 2018. *Cancer Res. Treat.* **2021**, *53*, 301–315. [CrossRef] [PubMed]
3. Mo, H.; Li, P.; Jiang, S. A novel nomogram based on cardia invasion and chemotherapy to predict postoperative overall survival of gastric cancer patients. *World J. Surg. Oncol.* **2021**, *19*, 256. [CrossRef] [PubMed]
4. Ko, G.H.; Go, S.I.; Lee, W.S.; Lee, J.H.; Jeong, S.H.; Lee, Y.J.; Hong, S.C.; Ha, W.S. Prognostic impact of Ki-67 in patients with gastric cancer-the importance of depth of invasion and histologic differentiation. *Medicine* **2017**, *96*, e7181. [CrossRef] [PubMed]
5. Go, S.I.; Ko, G.H.; Lee, W.S.; Kim, R.B.; Lee, J.H.; Jeong, S.H.; Lee, Y.J.; Hong, S.C.; Ha, W.S. CD44 Variant 9 Serves as a Poor Prognostic Marker in Early Gastric Cancer, But Not in Advanced Gastric Cancer. *Cancer Res. Treat.* **2016**, *48*, 142–152. [CrossRef] [PubMed]
6. Zhang, Y.; Yu, C. Development and validation of a Surveillance, Epidemiology, and End Results (SEER)-based prognostic nomogram for predicting survival in elderly patients with gastric cancer after surgery. *J. Gastrointest. Oncol.* **2021**, *12*, 278–296. [CrossRef] [PubMed]
7. Zhou, C.; Guo, Z.; Xu, L.; Jiang, H.; Sun, P.; Zhu, X.; Mu, X. PFND1 Predicts Poor Prognosis of Gastric Cancer and Promotes Cell Metastasis by Activating the Wnt/beta-Catenin Pathway. *Onco Targets Ther.* **2020**, *13*, 3177–3186. [CrossRef] [PubMed]
8. Liu, H.; Zhang, Z.; Han, Y.; Fan, A.; Liu, H.; Zhang, X.; Liu, Y.; Zhang, R.; Liu, W.; Lu, Y.; et al. The FENDRR/FOXC2 Axis Contributes to Multidrug Resistance in Gastric Cancer and Correlates With Poor Prognosis. *Front. Oncol.* **2021**, *11*, 634579. [CrossRef] [PubMed]
9. Guo, G.F.; Cai, Y.C.; Zhang, B.; Xu, R.H.; Qiu, H.J.; Xia, L.P.; Jiang, W.Q.; Hu, P.L.; Chen, X.X.; Zhou, F.F.; et al. Overexpression of SGLT1 and EGFR in colorectal cancer showing a correlation with the prognosis. *Med. Oncol.* **2011**, *28* (Suppl. S1), S197–S203. [CrossRef] [PubMed]
10. Park, S.J.; Gu, M.J.; Lee, D.S.; Yun, S.S.; Kim, H.J.; Choi, J.H. EGFR expression in pancreatic intraepithelial neoplasia and ductal adenocarcinoma. *Int. J. Clin. Exp. Pathol.* **2015**, *8*, 8298–8304.
11. Fard, S.S.; Saliminejad, K.; Sotoudeh, M.; Soleimanifard, N.; Kouchaki, S.; Yazdanbod, M.; Mahmoodzadeh, H.; Ghavamzadeh, A.; Malekzadeh, R.; Chahardouli, B.; et al. The Correlation between EGFR and Androgen Receptor Pathways: A Novel Potential Prognostic Marker in Gastric Cancer. *Anticancer Agents Med. Chem.* **2019**, *19*, 2097–2107. [CrossRef] [PubMed]
12. Al Zobair, A.A.; Al Obeidy, B.F.; Yang, L.; Yang, C.; Hui, Y.; Yu, H.; Zheng, F.; Yang, G.; Xie, C.; Zhou, F.; et al. Concomitant overexpression of EGFR and CXCR4 is associated with worse prognosis in a new molecular subtype of non-small cell lung cancer. *Oncol. Rep.* **2013**, *29*, 1524–1532. [CrossRef] [PubMed]
13. Uribe, M.L.; Marrocco, I.; Yarden, Y. EGFR in Cancer: Signaling Mechanisms, Drugs, and Acquired Resistance. *Cancers* **2021**, *13*, 2748. [CrossRef] [PubMed]
14. Yang, G.; Huang, L.; Jia, H.; Aikemu, B.; Zhang, S.; Shao, Y.; Hong, H.; Yesseyeva, G.; Wang, C.; Li, S.; et al. NDRG1 enhances the sensitivity of cetuximab by modulating EGFR trafficking in colorectal cancer. *Oncogene* **2021**, *40*, 5993–6006. [CrossRef] [PubMed]
15. Wang, Y.K.; Gao, C.F.; Yun, T.; Chen, Z.; Zhang, X.W.; Lv, X.X.; Meng, N.L.; Zhao, W.Z. Assessment of ERBB2 and EGFR gene amplification and protein expression in gastric carcinoma by immunohistochemistry and fluorescence in situ hybridization. *Mol. Cytogenet.* **2011**, *4*, 14.

16. Montalto, F.I.; De Amicis, F. Cyclin D1 in Cancer: A Molecular Connection for Cell Cycle Control, Adhesion and Invasion in Tumor and Stroma. *Cells* **2020**, *9*, 2648. [CrossRef] [PubMed]
17. Tchakarska, G.; Sola, B. The double dealing of cyclin D1. *Cell Cycle* **2020**, *19*, 163–178. [CrossRef]
18. Shan, Y.S.; Hsu, H.P.; Lai, M.D.; Hung, Y.H.; Wang, C.Y.; Yen, M.C.; Chen, Y.L. Cyclin D1 overexpression correlates with poor tumor differentiation and prognosis in gastric cancer. *Oncol. Lett.* **2017**, *14*, 4517–4526. [CrossRef]
19. He, Q.; Wu, J.; Liu, X.L.; Ma, Y.H.; Wu, X.T.; Wang, W.Y.; An, H.X. Clinicopathological and prognostic significance of cyclin D1 amplification in patients with breast cancer: A meta-analysis. *J. BUON* **2017**, *22*, 1209–1216.
20. Bachmann, K.; Neumann, A.; Hinsch, A.; Nentwich, M.F.; El Gammal, A.T.; Vashist, Y.; Perez, D.; Bockhorn, M.; Izbicki, J.R.; Mann, O. Cyclin D1 is a strong prognostic factor for survival in pancreatic cancer: Analysis of CD G870A polymorphism, FISH and immunohistochemistry. *J. Surg. Oncol.* **2015**, *111*, 316–323. [CrossRef]
21. Luangdilok, S.; Wanchaijiraboon, P.; Chantranuwatana, P.; Teerapakpinyo, C.; Shuangshoti, S.; Sriuranpong, V. Cyclin D1 expression as a potential prognostic factor in advanced KRAS-mutant non-small cell lung cancer. *Transl. Lung Cancer Res.* **2019**, *8*, 959–966. [CrossRef] [PubMed]
22. Gu, X.Y.; Jiang, Y.; Li, M.Q.; Han, P.; Liu, Y.L.; Cui, B.B. Over-expression of EGFR regulated by RARA contributes to 5-FU resistance in colon cancer. *Aging* **2020**, *12*, 156–177. [CrossRef] [PubMed]
23. Li, W.; Liu, Z.; Li, C.; Li, N.; Fang, L.; Chang, J.; Tan, J. Radionuclide therapy using (1)(3)(1)I-labeled anti-epidermal growth factor receptor-targeted nanoparticles suppresses cancer cell growth caused by EGFR overexpression. *J. Cancer Res. Clin. Oncol.* **2016**, *142*, 619–632. [CrossRef] [PubMed]
24. Ema, A.; Waraya, M.; Yamashita, K.; Kokubo, K.; Kobayashi, H.; Hoshi, K.; Shinkai, Y.; Kawamata, H.; Nakamura, K.; Nishimiya, H.; et al. Identification of EGFR expression status association with metastatic lymph node density (ND) by expression microarray analysis of advanced gastric cancer. *Cancer Med.* **2015**, *4*, 90–100. [CrossRef] [PubMed]
25. Tang, C.; Yang, L.; Wang, N.; Li, L.; Xu, M.; Chen, G.G.; Liu, Z.M. High expression of GPER1, EGFR and CXCR1 is associated with lymph node metastasis in papillary thyroid carcinoma. *Int. J. Clin. Exp. Pathol.* **2014**, *7*, 3213–3223. [PubMed]
26. Carlsson, J.; Shen, L.; Xiang, J.; Xu, J.; Wei, Q. Tendencies for higher co-expression of EGFR and HER2 and downregulation of HER3 in prostate cancer lymph node metastases compared with corresponding primary tumors. *Oncol. Lett.* **2013**, *5*, 208–214. [CrossRef] [PubMed]
27. Hashmi, A.A.; Hussain, Z.F.; Aijaz, S.; Irfan, M.; Khan, E.Y.; Naz, S.; Faridi, N.; Khan, A.; Edhi, M.M. Immunohistochemical expression of epidermal growth factor receptor (EGFR) in South Asian head and neck squamous cell carcinoma: Association with various risk factors and clinico-pathologic and prognostic parameters. *World J. Surg. Oncol.* **2018**, *16*, 118. [CrossRef] [PubMed]
28. Lerias, S.; Esteves, S.; Silva, F.; Cunha, M.; Cochicho, D.; Martins, L.; Felix, A. CD274 (PD-L1), CDKN2A (p16), TP53, and EGFR immunohistochemical profile in primary, recurrent and metastatic vulvar cancer. *Mod. Pathol.* **2020**, *33*, 893–904. [CrossRef] [PubMed]
29. Ben Brahim, E.; Ayari, I.; Jouini, R.; Atafi, S.; Koubaa, W.; Elloumi, H.; Chadli, A. Expression of epidermal growth factor receptor (EGFR) in colorectal cancer: An immunohistochemical study. *Arab. J. Gastroenterol.* **2018**, *19*, 121–124. [CrossRef] [PubMed]
30. Martins, W.P.; Zanardi, J.V. Subgroup analysis and statistical power. *Eur. J. Obstet. Gynecol. Reprod. Biol.* **2011**, *159*, 244–245. [CrossRef]
31. Shevra, C.R.; Ghosh, A.; Kumar, M. Cyclin D1 and Ki-67 expression in normal, hyperplastic and neoplastic endometrium. *J. Postgrad. Med.* **2015**, *61*, 15–20.
32. Kaufmann, C.; Kempf, W.; Mangana, J.; Cheng, P.; Emberger, M.; Lang, R.; Kaiser, A.K.; Lattmann, E.; Levesque, M.; Dummer, R.; et al. The role of cyclin D1 and Ki-67 in the development and prognostication of thin melanoma. *Histopathology* **2020**, *77*, 460–470. [CrossRef]
33. Go, S.I.; Ko, G.H.; Lee, W.S.; Lee, J.H.; Jeong, S.H.; Lee, Y.J.; Hong, S.C.; Ha, W.S. The Use of CD44 Variant 9 and Ki-67 Combination Can Predicts Prognosis Better Than Their Single Use in Early Gastric Cancer. *Cancer Res. Treat.* **2019**, *51*, 1411–1419. [CrossRef] [PubMed]
34. Zinczuk, J.; Zareba, K.; Guzinska-Ustymowicz, K.; Kedra, B.; Kemona, A.; Pryczynicz, A. Expression of chosen cell cycle and proliferation markers in pancreatic intraepithelial neoplasia. *Prz. Gastroenterol.* **2018**, *13*, 118–126. [CrossRef] [PubMed]
35. Gao, P.; Zhou, G.Y.; Liu, Y.; Li, J.S.; Zhen, J.H.; Yuan, Y.P. Alteration of cyclin D1 in gastric carcinoma and its clinicopathologic significance. *World J. Gastroenterol.* **2004**, *10*, 2936–2939. [CrossRef]
36. Patel, S.B.; Manjunatha, B.S.; Shah, V.; Soni, N.; Sutariya, R. Immunohistochemical evaluation of p63 and cyclin D1 in oral squamous cell carcinoma and leukoplakia. *J. Korean Assoc. Oral Maxillofac. Surg.* **2017**, *43*, 324–330. [CrossRef] [PubMed]
37. Cho, E.Y.; Han, J.J.; Choi, Y.L.; Kim, K.M.; Oh, Y.L. Comparison of Her-2, EGFR and cyclin D1 in primary breast cancer and paired metastatic lymph nodes: An immunohistochemical and chromogenic in situ hybridization study. *J. Korean Med. Sci.* **2008**, *23*, 1053–1061. [CrossRef]

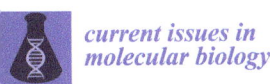

current issues in
molecular biology

MDPI

Article

PNU-74654 Suppresses TNFR1/IKB Alpha/p65 Signaling and Induces Cell Death in Testicular Cancer

Wen-Jung Chen [1,2,3,†], Wen-Wei Sung [1,2,3,†], Chia-Ying Yu [2], Yu-Ze Luan [2], Ya-Chuan Chang [2], Sung-Lang Chen [2,3,*,‡] and Tsung-Hsien Lee [1,4,5,*,‡]

1 Institute of Medicine, Chung Shan Medical University, Taichung 40201, Taiwan; mimic1024@gmail.com (W.-J.C.); flutewayne@gmail.com (W.-W.S.)
2 School of Medicine, Chung Shan Medical University, Taichung 40201, Taiwan; cyyu2015@gmail.com (C.-Y.Y.); vm6vul4@gmail.com (Y.-Z.L.); raptor7037@gmail.com (Y.-C.C.)
3 Department of Urology, Chung Shan Medical University Hospital, Taichung 40201, Taiwan
4 Department of Obstetrics and Gynecology, Chung Shan Medical University Hospital, Taichung 40201, Taiwan
5 Division of Infertility Clinic, Lee Women's Hospital, Taichung 40201, Taiwan
* Correspondence: cshy650@csh.org.tw (S.-L.C.); jackth.lee@gmail.com (T.-H.L.)
† These authors contributed equally to this work.
‡ These authors contributed equally to this work.

Abstract: Testicular cancer (TC) is a rare malignancy worldwide and is the most common malignancy in males aged 15–44 years. The Wnt/β-catenin signaling pathway mediates numerous essential cellular functions and has potentially important effects on tumorigenesis and cancer progression. The search for drugs to inhibit this pathway has identified a small molecule, PNU-74654, as an inhibitor of the β-catenin/TCF4 interaction. We evaluated the therapeutic role of PNU-74654 in two TC cell lines, NCCIT and NTERA2, by measuring cell viability, cell cycle transition and cell death. Potential pathways were evaluated by protein arrays and Western blots. PNU-74654 decreased cell viability and induced apoptosis of TC cells, with significant increases in the sub G1, Hoechst-stained, Annexin V-PI-positive rates. PNU-74654 treatment of both TC cell lines inhibited the TNFR1/IKB alpha/p65 pathway and the execution phase of apoptosis. Our findings demonstrate that PNU-74654 can induce apoptosis in TC cells through mechanisms involving the execution phase of apoptosis and inhibition of TNFR1/IKB alpha/p65 signaling. Therefore, small molecules such as PNU-74654 may identify potential new treatment strategies for TC.

Keywords: PNU-74654; TNF receptor-1; apoptosis; testicular cancer

Citation: Chen, W.-J.; Sung, W.-W.; Yu, C.-Y.; Luan, Y.-Z.; Chang, Y.-C.; Chen, S.-L.; Lee, T.-H. PNU-74654 Suppresses TNFR1/IKB Alpha/p65 Signaling and Induces Cell Death in Testicular Cancer. *Curr. Issues Mol. Biol.* **2022**, *44*, 222–232. https://doi.org/10.3390/cimb44010016

Academic Editor: Dumitru A. Iacobas

Received: 1 December 2021
Accepted: 2 January 2022
Published: 4 January 2022

1. Introduction

Testicular cancer is a rare malignancy in males worldwide and accounts for about 1% of newly diagnosed male cancers every year [1]. It also shows a specific age distribution, as it is most commonly encountered in males aged 15–44 years [2]. For unknown reasons, the incidence of this male cancer has been increasing in many developed countries in the last decades [3]. TC has a close relationship with cryptorchidism, but other known risk factors include prior TC, a family history of a father or brother with TC, increased adult height (per 5 cm increase), ethnicity and infertility [3,4]. Despite the increasing incidence, the overall five-year survival is as high as 97% with effective treatment [4]. However, the development of effective treatments depends on knowledge of the mechanisms that cause testicular tissues to become cancerous.

Current evidence now implicates the Wnt/β-catenin signaling pathway as part of the mechanism involved in the development of several malignancies. This pathway mediates numerous essential cellular functions, such as proliferation, differentiation, apoptosis and cell migration [5]. *INT1*, the first gene identified in the Wnt/β-catenin pathway, shows high similarity in both humans and mice; therefore, it is considered a highly conserved pathway evolutionarily among various species [6]. Wnt proteins are secreted extracellularly by cells and function through a receptor-mediated pathway [7]. By contrast, β-catenin is a component of the cell adhesion complex and plays an essential role in cell–cell adhesion [8]. In the canonical Wnt/β-catenin signaling pathway, WNT proteins are activated by binding to a transmembrane Frizzled family receptor and the LDL receptor-related protein (LRP) co-receptor and this inhibits the activity of glycogen synthase kinase-3β (GSK-3β). In this Wnt activation state, the β-catenin destruction complex is disrupted by activated Dishevelled protein (DVL) and the phosphorylation of β-catenin is prevented, thereby allowing β-catenin accumulation to occur within the cytoplasm. Then, β-catenin translocates into the nucleus, where it interacts with T-cell factor/lymphoid enhancer factor (TCF/Lef) to activate the TCF/Lef transcription complex [9–11]. In turn, the transcription of Wnt/β-catenin target genes, such as c-myc, cyclin D1 and Bcl-w, is activated, followed by subsequent cell proliferation that is closely related to tumorigenesis and cancer progression [12–14].

By contrast, in the absence of the extracellular Wnt needed for interaction with the receptors, β-catenin is degraded by a destruction complex composed of GSK-3β, adenomatous polyposis coli protein (APC), casein kinase 1α (CK1α) and Axin [15]. During the degradation process, β-catenin is initially phosphorylated by CK1α and GSK-3β, which makes it recognizable by a ubiquitin E3 ligase via β-transducin repeat-containing proteins (β-TrCP). Subsequent ubiquitination and proteasomal degradation of β-catenin occur, thereby maintaining a low intracellular level of β-catenin [16]. In this normal condition, TCF/Lef proteins are bound by repressors, such as CREB-binding protein or p300, and abnormal cell proliferation does not occur.

The Wnt/β-catenin signaling pathway is involved in the tumorigenesis of several malignancies, including colorectal cancers, non-colorectal gastrointestinal cancers, desmoid tumors, breast cancers, adrenocortical tumors, melanoma, glioblastoma multiforme, renal cell carcinoma, osteosarcoma and hematologic malignancies [5]. This pathway plays an important role in the transcriptional regulation of multiple oncogenes; therefore, it is essential for tumor occurrence and progression. Aberrant activation of this signaling pathway has been verified in various types of cancer [17]. For example, β-catenin-stabilizing mutations have been identified in colorectal cancers and have resulted in constitutive nuclear localization of β-catenin and transcription of its target genes [18].

In addition, multiple fusion transcripts, including recurrent gene fusions involving the R-spondin family members RSPO2 and RSPO3, have been identified in colon cancer cells [19]. R-spondins are proteins that enhance Wnt signaling and serve as oncogenic drivers in RSPO fusion tumors [19,20]. Hyperactivation of Wnt signaling has also been identified in hepatocellular carcinoma, which is mediated by loss of function and inactivation of the Wnt negative regulators AXIN1 and/or AXIN2 [21]. In breast and ovarian cancer cell lines, autocrine production of various Wnt ligands has been reported and has resulted in an increase in β-catenin stability [22].

Fortunately, in the last decade, many inhibitors, including biologics and small molecules, that focus on the Wnt pathway have been discovered. The biologics, which include antibodies and RNA interference molecules, mainly focus on targeting the Wnt protein itself and on the use of recombinant proteins that target extracellular modulators of this pathway [23–25]. By contrast, the small molecules fall into four categories based on their functional mechanism: (1) β-catenin/TCF interaction inhibitors; (2) antagonists of transcriptional co-activators, such as CBP and p300; (3) molecules binding to DVL; and (4) inhibitors based on the stabilization of Axin protein [26]. All these drugs show promising anticancer effects mediated by disrupting the signal transduction of the Wnt/β-catenin pathway.

One of the identified small molecule inhibitors of the β-catenin/TCF4 interaction is PNU-74654, a compound that prevents TCF4 from binding to β-catenin; therefore, it acts as a Wnt/β-catenin antagonist. This small molecule was discovered through virtual screening and confirmed by biophysical screening to disturb protein–protein interactions [27]. PNU-74654 competes with TCF4 for the binding site on β-catenin and its effective role as a

Wnt pathway antagonist has been proven by a luciferase activity assay for TCF transactivation [27].

TCF4 is the most important transcription factor in the TCF/Lef family, as it has the ability to bind DNA and modulate the transcription of many oncogenes. Therefore, a small molecule that can block the interaction between β-catenin and TCF4 can prevent the transcription of a variety of tumor-related proteins, such as c-myc, cyclin D1, Bcl-w, MDR1, IL-8 and ZEB1 [12–14,28–30]. PNU-74654 has been shown to significantly decrease cell proliferation, increase tumor cell apoptosis, decrease nuclear β-catenin accumulation and impair CTNNB1/β-catenin expression in adrenocortical tumors [31]. PNU-74654 has also shown antitumor efficacy in breast cancers, where it causes tumor shrinkage and increased tumor cell apoptosis [32]. To our knowledge, no study has yet explored the potential for using PNU-74654 as a treatment in TC. Therefore, the aim of the current study is to evaluate the antitumor activity of PNU74654 in TC cells.

2. Materials and Methods

2.1. Cell Culture

Two human testicular teratocarcinoma cell lines, NCCIT and NTERA2, were obtained from BCRC (Bioresource Collection and Research Center, Taiwan). The cells were cultured and stored according to the suppliers' instructions. NTERA-2 cells were maintained in high-glucose (4.5 g/L) Dulbecco's modified Eagle medium containing 10% fetal bovine serum, 1 mM sodium pyruvate, 100 U/mL of penicillin and 100 μg/mL of streptomycin. NCCIT cells were maintained in RPMI supplemented as described above. Cells were cultivated at 37 °C in a humidified 5% CO_2 atmosphere [33].

2.2. MTT Assay

The MTT assay was used to detect cytotoxicity and cell growth, as described previously [33]. Briefly, TC cells were seeded at a density of 1×10^4 cells in each well of 96-well plates and incubated overnight. Then, the cells were treated with PNU-74654 (MedChemExpress, Monmouth Junction, NJ, USA) at 50–250 μM for 24 h. Then, an MTT solution (0.5 mg/mL) was added to the wells and incubated for 3 h at 37 °C. The reaction was stopped by removing the supernatant, followed by dissolving the formazan product in DMSO. The absorbance at 570 nm was measured with an ELISA reader.

2.3. Flow Cytometry Analysis

A flow cytometer (FACSCanto II; BD Biosciences, San Jose, CA, USA) was used to determine the cell cycle population and apoptosis percentage, as described previously [33]. An Annexin V/PI apoptosis detection kit (Elabscience Biotechnology Inc., Houston, TX, USA) was used to mark apoptotic cells. Apoptosis rates were quantified by treating the cells with PNU-74654 (0, 50 and 200 μM) for 48 h, then staining them with FITC-Annexin V and PI in the dark at room temperature, according to the manufacturer's protocol. The fractions of cells in different phases of the cell cycle were determined by treating the cells with PNU-74654 (0, 50 and 200 μM) for 24 h. Then, the supernatant and adherent cells were harvested, washed with PBS and fixed overnight in 70% ice-cold ethanol at −20 °C. After washing, the cells were resuspended in 0.4 mL of PBS containing 4 μg/mL of propidium iodide and 0.5 of mg/mL RNase A and incubated for 30 min at 37 °C in the dark. Individual cell suspensions were analyzed by flow cytometry and cell profiles were analyzed using FlowJo software (BD Biosciences, USA).

2.4. Hoechst 33,342 Staining

Cells were plated in 6-well plates and incubated for 16 h. Different concentrations of PNU-74654 were added to each well and incubated for 24 h. The cells were harvested at 48 h and stained with Hoechst 33,342 (10 μg/mL) for 20 min at 37 °C. Images of the Hoechst 33,342 fluorescence were captured using a fluorescence microscope (ImageXpress PICO; excitation wavelength of 350–390 nm, emission wavelength of 420–480 nm) at

20× magnification. Apoptotic cells emitted blue fluorescence and exhibited morphological changes in the nuclei typical of apoptosis. The fluorescence staining percentage of positive cells was calculated based on cell counts made from five random visual fields, as described previously [33].

2.5. Human Apoptosis Array for Proteome Profiling

The Human Apoptosis Proteome Profiler™ array (R&D Systems, Minneapolis, MN, USA), which examines changes in 35 apoptosis-related proteins, was used in this study. Cells were treated with or without PNU-74654 (200 µM) for 24 h and 400 µg of total protein obtained after cell lysis was used for each array and analyzed according to the manufacturer's instructions. Membranes were imaged by chemiluminescence and the integrated density of the spots was quantified using Image J software, as described previously [33].

2.6. Protein Extraction from Cells and Western Blotting

The cells were treated with PNU-74654 (0, 50 and 200 µM) for 24 h, then lysed in RIPA buffer containing a protease inhibitor cocktail (Roche Molecular Biochemicals, Basel, Switzerland). The lysate was centrifuged for 20 min at 10,400 rcf at 4 °C and the supernatant was saved. The protein concentration was measured with the Bio-Rad Protein Assay (Bio-Rad Laboratories Inc., Hercules, CA, USA). Equal amounts of protein (15 µg) per sample were boiled in sample buffer, electrophoresed on sodium dodecyl sulfate–polyacrylamide gels, then transferred onto an Immobilon™-P transfer membrane (Millipore, Burlington, MA, USA). After blocking, the membranes were probed with antibodies at 4 °C for 16 h. The membranes were washed with Tris-buffered saline with 0.1% Tween 20 detergent (TTBS), then incubated for 1 h with horseradish peroxidase-conjugated secondary antibodies. The membrane was washed with TTBS buffer and the Western blots were visualized with ImmobilonTM-Western Chemiluminescent HRP Substrate (Millipore, USA). The results were displayed on an Amersham™ Imager 680 (GE Healthcare, Chicago, IL, USA) and the integrated density of the spots was quantified using Image J software, as described previously [33].

2.7. Statistical Analysis

The statistical analyses were conducted using IBM SPSS software (version 20.0). The data are presented as the mean ± S.D. The Student's t-test was used for continuous or discrete data analysis. All statistical tests were two-sided (SEM) and values of $p < 0.05$ were considered statistically significant (*$p < 0.05$; ** $p < 0.01$; *** $p < 0.001$).

3. Results

3.1. PNU-74654 Had a Cytotoxic Effect on TC Cells and Elicited Cell Death

We confirmed the cytotoxic effect of PNU-74654 treatment on both the NCCIT and NTERA2 testicular carcinoma cell lines (Figure 1A,B). The MTT assay revealed a dose-dependent decrease in the viability of PNU-74654-treated TC cells after a 24 h exposure. Therefore, we performed a cell cycle analysis by flow cytometry to seek the causes of this cytotoxicity. Figure 1C–E shows a dose-dependent increase in the sub G1 group of TC cells after 24 h treatment with 50 and 200 µM PNU-74654 compared with the untreated control groups (NCCIT: control vs. 50 µM and 200 µM, p value = 0.002 and 0.001; NTERA2: control vs. 50 µM and 200 µM, p value = 0.564 and 0.002). The proportions of NCCIT cells in the sub G1 phase rose from 1.39% ± 0.24% (control group) to 2.58% ± 0.13% (50 µM group) and 37.93% ± 2.80% (200 µM group). The proportions of NTERA2 cells in the sub G1 phase rose from 5.41% ± 1.38% (control group) to 5.92% ± 0.24% (50 µM group) and 23.03% ± 0.12% (200 µM group). Therefore, the cytotoxicity of PNU-74654 was confirmed to cause the death of both NCCIT and NTERA2 TC cells.

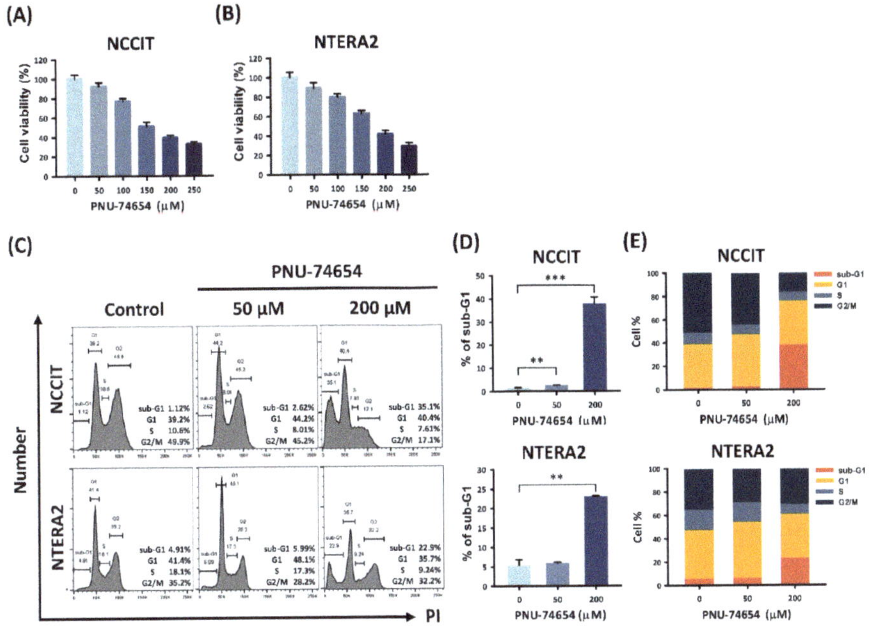

Figure 1. PNU-74654 caused cell death in NCCIT and NTERA2 testicular cancer cell lines. (**A**,**B**) The MTT assay determined reduced viabilities of TC cells after PNU-74654 treatment. (**C**) Flow cytometry showed increased proportions of TC cells in sub G1 phase after PNU-74654 treatment. (**D**) Columns illustrate the increased proportions of TC cells in sub G1 phase after PNU-74654 treatment. (**E**) Cell cycle distributions are shown for TC cells after PNU-74654 treatment. Data are shown as mean ± S.D (** $p < 0.01$; *** $p < 0.001$).

3.2. Apoptosis Induced by PNU-74654 in TC Cells

We further distinguished the types of cell death occurring in PNU-74654-treated TC cells by Hoechst 33,342 staining and Annexin V/PI double staining (Figure 2). The Hoechst 33,342 staining (Figure 2A,B) revealed that the proportions of apoptotic TC cells rose dose-dependently with PNU-74654 treatment compared to the control groups (NCCIT: control vs. 50 μM and 200 μM, p value = 0.001 and 0.001; NTERA2: control vs. 50 μM and 200 μM, p value = 0.001 and 0.001). The proportions of apoptotic NCCIT cells rose from 0.24% ± 0.33% (control group) to 2.99% ± 0.56% (50 μM group) and 4.54% ± 0.46% (200 μM group). The proportion of apoptotic NTERA2 cells rose from 1.69% ± 0.41% (control group) to 8.91% ± 0.91% (50 μM group) and 21.83% ± 1.11% (200 μM group).

The Annexin V/PI double staining (Figure 2C,D) showed dose-dependent rises in the proportion of early-apoptotic (Annexin V+/PI−) and late-apoptotic (Annexin V+/PI+) TC cells with PNU-74654 treatment compared to the control groups (NCCIT: control vs. 50 μM and 200 μM, p value = 0.053 and 0.008; NTERA2: control vs. 50 μM and 200 μM, p value = 0.001 and 0.001). The proportions of apoptotic NCCIT cells rose from 4.91% ± 0.60% (control group) to 10.92% ± 2.66% (50 μM group) and 50.13% ± 7.01% (200 μM group). The proportions of apoptotic NTERA2 cells rose from 7.14% ± 0.06% (control group) to 56.47% ± 3.26% (50 μM group) and 77.87% ± 3.12% (200 μM group). In brief, the results of Hoechst 33,342 staining and Annexin V/PI double staining confirmed apoptosis as the major type of cell death in PNU-74654-treated TC cells.

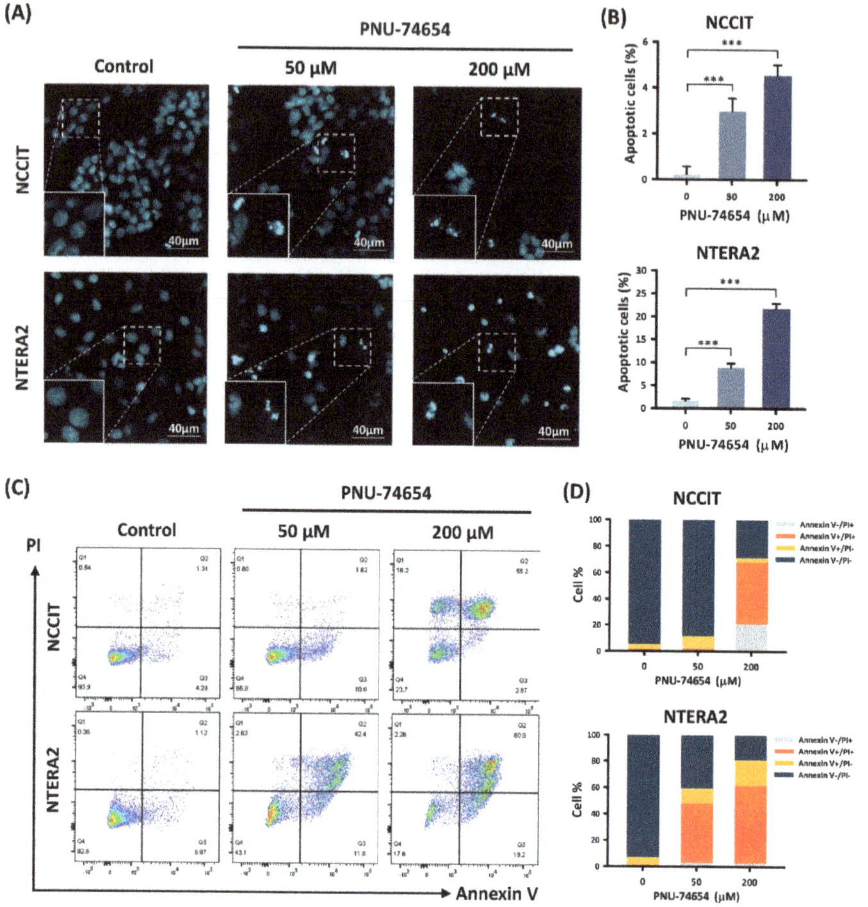

Figure 2. PNU-74654 caused apoptosis in TC cells. (**A**) Hoechst 33,342 staining showed apoptosis-associated morphological changes and increased numbers of apoptotic cells in TC cell lines after PNU-74654 treatment. (**B**) Bar chart shows increased proportions of apoptotic TC cells after PNU-74654 treatment. (**C**) Flow cytometry after Annexin V/PI double staining showed distributions of early apoptosis, late apoptosis and necrosis in TC cells after PNU-74654 treatment. (**D**) Columns show proportions of Annexin $V^{+/-}$/$PI^{+/-}$ TC cells after PNU-74654 treatment. Data are shown as mean ± S.D (*** $p < 0.001$).

3.3. *Inhibition of TNF R1 Triggered NF-κB Anti-Apoptotic Pathway and Execution Phase of Apoptosis Contributed to PNU-74654-Induced Apoptosis in TC Cells*

We explored the underlying mechanisms of PNU-74654-induced apoptosis in TC cells using apoptosis arrays. As shown in Figure 3A–D, after 24 h treatment with 200 μM PNU-74654, both TC cell lines showed notable upregulations of cleaved-caspase-3 and downregulations of claspin and survivin, but only NCCIT cells showed downregulations of TNF R1 and cIAP-1.

We sought a deeper understanding and ascertainment by conducting Western blots to test the expressions of proteins involved in the execution phase of apoptosis and in TNFR1/IKB alpha/p65 and apoptotic signaling (Figure 4A,B). We observed inhibition of the TNFR1-triggered pathway (Figure 4A) in the form of downregulations of TNFR1, phospho-IκBα and phosphor-p65. We observed inhibition of the execution phase of apoptosis

(Figure 4B) in the form of downregulations of $FLIP_L$ and survivin and upregulations of cleaved-caspase-3, cleaved-caspase-7 and cleaved-PARP.

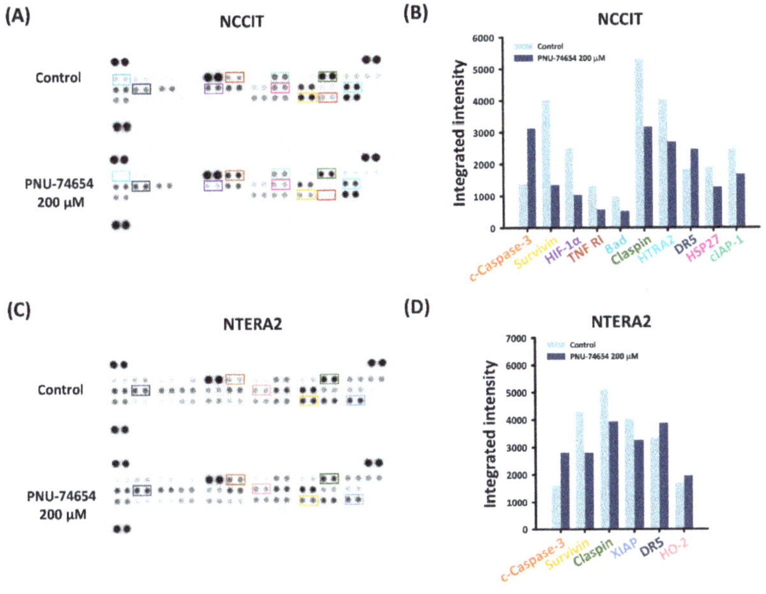

Figure 3. Mechanisms underlying PNU-74654-induced apoptosis in NCCIT and NTERA2 cells. (**A**,**C**) Apoptosis array testing expression of apoptosis-related proteins in TC cells after PNU-74654 treatment. (**B**,**D**) Quantitative analysis of proteins with apparent changes in integrated intensities in TC cells after PNU-74654 treatment, with cut-off >25% and <25%.

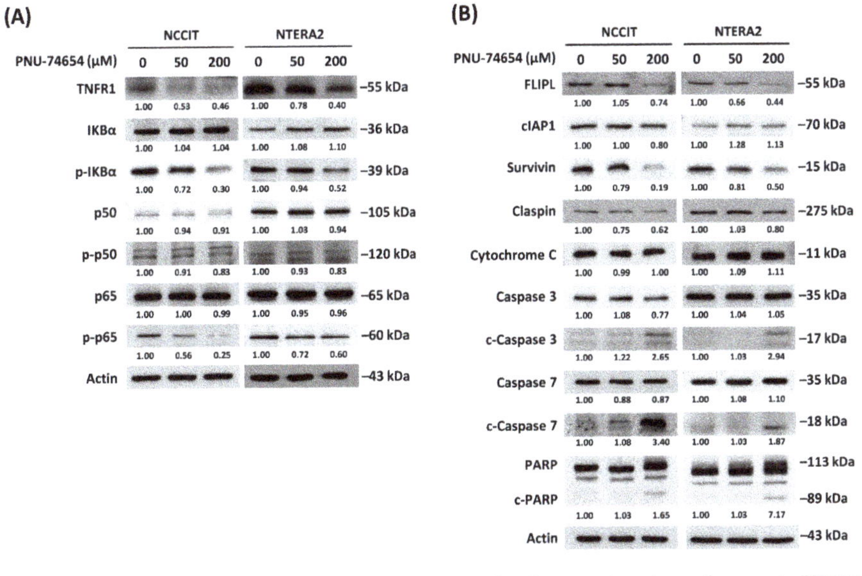

Figure 4. Western blots of proteins related to the execution phase of apoptosis. PNU-74654 treatment suppressed (**A**) TNFR1/IKB alpha/p65 signaling and induced (**B**) signaling pathway in NCCIT and NTERA2 TC cell lines.

4. Discussion

The findings of the current study confirmed that PNU-74654 can exert a cytotoxic effect on TC cells by inducing apoptosis. When treated with PNU-74654, TC cells showed a dose-dependent decrease in viability, as determined by the MTT assay. Our cell cycle analysis revealed a significant dose-dependent increase in the sub-G1 group of TC cells treated with PNU-74654, indicating increased cell death. Our Hoechst 33,342 and Annexin V/PI double staining results further confirmed apoptosis as the major type of cell death in PNU-74654-treated TC cells. Our further analyses implicated Wnt/β-catenin signaling in this response.

Wnt/β-catenin signaling has been implicated in both germ cell tumor (GCT) progression and in treatment resistance to cisplatin-based chemotherapy [34]. Another study focusing on the aberrations of the adenomatous polyposis coli (APC) tumor suppressor in GCT arising during childhood and adolescence discovered methylation of APC and loss of heterozygosity at 5q21-22 in yolk sac tumor and teratomas [35]. To our knowledge, the present study is the first to prove a direct cytotoxic ability of a small molecule inhibitor targeting the β-catenin/TCF4 interaction in this signaling pathway.

The anti-apoptotic effect of nuclear factor kappa-light-chain-enhancer of activated B cells (NF-κB) induced by tumor necrosis factor receptor 1 (TNF R1) has been confirmed for decades [36]. We also examined the mechanisms behind PNU-74654-induced apoptosis in NCCIT and NTERA2 cells by an apoptosis array and quantitative analysis. Upregulations of cleaved-caspase-3 in TC cells, downregulations of claspin and survivin in TC cells and downregulations of TNF R1 and cIAP-1 in NCCIT cells alone were disclosed. Caspase-3 serves as a marker of programmed cell death. It is cleaved and activated following the initiation of apoptosis and its antibodies are now used as strong indicators of cell death induction [37]. By contrast, claspin acts as an adaptor or scaffold protein that facilitates the ATR-dependent activation of Chk1 [38]. Claspin undergoes cleavage and degradation by caspases and the proteasome during apoptosis [39]. Conversely, survivin is a member of the inhibitor of apoptosis (IAP) family and functions as an inhibitor of the activation of caspases and as a negative regulator of programmed cell death [40].

Our Western blot data revealed the expression of proteins involved in the execution phase of apoptosis and in the TNF R1-triggered NF-κB anti-apoptotic pathway. For the execution phase of apoptosis, we noted upregulations of cleaved-caspase-3, cleaved-caspase-7 and cleaved-PARP and downregulations of survivin and claspin. For the inhibition of the TNF R1-triggered NF-κB anti-apoptotic pathway, we observed downregulations of TNF R1, phospho-IκBα, FLIPL, phospho-p50 and phospho-p65. Therefore, we concluded that the execution phase of apoptosis and the inhibition of the TNF R1-triggered NF-κB anti-apoptotic pathway seem to be responsible for the apoptosis induced by PNU-74654 treatment of TC cells.

As already mentioned, the Wnt/β-catenin signaling pathway plays an essential role in the regulation of the transcription of a variety of oncogenes and is involved in the tumorigenesis of several cancers. Numerous medications targeting this signaling pathway have been discovered in the past decade, including the small molecule inhibitor PNU-74654 used in the current study. PNU-74654 interferes with the β-catenin/TCF4 interaction, thereby blocking the ongoing transcription of various oncogenes. The evidence in our study indicates that TC cells treated with PNU-74654 initiated a dose-dependent programmed cell death through the execution phase of apoptosis and inhibition of the TNF R1-triggered NF-κB anti-apoptotic pathway.

The Wnt signal transduction cascade mediates numerous essential cellular functions, such as proliferation, differentiation, apoptosis and cell migration, across many species [5]. Wnts are also key factors for most types of tissue stem cells in mammals [41]. In normal tissue, the β-catenin/TCF4 pathway is maintained in the "Wnt off" state, where β-catenin is phosphorylated by the destruction complex and then undergoes proteasomal degradation. This maintains a low intracellular level of β-catenin and prevents further abnormal cell proliferation. PNU-74654 blocks the interaction between β-catenin and TCF4, thereby preventing the transcription of tumor-related proteins in the affected testis or in metastatic lesions. Since the β-catenin/TCF4 pathway maintains the "Wnt off" state in normal tissue, where the intracellular β-catenin is low, the β-catenin/TCF4 interaction is not turned on. Therefore, the negative impact of PNU-74654 on the contralateral healthy testis or normal tissue can be ignored.

This result is well explained by the process of Wnt/β-catenin signaling. In a breast cancer model, PNU-74654, either alone or in combination with fluorouracil (5-FU), was shown to suppress breast cancer cell growth and to synergistically enhance the antiproliferative activity of gemcitabine [32]. The PNU-74654/5-FU combination increased the percentage of cells in the S-phase and significantly induced apoptosis in PNU-7465-treated breast cancer cells. PNU-74654 also significantly decreased cell proliferation, increased early and late apoptosis and impaired CTNNB1/β-catenin expression in an adrenocortical cancer cell model [31]. In the setting of acute myeloid leukemia (AML), higher levels of β-catenin, Ser675-phospho-β-catenin and GSK-3α were found in AML cells from intermediate- or poor-risk patients and the patients presenting with a higher activity of Wnt/β-catenin showed a shorter progression-free survival [42]. The authors of that study tested the combinatorial treatment between Wnt inhibitors and classic anti-leukemia drugs, both in vitro and in vivo and found that in vitro administration of PNU-74654, niclosamide and LiCl significantly reduced the bone marrow leukemic burden, with a synergistic effect on Ara-C, thereby improving mouse survival. These results, combined with ours, demonstrate the ability of PNU-74654 to specifically target the Wnt pathway, interfere with cell proliferation, induce apoptosis and reduce tumor cell migration. These results indicate that PNU-74654 could represent a promising therapy against several cancers.

Due to the diversity of TC types, one limitation of the current study is our use of NCCIT and NTERA2 cells only, as these two lines cannot represent all cancer cell types associated with TC. In addition, the antibody array we used to select the possible pathway leading to apoptosis has the potential flaw that we might have missed some other interesting pathway. To compensate for this limitation, we performed Western blotting to examine other interesting proteins that might be involved in apoptosis. Another limitation of our work is the lack of experimentation on animals and the lack of evaluation of treatment dosage and toxicity. In the published research works, various doses were used for different types of cancer. Doses ranging from 10 to 200 μM were used in vitro and doses ranging from 0.5 to 30 mg/kg were used in mouse models [31,32,42–44]. More precise evaluations will be needed before considering clinical trials.

5. Conclusions

PNU-74654 can induce the apoptosis of NCCIT and NTERA2 TC cells through mechanisms involving the execution phase of apoptosis and the inhibition of TNFR1/IKB alpha/p65 signaling. Along with previous studies on the antitumor effects of PNU-74654, the present findings may provide preliminary evidence to support further experiments on the use of PNU-74654 as a potential new treatment strategy for patients with TC.

Author Contributions: Conceptualization, W.-J.C., W.-W.S. and S.-L.C.; C.-Y.Y., Y.-Z.L. and Y.-C.C. carried out the experiments; formal analysis, C.-Y.Y. and Y.-C.C.; resources, W.-J.C. and W.-W.S.; supervision, W.-W.S., S.-L.C. and T.-H.L.; writing—original draft, W.-J.C. and Y.-Z.L.; writing—review and editing, W.-W.S., S.-L.C. and T.-H.L. All authors have read and agreed to the published version of the manuscript.

Funding: This work was supported by the grants from Ministry of Science and Technology (MOST 106-2314-B-040-026-MY2 and MOST 108-2314-B-040-014-MY3) of Taiwan.

Data Availability Statement: All data analyzed are included in this article is available upon request.

Conflicts of Interest: The authors declare no conflict of interest.

References

1. Chen, W.J.; Huang, C.Y.; Huang, Y.H.; Wang, S.C.; Hsieh, T.Y.; Chen, S.L.; Sung, W.W.; Lee, T.H. Correlations between Mortality-to-Incidence Ratios and Health Care Disparities in Testicular Cancer. *Int. J. Env. Res. Public Health* **2019**, *17*, 130. [CrossRef]
2. Znaor, A.; Lortet-Tieulent, J.; Jemal, A.; Bray, F. International variations and trends in testicular cancer incidence and mortality. *Eur. Urol.* **2014**, *65*, 1095–1106. [CrossRef]
3. McGlynn, K.A.; Trabert, B. Adolescent and adult risk factors for testicular cancer. *Nat. Rev. Urol.* **2012**, *9*, 339–349. [CrossRef]
4. Baird, D.C.; Meyers, G.J.; Hu, J.S. Testicular Cancer: Diagnosis and Treatment. *Am. Fam. Physician* **2018**, *97*, 261–268.
5. Pai, S.G.; Carneiro, B.A.; Mota, J.M.; Costa, R.; Leite, C.A.; Barroso-Sousa, R.; Kaplan, J.B.; Chae, Y.K.; Giles, F.J. Wnt/beta-catenin pathway: Modulating anticancer immune response. *J. Hematol. Oncol.* **2017**, *10*, 101. [CrossRef]
6. van Ooyen, A.; Kwee, V.; Nusse, R. The nucleotide sequence of the human int-1 mammary oncogene; evolutionary conservation of coding and non-coding sequences. *EMBO J.* **1985**, *4*, 2905–2909. [CrossRef]
7. Niehrs, C. The complex world of WNT receptor signalling. *Nat. Rev. Mol. Cell Biol.* **2012**, *13*, 767–779. [CrossRef]
8. Kemler, R. From cadherins to catenins: Cytoplasmic protein interactions and regulation of cell adhesion. *Trends Genet.* **1993**, *9*, 317–321. [CrossRef]
9. Stadeli, R.; Hoffmans, R.; Basler, K. Transcription under the control of nuclear Arm/beta-catenin. *Curr. Biol.* **2006**, *16*, R378–R385. [CrossRef]
10. Molenaar, M.; van de Wetering, M.; Oosterwegel, M.; Peterson-Maduro, J.; Godsave, S.; Korinek, V.; Roose, J.; Destree, O.; Clevers, H. XTcf-3 transcription factor mediates beta-catenin-induced axis formation in Xenopus embryos. *Cell* **1996**, *86*, 391–399. [CrossRef]
11. Huber, O.; Korn, R.; McLaughlin, J.; Ohsugi, M.; Herrmann, B.G.; Kemler, R. Nuclear localization of beta-catenin by interaction with transcription factor LEF-1. *Mech. Dev.* **1996**, *59*, 3–10. [CrossRef]
12. He, T.C.; Sparks, A.B.; Rago, C.; Hermeking, H.; Zawel, L.; da Costa, L.T.; Morin, P.J.; Vogelstein, B.; Kinzler, K.W. Identification of c-MYC as a target of the APC pathway. *Science* **1998**, *281*, 1509–1512. [CrossRef]
13. Saegusa, M.; Hashimura, M.; Kuwata, T.; Hamano, M.; Okayasu, I. Beta-catenin simultaneously induces activation of the p53-p21WAF1 pathway and overexpression of cyclin D1 during squamous differentiation of endometrial carcinoma cells. *Am. J. Pathol.* **2004**, *164*, 1739–1749. [CrossRef]
14. Lapham, A.; Adams, J.E.; Paterson, A.; Lee, M.; Brimmell, M.; Packham, G. The Bcl-w promoter is activated by beta-catenin/TCF4 in human colorectal carcinoma cells. *Gene* **2009**, *432*, 112–117. [CrossRef]
15. Rao, T.P.; Kuhl, M. An updated overview on Wnt signaling pathways: A prelude for more. *Circ. Res.* **2010**, *106*, 1798–1806. [CrossRef]
16. MacDonald, B.T.; Tamai, K.; He, X. Wnt/beta-catenin signaling: Components, mechanisms, and diseases. *Dev. Cell* **2009**, *17*, 9–26. [CrossRef]
17. de Sousa, E.M.F.; Vermeulen, A.L. Wnt Signaling in Cancer Stem Cell Biology. *Cancers* **2016**, *8*, 60. [CrossRef]
18. Fearon, E.R. Molecular genetics of colorectal cancer. *Ann. Rev. Pathol.* **2011**, *6*, 479–507. [CrossRef]
19. Seshagiri, S.; Stawiski, E.W.; Durinck, S.; Modrusan, Z.; Storm, E.E.; Conboy, C.B.; Chaudhuri, S.; Guan, Y.; Janakiraman, V.; Jaiswal, B.S.; et al. Recurrent R-spondin fusions in colon cancer. *Nature* **2012**, *488*, 660–664. [CrossRef]
20. Storm, E.E.; Durinck, S.; de Sousa e Melo, F.; Tremayne, J.; Kljavin, N.; Tan, C.; Ye, X.; Chiu, C.; Pham, T.; Hongo, J.A.; et al. Targeting PTPRK-RSPO3 colon tumours promotes differentiation and loss of stem-cell function. *Nature* **2016**, *529*, 97–100. [CrossRef]
21. Satoh, S.; Daigo, Y.; Furukawa, Y.; Kato, T.; Miwa, N.; Nishiwaki, T.; Kawasoe, T.; Ishiguro, H.; Fujita, M.; Tokino, T.; et al. AXIN1 mutations in hepatocellular carcinomas, and growth suppression in cancer cells by virus-mediated transfer of AXIN1. *Nat. Genet.* **2000**, *24*, 245–250. [CrossRef]
22. Bafico, A.; Liu, G.; Goldin, L.; Harris, V.; Aaronson, S.A. An autocrine mechanism for constitutive Wnt pathway activation in human cancer cells. *Cancer Cell* **2004**, *6*, 497–506. [CrossRef]
23. He, B.; You, L.; Uematsu, K.; Xu, Z.; Lee, A.Y.; Matsangou, M.; McCormick, F.; Jablons, D.M. A monoclonal antibody against Wnt-1 induces apoptosis in human cancer cells. *Neoplasia* **2004**, *6*, 7–14. [CrossRef]
24. You, L.; He, B.; Xu, Z.; Uematsu, K.; Mazieres, J.; Mikami, I.; Reguart, N.; Moody, T.W.; Kitajewski, J.; McCormick, F.; et al. Inhibition of Wnt-2-mediated signaling induces programmed cell death in non-small-cell lung cancer cells. *Oncogene* **2004**, *23*, 6170–6174. [CrossRef]
25. Lavergne, E.; Hendaoui, I.; Coulouarn, C.; Ribault, C.; Leseur, J.; Eliat, P.A.; Mebarki, S.; Corlu, A.; Clément, B.; Musso, O. Blocking Wnt signaling by SFRP-like molecules inhibits in vivo cell proliferation and tumor growth in cells carrying active beta-catenin. *Oncogene* **2011**, *30*, 423–433. [CrossRef]
26. Takahashi-Yanaga, F.; Kahn, M. Targeting Wnt signaling: Can we safely eradicate cancer stem cells? *Clin Cancer Res* **2010**, *16*, 3153–3162. [CrossRef]
27. Trosset, J.Y. Inhibition of protein-protein interactions: The discovery of druglike beta-catenin inhibitors by combining virtual and biophysical screening. *Proteins* **2006**, *64*, 60–67. [CrossRef]
28. Lévy, L.; Neuveut, C.; Renard, C.A.; Charneau, P.; Branchereau, S.; Gauthier, F.; Van Nhieu, J.T.; Cherqui, D.; Petit-Bertron, A.F.; Mathieu, D.; et al. Transcriptional activation of interleukin-8 by beta-catenin-Tcf4. *J. Biol. Chem.* **2002**, *277*, 42386–42393. [CrossRef]

29. Yamada, T.; Takaoka, A.S.; Naishiro, Y.; Hayashi, R.; Maruyama, K.; Maesawa, C.; Ochiai, A.; Hirohashi, S. Transactivation of the multidrug resistance 1 gene by T-cell factor 4/beta-catenin complex in early colorectal carcinogenesis. *Cancer Res.* **2000**, *60*, 4761–4766.
30. Sanchez-Tillo, E.; de Barrios, O.; Siles, L.; Cuatrecasas, M.; Castells, A.; Postigo, A. beta-catenin/TCF4 complex induces the epithelial-to-mesenchymal transition (EMT)-activator ZEB1 to regulate tumor invasiveness. *Proc. Natl. Acad. Sci. USA* **2011**, *108*, 19204–19209. [CrossRef]
31. Leal, L.F.; Bueno, A.C.; Gomes, D.C.; Abduch, R.; de Castro, M.; Antonini, S.R. Inhibition of the Tcf/beta-catenin complex increases apoptosis and impairs adrenocortical tumor cell proliferation and adrenal steroidogenesis. *Oncotarget* **2015**, *6*, 43016–43032. [CrossRef]
32. Rahmani, F.; Amerizadeh, F.; Hassanian, S.M.; Hashemzehi, M.; Nasiri, S.N.; Fiuji, H.; Ferns, G.A.; Khazaei, M.; Avan, A. PNU-74654 enhances the antiproliferative effects of 5-FU in breast cancer and antagonizes thrombin-induced cell growth via the Wnt pathway. *J. Cell Physiol.* **2019**, *234*, 14123–14132. [CrossRef]
33. Wang, S.C.; Chang, Y.C.; Wu, M.Y.; Yu, C.Y.; Chen, S.L.; Sung, W.W. Intravesical Instillation of Azacitidine Suppresses Tumor Formation through TNF-R1 and TRAIL-R2 Signaling in Genotoxic Carcinogen-Induced Bladder Cancer. *Cancers* **2021**, *13*, 3933. [CrossRef]
34. Bagrodia, A.; Lee, B.H.; Lee, W.; Cha, E.K.; Sfakianos, J.P.; Iyer, G.; Pietzak, E.J.; Gao, S.P.; Zabor, E.C.; Ostrovnaya, I.; et al. Genetic Determinants of Cisplatin Resistance in Patients with Advanced Germ Cell Tumors. *J. Clin. Oncol.* **2016**, *34*, 4000–4007. [CrossRef]
35. Okpanyi, V.; Schneider, D.T.; Zahn, S.; Sievers, S.; Calaminus, G.; Nicholson, J.C.; Palmer, R.D.; Leuschner, I.; Borkhardt, A.; Schönberger, S. Analysis of the adenomatous polyposis coli (APC) gene in childhood and adolescent germ cell tumors. *Pediatr. Blood Cancer* **2011**, *56*, 384–391. [CrossRef]
36. Baichwal, V.R.; Baeuerle, P.A. Activate NF-kappa B or die? *Curr. Biol.* **1997**, *7*, R94–R96. [CrossRef]
37. Strasser, A.; Cory, S.; Adams, J.M. Deciphering the rules of programmed cell death to improve therapy of cancer and other diseases. *EMBO J.* **2011**, *30*, 3667–3683. [CrossRef]
38. Smits, V.A.J.; Cabrera, E.; Freire, R.; Gillespie, D.A. Claspin—Checkpoint adaptor and DNA replication factor. *FEBS J.* **2019**, *286*, 441–455. [CrossRef]
39. Semple, J.I.; Smits, V.A.; Fernaud, J.R.; Mamely, I.; Freire, R. Cleavage and degradation of Claspin during apoptosis by caspases and the proteasome. *Cell Death Differ.* **2007**, *14*, 1433–1442. [CrossRef]
40. Sah, N.K.; Khan, Z.; Khan, G.J.; Bisen, P.S. Structural, functional and therapeutic biology of survivin. *Cancer Lett.* **2006**, *244*, 164–171. [CrossRef]
41. Nusse, R.; Clevers, H. Wnt/β-Catenin Signaling, Disease, and Emerging Therapeutic Modalities. *Cell* **2017**, *169*, 985–999. [CrossRef]
42. Takam Kamga, P.; Dal Collo, G.; Cassaro, A.; Bazzoni, R.; Delfino, P.; Adamo, A.; Bonato, A.; Carbone, C.; Tanasi, I.; Bonifacio, M.; et al. Small Molecule Inhibitors of Microenvironmental Wnt/beta-Catenin Signaling Enhance the Chemosensitivity of Acute Myeloid Leukemia. *Cancers* **2020**, *12*, 2696. [CrossRef]
43. Zhang, F.; Li, P.; Liu, S.; Yang, M.; Zeng, S.; Deng, J.; Chen, D.; Yi, Y.; Liu, H.J.O. β-Catenin-CCL2 feedback loop mediates crosstalk between cancer cells and macrophages that regulates breast cancer stem cells. *Oncogene* **2021**, *40*, 5854–5865. [CrossRef]
44. Maria, A.G.; Silva Borges, K.; Lira, R.; Hassib Thomé, C.; Berthon, A.; Drougat, L.; Kiseljak-Vassiliades, K.; Wierman, M.E.; Faucz, F.R.; Faça, V.M.; et al. Inhibition of Aurora kinase A activity enhances the antitumor response of beta-catenin blockade in human adrenocortical cancer cells. *Mol. Cell. Endocrinol.* **2021**, *528*, 111243. [CrossRef]

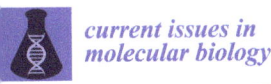

current issues in molecular biology

MDPI

Article

The Search for Cancer Biomarkers: Assessing the Distribution of INDEL Markers in Different Genetic Ancestries

Roberta B. Andrade [1,†], Giovanna C. Cavalcante [2,†], Marcos A. T. Amador [2], Fabiano Cordeiro Moreira [1], André S. Khayat [1], Paulo P. Assumpção [1], Ândrea Ribeiro-dos-Santos [1,2], Ney P. C. Santos [1] and Sidney Santos [1,2,*]

1 Center of Oncology Research, Graduate Program in Oncology and Medical Sciences, Federal University of Pará, Belém 66073-005, Brazil; robertaborgesandrade@hotmail.com (R.B.A.); fabiano.ufpa@gmail.com (F.C.M.); khayatas@gmail.com (A.S.K.); assumpcaopp@gmail.com (P.P.A.); akelyufpa@gmail.com (Â.R.-d.-S.); npcsantos@yahoo.com.br (N.P.C.S.)

2 Laboratory of Human and Medical Genetics, Graduate Program in Genetics and Molecular Biology, Federal University of Pará, Belém 66075-110, Brazil; giovannaccavalcante@gmail.com (G.C.C.); marcosmata52@yahoo.com.br (M.A.T.A.)

* Correspondence: sidneysantosufpa@gmail.com

† These authors contributed equally to this work.

Abstract: Cancer is a multifactorial group of diseases, being highly incident and one of the leading causes of death worldwide. In Brazil, there is a great variation in cancer incidence and impact among the different geographic regions, partly due to the genetic heterogeneity of the population in this country, composed mainly by European (EUR), Native American (NAM), African (AFR), and Asian (ASN) ancestries. Among different populations, genetic markers commonly present diverse allelic frequencies, but in admixed populations, such as the Brazilian population, data is still limited, which is an issue that might influence cancer incidence. Therefore, we analyzed the allelic and genotypic distribution of 12 INDEL polymorphisms of interest in populations from the five Brazilian geographic regions and in populations representing EUR, NAM, AFR, and ASN, as well as tissue expression in silico. Genotypes were obtained by multiplex PCR and the statistical analyses were done using R, while data of tissue expression for each marker was extracted from GTEx portal. We highlight that all analyzed markers presented statistical differences in at least one of the population comparisons, and that we found 39 tissues to be differentially expressed depending on the genotype. Here, we point out the differences in genotype distribution and gene expression of potential biomarkers for risk of cancer development and we reinforce the importance of this type of study in populations with different genetic backgrounds.

Keywords: INDEL; cancer; genetic ancestry; Amazon; biomarkers

Citation: Andrade, R.B.; Cavalcante, G.C.; Amador, M.A.T.; Moreira, F.C.; Khayat, A.S.; Assumpção, P.P.; Ribeiro-dos-Santos, Â.; Santos, N.P.C.; Santos, S. The Search for Cancer Biomarkers: Assessing the Distribution of INDEL Markers in Different Genetic Ancestries. *Curr. Issues Mol. Biol.* **2022**, *44*, 2275–2286. https://doi.org/10.3390/cimb44050154

Academic Editor: Dumitru A. Iacobas

Received: 22 April 2022
Accepted: 13 May 2022
Published: 19 May 2022

1. Introduction

Cancer is one of the leading causes of death worldwide [1], being considered a group of complex diseases that involve environmental, epigenetic, and genetic factors [2,3]. It is estimated that, in 2018, around 18 million new cases of cancer occurred in the world [1]. In Brazil, the National Cancer Institute (INCA) estimates that, for each year from 2020 to 2022, there were 625 thousand new cases, although there is a great variation in magnitude and in the cancer types among the different geographic regions of this country [4]. This occurs partly because Brazil has one of the most genetically heterogeneous populations in the world, composed mainly by Native American, European, and African contributions [5]. In addition, the biggest Japanese community outside Japan is in Brazil, estimated to be around 1.5 million people [6], which allows a certain degree of admixture between this population and the Brazilian population, mainly within the regions where this community is concentrated, North and Southeast of Brazil.

In the global literature, we may find several studies involving genetic markers related to cancer, mostly in case-control association studies, in which these are used to predict risk of development and/or prognosis of a certain type of cancer in different populations [7,8]. It is notable that, among different ethnic populations (also called continental populations), genetic markers commonly present diverse allelic frequencies [9]. However, in admixed populations, such as the Brazilian population, data on the distribution of this kind of markers are still limited.

In this work, we describe the allelic and genotypic distribution of 12 Insertion/Deletion (INDEL) polymorphisms, located in genes involved in important metabolic pathways associated with carcinogenesis, in populations from the five Brazilian geographic regions and in populations representing Europeans, Africans, Native Americans, and Asians. These genes and polymorphisms have been studied and associated with various types of cancer in different populations, such as bladder cancer [10], oral cancer [11], hepatocellular carcinoma [12], breast cancer [13–15], chronic lymphoblastic leukemia [16], colorectal cancer [17–21], thyroid cancer [22] and gastric cancer [23,24]. Thus, these markers were chosen based on the importance of each gene and their potential as an influence in tumor development.

The investigated markers were divided in three groups according to gene function: genomic stability and cell death (rs3834129, rs3730485, rs17878362, rs151264360, and rs3213239, respectively in *CASP8, MDM2, TP53, TYMS* and *XRCC1* genes); biometabolism and cell energy (rs8175347, rs28892005 and a 96 pb-insertion, respectively in *UGT1A1, CYP19A1* and *CYP2E1* genes); and immune response and inflammatory processes (rs3783553, rs79071878, rs28362491 and rs11267092, respectively in *IL1A, IL4, NFKB1* and *PAR1* genes).

2. Materials and Methods

This study included a population of 1411 non-related and cancer-free adult individuals, recruited in ten Brazilian states, in the years of 2009 and 2010, being 480 individuals from Pará (n = 360), Amazonas (n = 60) and Rondônia (n = 60) representing the North region; 370 individuals from Ceará (n = 135), Rio Grande do Norte (n = 175), Maranhão (n = 8) and Pernambuco (n = 52) representing the Northeast region; 186 individuals from Goiás (n = 101), Mato Grosso do Sul (n = 49) and Distrito Federal (n = 36) representing the Midwest region; 184 individuals from São Paulo representing the Southeast region; and 191 individuals from Rio Grande do Sul representing the South region. More details on the sampling approach may be found in previous studies [25,26].

In addition, we investigated a sample of 896 individuals representative of the main ethnic groups that contributed to the Brazilian population: 222 Native Americans (NAM) from nine tribes of the Brazilian Amazon (Tiriyó, Waiãpi, Zoé, Urubu-Kaapor, Awa-Guajá, Parakanã, Wai Wai, Gavião, Zoró) [27]; 211 Africans (AFR) from five different countries (Angola, Mozambique, Congo Republic, Cameroon, Ivory Coast) [28]; 270 Europeans (EUR) from two different countries (Portugal and Spain) [25,29]; and 193 Asians (ASN) from Japan [30]. By using a panel of ancestry informative markers (AIM), we have previously estimated the genomic ancestry of each group [30]. Informed consent for DNA analysis was obtained from all participants. Project approval was given by the Ethics Committee of Instituto de Ciências da Saúde, Universidade Federal do Pará.

2.1. DNA Extraction and Quantification

Samples of peripheral blood were collected from all individuals of the study and the DNA extraction was performed accordingly [31]. DNA quantification was performed with the NanoDrop 1000 spectrophotometer (Thermo Fisher Scientific, Wilmington, DE, USA).

2.2. Genotyping of Investigated Polymorphisms

Polymorphisms were genotyped by a single multiplex reaction with Master Mix QIAGEN® Multiplex PCR kit (Qiagen, Hilden, Germany) and the primers are described in Table 1. PCR preparation protocol was done as described by Cavalcante et al. [32]. All polymorphisms are functional and correspond to INDEL of DNA fragments.

Multiplex PCR products were separated and analyzed by capillary electrophoresis on the ABI 3130 Genetic Analyzer instrument, using GS-500 LIZ as a pattern of molecular weight, G5 virtual filter matrix and POP7 (instrument and reagents by Thermo Fisher Scientific). Then, samples were analyzed with GeneMapper®3.7 software (also by Thermo Fisher Scientific).

Table 1. Technical characterization of the investigated polymorphisms.

Gene	ID	Size (bp)	Primers	Amplicon (bp)
CASP8	rs3834129	6	F-5′CTCTTCAATGCTTCCTTGAGGT3′ R-5′CTGCATGCCAGGAGCTAAGTAT3′	249–255
CYP2E1	-	96	F-5′TGTCCCAATACAGTCACCTCTTT3′ R-5′GGCTTTTATTTGTTTTGCATCTG3′	397–493
CYP19A1	rs28892005	3	F-5′TGCATGAGAAAGGCATCATATT3′ R-5′AAAAGGCACATTCATAGACAAAAA3′	122–125
IL1A	rs3783553	4	F-5′TGGTCCAAGTTGTGCTTATCC3′ R-5′ACAGTGGTCTCATGGTTGTCA3′	230–234
IL4	rs79071878	70 (1–3 repeats)	F-5′AGGGTCAGTCTGGCTACTGTGT3′ R-5′CAAATCTGTTCACCTCAACTGC3′	147/217/287
MDM2	rs3730485	40	F-5′GGAAGTTTCCTTTCTGGTAGGC3′ R-5′TTTGATGCGGTCTCATAAATTG3′	192–232
NFKB1	rs28362491	4	F-5′TATGGACCGCATGACTCTATCA3′ R-5′GGCTCTGGCATCCTAGCAG3′	366–370
PAR1	rs11267092	13	F-5′AAAACTGAACTTTGCCGGTGT3′ R-5′GGGCCTAGAAGTCCAAATGAG	265–277
TP53	rs17878362	16	F-5′GGGACTGACTTTCTGCTCTTGT3′ R-5′GGGACTGTAGATGGGTGAAAAG3′	135–141
TYMS	rs151264360	6	F-5′ATCCAAACCAGAATACAGCACA3′ R-5′CTCAAATCTGAGGGAGCTGAGT3′	148–164
UGT1A1	rs8175347	2 (5–8 repeats)	F-5′CTCTGAAAGTGAACTCCCTGCT3′ R-5′AGAGGTTCGCCCTCTCCTAT3′	133/135/137/139
XRCC1	rs3213239	4	F-5′GAACCAGAATCCAAAAGTGACC3′ R-5′AGGGGAAGAGAGAGAAGGAGAG3′	243–247

2.3. Data Analyses

Allelic and genotypic frequencies were obtained by direct counting. Hardy-Weinberg Equilibrium (HWE) deviations were tested in Arlequin 3.1 software [33] and corrected by Bonferroni method. Differences in genotypic frequencies among Brazilian regions and parental populations were measured by chi-squared test (χ^2 test, df = 2). FDR (False Discovery Rate) method was used to correct multiple analyses. All statistical analyses were performed in the statistical package R [34]. *p*-Value was considered significant if equal or lower than 0.05. In addition, to infer possible influences on cancer development, we assessed the Genotype-Tissue Expression (GTEx) Portal (https://gtexportal.org/home/, accessed on 1 May 2022) [35] to obtain the expression of each variant in different tissues.

3. Results

The observed allele and genotype frequencies for the 12 markers investigated in the Brazilian population and the continental populations (AFR, NAM, EUR and ASN) are shown in Tables S1–S3; and the distribution of the genotypes is plotted in Figure 1.

When assessing HWE with correction for multiple testing for all markers in the different populations, we did not find any deviation from HWE in the admixed populations from Brazil. However, the markers in *CASP8*, *TP53* and *XRCC1* genes in Amerindian, *UGT1A1* gene in African, *NFKB1* gene in European, as well as *MDM2* and *IL4* genes in

Asian populations, presented HWE deviation, indicating the distribution of these markers in such populations is not normalized according to HWE principles.

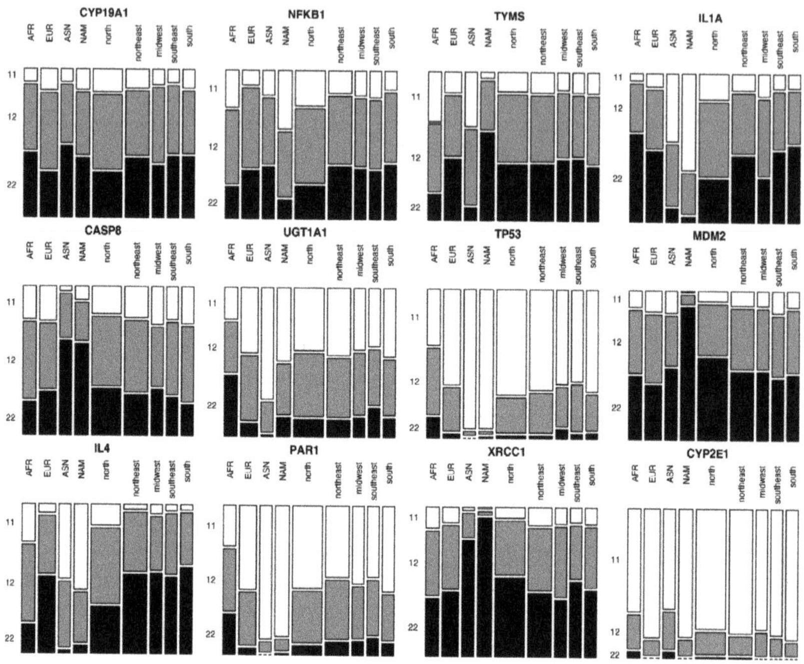

Figure 1. Genotype distribution of the markers across the studied populations. Genotype distribution of each marker in all analyzed populations: African, European, Asian, Native American and the five regions of Brazil (North, Northeast, Midwest, Southeast and South). 11, deletion/deletion; 12, deletion/insertion; 22, insertion/insertion.

We then compared the genotypic distribution of the 12 markers among continental populations and the following results should be highlighted. Regarding biometabolism and cell energy markers (in *UGT1A1*, *CYP19A1* and *CYP2E1* genes), *CYP19A1* e *UGT1A1* did not present differences among populations in most comparisons, only in EUR vs. NAM, and *CYP2E1* also did not differ in the comparisons, except for AFR vs. ASN. As for genomic stability and cell death markers (*TYMS*, *XRCC1*, *CASP8*, *MDM2* and *TP53*), *XRCC1* and *CASP8* were significantly different in all populations, except for AFR vs. EUR; *TYMS* and *TP53* only presented statistical difference in AFR vs. ASN and NAM vs. ASN comparisons, respectively; and marker *MDM2* presented differences in the comparisons between NAM and all the other groups, but not in the other comparisons. Concerning markers of immune response and inflammatory processes (*IL1A*, *IL4*, *NKFB1* and *PAR1*), we observed significant differences for both *IL4* and *PAR1* in all comparisons; *IL1A* was different in all comparisons, but not in AFR vs. EUR; and *NFKB1* was only different in AFR vs. EUR and NAM vs. EUR. Due to the table size, *p*-values for these comparative analyses are shown in Table S4.

Moreover, we measured and analyzed δ (delta) values or mean frequencies among continental populations (Table 2). Among the investigated markers, the difference of δ values between NAM and AFR was 32%, between NAM and EUR was 23% and between EUR and AFR was 19%. In the comparisons involving ASN, an average delta value of 14% was estimated between ASN and NAM; 21% between ANS and EUR; and 26% between ASN and AFR.

Table 2. Frequency of the shortest alleles of the 12 polymorphisms in the AFR, EUR, NAM and ASN populations, and the mean difference in frequency among populations (δ values).

	Frequencies				δ					
Markers	**EUR**	**ASN**	**AFR**	**NAM**	**EUR/ASN**	**AFR/ASN**	**NAM/ASN**	**EUR/AFR**	**EUR/NAM**	**AFR/NAM**
IL1A	0.3	0.69	0.22	0.82	0.39	0.47	0.13	0.08	0.52	0.6
IL4	0.26	0.69	0.58	0.77	0.43	0.11	0.08	0.32	0.51	0.19
NFKB1	0.38	0.10	0.52	0.64	0.02	0.12	0.24	0.14	0.26	0.12
PAR1	0.77	0.96	0.5	0.95	0.19	0.46	0.01	0.27	0.18	0.45
UGT1A1	0.68	0.89	0.45	0.70	0.21	0.44	0.19	0.23	0.02	0.25
CYP19A1	0.42	0.3	0.32	0.37	0.12	0.02	0.07	0.1	0.06	0.05
CYP2E1	0.94	0.8	0.83	0.95	0.14	0.03	0.15	0.11	0.01	0.12
CASP8	0.47	0.18	0.5	0.23	0.29	0.32	0.05	0.03	0.24	0.27
TYMS	0.36	0.65	0.58	0.21	0.29	0.07	0.44	0.22	0.15	0.37
XRCC1	0.33	0.1	0.83	0.02	0.23	0.73	0.08	0.5	0.31	0.81
MDM2	0.38	0.33	0.33	0.04	0.05	0.00	0.29	0.05	0.34	0.29
TP53	0.82	0.99	0.62	0.98	0.17	0.37	0.01	0.2	0.16	0.36
Average					0.21	0.26	0.14	0.19	0.23	0.32

In the comparison of geographic regions, the marker in IL4 was significantly different between North and the other populations of Brazil. Additionally, distribution of the marker in IL1A was significantly different between North and the regions South, Southeast and Northeast, and between Midwest and South. As for the polymorphism in NFKB1, it showed statistically significant difference between North and regions South and Southeast, but it was similar in all other comparisons.

Furthermore, in the GTEx analysis—performed to infer possible influences on cancer development, 39 tissues were found to be differentially expressed depending on the genotype in six of the 12 variants here investigated (Table 3). The variant that was differentially expressed in the highest number of tissues was rs3213239 (XRCC1), 30 tissues; followed by rs3834129 (CASP8), 15 tissues; rs3783553 (IL1A) and rs28362491 (NFKB1), six tissues each; rs151264360 (TYMS), three tissues; and rs11575899 (CYP19A1), two tissues.

Table 3. Differentially expressed variants in diverse tissues from GTEx. NES, Normalized Effect Size.

Gene	Variant ID	*p*-Value	NES	Tissue
CASP8	rs3834129	4.9×10^{-13}	0.28	Cells-Cultured fibroblasts
		1.1×10^{-9}	0.18	Esophagus-Mucosa
		1.5×10^{-7}	−0.27	Pituitary gland
		2.6×10^{-7}	−0.50	Brain-Cerebellum
		6.9×10^{-7}	0.16	Adipose-Visceral (Omentum)
		7.3×10^{-7}	−0.48	Brain-Frontal Cortex (BA9)
		0.0000019	−0.16	Thyroid
		0.0000030	0.18	Skin-Sun Exposed (Lower leg)
		0.0000035	0.22	Breast-Mammary Tissue
		0.000011	0.24	Spleen
		0.000024	−0.38	Brain-Cortex
		0.000055	0.14	Adipose-Subcutaneous
		0.00012	0.17	Skin-Not Sun Exposed (Suprapubic)
		0.00012	0.12	Muscle-Skeletal
		0.00023	−0.12	Nerve-Tibial
CYP19A1	rs28892005	1.1×10^{-9}	0.20	Skin-Sun Exposed (Lower leg)
		0.00020	0.14	Skin-Not Sun Exposed (Suprapubic)

Table 3. *Cont.*

Gene	Variant ID	*p*-Value	NES	Tissue
IL1A	rs3783553	3.9×10^{-14}	0.35	Skin-Not Sun Exposed (Suprapubic)
		2.6×10^{-11}	0.30	Skin-Sun Exposed (Lower leg)
		2.0×10^{-7}	0.37	Spleen
		0.0000018	0.19	Testis
		0.000028	0.16	Esophagus-Mucosa
		0.00016	0.19	Thyroid
NFKB1	rs28362491	9.9×10^{-14}	0.15	Muscle-Skeletal
		6.0×10^{-8}	0.12	Cells-Cultured fibroblasts
		2.4×10^{-7}	82	Whole Blood
		0.0000018	−0.12	Testis
		0.000099	−0.12	Heart-Atrial Appendage
		0.00021	−77	Skin-Not Sun Exposed (Suprapubic)
TYMS	rs151264360	6.6×10^{-34}	0.66	Esophagus-Muscularis
		1.7×10^{-22}	−0.28	Testis
		8.1×10^{-17}	0.53	Esophagus-Gastroesophageal Junction
XRCC1	rs3213239	4.4×10^{-45}	−0.43	Thyroid
		5.4×10^{-42}	−0.53	Pancreas
		2.0×10^{-26}	−0.49	Testis
		2.0×10^{-24}	−0.58	Adrenal Gland
		3.8×10^{-19}	−0.23	Muscle-Skeletal
		6.3×10^{-15}	−0.32	Pituitary
		4.1×10^{-13}	0.20	Nerve-Tibial
		4.2×10^{-13}	−0.46	Brain-Hypothalamus
		8.7×10^{-11}	−0.23	Colon-Sigmoid
		1.5×10^{-10}	−0.54	Brain-Anterior cingulate cortex (BA24)
		4.8×10^{-10}	−0.38	Brain-Caudate (basal ganglia)
		6.8×10^{-10}	−0.23	Stomach
		1.1×10^{-9}	−0.34	Ovary
		1.7×10^{-9}	−0.24	Heart-Left Ventricle
		1.2×10^{-8}	−0.41	Brain-Hippocampus
		2.0×10^{-8}	−0.33	Brain-Nucleus accumbens (basal ganglia)
		3.7×10^{-8}	−0.28	Liver
		3.8×10^{-8}	−0.14	Colon-Transverse
		1.7×10^{-7}	−0.29	Brain-Frontal Cortex (BA9)
		1.9×10^{-7}	−0.29	Brain-Cortex
		2.0×10^{-7}	−0.48	Brain-Amygdala
		4.5×10^{-7}	−0.15	Esophagus-Mucosa
		0.0000022	−0.40	Minor Salivary Gland
		0.0000028	−0.55	Brain-Substantia nigra
		0.0000045	−0.30	Brain-Putamen (basal ganglia)
		0.000011	−84	Whole Blood
		0.000018	−78	Cells-Cultured fibroblasts
		0.000031	−0.24	Prostate
		0.000056	−0.15	Heart-Atrial Appendage
		0.00012	−94	Artery-Tibial

Of all the found tissues, 12 seem to be regulated by more than one of the studied variants: cells—cultured fibroblasts (variants in *CASP8*, *NFKB1*, *XRCC1*), esophagus—mucosa (*CASP8*, *IL1A*, *XRCC1*), heart—atrial appendage (*NFKB1*, *XRCC1*), muscle—skeletal (*CASP8*, *NFKB1*, *XRCC1*), nerve—tibial (*CASP8*, *XRCC1*), pituitary gland (*CASP8*, *XRCC1*), skin—not sun exposed (NSE, suprapubic; *CASP8*, *CYP19A1*, *IL1A*, *NFKB1*), skin—sun exposed (SE, lower leg; *CASP8*, *CYP19A1*, *IL1A*), spleen (*CASP8*, *IL1A*), testis (*IL1A*, *NFKB1*, *TYMS*, *XRCC1*), thyroid (*CASP8*, *IL1A*, *XRCC1*) and whole blood (*NFKB1*, *XRCC1*).

4. Discussion

This study aimed to investigate and describe the frequencies of markers of interest (located in genes involved in important metabolic pathways associated with carcinogenesis) in populations from the five geographic populations of Brazil and in populations representing European, African, Native American, and Asian ancestries. These markers were divided according to gene functions.

In a previous study [5], the description of the group of markers of immune response and inflammatory processes was performed in the same populations investigated here, except for the Asian population. Regarding this group of markers, in addition to the results presented in that paper, it is possible to highlight that ASN population was different from all other continental populations for the investigated markers in *IL1A*, *IL4* and *PAR1*. As for the marker in *NFKB1*, it only differed between AFR and EUR and between NAM and EUR.

In the comparisons of geographic regions, marker *IL4* was significantly different between North and the other Brazilian populations. Besides that, our analysis also showed the distribution of the *IL1A* marker with statistical differences between North and the South, Southeast and Northeast regions, as well as between Midwest and South regions. The polymorphism in *NFKB1* was also significantly different between North and the regions South and Southeast. All other distributions of this group of markers were similar among these regions.

Regarding the investigated variants in genes of biometabolism and cell energy, not many studies can be currently found analyzing their distribution in populations of different genetic ancestries. However, a study by Fritsche et al. [36] compared genotypes of the 96-bp INDEL in *CYP2E1* gene in samples from individuals with African (African-American), European (European-American) and Asian (Taiwanese) genetic backgrounds and observed statistically significant differences between Europeans and Asians and between Europeans and Africans, but none between Asians and Africans, which corroborates our findings here. Concerning variant rs8175347 in *UGT1A1* gene, allele frequency of this variant has been reported as different when compared in groups of European, African, and Asian (including Japanese) ancestries [37]. No studies were found with the rs28892005 variant in *CYP19A1*.

As for variants in the group of genomic stability and cell death, there are some papers discussing their distribution in different populations in the global literature. For instance, a previous study by our research group compared the allele distribution of rs17878362 in *TP53* gene in populations of European, African, and Asian ancestry from 1000 genomes database [9], as well as a population from Northern Brazil, and observed statistical differences in all comparisons, with the exception of the one between Northern Brazil and European populations, which could be expected given the high contribution of European ancestry in this region [38]. However, it is notable that these frequencies significantly vary among different genetic backgrounds.

Similarly, the variant rs3730485 (also known as del1518) in *MDM2* gene has been investigated in different populations, particularly in connection with cancer development. For example, two independent studies involving different types of cancer in Chinese cohorts have reported a frequency of 30% of DEL allele in both groups of controls [12,39]. The study by Gansmo et al. [39] also investigated this variant in other populations, indicating the presence of the same allele in 38% and 42% in the African American and the Norwegian controls, respectively. Here, we found this allele in 33%, 38% and 33% of the African, European, and Asian groups, respectively, which seem to be close to the corresponding frequencies in these previous reports. To the best of our knowledge, this is the first study investigating this variant in NAM populations from the Brazilian Amazon.

Variant rs151264360 in *TYMS* gene has also been widely studied regarding cancer treatment in different regions. In this context, it has been associated with response to chemotherapy for colorectal cancer in a Mexican cohort, highlighting the importance of *TYMS* to cancer treatment in Latin American populations [40]. In that study, DEL allele was present in 33.0% of the participants, which was similar to the frequency of this allele in a study carried in a Slovak population (37.5%) [41]. A study by Summers et al. [42] reported a significant difference in the distribution of rs151264360 between African-Americans (DEL 53.75%) and Europeans (DEL 33.3%). These frequencies are like the ones observed here for African (DEL 58.0%) and European (DEL 36.0%) ancestries, which also showed significant differences.

Likewise, polymorphism rs3834129 in *CASP8* gene has been broadly studied regarding cancer, particularly cancer development. For instance, in a study by Pardini et al. [43], DEL allele of this variant was suggested as a protective effect to colorectal cancer in the multiple populations investigated, mostly from European countries. In these populations, the presence of DEL allele ranged from 45% to 52%. In addition, two independent studies investigating the distribution of this polymorphism in British cohorts in association to different diseases have reported the presence of DEL allele as 50% in the controls [44,45]. Similarly, a study by Chatterjee et al. [46] investigated the association of this marker with HPV infection and cervical cancer in South Africa and showed the presence of DEL allele in around 52% of the controls. Here, we found this allele in 50% and 47% of the African and European groups, respectively, which are also similar frequencies and corroborate these studies.

On the other hand, there are not many studies with the variant rs3213239 (*XRCC1* gene) in the specialized literature. Two studies by our research group have investigated this variant regarding cancer susceptibility in Northern Brazil, reporting association with acute lymphoblastic leukemia (ALL) [47], but not with gastric cancer or colorectal cancer [32]. Curiously, in the study by Carvalho et al. [47], not only the DEL/DEL genotype of this variant was associated with ALL, but also the genetic ancestry: NAM and EUR ancestries were associated with increased and decreased risk of developing ALL, respectively, highlighting the importance of investigating this variant in different populations.

Moreover, in the GTEx analysis, it is notable that the studied variants in *CASP8* and *XRCC1* appeared in most of the tissues showing more than one differentially expressed gene, nine each, of which six presented both markers: (i) cells—cultured fibroblasts, (ii) esophagus—mucosa, (iii) muscle—skeletal, (iv) nerve—tibial, (v) pituitary gland and (vi) thyroid. This suggests that these tissues are likely to be regulated by the variants in such genes, which are related to genomic stability and cell death.

Even though we did not find any works involving both *CASP8* and *XRCC1* and such tissues, there are a few studies in the global literature on the possible association of these genes and the development of different types of cancer, such as lung adenocarcinoma, breast cancer, gallbladder cancer, acute lymphoblastic leukemia, as well as gastric and colorectal cancers [32,47–52].

It is also notable that both skin tissues (SE and NSE) presented differential expression in *CASP8*, *CYP19A1* and *IL1A* and that NSE skin also presented this difference for *NFKB1*. This finding suggests a possible influence of this variant and gene on skin cancer development and reinforces previous studies that have reported the association of INS/INS genotype of this variant in *NFKB1* with an increased risk of developing melanoma in a Swedish and in a Brazilian population [53,54].

In addition to NSE skin, testis tissue also presented four differentially expressed variants (in *IL1A*, *NFKB1*, *TYMS* and *XRCC1* genes), the highest number of variants per tissue in this analysis. No studies were found about these specific variants in testis or these genes in testicular cancer, but the role of *IL1A*, *NFKB1* and *XRCC1* have been reported in Sertoli cells and other essential factors for spermatogenesis [55–60]. Hence, given their importance in testis function, these genes might also be involved in carcinogenesis in this tissue.

In summary, here we thoroughly analyzed the distribution of 12 polymorphisms in diverse populations (groups from European, African, Native American, and Asian populations, as well as groups from the five admixed Brazilian geographical regions) and tissue expression. All analyzed markers presented statistical differences in at least one of the population comparisons, and we found 39 tissues to be differentially expressed depending on the genotype, suggesting these markers might play a role in cancer distribution in different populations. Thus, we recommend future studies with larger cohorts to explore these novel observations, as this was the first study to investigate some of these markers in these populations. Based on our findings, we point out some potential biomarkers for risk of cancer development and we highlight the importance of this type of study in populations with different genetic backgrounds.

Supplementary Materials: The following supporting information can be downloaded at: https://www.mdpi.com/article/10.3390/cimb44050154/s1, Table S1. Allelic and genotypic frequencies of markers in Biometabolism and Cell Energy group. Table S2. Allelic and genotypic frequencies of markers in Genomic Stability and Cell Death group. Table S3. Allelic and genotypic frequencies of markers in Immune Response and Inflammatory Processes group. Table S4. *p*-Values of the comparative analyses of frequencies among all populations for each marker.

Author Contributions: Conceptualization: N.P.C.S. and S.S.; Data curation: R.B.A., G.C.C. and M.A.T.A.; Formal analysis: F.C.M.; Funding acquisition: A.S.K., P.P.A. and Â.R.-d.-S.; Investigation: R.B.A., G.C.C. and M.A.T.A.; Methodology: N.P.C.S. and S.S.; Resources: A.S.K., P.P.A. and Â.R.-d.-S.; Supervision: N.P.C.S. and S.S.; Writing–Original draft preparation: R.B.A. and G.C.C.; Writing–Review & editing: R.B.A. and G.C.C. All authors have read and agreed to the published version of the manuscript.

Funding: This research was funded by CNPq (Conselho Nacional do Desenvolvimento Científico e Tecnológico), CAPES (Coordenação de Aperfeiçoamento de Pessoal de Nível Superior), HYDRO/UFPA/FADESP (Fundação Amparo e Desenvolvimento da Pesquisa) and PROPESP (Pró-Reitoria de Pesquisa da Universidade Federal do Pará. It is part of Rede de Pesquisa em Genômica Populacional Humana (Biocomputacional—Protocol no. 3381/2013/CAPES).

Institutional Review Board Statement: The study was conducted in accordance with the Declaration of Helsinki, and approved by the Ethics Committee of Instituto de Ciências da Saúde, Universidade Federal do Pará.

Informed Consent Statement: Informed consent was obtained from all subjects involved in the study.

Data Availability Statement: The data presented in this study are available within the article or supplementary material.

Acknowledgments: In this section, you can acknowledge any support given which is not covered by the author contribution or funding sections. This may include administrative and technical support, or donations in kind (e.g., materials used for experiments).

Conflicts of Interest: The authors declare no conflict of interest. The funders had no role in the design of the study; in the collection, analyses, or interpretation of data; in the writing of the manuscript, or in the decision to publish the results.

References

1. Globocan Cancer Today. Available online: http://gco.iarc.fr/today/home (accessed on 6 March 2019).
2. Hanahan, D.; Weinberg, R.A. Hallmarks of Cancer: The next Generation. *Cell* **2011**, *144*, 646–674. [CrossRef] [PubMed]
3. Giampazolias, E.; Tait, S.W.G. Mitochondria and the Hallmarks of Cancer. *FEBS J.* **2016**, *283*, 803–814. [CrossRef] [PubMed]
4. INCA. Estimativa 2020: Incidência de Câncer No Brasil | INCA—Instituto Nacional de Câncer. Available online: https://www.inca.gov.br/publicacoes/livros/estimativa-2020-incidencia-de-cancer-no-brasil (accessed on 17 October 2020).
5. Amador, M.A.T.; Cavalcante, G.C.; Santos, N.P.C.; Gusmão, L.; Guerreiro, J.F.; Ribeiro-dos-Santos, Â.; Santos, S. Distribution of Allelic and Genotypic Frequencies of IL1A, IL4, NFKB1 and PAR1 Variants in Native American, African, European and Brazilian Populations. *BMC Res. Notes* **2016**, *9*, 101. [CrossRef] [PubMed]
6. Suarez-Kurtz, G. Pharmacogenetics in the Brazilian Population. *Front. Pharm.* **2010**, *1*, 118. [CrossRef]
7. Yuan, F.; Sun, R.; Li, L.; Jin, B.; Wang, Y.; Liang, Y.; Che, G.; Gao, L.; Zhang, L. A Functional Variant Rs353292 in the Flanking Region of MiR-143/145 Contributes to the Risk of Colorectal Cancer. *Sci. Rep.* **2016**, *6*, 30195. [CrossRef]
8. Wang, C.; Zhao, H.; Zhao, X.; Wan, J.; Wang, D.; Bi, W.; Jiang, X.; Gao, Y. Association between an Insertion/Deletion Polymorphism within 3′UTR of SGSM3 and Risk of Hepatocellular Carcinoma. *Tumour Biol.* **2014**, *35*, 295–301. [CrossRef]
9. 1000 Genomes Project Consortium; Auton, A.; Brooks, L.D.; Durbin, R.M.; Garrison, E.P.; Kang, H.M.; Korbel, J.O.; Marchini, J.L.; McCarthy, S.; McVean, G.A.; et al. A Global Reference for Human Genetic Variation. *Nature* **2015**, *526*, 68–74. [CrossRef]

10. Ahirwar, D.; Kesarwani, P.; Manchanda, P.K.; Mandhani, A.; Mittal, R.D. Anti- and Proinflammatory Cytokine Gene Polymorphism and Genetic Predisposition: Association with Smoking, Tumor Stage and Grade, and Bacillus Calmette-Guérin Immunotherapy in Bladder Cancer. *Cancer Genet. Cytogenet.* **2008**, *184*, 1–8. [CrossRef]
11. Yang, C.-M.; Chen, H.-C.; Hou, Y.-Y.; Lee, M.-C.; Liou, H.-H.; Huang, S.-J.; Yen, L.-M.; Eng, D.-M.; Hsieh, Y.-D.; Ger, L.-P. A High IL-4 Production Diplotype Is Associated with an Increased Risk but Better Prognosis of Oral and Pharyngeal Carcinomas. *Arch. Oral Biol.* **2014**, *59*, 35–46. [CrossRef]
12. Dong, D.; Gao, X.; Zhu, Z.; Yu, Q.; Bian, S.; Gao, Y. A 40-Bp Insertion/Deletion Polymorphism in the Constitutive Promoter of MDM2 Confers Risk for Hepatocellular Carcinoma in a Chinese Population. *Gene* **2012**, *497*, 66–70. [CrossRef]
13. Zhang, Y.J.; Zhong, X.P.; Chen, Y.; Liu, S.R.; Wu, G.; Liu, Y.F. Association between CASP-8 Gene Polymorphisms and Cancer Risk in Some Asian Population Based on a HuGE Review and Meta-Analysis. *Genet. Mol. Res.* **2013**, *12*, 6466–6476. [CrossRef] [PubMed]
14. Ramalhinho, A.C.M.; Fonseca-Moutinho, J.A.; Breitenfeld Granadeiro, L.A.T.G. Positive Association of Polymorphisms in Estrogen Biosynthesis Gene, CYP19A1, and Metabolism, GST, in Breast Cancer Susceptibility. *DNA Cell Biol.* **2012**, *31*, 1100–1106. [CrossRef] [PubMed]
15. Kuhlmann, J.D.; Bankfalvi, A.; Schmid, K.W.; Callies, R.; Kimmig, R.; Wimberger, P.; Siffert, W.; Bachmann, H.S. Prognostic Relevance of Caspase 8 -652 6N InsDel and Asp302His Polymorphisms for Breast Cancer. *BMC Cancer* **2016**, *16*, 618. [CrossRef] [PubMed]
16. Karakosta, M.; Kalotychou, V.; Kostakis, A.; Pantelias, G.; Rombos, I.; Kouraklis, G.; Manola, K.N. UGT1A1*28 Polymorphism in Chronic Lymphocytic Leukemia: The First Investigation of the Polymorphism in Disease Susceptibility and Its Specific Cytogenetic Abnormalities. *Acta Haematol.* **2014**, *132*, 59–67. [CrossRef]
17. Bajro, M.H.; Josifovski, T.; Panovski, M.; Jankulovski, N.; Nestorovska, A.K.; Matevska, N.; Petrusevska, N.; Dimovski, A.J. Promoter Length Polymorphism in UGT1A1 and the Risk of Sporadic Colorectal Cancer. *Cancer Genet.* **2012**, *205*, 163–167. [CrossRef]
18. Jiang, O.; Zhou, R.; Wu, D.; Liu, Y.; Wu, W.; Cheng, N. CYP2E1 Polymorphisms and Colorectal Cancer Risk: A HuGE Systematic Review and Meta-Analysis. *Tumor Biol.* **2013**, *34*, 1215–1224. [CrossRef]
19. Jirásková, A.; Novotný, J.; Novotný, L.; Vodicka, P.; Pardini, B.; Naccarati, A.; Schwertner, H.A.; Hubácek, J.A.; Puncochárová, L.; Šmerhovský, Z.; et al. Association of Serum Bilirubin and Promoter Variations in HMOX1 and UGT1A1 Genes with Sporadic Colorectal Cancer. *Int. J. Cancer* **2012**, *131*, 1549–1555. [CrossRef]
20. Morita, M.; Le Marchand, L.; Kono, S.; Yin, G.; Toyomura, K.; Nagano, J.; Mizoue, T.; Mibu, R.; Tanaka, M.; Kakeji, Y.; et al. Genetic Polymorphisms of CYP2E1 and Risk of Colorectal Cancer: The Fukuoka Colorectal Cancer Study. *Cancer Epidemiol. Biomark. Prev.* **2009**, *18*, 235–241. [CrossRef]
21. Silva, T.D.; Felipe, A.V.; Pimenta, C.A.; Barão, K.; Forones, N.M. CYP2E1 RsaI and 96-Bp Insertion Genetic Polymorphisms Associated with Risk for Colorectal Cancer. *Genet. Mol. Res.* **2012**, *11*, 3138–3145. [CrossRef]
22. Santoro, A.B.; Vargens, D.D.; Barros Filho Mde, C.; Bulzico, D.A.; Kowalski, L.P.; Meirelles, R.M.R.; Paula, D.P.; Neves, R.R.S.; Pessoa, C.N.; Struchine, C.J.; et al. Effect of UGT1A1, UGT1A3, DIO1 and DIO2 Polymorphisms on L-Thyroxine Doses Required for TSH Suppression in Patients with Differentiated Thyroid Cancer. *Br. J. Clin. Pharm.* **2014**, *78*, 1067–1075. [CrossRef]
23. Fujimoto, D.; Hirono, Y.; Goi, T.; Katayama, K.; Matsukawa, S.; Yamaguchi, A. The Activation of Proteinase-Activated Receptor-1 (PAR1) Promotes Gastric Cancer Cell Alteration of Cellular Morphology Related to Cell Motility and Invasion. *Int. J. Oncol.* **2013**, *42*, 565–573. [CrossRef] [PubMed]
24. Qiao, W.; Wang, T.; Zhang, L.; Tang, Q.; Wang, D.; Sun, H. Association Study of Single Nucleotide Polymorphisms in XRCC1 Gene with the Risk of Gastric Cancer in Chinese Population. *Int. J. Biol. Sci.* **2013**, *9*, 753–758. [CrossRef] [PubMed]
25. Santos, N.P.C.; Ribeiro-Rodrigues, E.M.; Ribeiro-dos-Santos, Â.K.C.; Pereira, R.; Gusmão, L.; Amorim, A.; Guerreiro, J.F.; Zago, M.A.; Matte, C.; Hutz, M.H.; et al. Assessing Individual Interethnic Admixture and Population Substructure Using a 48-Insertion-Deletion (INSEL) Ancestry-Informative Marker (AIM) Panel. *Hum. Mutat.* **2010**, *31*, 184–190. [CrossRef] [PubMed]
26. Ramos, B.R.D.A.; Mendes, N.D.; Tanikawa, A.A.; Amador, M.A.T.; dos Santos, N.P.C.; dos Santos, S.E.B.; Castelli, E.C.; Witkin, S.S.; da Silva, M.G. Ancestry Informative Markers and Selected Single Nucleotide Polymorphisms in Immunoregulatory Genes on Preterm Labor and Preterm Premature Rupture of Membranes: A Case Control Study. *BMC Pregnancy Childbirth* **2016**, *16*, 30. [CrossRef] [PubMed]
27. Ribeiro-Rodrigues, E.M.; Palha, T.D.J.B.F.; Bittencourt, E.A.; Ribeiro-Dos-Santos, A.; Santos, S. Extensive Survey of 12 X-STRs Reveals Genetic Heterogeneity among Brazilian Populations. *Int. J. Leg. Med* **2011**, *125*, 445–452. [CrossRef] [PubMed]
28. Silva, W.A.; Bortolini, M.C.; Schneider, M.P.C.; Marrero, A.; Elion, J.; Krishnamoorthy, R.; Zago, M.A. MtDNA Haplogroup Analysis of Black Brazilian and Sub-Saharan Populations: Implications for the Atlantic Slave Trade. *Hum. Biol.* **2006**, *78*, 29–41. [CrossRef] [PubMed]
29. Palha, T.; Gusmão, L.; Ribeiro-Rodrigues, E.; Guerreiro, J.F.; Ribeiro-dos-Santos, Â.; Santos, S. Disclosing the Genetic Structure of Brazil through Analysis of Male Lineages with Highly Discriminating Haplotypes. *PLoS ONE* **2012**, *7*, e40007. [CrossRef]
30. Andrade, R.B.; Amador, M.A.T.; Cavalcante, G.C.; Leitão, L.P.C.; Fernandes, M.R.; Modesto, A.A.C.; Moreira, F.C.; Khayat, A.S.; Assumpção, P.P.; Ribeiro-dos-Santos, Â.; et al. Estimating Asian Contribution to the Brazilian Population: A New Application of a Validated Set of 61 Ancestry Informative Markers. *G3 Genes Genomes Genet.* **2018**, *8*, 3577–3582. [CrossRef]

31. Sambrook, J.; Fritsch, E.F.; Maniatis, T. *Molecular Cloning: A Laboratory Manual*; Cold Spring Harbor Laboratory Press: Long Island, NY, USA, 1989.
32. Cavalcante, G.C.; Amador, M.A.; Ribeiro Dos Santos, A.M.; Carvalho, D.C.; Andrade, R.B.; Pereira, E.E.; Fernandes, M.R.; Costa, D.F.; Santos, N.P.; Assumpção, P.P.; et al. Analysis of 12 Variants in the Development of Gastric and Colorectal Cancers. *World J. Gastroenterol.* **2017**, *23*, 8533–8543. [CrossRef]
33. Excoffier, L.; Lischer, H.E.L. Arlequin Suite Ver 3.5: A New Series of Programs to Perform Population Genetics Analyses under Linux and Windows. *Mol. Ecol. Resour.* **2010**, *10*, 564–567. [CrossRef]
34. R Core Team. *R: A Language and Environment for Statistical Computing*; R Core Team: Vienna, Austria, 2014.
35. GTEx Consortium. The Genotype-Tissue Expression (GTEx) Project. *Nat. Genet.* **2013**, *45*, 580–585. [CrossRef] [PubMed]
36. Fritsche, E.; Pittman, G.S.; Bell, D.A. Localization, Sequence Analysis, and Ethnic Distribution of a 96-Bp Insertion in the Promoter of the Human CYP2E1 Gene. *Mutat. Res.* **2000**, *432*, 1–5. [CrossRef]
37. Shin, H.J.; Kim, J.Y.; Cheong, H.S.; Na, H.S.; Shin, H.D.; Chung, M.W. Functional Study of Haplotypes in UGT1A1 Promoter to Find a Novel Genetic Variant Leading to Reduced Gene Expression. *Drug Monit.* **2015**, *37*, 369–374. [CrossRef] [PubMed]
38. Cavalcante, G.C.; Ribeiro-Dos-Santos, A.M.; Carvalho, D.C.D.; Silva, E.M.D.; Assumpção, P.P.D.; Ribeiro-Dos-Santos, Â.; Santos, S. Investigation of Potentially Deleterious Alleles for Response to Cancer Treatment with 5-Fluorouracil. *Anticancer Res.* **2015**, *7*, 6971–6977.
39. Gansmo, L.B.; Vatten, L.; Romundstad, P.; Hveem, K.; Ryan, B.M.; Harris, C.C.; Knappskog, S.; Lønning, P.E. Associations between the MDM2 Promoter P1 Polymorphism Del1518 (Rs3730485) and Incidence of Cancer of the Breast, Lung, Colon and Prostate. *Oncotarget* **2016**, *7*, 28637–28646. [CrossRef] [PubMed]
40. Castro-Rojas, C.A.; Esparza-Mota, A.R.; Hernandez-Cabrera, F.; Romero-Diaz, V.J.; Gonzalez-Guerrero, J.F.; Maldonado-Garza, H.; Garcia-Gonzalez, I.S.; Buenaventura-Cisneros, S.; Sanchez-Lopez, J.Y.; Ortiz-Lopez, R.; et al. Thymidylate Synthase Gene Variants as Predictors of Clinical Response and Toxicity to Fluoropyrimidine-Based Chemotherapy for Colorectal Cancer. *Drug Metab. Pers.* **2017**, *32*, 209–218. [CrossRef]
41. Pastorakova, A.; Chandogova, D.; Chandoga, J.; Luha, J.; Bohmer, D.; Malova, J.; Braxatorisova, T.; Juhosova, M.; Reznakova, S.; Petrovic, R. Distribution of the Most Common Polymorphisms in TYMS Gene in Slavic Population of Central Europe. *Neoplasma* **2017**, *64*, 962–970. [CrossRef]
42. Summers, C.M.; Cucchiara, A.J.; Nackos, E.; Hammons, A.L.; Mohr, E.; Whitehead, A.S.; Von Feldt, J.M. Functional Polymorphisms of Folate-Metabolizing Enzymes in Relation to Homocysteine Concentrations in Systemic Lupus Erythematosus. *J. Rheumatol.* **2008**, *35*, 2179–2186. [CrossRef]
43. Pardini, B.; Verderio, P.; Pizzamiglio, S.; Nici, C.; Maiorana, M.V.; Naccarati, A.; Vodickova, L.; Vymetalkova, V.; Veneroni, S.; Daidone, M.G.; et al. Association between CASP8 –652 6N Del Polymorphism (Rs3834129) and Colorectal Cancer Risk: Results from a Multi-Centric Study. *PLoS ONE* **2014**, *9*, e85538. [CrossRef]
44. Pittman, A.M.; Broderick, P.; Sullivan, K.; Fielding, S.; Webb, E.; Penegar, S.; Tomlinson, I.; Houlston, R.S. CASP8 Variants D302H and −652 6N Ins/Del Do Not Influence the Risk of Colorectal Cancer in the United Kingdom Population. *Br. J. Cancer* **2008**, *98*, 1434–1436. [CrossRef]
45. Brown, K.L.; Seale, K.B.; El Khoury, L.Y.; Posthumus, M.; Ribbans, W.J.; Raleigh, S.M.; Collins, M.; September, A.V. Polymorphisms within the COL5A1 Gene and Regulators of the Extracellular Matrix Modify the Risk of Achilles Tendon Pathology in a British Case-Control Study. *J. Sports Sci.* **2017**, *35*, 1475–1483. [CrossRef] [PubMed]
46. Chatterjee, K.; Williamson, A.-L.; Hoffman, M.; Dandara, C. CASP8 Promoter Polymorphism Is Associated with High-Risk HPV Types and Abnormal Cytology but Not with Cervical Cancer. *J. Med. Virol.* **2011**, *83*, 630–636. [CrossRef] [PubMed]
47. Carvalho, D.C.; Wanderley, A.V.; Amador, M.A.T.; Fernandes, M.R.; Cavalcante, G.C.; Pantoja, K.B.C.C.; Mello, F.A.R.; de Assumpção, P.P.; Khayat, A.S.; Ribeiro-dos-Santos, Â.; et al. Amerindian Genetic Ancestry and INDEL Polymorphisms Associated with Susceptibility of Childhood B-Cell Leukemia in an Admixed Population from the Brazilian Amazon. *Leuk. Res.* **2015**, *39*, 1239–1245. [CrossRef]
48. Sanjari Moghaddam, A.; Nazarzadeh, M.; Sanjari Moghaddam, H.; Bidel, Z.; Keramatinia, A.; Darvish, H.; Mosavi-Jarrahi, A. XRCC1 Gene Polymorphisms and Breast Cancer Risk: A Systematic Review and Meta- Analysis Study. *Asian Pac. J. Cancer Prev.* **2016**, *17*, 323–330. [CrossRef] [PubMed]
49. Fehringer, G.; Kraft, P.; Pharoah, P.D.; Eeles, R.A.; Chatterjee, N.; Schumacher, F.R.; Schildkraut, J.M.; Lindström, S.; Brennan, P.; Bickeböller, H.; et al. Cross-Cancer Genome-Wide Analysis of Lung, Ovary, Breast, Prostate, and Colorectal Cancer Reveals Novel Pleiotropic Associations. *Cancer Res.* **2016**, *76*, 5103–5114. [CrossRef] [PubMed]
50. Hashemi, M.; Aftabi, S.; Moazeni-Roodi, A.; Sarani, H.; Wiechec, E.; Ghavami, S. Association of CASP8 Polymorphisms and Cancer Susceptibility: A Meta-Analysis. *Eur. J. Pharm.* **2020**, *881*, 173201. [CrossRef]
51. Marques, D.; Ferreira-Costa, L.R.; Ferreira-Costa, L.L.; da Silva Correa, R.; Borges, A.M.P.; Ito, F.R.; de Oliveira Ramos, C.C.; Bortolin, R.H.; Luchessi, A.D.; Ribeiro-Dos-Santos, Â.; et al. Association of Insertion-Deletions Polymorphisms with Colorectal Cancer Risk and Clinical Features. *World J. Gastroenterol.* **2017**, *23*, 6854–6867. [CrossRef]
52. Cătană, A.; Pop, M.; Hincu, B.D.; Pop, I.V.; Petrişor, F.M.; Porojan, M.D.; Popp, R.A. The XRCC1 Arg194Trp Polymorphism Is Significantly Associated with Lung Adenocarcinoma: A Case-Control Study in an Eastern European Caucasian Group. *Onco Targets* **2015**, *8*, 3533–3538. [CrossRef]

53. Bu, H.; Rosdahl, I.; Sun, X.-F.; Zhang, H. Importance of Polymorphisms in NF-KappaB1 and NF-KappaBIalpha Genes for Melanoma Risk, Clinicopathological Features and Tumor Progression in Swedish Melanoma Patients. *J. Cancer Res. Clin. Oncol.* **2007**, *133*, 859–866. [CrossRef]
54. Escobar, G.F.; Arraes, J.A.A.; Bakos, L.; Ashton-Prolla, P.; Giugliani, R.; Callegari-Jacques, S.M.; Santos, S.; Bakos, R.M. Polymorphisms in CYP19A1 and NFKB1 Genes Are Associated with Cutaneous Melanoma Risk in Southern Brazilian Patients. *Melanoma Res.* **2016**, *26*, 348–353. [CrossRef]
55. Chojnacka, K.; Bilinska, B.; Mruk, D.D. Interleukin 1alpha-Induced Disruption of the Sertoli Cell Cytoskeleton Affects Gap Junctional Communication. *Cell. Signal.* **2016**, *28*, 469–480. [CrossRef] [PubMed]
56. Griswold, M.D. The Central Role of Sertoli Cells in Spermatogenesis. *Semin. Cell Dev. Biol.* **1998**, *9*, 411–416. [CrossRef] [PubMed]
57. Gutti, R.K.; Tsai-Morris, C.-H.; Dufau, M.L. Gonadotropin-Regulated Testicular Helicase (DDX25), an Essential Regulator of Spermatogenesis, Prevents Testicular Germ Cell Apoptosis. *J. Biol. Chem.* **2008**, *283*, 17055–17064. [CrossRef]
58. O'Bryan, M.K.; Hedger, M.P. Inflammatory Networks in the Control of Spermatogenesis: Chronic Inflammation in an Immunologically Privileged Tissue? *Adv. Exp. Med. Biol.* **2008**, *636*, 92–114. [CrossRef] [PubMed]
59. Singh, V.; Kumar Mohanty, S.; Verma, P.; Chakraborty, A.; Trivedi, S.; Rajender, S.; Singh, K. XRCC1 Deficiency Correlates with Increased DNA Damage and Male Infertility. *Mutat. Res. Toxicol. Environ. Mutagen.* **2019**, *839*, 1–8. [CrossRef]
60. Walter, C.A.; Trolian, D.A.; McFarland, M.B.; Street, K.A.; Gurram, G.R.; McCarrey, J.R. Xrcc-1 Expression during Male Meiosis in the Mouse. *Biol. Reprod.* **1996**, *55*, 630–635. [CrossRef]

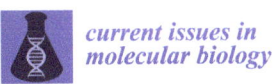

current issues in molecular biology

MDPI

Review

Nematode-Applied Technology for Human Tumor Microenvironment Research and Development

Eric di Luccio [1], Satoru Kaifuchi [1], Nobuaki Kondo [1], Ryota Chijimatsu [2], Andrea Vecchione [3], Takaaki Hirotsu [1,*] and Hideshi Ishii [2,*]

1 Hirotsu Bio Science Inc., Chiyoda-Ku, Tokyo 102-0094, Japan; e.diluccio@hbio.jp (E.d.L.); kaifuchi@hbio.jp (S.K.); kondo@hbio.jp (N.K.)
2 Department of Medical Data Science, Center of Medical Innovation and Translational Research, Osaka University Graduate School of Medicine, Suita, Yamadaoka 2-2, Osaka 565-0871, Japan; rchijimatsu@cfs.med.osaka-u.ac.jp
3 Department of Clinical and Molecular Medicine, Santo Andrea Hospital, University of Rome "Sapienza", Via di Grottarossa, 00189 Rome, Italy; andrea.vecchione@uniroma1.it
* Correspondence: hirotsu@hbio.jp (T.H.); hishii@gesurg.med.osaka-u.ac.jp (H.I.); Tel.: +81-(0)3-6277-8902 (T.H.); +81-(0)6-6210-8406 (H.I.); Fax: +81-(0)6-6210-8407 (H.I.)

Abstract: Nematodes, such as *Caenorhabditis elegans*, have been instrumental to the study of cancer. Recently, their significance as powerful cancer biodiagnostic tools has emerged, but also for mechanism analysis and drug discovery. It is expected that nematode-applied technology will facilitate research and development on the human tumor microenvironment. In the history of cancer research, which has been spurred by numerous discoveries since the last century, nematodes have been important model organisms for the discovery of cancer microenvironment. First, microRNAs (miRNAs), which are noncoding small RNAs that exert various functions to control cell differentiation, were first discovered in *C. elegans* and have been actively incorporated into cancer research, especially in the study of cancer genome defects. Second, the excellent sense of smell of nematodes has been applied to the diagnosis of diseases, especially refractory tumors, such as human pancreatic cancer, by sensing complex volatile compounds derived from heterogeneous cancer microenvironment, which are difficult to analyze using ordinary analytical methods. Third, a nematode model system can help evaluate invadosomes, the phenomenon of cell invasion by direct observation, which has provided a new direction for cancer research by contributing to the elucidation of complex cell–cell communications. In this cutting-edge review, we highlight milestones in cancer research history and, from a unique viewpoint, focus on recent information on the contributions of nematodes in cancer research towards precision medicine in humans.

Keywords: nematode; microenvironment; cancer; research

Citation: di Luccio, E.; Kaifuchi, S.; Kondo, N.; Chijimatsu, R.; Vecchione, A.; Hirotsu, T.; Ishii, H. Nematode-Applied Technology for Human Tumor Microenvironment Research and Development. *Curr. Issues Mol. Biol.* **2022**, *44*, 988–997. https://doi.org/10.3390/cimb44020065

Academic Editor: Dumitru A. Iacobas

Received: 24 January 2022
Accepted: 17 February 2022
Published: 21 February 2022

1. Introduction

Most nematode species live nonparasitic lives in soil and the ocean; however, many parasitic nematodes are also present [1]. Several nematodes, including human parasites, are closely related to human life, and while research on them has advanced, research on free-living animals has tended to be postponed. Enormous populations of nematodes are present in the soil, and account for 15% of earth biomass [2]. Cancer research began by considering the effects of its interaction with the environment before being deepened, and its complex mechanisms unraveled. Among them, nematodes have appeared in various points as research subjects or supporters that provide clues to cancer research. Particularly, as an extension of nematode research, these are: (1) the discovery of RNA interference, especially microRNAs, in cancer; (2) smell research objects and medical applications in cancer research; and (3) innovative applied methods for examining cell–cell interactions in the tumor microenvironment, all of which are discussed in this review article. We noted the milestones in cancer research and then focused on the advantages and discussed the usefulness of nematodes in the study of tumor microenvironment.

2. Milestones in Cancer Research

The cell theory was first described by Schleiden, Schwann, and Virchow [3]. Given that Rudolf Ludwig Karl Virchow (1821–1902), the founder of cellular pathology, who laid the foundations for cytopathology, comparative pathology (as a comparison of diseases common to humans with those common to animals), and anthropology, advocated his Latin motto "omnis cellula e cellula", which means that every cell originates from a cell—the concept has been considered by many other current researchers that alterations in cell organization were the basis of disease [3,4]. He discovered the concept that only certain cells or groups of cells become sick, not the entire living body [3–5]. In the 1900s, the concept that tumors originate from other body parts was beginning to be debated. Considering the discussion that scrotal cancer, which was seen in factory chimney sweepers, is presumably due to repeated stimulation under the influence of the Industrial Revolution in Western countries. It was first described by Percival Pott in 1775 [6], and since then Virchow's repetitive stimulus theory regarding cancer has emerged, and Fibiger's work was a strong proof of Virchow's theory; Fibiger received the Nobel Prize in Physiology and Medicine in 1926 because of his parasite carcinogenesis theory, for which he studied a type of parasite nematode called Spiroptera carcinoma (*Gongylonema neoplasticum*) [7]. Although, Fibiger's Nobel Prize-winning parasite carcinogenesis theory is now believed to have been false, and a 2004 document investigating the 1926 Physiology and Medicine Award selection process has stated that it is easy to conclude that Fibiger's Nobel Prize was wrong today; historically, it is invalid [7]. In this way, the royal road to the truth is to accumulate information, which has not been changed until modern life science. At that time, the main theories included "stimulation theory" and "predisposition theory," which are discussed as the cause of cancer. Katsusaburo Yamagiwa succeeded in developing artificial cancer in 1915 by conducting experiments on the steady process of continuously rubbing coal tar on the ears of rabbits for over 3 years [7]. It remains supported that repetitive stimuli, especially the importance of inflammation, will contribute closely to the initiation, progression, and development of cancer, which suggests the importance of the cancer microenvironment. In 1931, Otto Warburg was awarded the Nobel Prize in Physiology and Medicine for his study on tumor metabolism and cell respiration, especially cancer cells [8]. The concept of metabolic reprogramming is now a hallmark of cancer [9,10]. DNA-sequencing techniques, which are now commonplace and incorporated into standard medical practice, began with the discovery of the Watson–Crick structure of DNA (double-helix structure) in the late 20th century [11], for which Watson and Crick received the 1962 Nobel Prize in Physiology and Medicine. Understanding the cancer microenvironment and communications between cells has become indispensable in understanding cancer overall. The study of the cancer genome has made great strides.

Rous has found non-epithelial malignancies that infect and develop not only when cancer cells are transplanted but also when substances extracted from cancer cells are injected [12]. This finding brought about the viral theory of cancer development. His work was ridiculed at the time; however, subsequent experiments proved his claim. He was awarded the 1966 Nobel Prize in Physiology and Medicine with Huggins, who discovered that hormones suppress the metastasis of certain cancers and showed for the first time that cancer can be controlled by chemicals [12]. Temin discovered reverse transcriptase in the 1970s and, along with Dulbecco and Baltimore, received the 1975 Nobel Prize in Physiology and Medicine [13]. Temin clarified how oncoviruses use reverse transcriptase to rewrite the genetic information of host cells. The discovery also urged a revision of the widely believed concept of Central Dogma, advocated by Watson and Crick, since other molecular biologists at the time believed that genetic information flows in only one direction, from DNA via RNA to protein. However, Temin demonstrated that in a type of tumor virus, reverse transcriptase is essential for transmitting genetic information toward

DNA [13]. Varmus was awarded the Nobel Prize in Physiology and Medicine with Bishop for discovering the proto-oncogene tyrosine-protein kinase (c-Src), a human oncogene. Moreover, Varmus discovered that the cancer gene of a retrovirus has a cellular origin [14]. With the application of induced pluripotent stem (iPS) cells, Yamanaka was awarded the 2012 Nobel Prize in Physiology and Medicine, with co-winner Gurdon, for discovering that mature cells are reprogrammed and pluripotent. After this innovation of iPS technology [15], several studies have been conducted on the concept of reprogramming the properties of cells to regenerative medicine. Those concepts, i.e., metabolic reprogramming, which changes the metabolic mechanism from anaerobic to aerobic, and epigenetic reprogramming, which controls cell differentiation by regulating gene expression, have been applied in the diagnosis and treatment of human diseases. Recently, Honjo received the Nobel Prize in Physiology and Medicine in 2018 with Allison for the discovery of immune checkpoint inhibitors and their application to cancer treatment [16,17]. Honjo has been recognized for his seminal publication in 1992 describing a new molecule, which he termed programmed death-1 (PD-1), based on its functional role in mediating classical apoptosis in a T-cell hybridoma and hematopoietic progenitor cells [18]. Although tumor tissues contain carious components, including epithelial cancer cells, mesenchymal fibroblasts, blood vessels, and immune cells, given that cancer cells, but not noncancerous cells, harbor genetic alterations, much emphasis has been placed on the study of genetic alterations that can attenuate the function of tumor suppressor genes or induce the activation of tumor-promoting oncogenes [19]. Taken together, the study of genomic losses in cancer cells allowed the identification of microRNAs (miRNAs), which followed the discovery of miRNAs in nematodes. *C. elegans* has been crucial for miRNA research, that helped to unravel the role of miRNAs in cancer. Therefore, nematodes can be considered as a very useful tool to human cancer research (Figure 1).

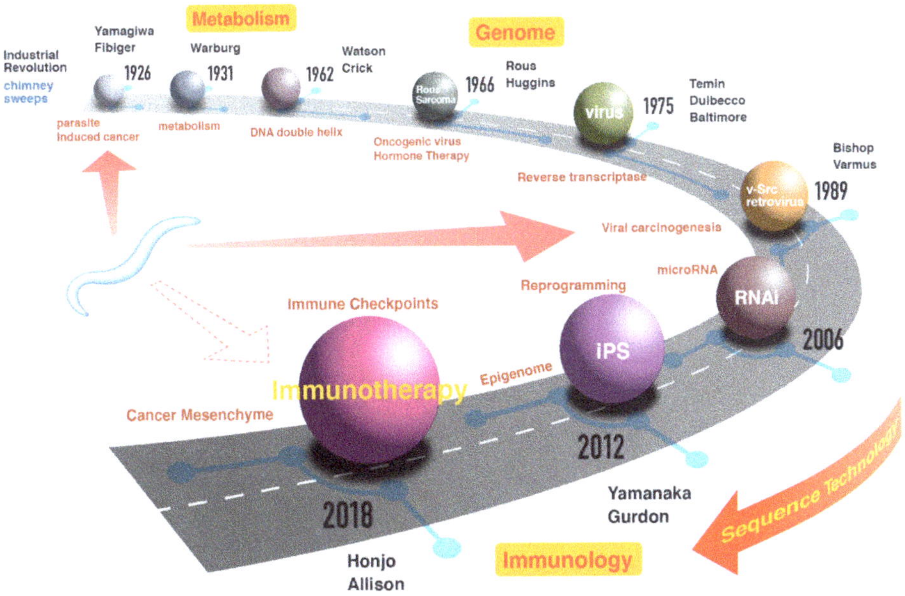

Figure 1. Historical overview of cancer research. Based on Virchow's accomplishments on cellular pathology, the modern state of cancer research has been developed for over a century. With the background of the Industrial Revolution, cancer induction was studied by involving various stimuli,

such as parasitic infections, nematode infections, or chemical substances, which were believed to induce inflammation in the epithelium. In 1962, DNA's structure was elucidated, which opened the avenue to the current genome sequencing technology. Meanwhile, important discoveries were accumulated regarding viruses and their biochemistry. As a result of the discovery of *C. elegans* in the 21st century, miRNAs were discovered in human cancer. The iPS technology in regenerative medicine facilitated the study of reprogramming in cancer research. Immunotherapy is a current rewiring cancer treatment targeting the cancer microenvironment. The control of cell–cell communications in the cancer microenvironment is a critical issue in nematode technology. In a schema, sequential discoveries were illustrated according to the Nobel Prize in the field of cancer research. In the schema, the knowledge of nematode study induced innovation, which are depicted by three arrows. Detailed events are described in the text.

3. MicroRNA

Fire and Mello were awarded the Nobel Prize in Physiology in 2006 for their discovery of RNA interference and gene silencing by double-stranded RNA [20]. Fire and Mellow, along with colleagues Xu, Montgomery, Kostas, and Sam Driver, translated small double-stranded RNAs (dsRNA) into proteins by disrupting mRNA with complementary sequences, which led to suppressing a specific gene. They found that dsRNA suppresses gene expression more efficiently than the previously reported RNA interference by single-stranded RNA. Since they needed short dsRNA, they suggested involving a catalytic process, and this hypothesis was substantiated by later studies [21,22]. The microRNAs lin-4 and lin-14 were discovered using genetic analysis of developmental timing mutants in *C. elegans* [23,24]. The study of the developmental timing pathway was pioneered by Brenner, Sulston and Horvitz on the genetics and cell lineage of *C. elegans* [25,26]. Horvitz defined several components of programmed cell death. *C. elegans* was key in understanding the general features of miRNA biology, which brings about the evaluation of the function of miRNAs [27]. In contrast, the microRNAs miR-15 and miR16 in cancer were discovered first in humans by positional cloning tumor suppressor genes of hematopoietic malignancies in cancer research of chromosome 13 [28], and the expression of these microRNAs was associated with the progression of chronic lymphocytic leukemia [29]. A study on lung cancer has shown that the expression of the microRNA let-7 in human solid tumors, such as lung cancers, was decreased and associated with shortened postoperative survival [30]. The study has indicated that the expression of miRNAs successfully classifies poorly differentiated tumors using miRNA expression profiles, whereas messenger RNA profiles are highly inaccurate, suggesting the potential of miRNA profiling in cancer diagnosis [31].

4. Smell Research

It is common knowledge that some animals have sensory abilities superior to those of humans, such as smelling, and animal use will fill the gap in achieving detection of phenomena via smelling [32]. In a study by Lo et al., memory tests of *Canis familiaris*, *Rattus norvegicus*, and *Homo sapiens* indicated that dogs were superior to rats and that dogs and rats were superior to humans [33]. The study has suggested that the relatively poor performance of humans contrasts with high recognition memory for odors, suggesting that humans complement their low sensory abilities with intelligence and emotions [33]. Therefore, using animals will be better for objective testing. Incidentally, there have been some subjective opinions that patients with cancer have a peculiar odor; however, there was no way to objectively investigate this theory. Attempts were made to use animals to test this theory as objectively as possible. To examine the ability of beagle dogs to discriminate fresh biopsy and discharge samples from patients with cervical cancer, which is based on the impression of clinical doctors that cervical cancer with discharges might express any odors, a double-blinded procedure was performed. The results indicated that trained dogs seemed useful as a noninvasive alternative method for identifying patients with cervical cancer [34]. Moreover, another study has indicated that canine olfaction can detect liquid samples from breast cancer and colorectal cancer cell cultures, although dogs could not discriminate the odor of metabolic wastes between breast and colorectal cancers [35], suggesting that such animals sense some odors that have not been characterized so far. Although dogs may be considered good candidates for scent test detectors, the cost for training and maintenance is relatively high, which inhibits repeats and requires several examinations a day to confirm its reproducibility, which humps a large-scale study [32].

Nematodes may be an ideal tool to assess odorants from samples, such as urines of patients with cancer, and study uncharacterized mechanisms that will reflect tumor microenvironments. Studies have indicated that *C. elegans* could discriminate urine samples from patients with cancer from those obtained from healthy individuals [36]. The receiver operating characteristic (ROC) analysis has indicated that tests using *C. elegans* had a higher diagnostic ability than those using classical tumor markers; moreover, *C. elegans* showed a significant difference in behavior before and after tumor removal, suggesting that the *C. elegans* test will be useful in monitoring patients postoperatively [37]. Moreover, a relatively large study involving 180 urine samples from patients with gastrointestinal cancer and 76 samples from healthy participants has demonstrated that gastrointestinal cancer screening test has a high sensitivity, with a significant value of 0.80 in the ROC analysis, even in early-stage cancers [38]. Furthermore, a nationwide study group comprising high-volume centers throughout Japan to collect patients with pancreatic cancer reported that an open-label study involving 83 cases (stage 0–IV) of pancreatic cancer showed the efficacy of the *C. elegans* test to detect pancreatic cancer; a blinded study on 28 cases conducted by comparing patients with very early stage pancreatic cancer indicated that preoperative urine samples had a significantly higher chemotaxis index than postoperative samples in patients with pancreatic cancer; using the changes in the preoperative and postoperative chemotaxis index, this method had a higher sensitivity for detecting early pancreatic cancer than existing diagnostic markers, suggesting the rationales for the clinical application of *C. elegans* in the early diagnosis of pancreatic cancer [39]. In contrast, a study on the mechanism of genetically engineered mice indicated that the *C. elegans* test detected the urine of oncogenic KrasG12D mouse model, which is frequently mutated and activated in pancreatic cancer in humans, whereas the role of mouse c-Met, a receptor of hepatocyte growth factor, was not detected [40], suggesting that the downstream products of mouse KrasG12D is involved in the chemotaxis or olfactory behavior response alteration in *C. elegans*.

Many parasitic nematodes actively search for hosts to infect by using volatile chemical cues [41]. By understanding the olfactory signals of free-living nematodes as conventional research tools, we will be able to apply the knowledge to prevent infection by parasitic nematodes in humans. Eventually, the study of circuit mechanisms has allowed the identification of substances, including odorants, gases, and pheromones, that *C. elegans* respond to [41]. It shows that chemosensory neurons of *C. elegans* include: amphid wing C (AWC), which functions as an attraction by sensing odors, temperatures, carbon dioxide, salt, osmotic pressure, and pH; AWA olfactory neuron, which functions as an attraction by sensing odors; ASH sensory neuron, which mediates avoidance by sensing odors, soluble chemicals, and mechanical and osmotic stimuli; BAG neuron, which functions as avoidance (adults) or attraction (daughters) by sensing carbon dioxide and oxygen; and ADL neuron, which functions as avoidance by sensing odors and pheromones [41] (https://www.wormatlas.org) (accessed on 1 November 2021). The proposed models of microcircuit motifs present in the olfactory system of *C. elegans* indicate two stages. First, the feedback-inhibition regulatory system can elicit odor adaptation [42]. In the absence of an odor, AWC olfactory neurons will release neuropeptide-like protein 1 (NLP-1), which binds the neuropeptide receptor resemblance-11 (NPR-11) on the surface of AIA interneurons to inhibit their activity. In contrast, in the presence of an odor, AWC activity is suppressed, resulting in a decrease in NLP-1 signaling and leads AIA to release insulin-related 1 (ins-1), which inhibits AWC [42]. Second, the reciprocal inhibition system can modulate feedings in an odor environment [43]. In the presence of attractive odors, nematodes increase their feeding. As a mechanism, attractive odorants, such as diacetyl, are sensed by AWA neurons and cause the release of serotonin (5-hydroxytryptamine, 5-HT) from NSM neurons. 5-HT binds MOD-1 (Modulation Of locomotion Defective), a serotonin-gated chloride channel on RIM and RIC interneurons, resulting in inhibition and an increase in feeding. In contrast, the presence of repulsive odors decreases feeding caused by repellents, such as quinine, or high concentrations of isoamyl alcohol, which are sensed by ASH neurons and promote the release of octopamine and tyramine from RIM and RIC. Octopamine and tyramine bind to the tyramine receptor (SER-2) on NSM neurons and inhibit serotonin release [43]. As such, uncharacterized substances, including some volatiles, may be involved in the response to stimuli in nematodes [36,39,44]. A study on urine samples from patients with pancreatic cancer showed unique patterns of volatile organic compounds, suggesting that they are useful in distinguishing between cancer and inflammation in the pancreas [45]. Moreover, a study on pancreatic ductal adenocarcinoma has indicated that acetone, 2-pantanone, 4-methyl-2-heptanone, D-limonene, and levomenthol were possible volatile organic compounds and metabolite biomarkers in urine, though both chronic pancreatitis and pancreatic ductal adenocarcinoma were investigated [46]. Furthermore, another study has suggested several candidate volatile organic compounds, including 2-octonone and pentanal, as these compounds increased in the urine of patients with prostate cancer compared with those in healthy controls [47].

Recent studies of lung cancer have indicated that volatile organic compounds in breath are potentially associated with disease progression, suggesting its usefulness as a biomarker [48–53]. A study on the linkage between volatile organic compounds and gene mutations of KRASV12 and TP53 has indicated that genetic changes lead to detectable differences in levels of specific volatile organic compounds in cell culture experiments, suggesting that breath analysis can be used for detecting cancers [54]. Volatile organic compounds may be involved in other mutations observed in cancer [55].

The gene mutation-related mechanism by which nematodes sense the smell of cancer is interesting. Substances that can stimulate nematode nerve cells may be released from cancerous tissues under the control of KrasG12D, but not Met activation, which elicited a response from nematode nerve cells according to animal experiments [40]. Metabolites located downstream of KRASG12D may be involved. However, a clinical sequence study has indicated that KRAS is mutated in more than 90% of cases of pancreatic cancer, with frequent associations with other mutations, such as mothers against decapentaplegic homolog 1 (SMAD1) family in the transforming growth factor beta 1 (TGF-β) pathway, and tumor suppressor genes, including tumor protein P53 (TP53) and cyclin-dependent kinase inhibitor 2A (p16Ink4a) [56]. Recently, a study on the metabolism in pancreatic cancer has demonstrated that pancreatic cancer cells rely on the distinct pathway in which glutamine supports pancreatic cancer growth through a KRAS-regulated metabolic pathway [57]. Glutamine is converted to oxaloacetate by aspartate transaminase (GOT1), and oxaloacetate is converted further into malate and then pyruvate, and this metabolic pathway is associated with an increase in the NADPH/NADP+ ratio, resulting in the maintenance of the cellular redox state [57]. Taken together, it appears that various metabolic pathway abnormalities occur downstream of the KRAS mutation in pancreatic cancer, which results in the generation of substances that affect the odorant behaviors of nematodes. Further studies undoubtedly will be necessary to further understand the mechanism of *C. elegans* sensing to develop an efficient innovative tool (Figure 2).

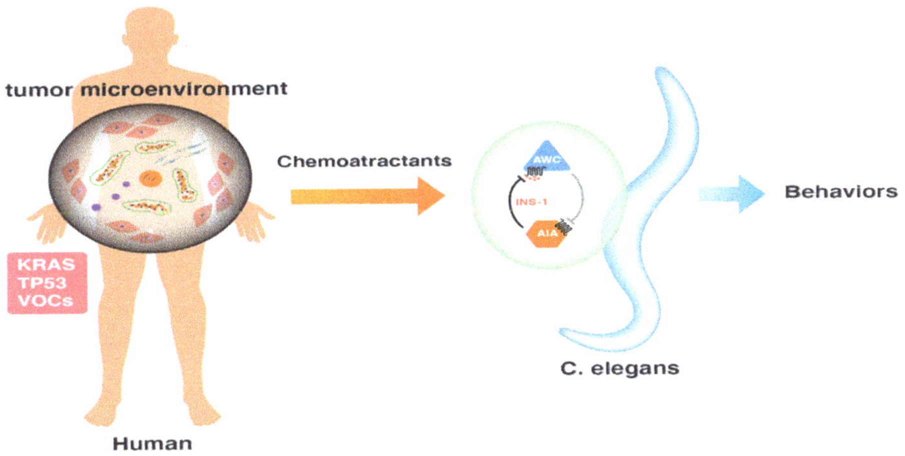

Figure 2. Schema for the nematode scent test of cancer. Given that cancer is a genetic disease harboring the accumulation of several mutations of malignant phenotype-promoting oncogenes and tumor suppressor genes. In pancreatic cancer, mutations in KRAS and TP53 occur frequently, which stimulate the downstream signals in cancer cells, influencing the surrounding mesenchymal fibroblasts, vessels, neural cells, and immune cells in the tumor microenvironment. Studies have indicated that *C. elegans* respond differentially to the presence or absence of tumors by sensing liquid samples, such as urine from patients with cancer. As a mechanism, volatile organic compounds, the production of which was elicited in the influence of genetic mutations of KRAS and TP53, may be involved in the behavior reaction by stimulating the neural system in nematodes. Detailed events are described in the text.

5. Innovative Method for Studying the Tumor Microenvironment

Recently, animal models, including nematodes, are used for basic studies for medical applications, such as mechanism studies and drug discovery, as summarized in [32]. Here, we focused on another application to study the tumor microenvironment using *C. elegans*. Recent studies have reported, in general, the importance of invadosomes, including podosomes and invadopodia, which are involved in cell–cell interactions via specialized F-actin-based adhesive structures formed as cell protrusions at sites of cell–extracellular matrix contacts on the ventral membrane of various cell types in tumor tissues [58]. Invadosomes are referred to as podosomes when they are found in normal cells and invadopodia when they are found in cancer cells. In this review, we discussed both, considering that common mechanisms are shared between them [59,60]. In vivo invadosome homologs have been reported in developmental model systems, including *C. elegans* [61]. The phenomenon of invasion occurs during both physiological and pathological processes. The formation of invadosomes is observed in various cells, including vascular cells, monocytic cells, osteoclasts, cancer cells, fibroblasts, and cancer-associated fibroblasts, which are transformed by oncogenic signals on almost all life processes in different stages of embryonic and tissue development, wound-healing, inflammation, and cancer invasion and metastasis, which are characteristics of the tumor microenvironment [60,62]. The structures of invadosomes were first discovered in a study of chicken embryo fibroblasts transformed using v-Src, a viral oncogene found in Rous sarcoma virus (RSV) [63]. The small size and transparent nature of *C. elegans* offer an important feature of being able to visualize invasive protrusions in vivo, which can address the issues in observing in higher organisms. Thus, *C. elegans* is often used as a model system in studies of developmental processes [61]. The genome of *C. elegans* encodes orthologs of most components implicated in invadosome formation or function, including Src [64,65]. Studies have reported that one exception to the structural components observed in *C. elegans* is cortactin, a key regulator of invadosome formation in cancer cells in vitro, suggesting that the common mechanism is shared over species [64,66,67], showing the valuable significance of nematode application to human life science. The first study that has used *C. elegans* has indicated that in vivo screening of genes regulating invadopodia allowed the identification of genes promoting invadopodia function in vivo—cell division control protein 42 homolog (CDC42) and Rab GDP dissociation inhibitor 1 (Gdi1)—which are involved in the direct control of invadopodia formation. The aforementioned results clarified the notion that invadopodia formation requires the integration of distinct cellular processes coordinated by an extracellular cue [68]. For the screening of cell–cell interactions and cytokines and chemokines, especially volatile organic compounds secreted and contained in the cancer microenvironment and expectedly sensed by nematodes as the cancer screening, the nematode system is expected to produce new results that have never been seen.

Author Contributions: E.d.L., T.H. and H.I. conceptualized the study objectives. E.d.L., S.K., N.K., R.C., A.V., T.H. and H.I. wrote the manuscript and outlined the content. All authors have read and agreed to the published version of the manuscript.

Funding: This work was supported in part by a Grant-in-Aid for Scientific Research from the Ministry of Education, Culture, Sports, Science and Technology (20H00541; 21K19526) to H.I. Partial support was received from Princess Takamatsu Cancer Research Fund, Senshin Medical Research Foundation, and Mitsubishi Foundation, to H.I.

Acknowledgments: We thank Otsuka, C., Hamano, Y., Nakayama, M., and Arao, Y., for fruitful discussion and excellent support of preparation of our research manuscript.

Conflicts of Interest: Partial institutional endowments were received from Taiho Pharmaceutical Co., Ltd. (Tokyo, Japan), Hirotsu Bio Science Inc. (Tokyo, Japan), Kinshu-kai Medical Corporation (Osaka, Japan), Kyowa-kai Medical Corporation (Osaka, Japan), IDEA Consultants Inc. (Tokyo, Japan), and Unitech Co., Ltd. (Chiba, Japan). E.d.L., S.K. and N.K. are employees and T.H. is a Chief Executive Officer of Hirotsu Bio Science Inc. (Tokyo, Japan).

References

1. Makarova, A.A.; Polilov, A.A.; Chklovskii, D.B. Small brains for big science. *Curr. Opin. Neurobiol.* **2021**, *71*, 77–83. [CrossRef] [PubMed]
2. Tuli, M.A.; Daul, A.; Schedl, T. Caenorhabditis nomenclature. *WormBook* **2018**, *2018*, 1–14. [CrossRef] [PubMed]
3. Ribatti, D. Rudolf Virchow, the founder of cellular pathology. *Rom. J. Morphol. Embryol.* **2019**, *60*, 1381–1382. [PubMed]
4. Listed, N.A. Rudolf Ludwig Karl Virchow (1821–1902) "Omnis Cellula E Cellula". *Minn. Med.* **1966**, *49*, 533–534.
5. Brotman, D.J.; Deitcher, S.R.; Lip, G.Y.; Matzdorff, A.C. Virchow's triad revisited. *South. Med. J.* **2004**, *97*, 213–214. [CrossRef] [PubMed]
6. Urbach, F. Environmental risk factors for skin cancer. In *Skin Carcinogenesis in Man and in Experimental Models*; Springer: Berlin/Heidelberg, Germany, 1993; Volume 128, pp. 243–262.
7. Stolt, C.-M.; Klein, G.; Jansson, A.T. An Analysis of a Wrong Nobel Prize—Johannes Fibiger, 1926: A Study in the Nobel Archives. *Adv. Cancer Res.* **2004**, *92*, 1–12. [CrossRef] [PubMed]
8. Urbano, A.M. Otto Warburg: The journey towards the seminal discovery of tumor cell bioenergetic reprogramming. *Biochim. Biophys. Acta (BBA)-Mol. Basis Dis.* **2021**, *1867*, 165965. [CrossRef] [PubMed]
9. Hanahan, D.; Weinberg, R.A. The hallmarks of cancer. *Cell* **2000**, *100*, 57–70. [CrossRef]
10. Hanahan, D.; Weinberg, R.A. Hallmarks of Cancer: The Next Generation. *Cell* **2011**, *144*, 646–674. [CrossRef]
11. Watson, J.D.; Crick, F.H.C. Molecular Structure of Nucleic Acids: A Structure for Deoxyribose Nucleic Acid. *Nature* **1953**, *171*, 737–738. [CrossRef]
12. Raju, T.N. The Nobel chronicles. 1966: Francis Peyton Rous (1879–1970) and Charles Brenton Huggins (1901–1997). *Lancet* **1999**, *354*, 520. [CrossRef]
13. Raju, T.N. The Nobel chronicles. 1975: Renato Dulbecco (b 1914), David Baltimore (b 1938), and Howard Martin Temin (1934–1994). *Lancet* **1999**, *354*, 1308. [CrossRef]
14. Raju, T.N. The Nobel Chronicles. 1989: John Michael Bishop (b 1936) and Harold Eliot Varmus (b 1939). *Lancet* **2000**, *355*, 1106. [CrossRef]
15. Takahashi, K.; Yamanaka, S. Induction of Pluripotent Stem Cells from Mouse Embryonic and Adult Fibroblast Cultures by Defined Factors. *Cell* **2006**, *126*, 663–676. [CrossRef] [PubMed]
16. Wolchok, J. Putting the Immunologic Brakes on Cancer. *Cell* **2018**, *175*, 1452–1454. [CrossRef] [PubMed]
17. Ishida, Y. PD-1: Its Discovery, Involvement in Cancer Immunotherapy, and Beyond. *Cells* **2020**, *9*, 1376. [CrossRef] [PubMed]
18. Ishida, Y.; Agata, Y.; Shibahara, K.; Honjo, T. Induced expression of PD-1, a novel member of the immunoglobulin gene superfamily, upon programmed cell death. *EMBO J.* **1992**, *11*, 3887–3895. [CrossRef]
19. Weinberg, R. Tumor suppressor genes. *Science* **1991**, *254*, 1138–1146. [CrossRef]

20. Fire, A.; Xu, S.; Montgomery, M.K.; Kostas, S.A.; Driver, S.E.; Mello, C.C. Potent and specific genetic interference by double-stranded RNA in *Caenorhabditis elegans*. *Nature* **1998**, *391*, 806–811. [CrossRef]
21. Gregory, R.I.; Chendrimada, T.P.; Cooch, N.; Shiekhattar, R. Human RISC Couples MicroRNA Biogenesis and Posttranscriptional Gene Silencing. *Cell* **2005**, *123*, 631–640. [CrossRef]
22. Zhang, H.; Kolb, F.A.; Jaskiewicz, L.; Westhof, E.; Filipowicz, W. Single Processing Center Models for Human Dicer and Bacterial RNase III. *Cell* **2004**, *118*, 57–68. [CrossRef] [PubMed]
23. Lee, R.C.; Feinbaum, R.L.; Ambros, V. The *C. elegans* heterochronic gene lin-4 encodes small RNAs with antisense complementarity to lin-14. *Cell* **1993**, *75*, 843–854. [CrossRef]
24. Wightman, B.; Ha, I.; Ruvkun, G. Posttranscriptional regulation of the heterochronic gene lin-14 by lin-4 mediates temporal pattern formation in *C. elegans*. *Cell* **1993**, *75*, 855–862. [CrossRef]
25. Sulston, J.; Horvitz, H. Post-embryonic cell lineages of the nematode, *Caenorhabditis elegans*. *Dev. Biol.* **1977**, *56*, 110–156. [CrossRef]
26. Sulston, J.; Schierenberg, E.; White, J.; Thomson, J. The embryonic cell lineage of the nematode *Caenorhabditis elegans*. *Dev. Biol.* **1983**, *100*, 64–119. [CrossRef]
27. Kaufman, E.J.; Miska, E.A. The microRNAs of *Caenorhabditis elegans*. *Semin. Cell Dev. Biol.* **2010**, *21*, 728–737. [CrossRef]
28. Calin, G.A.; Dumitru, C.D.; Shimizu, M.; Bichi, R.; Zupo, S.; Noch, E.; Aldler, H.; Rattan, S.; Keating, M.; Rai, K.; et al. Frequent deletions and down-regulation of micro- RNA genes *miR15* and *miR16* at 13q14 in chronic lymphocytic leukemia. *Proc. Natl. Acad. Sci. USA* **2002**, *99*, 15524–15529. [CrossRef] [PubMed]
29. Calin, G.; Ferracin, M.; Cimmino, A.; Di Leva, G.; Shimizu, M.; Wojcik, S.E.; Iorio, M.; Visone, R.; Sever, N.I.; Fabbri, M.; et al. A MicroRNA Signature Associated with Prognosis and Progression in Chronic Lymphocytic Leukemia. *N. Engl. J. Med.* **2005**, *353*, 1793–1801. [CrossRef]
30. Takamizawa, J.; Konishi, H.; Yanagisawa, K.; Tomida, S.; Osada, H.; Endoh, H.; Harano, T.; Yatabe, Y.; Nagino, M.; Nimura, Y.; et al. Reduced Expression of the let-7 MicroRNAs in Human Lung Cancers in Association with Shortened Postoperative Survival. *Cancer Res.* **2004**, *64*, 3753–3756. [CrossRef]
31. Lu, J.; Getz, G.; Miska, E.A.; Alvarez-Saavedra, E.; Lamb, J.; Peck, D.; Sweet-Cordero, A.; Ebert, B.L.; Mak, R.H.; Ferrando, A.A.; et al. MicroRNA expression profiles classify human cancers. *Nature* **2005**, *435*, 834–838. [CrossRef]
32. Konno, M.; Asai, A.; Kitagawa, T.; Yabumoto, M.; Ofusa, K.; Arai, T.; Hirotsu, T.; Doki, Y.; Eguchi, H.; Ishii, H. State-of-the-Art Technology of Model Organisms for Current Human Medicine. *Diagnostics* **2020**, *10*, 392. [CrossRef] [PubMed]
33. Lo, G.K.-H.; Macpherson, K.; MacDonald, H.; Roberts, W.A. A comparative study of memory for olfactory discriminations: Dogs (*Canis familiaris*), rats (*Rattus norvegicus*), and humans (*Homo sapiens*). *J. Comp. Psychol.* **2020**, *134*, 170–179. [CrossRef] [PubMed]
34. Guerrero-Flores, H.; Apresa-García, T.; Garay-Villar, Ó.; Sánchez-Pérez, A.; Flores-Villegas, D.; Bandera-Calderón, A.; García-Palacios, R.; Rojas-Sánchez, T.; Romero-Morelos, P.; Sánchez-Albor, V.; et al. A non-invasive tool for detecting cervical cancer odor by trained scent dogs. *BMC Cancer* **2017**, *17*, 79. [CrossRef]
35. Seo, I.-S.; Lee, H.-G.; Koo, B.; Koh, C.S.; Park, H.-Y.; Im, C.; Shin, H.-C. Cross detection for odor of metabolic waste between breast and colorectal cancer using canine olfaction. *PLoS ONE* **2018**, *13*, e0192629. [CrossRef] [PubMed]
36. Hirotsu, T.; Sonoda, H.; Uozumi, T.; Shinden, Y.; Mimori, K.; Maehara, Y.; Ueda, N.; Hamakawa, M. A Highly Accurate Inclusive Cancer Screening Test Using *Caenorhabditis elegans* Scent Detection. *PLoS ONE* **2015**, *10*, e0118699. [CrossRef] [PubMed]
37. Kusumoto, H.; Tashiro, K.; Shimaoka, S.; Tsukasa, K.; Baba, Y.; Furukawa, S.; Furukawa, J.; Suenaga, T.; Kitazono, M.; Tanaka, S.; et al. Behavioural Response Alteration in *Caenorhabditis elegans* to Urine After Surgical Removal of Cancer: Nematode-NOSE (N-NOSE) for Postoperative Evaluation. *Biomark. Cancer* **2019**, *11*, 1179299x19896551. [CrossRef] [PubMed]
38. Kusumoto, H.; Tashiro, K.; Shimaoka, S.; Tsukasa, K.; Baba, Y.; Furukawa, S.; Furukawa, J.; Niihara, T.; Hirotsu, T.; Uozumi, T. Efficiency of Gastrointestinal Cancer Detection by Nematode-NOSE (N-NOSE). *In Vivo* **2019**, *34*, 73–80. [CrossRef]
39. Asai, A.; Konno, M.; Ozaki, M.; Kawamoto, K.; Chijimatsu, R.; Kondo, N.; Hirotsu, T.; Ishii, H. Scent test using *Caenorhabditis elegans* to screen for early-stage pancreatic cancer. *Oncotarget* **2021**, *12*, 1687–1696. [CrossRef]
40. Ueda, Y.; Kawamoto, K.; Konno, M.; Noguchi, K.; Kaifuchi, S.; Satoh, T.; Eguchi, H.; Doki, Y.; Hirotsu, T.; Mori, M.; et al. Application of *C. elegans* cancer screening test for the detection of pancreatic tumor in genetically engineered mice. *Oncotarget* **2019**, *10*, 5412–5418. [CrossRef]
41. Rengarajan, S.; Hallem, E.A. Olfactory circuits and behaviors of nematodes. *Curr. Opin. Neurobiol.* **2016**, *41*, 136–148. [CrossRef]
42. Chalasani, S.H.; Kato, S.; Albrecht, D.R.; Nakagawa, T.; Abbott, L.F.; Bargmann, C.I. Neuropeptide feedback modifies odor-evoked dynamics in *Caenorhabditis elegans* olfactory neurons. *Nat. Neurosci.* **2010**, *13*, 615–621. [CrossRef] [PubMed]
43. Li, Z.; Li, Y.; Yi, Y.; Huang, W.; Yang, S.; Niu, W.; Zhang, L.; Xu, Z.; Qu, A.; Wu, Z.; et al. Dissecting a central flip-flop circuit that integrates contradictory sensory cues in *C. elegans* feeding regulation. *Nat. Commun.* **2012**, *3*, 776. [CrossRef] [PubMed]
44. Hirotsu, T.; Saeki, S.; Yamamoto, M.; Iino, Y. The Ras-MAPK pathway is important for olfaction in *Caenorhabditis elegans*. *Nature* **2000**, *404*, 289–293. [CrossRef] [PubMed]
45. Nissinen, S.I.; Roine, A.; Hokkinen, L.; Karjalainen, M.; Venäläinen, M.; Helminen, H.; Niemi, R.; Lehtimäki, T.; Rantanen, T.; Oksala, N. Detection of Pancreatic Cancer by Urine Volatile Organic Compound Analysis. *Anticancer Res.* **2018**, *39*, 73–79. [CrossRef]
46. Daulton, E.; Wicaksono, A.N.; Tiele, A.; Kocher, H.M.; Debernardi, S.; Crnogorac-Jurcevic, T.; Covington, J.A. Volatile organic compounds (VOCs) for the non-invasive detection of pancreatic cancer from urine. *Talanta* **2021**, *221*, 121604. [CrossRef]
47. Thompson, M.; Feria, N.S.; Yoshioka, A.; Tu, E.; Civitci, F.; Estes, S.; Wagner, J.T. A *Caenorhabditis elegans* behavioral assay distinguishes early stage prostate cancer patient urine from controls. *Biol. Open* **2021**, *10*, bio057398. [CrossRef]

48. Chen, X.; Xu, F.; Wang, Y.; Pan, Y.; Lu, D.; Wang, P.; Ying, K.; Chen, E.; Zhang, W. A study of the volatile organic compounds exhaled by lung cancer cells in vitro for breath diagnosis. *Cancer* **2007**, *110*, 835–844. [CrossRef]
49. Phillips, M.; Altorki, N.; Austin, J.H.; Cameron, R.B.; Cataneo, R.N.; Kloss, R.; Maxfield, R.A.; Munawar, M.I.; Pass, H.; Rashid, A.; et al. Detection of lung cancer using weighted digital analysis of breath biomarkers. *Clin. Chim. Acta* **2008**, *393*, 76–84. [CrossRef]
50. Dragonieri, S.; Annema, J.T.; Schot, R.; van der Schee, M.P.; Spanevello, A.; Carratù, P.; Resta, O.; Rabe, K.F.; Sterk, P.J. An electronic nose in the discrimination of patients with non-small cell lung cancer and COPD. *Lung Cancer* **2009**, *64*, 166–170. [CrossRef]
51. Mazzone, P.J.; Wang, X.-F.; Xu, Y.; Mekhail, T.; Beukemann, M.C.; Na, J.; Kemling, J.W.; Suslick, K.; Sasidhar, M. Exhaled Breath Analysis with a Colorimetric Sensor Array for the Identification and Characterization of Lung Cancer. *J. Thorac. Oncol.* **2012**, *7*, 137–142. [CrossRef]
52. Amann, A.; Corradi, M.; Mazzone, P.; Mutti, A. Lung cancer biomarkers in exhaled breath. *Expert Rev. Mol. Diagn.* **2011**, *11*, 207–217. [CrossRef] [PubMed]
53. Hakim, M.; Broza, Y.Y.; Barash, O.; Peled, N.; Phillips, M.; Amann, A.; Haick, H. Volatile Organic Compounds of Lung Cancer and Possible Biochemical Pathways. *Chem. Rev.* **2012**, *112*, 5949–5966. [CrossRef] [PubMed]
54. Davies, M.P.A.; Barash, O.; Jeries, R.; Peled, N.; Ilouze, M.; Hyde, R.; Marcus, M.W.; Field, J.; Haick, H. Unique volatolomic signatures of TP53 and KRAS in lung cells. *Br. J. Cancer* **2014**, *111*, 1213–1221. [CrossRef] [PubMed]
55. Peled, N.; Barash, O.; Tisch, U.; Ionescu, R.; Broza, Y.Y.; Ilouze, M.; Mattei, J.; Bunn, P.A., Jr.; Hirsch, F.R.; Haick, H. Volatile fingerprints of cancer specific genetic mutations. *Nanomedicine* **2013**, *9*, 758–766. [CrossRef] [PubMed]
56. Makohon-Moore, A.; Iacobuzio-Donahue, C.A. Pancreatic cancer biology and genetics from an evolutionary perspective. *Nat. Rev. Cancer* **2016**, *16*, 553–565. [CrossRef] [PubMed]
57. Son, J.; Lyssiotis, C.A.; Ying, H.; Wang, X.; Hua, S.; Ligorio, M.; Perera, R.M.; Ferrone, C.R.; Mullarky, E.; Shyh-Chang, N.; et al. Glutamine supports pancreatic cancer growth through a KRAS-regulated metabolic pathway. *Nature* **2013**, *496*, 101–105. [CrossRef] [PubMed]
58. Murphy, D.A.; Courtneidge, S.A. The 'ins' and 'outs' of podosomes and invadopodia: Characteristics, formation and function. *Nat. Rev. Mol. Cell Biol.* **2011**, *12*, 413–426. [CrossRef]
59. Hoshino, D.; Branch, K.M.; Weaver, A.M. Signaling inputs to invadopodia and podosomes. *J. Cell Sci.* **2013**, *126*, 2979–2989. [CrossRef]
60. Schwab, M. *Cathepsins, Encyclopedia of Cancer*; Springer: Berlin/Heidelberg, Germany, 2017. [CrossRef]
61. Génot, E.; Gligorijevic, B. Invadosomes in their natural habitat. *Eur. J. Cell Biol.* **2014**, *93*, 367–379. [CrossRef]
62. Goicoechea, S.; Garcia-Mata, R.; Staub, J.; Valdivia, A.; Sharek, L.; McCulloch, C.G.; Hwang, R.F.; Urrutia, R.P.; Yeh, J.J.; Kim, H.J.; et al. Palladin promotes invasion of pancreatic cancer cells by enhancing invadopodia formation in cancer-associated fibroblasts. *Oncogene* **2013**, *33*, 1265–1273. [CrossRef]
63. David-Pfeuty, T.; Singer, S.J. Altered distributions of the cytoskeletal proteins vinculin and α-actinin in cultured fibroblasts transformed by Rous sarcoma virus. *Proc. Natl Acad. Sci. USA* **1980**, *77*, 6687–6691. [CrossRef] [PubMed]
64. Shaye, D.D.; Greenwald, I. OrthoList: A Compendium of *C. elegans* Genes with Human Orthologs. *PLoS ONE* **2011**, *6*, e20085. [CrossRef] [PubMed]
65. Saltel, F.; Daubon, T.; Juin, A.; Ganuza, I.E.; Veillat, V.; Génot, E. Invadosomes: Intriguing structures with promise. *Eur. J. Cell Biol.* **2011**, *90*, 100–107. [CrossRef] [PubMed]
66. Artym, V.V.; Zhang, Y.; Seillier-Moiseiwitsch, F.; Yamada, K.; Mueller, S.C. Dynamic Interactions of Cortactin and Membrane Type 1 Matrix Metalloproteinase at Invadopodia: Defining the Stages of Invadopodia Formation and Function. *Cancer Res.* **2006**, *66*, 3034–3043. [CrossRef] [PubMed]
67. Ayala, I.; Giacchetti, G.; Caldieri, G.; Attanasio, F.; Mariggiò, S.; Tetè, S.; Polishchuk, R.; Castronovo, V.; Buccione, R. Faciogenital dysplasia protein Fgd1 regulates invadopodia biogenesis and extracellular matrix degradation and is up-regulated in prostate and breast cancer. *Cancer Res.* **2009**, *69*, 747–752. [CrossRef]
68. Lohmer, L.L.; Clay, M.R.; Naegeli, K.M.; Chi, Q.; Ziel, J.W.; Hagedorn, E.J.; Park, J.E.; Jayadev, R.; Sherwood, D.R. A Sensitized Screen for Genes Promoting Invadopodia Function In Vivo: CDC-42 and Rab GDI-1 Direct Distinct Aspects of Invadopodia Formation. *PLoS Genet.* **2016**, *12*, e1005786. [CrossRef]

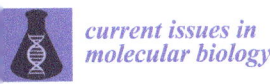

current issues in molecular biology

MDPI

Article

Lessons from a Single Amino Acid Substitution: Anticancer and Antibacterial Properties of Two Phospholipase A_2-Derived Peptides

José R. Almeida [1], Bruno Mendes [1,2], Marcelo Lancellotti [2,3], Gilberto C. Franchi, Jr. [4], Óscar Passos [5], Maria J. Ramos [5], Pedro A. Fernandes [5], Cláudia Alves [5], Nuno Vale [6,7,*], Paula Gomes [5] and Saulo L. da Silva [1,5,8]

1 Universidad Regional Amazónica Ikiam, Km 7 Via Muyuna, Tena 150150, Ecuador; rafael.dealmeida@ikiam.edu.ec (J.R.A.); bruno000mendes@gmail.com (B.M.); biomol2@hotmail.com (S.L.d.S.)
2 Departmento de Bioquímica e Biologia Tecidual, Instituto de Biologia, Universidade Estadual de Campinas (UNICAMP), Campinas 13083-862, SP, Brazil; marcelo.lancellotti@fcf.unicamp.br
3 Faculdade de Ciências Farmacêuticas, Universidade Estadual de Campinas (UNICAMP), Campinas 13083-871, SP, Brazil
4 Centro Integrado de Pesquisas Oncohematologicas da Infancia, Universidade Estadual de Campinas (UNICAMP), Campinas 13083-881, SP, Brazil; gfran@unicamp.br
5 LAQV/REQUIMTE, Departamento de Química e Bioquímica, Faculdade de Ciências, Universidade do Porto, Rua Campo Alegre s/n, 4169-007 Porto, Portugal; oscar.passos@fc.up.pt (Ó.P.); mjramos@fc.up.pt (M.J.R.); pafernan@fc.up.pt (P.A.F.); claudialves05@gmail.com (C.A.); pgomes@fc.up.pt (P.G.)
6 OncoPharma Research Group, Center for Health Technology and Services Research (CINTESIS), Rua Doutor Plácido da Costa, 4200-450 Porto, Portugal
7 Department of Community Medicine, Information and Health Decision Sciences (MEDCIDS), Faculty of Medicine, University of Porto, Alameda Professor Hernâni Monteiro, 4200-319 Porto, Portugal
8 Faculty of Chemical Sciences, University of Cuenca, Cuenca 010107, Ecuador
* Correspondence: nunovale@med.up.pt

Citation: Almeida, J.R.; Mendes, B.; Lancellotti, M.; Franchi, G.C., Jr.; Passos, Ó.; Ramos, M.J.; Fernandes, P.A.; Alves, C.; Vale, N.; Gomes, P.; et al. Lessons from a Single Amino Acid Substitution: Anticancer and Antibacterial Properties of Two Phospholipase A_2-Derived Peptides. *Curr. Issues Mol. Biol.* **2022**, *44*, 46–62. https://doi.org/10.3390/cimb44010004

Academic Editor: Dumitru A. Iacobas

Received: 6 December 2021
Accepted: 15 December 2021
Published: 22 December 2021

Abstract: The membrane-active nature of phospholipase A_2-derived peptides makes them potential candidates for antineoplastic and antibacterial therapies. Two short 13-mer C-terminal fragments taken from snake venom Lys49-PLA_2 toxins (p-AppK and p-Acl), differing by a leucine/phenylalanine substitution, were synthesized and their bioactivity was evaluated. Their capacity to interfere with the survival of Gram-positive and Gram-negative bacteria as well as with solid and liquid tumors was assessed in vitro. Toxicity to red blood cells was investigated via in silico and in vitro techniques. The mode of action was mainly studied by molecular dynamics simulations and membrane permeabilization assays. Briefly, both peptides have dual activity, i.e., they act against both bacteria, including multidrug-resistant strains and tumor cells. All tested bacteria were susceptible to both peptides, *Pseudomonas aeruginosa* being the most affected. RAMOS, K562, NB4, and CEM cells were the main leukemic targets of the peptides. In general, p-Acl showed more significant activity, suggesting that phenylalanine confers advantages to the antibacterial and antitumor mechanism, particularly for osteosarcoma lines (HOS and MG63). Peptide-based treatment increased the uptake of a DNA-intercalating dye by bacteria, suggesting membrane damage. Indeed, p-AppK and p-Acl did not disrupt erythrocyte membranes, in agreement with in silico predictions. The latter revealed that the peptides deform the membrane and increase its permeability by facilitating solvent penetration. This phenomenon is expected to catalyze the permeation of solutes that otherwise could not cross the hydrophobic membrane core. In conclusion, the present study highlights the role of a single amino acid substitution present in natural sequences towards the development of dual-action agents. In other words, dissecting and fine-tuning biomembrane remodeling proteins, such as snake venom phospholipase A_2 isoforms, is again demonstrated as a valuable source of therapeutic peptides.

Keywords: *Agkistrodon*; leucine; membrane; phenylalanine; venom peptides

1. Introduction

Bioactive peptides have opened a new horizon in drug discovery and are currently considered a cornerstone in developing therapies for cancer and bacterial infections [1]. In the last years, more than 7% of Food and Drug Administration-approved drugs are peptide-based entities [2]. Yet, the design and refinement of these active structures are highly challenging [3], and several prediction models and tools are being proposed and have been made available in recent years [4,5]. However, despite these bioinformatics and statistical advances, the isolation of peptides from natural sources, the identification by genomic and transcriptomic investigations, or the synthesis of molecular region mimics of target proteins remain useful both to discover new structures and to understand functional aspects that are crucial for the selection of more potent and selective molecules [6,7].

Toxins are natural products characterized by experiencing critical selective pressures for their action in pathways and cell structures of biomedical interest [8]. In snake venoms, a diversity of protein isoforms is found [9]. This repertoire of molecules with subtle differences in the primary structure offers ample opportunity to find therapeutic applications and obtain relevant information to elaborate predictive models and synthesize more efficient molecules. Biomedical products derived from toxins are a classical example of the pivotal contribution of natural products in the drug discovery process. An important precedent is Captopril, an antihypertensive synthetic analogue of one short molecule from *Bothrops jaracaca*, which is considered the first member of angiotensin-converting enzyme inhibitor medications [10]. In this connection, snake venom phospholipases A_2 (PLA_2) are a diverse group of great pharmaceutical relevance [11]. The PLA_2 family is extremely capable of interacting, modifying, and disrupting membrane lipids [12,13]. Additionally, the membranes of pathogens and cancer cells are an attractive target for low-resistance therapeutic approaches [14]. Thus, many lytic PLA_2-derived peptides with antitumor [15,16], leishmanicidal [17], and antibacterial [18,19] effects have been synthesized to reproduce biological interactions between the C-terminal of the protein templates and biomembranes. The peptide p-AppK (KKYKAYFKLKCKK) is a specific example of that, inspired in the myotoxic C-terminal region of a Lys49-PLA_2 derived from the snake *Agkistrodon piscivorus piscivorus*. This short synthetic molecule has antitumoral activity against murine tumor cell lines, such as sarcoma, melanoma, mammary carcinoma, and others [15]. p-AppK is amongst the PLA_2-mimicking peptides with highest cytotoxic activity, higher than that of pEM-2 that is recognized by its in vivo action [15]. Curiously, the 115–129 C-terminal region from the snake *Agkistrodon contortrix laticinctus*, corresponding to peptide p-Acl (KKYKAYFKFKCKK), has high sequence identity with p-AppK, differing only in the amino acid residue at position 9 (where p-Acl has a phenylalanine in place of the leucine present in p-AppK). Peptide p-Acl has not been previously evaluated against either bacteria or cancer cells, whereas some authors have reported that the presence of phenylalanine in certain peptide sequences is a determinant factor for the display of anticancer and antibacterial activities. Phenylalanine-containing peptides selectively recognize and fuse to certain membranes [20,21]. In view of this, we synthesized p-AppK and p-Acl for comparative evaluation of their cytotoxic and antibacterial activities, in order to establish structure–activity relationships (SAR) that can guide the bioengineering of novel peptides with dual therapeutic activity. These were further simulated in silico, to obtain a molecular-level picture of the effects induced by the peptides upon interaction with bacterial and eukaryotic (healthy and cancerous) cell membranes. Results herein reported confirm both peptides as dual-action therapeutic leads, where the Leu→Phe substitution, from p-AppK to p-Acl, apparently leads to an increase in peptide's antibacterial and antiproliferative potency.

2. Materials and Methods

2.1. Sequence Analysis

Online bioinformatics tools were used to perform amino acid sequence analysis of the peptides. A functional study was first run using CAMPR3 (http://www.camp.bicnirrh.res.in/prediction.php, accessed on 1 December 2021). AntiCP (https://webs.iiitd.edu.in/raghava/

anticp/index.html, accessed on 1 December 2021), ACPred (http://codes.bio/acpred/, accessed on 1 December 2021), and CPPred-RF (http://server.malab.cn/CPPred-RF/index.jsp, accessed on 1 December 2021) were also used to predict the antimicrobial, anticancer, and cell-penetrating properties of both peptides, respectively. The Rational Design of Antimicrobial Peptides tool in CAMPR3 (http://www.camp.bicnirrh.res.in/predict_c21/, accessed on 1 December 2021) was next employed to generate peptide analogues to confirm whether the exchange of amino acids found in nature is predicted as favorable to the antimicrobial activity by different software algorithms (Support Vector Machine, Random Forest, and Discriminant Analysis classifiers). Finally, PepDraw (http://www.tulane.edu/~biochem/WW/PepDraw/, accessed on 1 December 2021) was used to calculate the peptides' total net charge, PI and hydrophobicity, taking into account the C-terminal amidations.

2.2. Peptide Synthesis

C-terminally amidated p-AppK and p-Acl peptides were assembled in a CEM Liberty 1 system employing a rapid MW-SPPS protocol on Rink amide-MBHA resin. Once the elongation of the polypeptide chain ended, the peptide was fully deprotected and cleaved from the resin by a 2 h acidolytic treatment, at room temperature (RT), using a cocktail containing 95% trifluoroacetic acid (TFA), 2.5% triisopropylsilane (TIS), and 2.5% water. The crude peptides thus obtained were precipitated with cold diethyl ether, centrifuged, and the peptide pellets collected, re-dissolved in 10% aqueous acetic acid, and lyophilized. The peptides were then purified by RP-MPLC using the same conditions described by Almeida et al. (2018) [19]. The purity degree of peptides was determined by reverse-phase high performance liquid chromatography (RP-HPLC), and the purification step repeated, if necessary, until a minimum purity of 95% was reached. Finally, the molecular identity of the pure peptides was confirmed by ESI-IT MS, using an LTQ OrbitrapTM XL hybrid mass spectrometer.

2.3. Hemolysis Assays

The erythrocyte-lysing potential of the peptides was analyzed by in silico and in vitro approaches. The hemolytic peptide modules of the following bioinformatics tools: HLPred-Fuse (http://thegleelab.org/HLPpred-Fuse/, accessed on 1 December 2021), HAPPENN (https://research.timmons.eu/happen, accessed on 1 December 2021), HEmoPI (https://webs.iiitd.edu.in/raghava/hemopi/, accessed on 1 December 2021), and HempPImod (https://webs.iiitd.edu.in/raghava/hemopimod/ter_str.php, accessed on 1 December 2021) were used to predict the effect on the RBC. Quantitative in vitro hemolytic activity assays against heparinized collected human RBC (type AB Rh-) were performed as previously reported by Proaño-Bolanos et al., 2019 [22]. Briefly, a suspension of the mammalian cells was incubated with different concentrations of peptides dissolved in PBS buffer for 2 h at 37 °C. After this procedure, the material was centrifuged, and the supernatant aliquoted to a 96-well microplate. The percentage of disrupted RBC was quantitated by spectrophotometry, based on absorbance readings at 405 nm using a VersaMax multiwell plate reader (Molecular Devices, Sunnyvale, CA, USA). Erythrocytes incubated with Triton 2%, a known membrane-damage agent, were taken as positive control (100% hemolysis). Hemolytic assays were performed in triplicates at least three times.

2.4. Antibacterial Activity

The broth microdilution method was used to determine whether p-AppK and p-Acl display antibacterial properties. This evaluation was performed on three Gram-negative (*Pseudomonas aeruginosa* 31NM, *P. aeruginosa* ATCC 27853, and *Escherichia coli* ATCC 25922) and two Gram-positive (*Staphyloccocus aureus* BEC9393 and *S. aureus* Rib1) bacterial strains. The bacterial growth inhibition protocols were as detailed in a previous report from our group [19]. Briefly, the mid-log phase bacterial growth suspensions were incubated for 24 h with the synthetic peptides in a 96-well microplate. After this period, bacterial growth was determined, considering the absorbance values at 595 nm. Optical density measurements of bacterial suspensions incubated in medium only were taken as 100% growth. Three independent experiments were carried out in triplicate.

2.5. Membrane Damage

The effect of p-AppK and p-Acl on the membranes of Gram-positive (*S. aureus* BEC9393) and Gram-negative (*P. aeruginosa* ATCC 27853) bacteria was investigated using propidium iodide (PI) uptake assays. The bacteria were cultured and exposed to 100 μM peptides and PI (15 μg/mL). Aliquots (200 μL) of the test solutions were added to 96-well microplates. The time-dependent changes in PI fluorescence were quantified and recorded at excitation and emission wavelengths of 580 and 620 nm, respectively, for 30 min, using a VersaMax fluorescence microplate reader (Molecular Devices, Sunnyvale, CA, USA). The membrane damage was assessed by three independent assays performed in triplicate.

2.6. Cytotoxicity

In vitro cell viability assays were conducted to screen the anticancer properties of peptides p-AppK and p-Acl. The assays were based on colorimetric detection of the enzymatic reduction of 3-[4,5-dimethylthiazole-2-yl]-2,5-diphenyltetrazolium bromide (MTT) that occurs in viable cells. A vast repertoire of 13 solid tumors and 10 leukemia cell lines from the cell bank of the Integrated Center for Childhood Onco-Hematological Investigation (UNICAMP, Brazil) was used. Namely, the following solid malignant neoplasms were used: OVCAR (human ovarian adenocarcinoma), MACL1 and MGSO3 (human primary breast cancer), MCF7 (human breast adenocarcinoma), VW473 (human medulloblastoma), MG63 (human osteosarcoma), SHSY5S (human neuroblastoma), NCI (non-small cell lung cancer), U138 (human neuronal glioblastoma), U87 (glioblastoma), PC3 (human prostate cancer), H1299 (human non-small cell lung carcinoma), and HOS (human osteosarcoma). Leukemia cell lines included K562 (chronic myeloid/chronic myelogenous human), NB4 (human acute promyelocytic leukemia), HL60 (acute promyelocytic leukemia), RAMOS (human Burkitt lymphoma), NALM6 (acute B lymphoblastoma), B15 (human B cell precursor leukemia), REH (acute lymphocytic leukemia non-T; non-B), JURKAT (human T cell lymphoblast-like), TALL (acute lymphoblastic T-cell leukemia), and CEM (human acute lymphoblast leukemia). All procedures were run in automated mode on a Liquid Handling Workstation epMotion 5070 (Eppendorf). Firstly, the cell lines used in this investigation were cultured in plastic bottles (75 mL) in the presence of RPMI 1640 (Sigma R6504) medium supplemented with 10% fetal calf serum, 1% penicillin (IU/mL), and streptomycin (10 mg/mL) at 37 °C under humidified air with 5% CO_2. The tumor cells were distributed in 96-well microplates and were exposed to 100 μM solutions of the synthetic peptides. The metabolic activity of cells was measured by spectrophotometric monitoring of the absorbance at 570 nm on a Synergy ELISA plate reader (Bio Tek Instruments). Two reference antineoplastic drugs, paclitaxel and vincristine, were used as positive controls, whereas cells growing in medium alone was taken as 100% viability (negative control). The results were expressed as cell viability inhibition and represent three independent experiments run in triplicate.

2.7. Molecular Dynamics Simulations

2.7.1. Membrane Model

We modelled six different membranes of pure and mixed composition, each with either p-Acl or p-AppK peptides inserted across them, generating a total of twelve different systems. The membranes were inserted in rectangular boxes of water extending by 32 Å beyond the membrane headgroups. Counter-ions (K^+ and Cl^-) were used to neutralize the system (Supplementary Materials).

The number of phospholipids per membrane was 140; the total number of atoms was 37,447–48,190, depending on the specific membrane simulated. The six membrane models were composed of: 100% 2,3-dioleoyl-d-glycero-1-phosphatidylserine (DOPS, membrane 1, a totally negative membrane, which might exist in small, specific regions of the eukaryotic membrane, and for which the peptides should have maximal affinity); 100% 2,3 dipalmitoyl-d-glycero-1-phosphatidylcholine (DPPC, membrane 2, representing the dominant zwitterionic composition of eukaryotic membranes); 86% 1-palmytoil-2-cis-9,10-methylenehexadecanoyl-phosphorylethanolamine (PMPE); 4% palmitic acid (PAL); 4% palmitoleic acid (PALO); 4% oleic acid (OLE); 2% myristic acid (MYR, membrane 3, whose composition was taken from Almeida et al., 2018); 60% PMPE; 31% 1-palmitoyl-2-oleoyl-sn-glycero-3-[phospho-rac-(1-glycerol) (POPG); 6% cardiolipin with head group charge −2 (PVCL2); and 3% tetraoleoyl cardiolipin with head group charge −2 (TLCL2, membrane 4, whose composition was taken from Sahoo et al., 2017). Thymocytes-like membrane constituted by 13 different lipidic components (membrane 5, whose composition was taken from van Blitterswijk et al. [23] and is detailed in Supplementary Materials Table S1); and finally, a leukemia-like membrane constituted by 15 different lipidic components (membrane 6, composition taken from Van Blitterswijk et al. and detailed at Supplementary Materials Table S2). In summary, membranes 1 and 2 represent the extremes of the spectrum of the composition of eukaryotic membranes, which have a mix of anionic and zwitterionic phospholipids in different percentages at different regions; membranes 3 and 4 model the *E. coli* membrane; and membranes 5 and 6 are models for thymus and leukemia cancer cell membranes, respectively.

2.7.2. Molecular Dynamics Simulations

A series of molecular dynamics simulations were carried out within the GROMACS 2018 software package [24]. We conducted a 5000-step minimization for each membrane–peptide system to remove eventual tensions or clashes in the system's structure. Subsequently, the systems were gradually heated to 303.5 K in a 100 ps simulation using the Berendsen thermostat [25]. The density and interactions of the system were equilibrated in a 1 ns long simulation in the isothermal-isobaric ensemble (NPT) with decreasing restraints forces. At last, a production run of 200 ns was conducted in the NPT ensemble with Berendsen barostat [25]. The cut-off for the Lennard-Jones interactions was set to 12 Å. Coulomb interactions were treated using the Particle Mesh-Ewald (PME) summation method [26], with a cut-off of 12 Å for the real part of the sum. The time step of numerical integration was 2 fs (details in Supplementary Materials Table S3).

3. Discussion

Peptide drugs constitute an emerging therapy occupying a prominent market niche in the pharmaceutical industry trends [27,28]. The peptide therapeutics market is progressing with very favorable prospects, recognized potential, recent successful approvals by regulatory agencies and economic expansion [2]. The years of evolution of these compounds reveal clues for the selection of new hits, and simultaneously, new data to refine the statistical and computational models of prediction [29,30]. The bioengineering and optimization of antibacterial and antitumor peptides have been automated today [31,32]. The development of tools, methods, and computational algorithms has allowed significant advances in the peptide design and the understanding of critical interactions for the mechanism of action of these molecules [33,34]. However, in parallel, screening from natural resources, such as animal venoms and secretions, continues to be of great interest to the scientific and pharmaceutical community [35]. Motivated by this, in the present work, we combined computational and experimental approaches to evaluate the potential dual effects of structurally similar peptides derived from Lys49 toxins from snake venoms.

p-AppK was one of the first peptides mimicking PLA_2s proteins reported to display antitumor properties. This peptide is active against B16 (melanoma), EMT-6 (mammary carcinoma), S-180 (sarcoma), and P3X (myeloma), with IC50 values between 56 μM and 156 μM, being more potent against melanoma [15]. The present work extends the understanding of the spectrum of bioactivity of this peptide by investigating its capacity to inhibit bacterial growth, not characterized previously. This enabled confirmation of the dual-action of p-AppK, which is proven efficient against Gram-positive and Gram-negative bacteria, including the multidrug-resistant (MDR) clinical isolates *P. aeruginosa* 31NM, *S. aureus* rib1 and *S. aureus* BEC93. The present study additionally covered the characterization of peptide p-Acl, which differs from p-AppK in a single amino acid residue (Leu→Phe). Both peptides recapitulate short fragments derived from the C-terminal end of snake venom PLA_2 isoenzymes, and p-Ac1 was found to retain the antibacterial activity profile of p-AppK. Relevantly, the parent PLA_2s from which both peptides derive are catalytically inactive isoforms. Yet, they retain the ability to disturb bacterial cell membranes by a mechanism independent of the enzymatic hydrolysis dictated primarily by the leading role of cationic and hydrophobic residues in the protein C-terminal region [36]. Hence, it is herein demonstrated that such ability is conserved by the synthetic peptides mimicking this region, as a rapid influx of PI into Gram-positive and Gram-negative cells was detected upon peptide treatment. The uptake of this cell-impermeable DNA dye reveals that the structural and dynamic properties of bacterial membranes were significantly compromised in the presence of synthetic peptides, which was further supported by molecular dynamics simulations. Thus, the alteration of the integrity of specific layers of membranes induced by the 13-mer peptides should have functional consequences that can lead to loss of intracellular content and affect the survival of bacteria. Won and Ianoul (2009) have demonstrated the ability of pEM-2, a modified peptide also derived from a PLA_2, to interact with membrane phospholipids and induce cell lysis, which has been reported to be dependent on the lipid composition of the bacterial membrane and directly affected by hydrophobic interactions [36,37]. In general, the detergent and membrane disruptive action has been widely used as a strategy by many antimicrobial peptides and is seen as a promise to transform such peptides into therapeutics based on targeting bacterial membrane phospholipids [38,39].

Low toxicity is a keystone for the translational success of candidate peptides. Importantly, the biomimetic peptides herein studied did not promote lysis of the erythrocyte membrane in biologically active concentrations, suggesting selectivity and confirming the functional predictions determined by bioinformatics analyses. In earlier works addressing PLA_2 peptide mimics, similar results were found, confirming this type of synthetic peptides as non-hemolytic [18,40]. Although red blood cells are considered a classical model for the initial assessment of peptide toxicity, further works should consider the use of other healthy cells. The tunability and simple peptide synthesis enable an easy fine-tuning, which contributes to selectivity and generation of safer and efficient analogues [28].

A selective molecular recognition ability was also suggested by the cytotoxicity assays. Despite promising activity in a wide variety of cancer cells, both peptides were inactive or showed low activity against some solid tumor cell lines, such as U138, MGSO3 and VW473. In agreement with this selective behavior, different profiles were obtained for leukemic cells as both peptides were strongly active against leukemic cells (particularly NB4, RAMOS and CEM). This selectivity is probably due to the diversity in the complex composition of membrane lipids, which are often suggested as primary targets for bioactive peptides [41], especially those derived from, or inspired in, membrane binding toxins. Cells that are similarly affected must have an analogous membrane lipid composition: for instance, p-Acl is more active in two lines of osteosarcoma cells (HOS and MG63), in line with the higher structural similarity between their respective membranes [42]. In view of this, the knowledge of the lipidome profile of different tumor cells might help to develop anticancer pharmaceuticals [43,44] acting via selective tumor cell membrane permeation. The molecular basis behind antitumor effects detailed here has not been experimentally accessed. However, it is likely based on lytic activity as evidenced for antibacterial action and reported for other anticancer peptides [45,46]. Additionally, our in silico approach suggested a possible cell-penetrating capacity, which may be in line with the antitumor activity and should be confirmed in the future. The molecular dynamics simulations shown here provide further support for this hypothesis, as the peptides induce rapid and spontaneous membrane permeability. These observations can lead to applying these peptides as tumor-cell-targeting carriers, facilitating the transport of impermeable drugs into the cell, hence enabling their action and clinical application. Combining these anticancer peptides with clinical drugs can also be an exciting route for the generation of combined therapies that potentiate antineoplastic effects and allow a fast and efficient clinical response. Oncolytic peptides have shown promising in vivo effects, being of relevance for the clinical development of a new generation of antineoplastic drugs [47]. Further studies should access the mode of action of p-AppK and p-Acl and evaluate the activity of scrambled versions to understand better the differences in cell recognition, especially in solid tumor cells.

Overall, this work confirms dual functionality for both p-AppK and p-Ac1, which highlights how membrane-binding toxins such as PLA_2s are potential starting points to obtain short molecules capable of targeting cancer and bacterial cell membranes. Other previous studies have also reported potent antibacterial and antineoplastic peptides structurally related to p-AppK and p-Acl, such as pEM-2 [15], pep-MTII [16], pBmje [48] and pepBthTX-I [49]. Interestingly, these peptides were inspired in isoforms of membrane-remodeling proteins whose C-termini are highly homologous to those of our template proteins. As expected, p-AppK and p-Acl exhibit similar physicochemical properties, which also closely resemble those of other anticancer and/or antimicrobial peptides derived from PLA_2s, namely the synthetic peptides mentioned above. Thus, all these peptides are highly cationic, possessing high pI values and several hydrophobic residues. Commonly, peptide charge and hydrophobicity have been highlighted as flagships in the development of antibacterial agents, taking into account the specific features of bacterial membranes [50]. Still, p-Acl generally demonstrated more significant activity against Gram-positive bacteria and some solid tumor cell lines, especially MG63 and HOS. Therefore, a single leucine→phenylalanine substitution translates into significant differences in the peptides' functionality, particularly regarding effects against osteosarcoma cell lines. This type of cancer is one of the prominent malignant bone tumors in children [51]. Furthermore, bacterial infections, including those caused by MDR pathogens, are a relevant complication in the clinical outcome for patients with osteosarcoma, which defies current chemotherapy [52]. Therefore, the potent dual-action of p-Acl emerges as a cornerstone to develop therapeutics able to address both problems.

This work gives us an additional lesson on how a single amino acid substitution may be of relevance. In particular, the role of leucine and phenylalanine residues has been the subject of other studies involving therapeutic applications of peptides [53–55]. In line with this, Sahoo et al. (2017) demonstrated that the substitution of leucine for phenylalanine significantly increases the toxic effect of cathelicidin-5(1-18) against cancer and bacterial cells [56]. Studies have also demonstrated the interaction of these amino acid residues with biomembranes, which is strongly dependent on the specific membrane components [54,57]. Although it is not a general principle, as evidenced by our results (for some cells, the same effect is observed for p-AppK and p-Acl, or higher activity is induced by p-AppK), this amino acid substitution may increase the capacity for targeting the membranes of certain cancer cells. Consequently, tailored Leu→Phe replacements may be considered as a useful strategy for the design and selection of therapeutic candidates, especially for osteosarcoma. However, these structure–function relationships must be understood in a holistic manner considering the diversity of the membranes and the unique organization of the peptides so that optimal peptide–membrane interactions are promoted. In this context, future investigations involving alanine scanning libraries should be useful to clarify the role of phenylalanine for the cytotoxic activity. Additionally, this antiproliferative pattern must also be evaluated at lower concentrations.

Dual targeted peptide therapy constitutes a promising treatment approach for different diseases [58]. Currently, some small molecules with antimicrobial and/or anticancer properties have been tested in clinical trials. For example, the peptide-based pharmaceuticals ANG-1005, GRN-1201, C16G2 and NP108 were evaluated by AngioChem, Green Peptide, Chengdu Sen Nuo, Wei Biotechnology and NovaBiotics, respectively [58,59]. This combination of functional roles increases the interest of the pharmaceutical market in the possible applications of the peptides [60]. For example, in oncological patients, bacterial infections are common [61]. Complications due to these infections are widespread in children diagnosed with osteosarcoma [52]. Because of this, dual peptides such as PLA_2s-derived peptides are considered potential therapeutic candidates to this end.

4. Results

4.1. Peptide Design and Sequence Analysis

Two venom toxin-derived peptides, p-AppK and p-Acl, were selected for this study. Both are based on Lys49-PLA_2s isoforms from *Agkistrodon* spp. Peptide p-AppK is inspired in a membrane disrupting protein from *Agkistrodon piscivorus piscivorus*, and has been previously reported to exert cytotoxic effects on tumors. Peptide p-Acl is also derived from a membrane-damaging toxin, a Lys49-PLA_2 from *Agkistrodon contortrix laticinctus*. The main characteristics of both peptides and their parent toxins (used as templates) are detailed in Table 1.

Table 1. The biomimetic peptides were designed to replicate the C-terminal region (residues 115–129) of two membrane-damaging Lys49-PLA_2 from *Agkistrodon* spp. venoms. The molecular weight was estimated by PepCalc (https://pepcalc.com/, accessed on 1 December 2021), and the sequence identity was calculated by the Sequence Identity and Similarity tool (SIAS, http://imed.med.ucm.es/Tools/sias.html, accessed on 1 December 2021).

Peptides	Length	Molecular Weight (Da)	Parent Protein	UniProtKB	Snake Species	% Identity
p-AppK	13	1675.14	App toxin	P04361	*A. piscivorus piscivorus*	100
p-Acl	13	1709.15	Acl toxin	P49121	*A. contortrix laticinctus*	92.37

The bioinformatics tools employed predicted both phospholipase A_2-derived peptides to possess antibacterial, antitumor and cell-penetrating properties. The main physicochemical parameters of both peptides, as calculated by PepDraw, are shown in Table 2. Expectedly, p-AppK and p-Acl are structurally similar and share most physicochemical characteristics, such as charge and pI. Still, they slightly differ in hydrophobicity, hydropathicity, molecular volume, and molecular weight due to leucine substitution by phenylalanine.

Table 2. Functional predictions and physicochemical parameters of the phospholipase A_2-derived peptides. In silico tools described in the methodology section were used to predict whether the peptides had (+) or not (-) anticancer, antibacterial, and cell-penetrating peptide (CPP) properties. PepDraw was used to estimate physicochemical parameters (charge, pI, and hydrophobicity).

Peptides	Anticancer Properties	Antibacterial Properties	CPP Properties	Charge	pI	Hydrophobicity
p-AppK	+	+	+	+8	10.76	23.60
p-Acl	+	+	+	+8	10.76	23.14

4.2. Peptide Synthesis

Both peptides were successfully produced and purified by microwave-assisted solid-phase peptide synthesis (MW-SPPS) and reverse-phase medium pressure liquid chromatography (RP-MPLC), respectively. Final purity degrees were greater than 95.5%, and the expected molecular weights were confirmed by electrospray ionization-ion trap mass spectrometry (ESI-IT MS) (Supplementary Materials, Figure S1).

4.3. Hemolytic Character

Functional analysis using different algorithms predicted that p-AppK and p-Acl are non-toxic peptides, with a very low probability of exerting harmful effects on red blood cells (RBCs). Data obtained in silico were corroborated by an in vitro quantitative hemolytic assay at a peptide concentration range of 2.25–176 μM (Figure 1). Triton X-100 was used as the positive control for 100% hemolysis. Minor RBC lysis was induced by the peptides, namely, below 7.0% at the highest peptide concentration tested, for both synthetic molecules. Thus, the leucine/phenylalanine substitution does not affect toxicity, suggesting a favorable safety profile for both peptides.

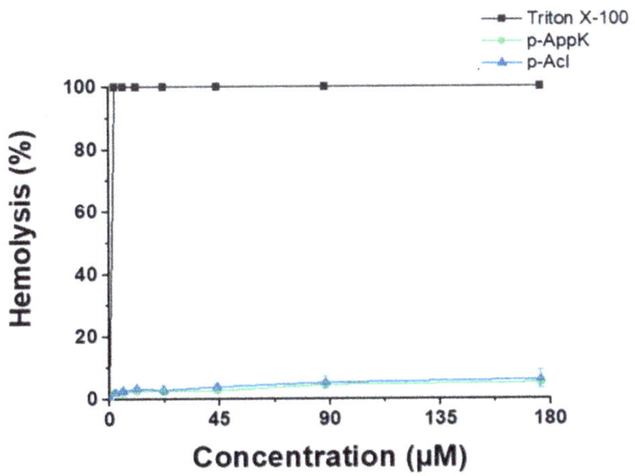

Figure 1. Evaluation of erythrocyte membrane disruption caused by the synthetic peptides. Red blood cells were incubated with 7 different concentrations of p-AppK (blue) and p-Acl (green). The percentage of hemolysis was measured in relation to the effect caused by a hemolytic surfactant, Triton X-100.

4.4. Antibacterial Activity

Peptides p-AppK and p-Acl were capable of inhibiting in vitro the growth of both Gram-positive and Gram-negative bacteria, including clinical isolates (Figure 2). *P. aeruginosa* strains (31NM and ATCC) were the most susceptible to both peptides, which showed similar antibacterial potency. The two peptides also significantly inhibited growth of *S. aureus*, but in this case p-Acl revealed a more significant effect. Interestingly, when the primary structure of p-Appk was analyzed by the Rational Design of Antimicrobial Peptides module of the CAMPR3 webserver, the sequence of p-Acl was suggested as a "potential analogous sequence".

4.5. Membrane Damage

The peptides' effects on the membranes of two bacterial species were analyzed over time in terms of percentage of PI uptake. PI uptake remained negligible in both bacterial cultures grown in the absence of the test peptides. In contrast, when *S. aureus* and *P. aeruginosa* were grown in the presence of the peptides (100 μM), a rapid and significant increase in PI uptake was observed (Figure 3). Based on the inability of this fluorescent dye to cross intact membranes, these findings demonstrate that the peptides caused membrane damage of both Gram-positive and Gram-negative bacteria, enabling the interaction of PI with DNA.

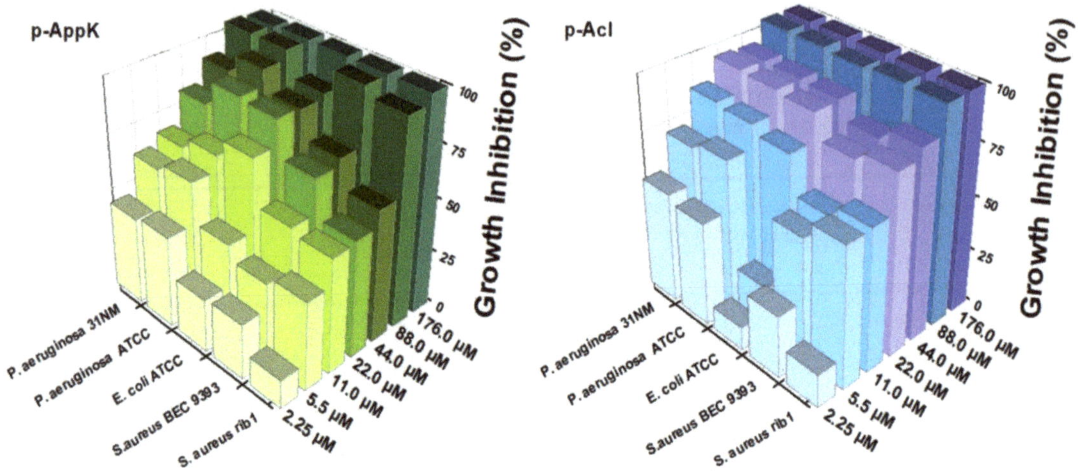

(**a**) (**b**)

Figure 2. Dose-dependent inhibition of bacterial growth caused by the synthetic peptides using broth microdilution assay. (Left—(**a**)) p-AppK (green) and (Right—(**b**)) p-Acl (blue) reduced the bacterial viability (*P. aeruginosa* 31NM, *P. aeruginosa* ATCC, *E. coli* ATCC, *S. aureus* BEC 9393, and *S. aureus* rib1) after 24 h as a function of their concentrations. The growth inhibition was calculated considering the maximum optical density of the negative control as the reference. The experiments were performed in triplicate.

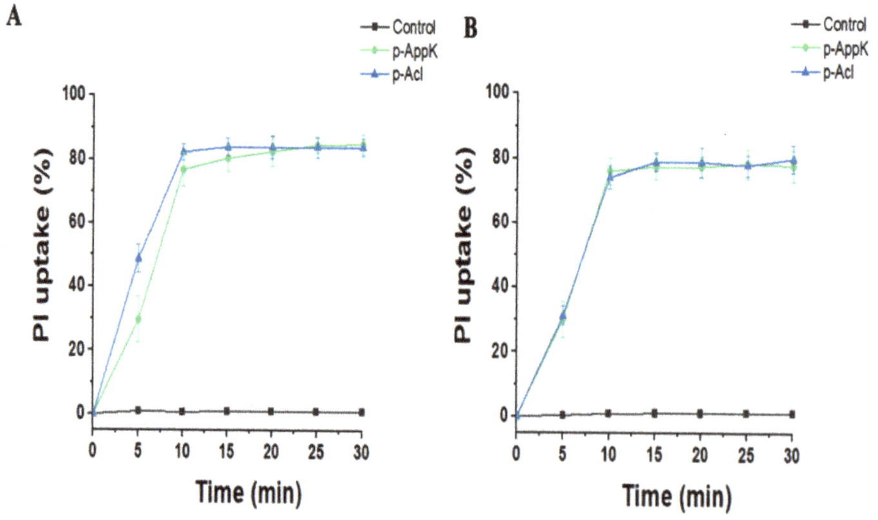

Figure 3. In vitro membrane-disruptive activity provoked by p-AppK (green) and p-Acl (blue). The membrane integrity evaluation of (**A**) *P. aeruginosa* ATCC and (**B**) *S. aureus* BEC9393 incubated with 100 μM peptides was determined as a function of fluorescent dye uptake.

4.6. Cytotoxicity

The microculture screening assay agreed with the cytotoxicity predicted by the sequence analysis software. Furthermore, at a concentration of 100 µM, the peptides showed a generalized and equipotent antiproliferative effect against most of the 10 leukemic cell lines tested (Figure 4A). Thus, the biomimetic p-AppK and p-Acl peptides were confirmed as quite promising candidates for anti-leukemic therapy, being highly toxic (≥75% cell viability inhibition) to K562, NB4, RAMOS and CEM cells.

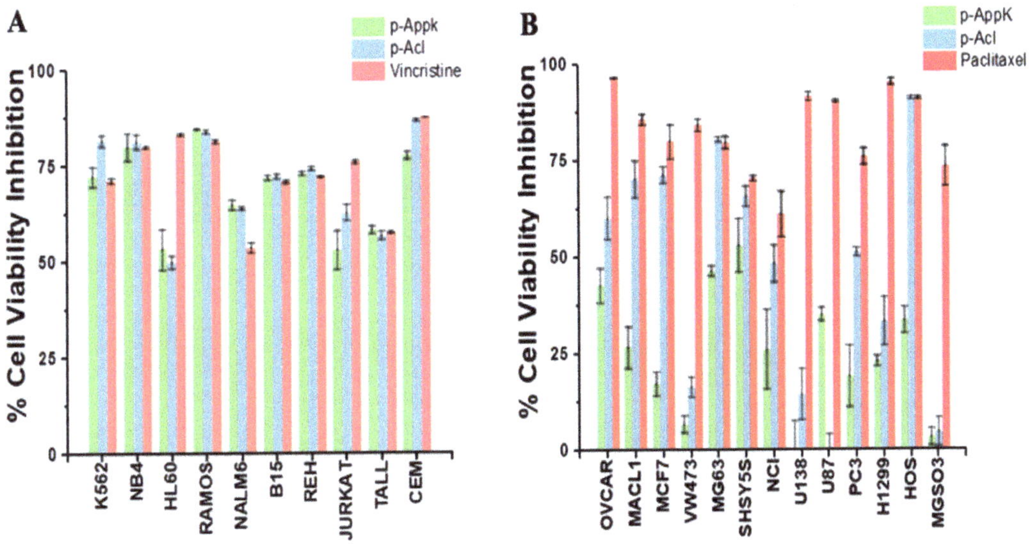

Figure 4. In vitro cytotoxicity of p-AppK (green) and p-Acl (blue). (**A**) Leukemia cell lines and (**B**) solid tumors were exposed to a concentration of 100 µM of both peptides. A colorimetric assay determined the cell viability inhibition after 24 h of peptide treatment. Cells cultured in a growth medium without the peptides were considered a positive control with 100% viability. The positive control (reference antineoplastic drug) is represented in salmon. The data represent the mean ± SD.

Interestingly, though generally less potent than the reference drug against solid tumors, both peptides showed higher selectivity between the cancer cell lines tested (Figure 4B). p-Acl was consistently more efficient than p-AppK in inhibiting the viability of most cell lines. Among the cell lines susceptible to p-Acl, the reduction in the mitochondrial metabolism of human osteosarcoma lines (HOS and MG63) stands out, reaching a viability inhibition greater than 75% and like paclitaxel. These results suggest a functional impact of the p-Acl single amino acid substitution in relation to p-AppK to recognize and treat osteosarcoma. On the other hand, the effect of p-AppK is significantly lower, displaying low toxicity on these solid tumor cell lines (≤50% inhibition of viability).

4.7. Molecular Dynamics Simulations

Six different model membranes representative of healthy, leukemia, and thymus human cancer cells were built. In each model, peptides p-AppK and p-Acl were separately inserted, generating twelve distinct membrane–peptide systems. Interestingly, the destabilizing effect of the two peptides in the six membranes was qualitatively similar, pointing to a common permeabilization mechanism. The p-Acl-DOPS system (membrane 1) is herein described in more detail, as this was the system where the destabilizing effects were most notorious. The similar but less emphatic results for the other eleven systems are given in Supplementary Materials Figures S2 and S3. From a phenomenological point of view, both peptides induced the same kind of membrane-disturbing effect. However, the effect of peptide p-Acl was shown to be quantitatively more pronounced.

The peptides fit very well within the membrane, parallel to the phospholipids, spanning the whole membrane width. The first and last two lysines (in blue) of both peptides (KKYKAYFKLKCKK and KKYKAYFKFKCKK) established salt bridges with the phospholipid phosphate groups. The side chains of the three more central lysines (in green), place the hydrophobic region grossly parallel to the lipidic tails of the phospholipids and the ammonium group near the phosphate headgroups (Figure 5).

(**a**)

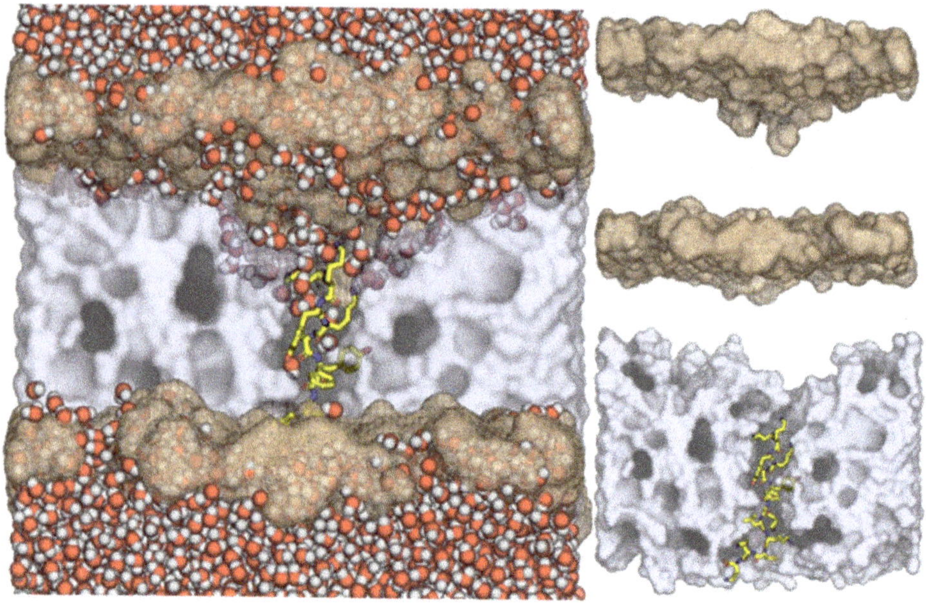

(**b**)

(**c**)

Figure 5. p-Acl peptide interaction with the DOPS membrane bilayer (one-half of the membrane and water molecules were removed for better visualization; the inner hydrophobic core of the membrane is shown in grey; the positions of the phosphorus atoms in the headgroups are shown in salmon). Left (**a**): a large deformation in the position of the phospholipid headgroups is visible, with the phosphate moieties penetrating deeply into the membrane core, more pronounced in the upper leaflet; in addition, the penetration of water molecules deep into the membrane is visible, confirming that the peptide induces a membrane-permeabilization effect. Top-right (**b**): only membrane inner hydrophobic core and phosphorus positions are shown, for clarity; the deformation of the membrane headgroups towards the inner part of the membrane is evident. Lower-right (**c**): insertion of peptide p-Acl (stick model) into the membrane inner hydrophobic core, showing the peptide's perfect structural fitness to span the whole width of the membrane.

In all membranes, the peptides induced a clear deformation in the bilayer, with a negative curvature emerging at both leaflets, but more pronounced in the upper one, resulting in a significant membrane thinning. For instance, the DOPS membrane width decreased by 30 Å in the center of the depression (Figure 5). This effect was the most pronounced deformation within all systems studied, as in all other cases, membrane thinning was between 15–27 Å.

The density of the system was calculated as a function of the z-axis perpendicular to the bilayer. The origin of the axis was defined at the bilayer center (Figure 6). The profile (Figure 6a, top) showed that the distance between the phosphorus atoms and the lysine nitrogen atoms was ~5–10 Å, placing the lysine ammonium groups near or within the headgroup region, with their hydrogen atoms at close- to medium-range from the phosphate oxygen atoms. The water density inside the bilayer was small as it was averaged out throughout the whole system, not only around peptide p-Acl. The number of water molecules along the p-Acl residues is also displayed in Figure 6b (bottom). All water molecules whose oxygen atom was within 3.0 Å from the peptide heteroatoms were accounted for. The figure shows that water penetrated deep into the bilayer, with more than two water molecules close to the peptide, on average, almost up to the leaflet separation, where a stretch of three hydrophobic residues (Ala-Tyr-Phe) almost broke the water molecule chain. There was no place in the simulated system where water molecules penetrated deeply in the membrane that was not around the peptide. Hence, the increased permeability of the bilayer around the peptide was unquestionable. In the other membrane systems, an evident water penetration across the membrane was visible as well, even though it did not reach the membrane core during the simulation time (details in Supplementary Materials).

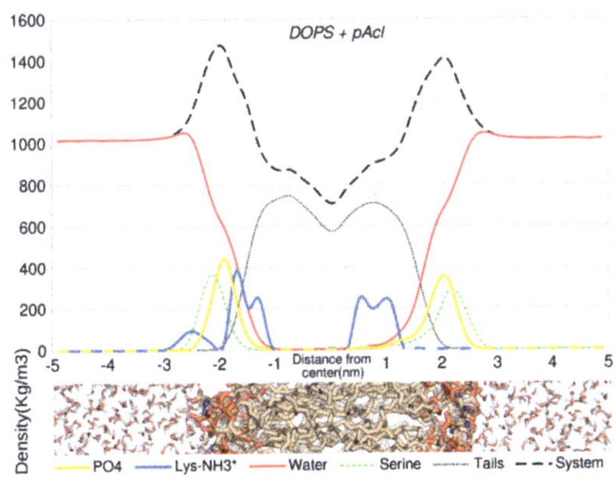

(a)

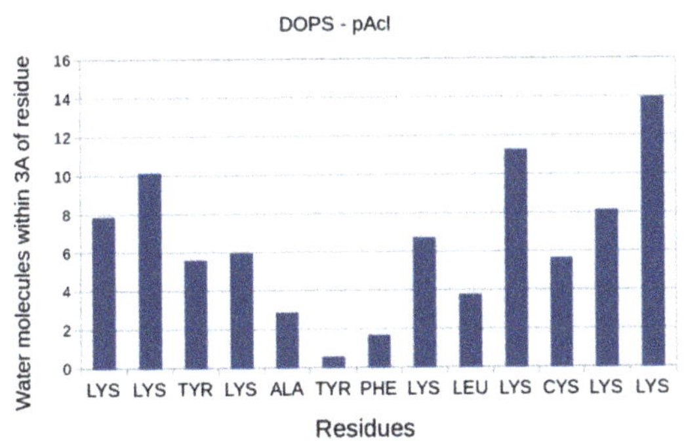

(b)

Figure 6. Density of the main constituents of the system as a function of the bilayer normal. Top (**a**): the center of the bilayer is located at z = 0. Bottom (**b**): the number of water molecules whose oxygen atom is within 3.0 Å of any heteroatom of the p-Acl peptide.

Within the sub-millisecond timescale of the simulation, the rapid penetration of water was most likely just the beginning of a chain of highly destabilizing events, which likely underpin the experimentally observed membrane permeability. Hence, the power of both peptides to induce extensive membrane weakening and permeation is clearly demonstrated both in vitro and in silico.

5. Conclusions

In summary, p-AppK and p-Acl showed a broad spectrum of in vitro effects, including antineoplastic and antibacterial activities in the absence of hemolysis. In general, phenylalanine favors the biological action of peptides in osteosarcoma and *S. aureus* isolates, as compared to leucine in an analogous peptide sequence. Therefore, isoforms of enzymes found in nature provide valuable information that can guide the customization of peptide drug candidates. Thus, our results validate PLA_2 toxins with membrane-disturbing activities as a source of promising dual-target therapeutic peptides that might be useful in the development of new treatments for cancer and bacterial infections, including those caused by MDR strains.

Supplementary Materials: The following are available online at https://www.mdpi.com/article/10.3390/cimb44010004/s1, Figure S1: Chromatographic homogeneity and mass spectrometry analysis of the synthetic peptides, Figure S2: Number of water molecules whose oxygen atom is within 3.0 Å of any heteroatom of the p-Acl and p-AppK peptides in membrane models 1-3, Figure S3: Number of water molecules whose oxygen atom is within 3.0 Å of any heavy atom of the p-Acl and p-AppK peptides in membranes 4-6, Figure S4: Representation of the DOPS model (membrane 1) used in this work with the p-Acl peptide located in the center of the bilayer (half of the membrane and water molecules were removed to visualize the peptide better), Table S1: Composition of Membrane 5 (thymocytes-like membrane model), Table S2: Composition of Membrane 6 (leukemia-like membrane model) and Table S3: Molecular dynamics minimizations and equilibrations.

Author Contributions: Conceptualization, J.R.A., P.G., N.V. and S.L.d.S.; methodology, J.R.A., B.M., M.L., Ó.P., G.C.F.J., N.V. and C.A.; formal analysis, J.R.A., P.A.F., M.J.R. and S.L.d.S.; writing—original draft preparation, J.R.A., B.M., Ó.P., M.J.R. and P.A.F.; writing—review and editing, J.R.A., P.G., N.V., P.A.F., M.J.R., M.L., G.C.F.J., and S.L.d.S.; supervision, S.L.d.S., N.V., P.G. and P.A.F.; project administration, S.L.d.S. and P.G.; funding acquisition, P.G. and N.V. All authors have read and agreed to the published version of the manuscript.

Funding: This research was funded by FCT/MCTES, grant number UIBD/50006/2020, PTDC/QUI-OUT/1401/2020, and CIRCNA/BRB/0281/2019.

Institutional Review Board Statement: Not applicable.

Informed Consent Statement: Not applicable.

Data Availability Statement: Not applicable.

Acknowledgments: The authors gratefully acknowledge the Coordination for the Improvement of Higher Education Personnel (CAPES) for doctoral scholarship and the National Council of Technological and Scientific Development (CNPq) for doctoral-sandwich scholarship and financial support.

Conflicts of Interest: The authors declare no conflict of interest.

References

1. Tornesello, A.L.; Borrelli, A.; Buonaguro, L.; Buonaguro, F.M.; Tornesello, M.L. Antimicrobial peptides as anticancer agents: Functional properties and biological activities. *Molecules* **2020**, *25*, 2850. [CrossRef] [PubMed]
2. De La Torre, B.G.; Albericio, F. Peptide therapeutics 2.0. *Molecules* **2020**, *25*, 2293. [CrossRef] [PubMed]
3. Lee, A.C.; Harris, J.L.; Khanna, K.K.; Hong, J.H. A comprehensive review on current advances in peptide drug development and design. *Int. J. Mol. Sci.* **2019**, *20*, 2383. [CrossRef] [PubMed]
4. Torrent, M.; Nogués, M.V.; Boix, E. Discovering new in silico tools for antimicrobial peptide prediction. *Curr. Drug Targets* **2012**, *13*, 1148–1157. [CrossRef]
5. Ali, N.; Shamoon, A.; Yadav, N.; Sharma, T. Peptide combination generator: A tool for generating peptide combinations. *ACS Omega* **2020**, *5*, 5781–5783. [CrossRef]

6. Kang, N.J.; Jin, H.-S.; Lee, S.-E.; Kim, H.J.; Koh, H.; Lee, D.-W. New approaches towards the discovery and evaluation of bioactive peptides from natural resources. *Crit. Rev. Environ. Sci. Technol.* **2020**, *50*, 72–103. [CrossRef]
7. Porosk, L.; Gaidutšik, I.; Langel, Ü. Approaches for the discovery of new cell-penetrating peptides. *Expert Opin. Drug Discov.* **2020**, *16*, 553–565. [CrossRef]
8. Zhang, Y. Why do we study animal toxins? *Zool. Res.* **2015**, *36*, 183–222.
9. Mouchbahani-Constance, S.; Sharif-Naeini, R. Proteomic and transcriptomic techniques to decipher the molecular evolution of venoms. *Toxins* **2021**, *13*, 154. [CrossRef]
10. Almeida, J.R.; Resende, L.M.; Watanabe, R.K.; Carregari, V.C.; Huancahuire-Vega, S.; da Caldeira, S.C.A.; Coutinho-Neto, A.; Soares, A.M.; Vale, N.; de Gomes, C.P.A.; et al. Snake venom peptides and low mass proteins: Molecular tools and therapeutic agents. *Curr. Med. Chem.* **2017**, *24*, 3254–3282. [CrossRef]
11. Lomonte, B.; Gutiérrez, J.M. Phospholipases A_2 from viperidae snake venoms: How do they induce skeletal muscle damage? *Acta Chim. Slov.* **2011**, *58*, 647–659.
12. Lomonte, B.; Angulo, Y.; Moreno, E. Synthetic peptides derived from the C-terminal region of Lys49 phospholipase A_2 homologues from viperidae snake venoms: Biomimetic activities and potential applications. *Curr. Pharm. Des.* **2010**, *16*, 3224–3230. [CrossRef]
13. Almeida, J.R.; Resende, L.M.; Silva, A.G.; Ribeiro, R.I.; Stabeli, R.G.; Soares, A.M.; Calderon, L.A.; Marangoni, S.; Da Silva, S.L. Biochemical and functional studies of ColTx-I, a new myotoxic phospholipase A_2 isolated from *Crotalus oreganus lutosus* (Great Basin rattlesnake) snake venom. *Toxicon* **2016**, *117*, 1–12. [CrossRef]
14. Ghosh, C.; Haldar, J. Membrane-active small molecules: Designs inspired by antimicrobial peptides. *ChemMedChem* **2015**, *10*, 1606–1624. [CrossRef]
15. Araya, C.; Lomonte, B. Antitumor effects of cationic synthetic peptides derived from Lys49 phospholipase A_2 homologues of snake venoms. *Cell Biol. Int.* **2007**, *31*, 263–268. [CrossRef]
16. Costa, T.R.; Menaldo, D.L.; Oliveira, C.Z.; Santos-Filho, N.A.; Teixeira, S.S.; Nomizo, A.; Fuly, A.L.; Monteiro, M.C.; de Souza, B.M.; Palma, M.S.; et al. Myotoxic phospholipases A_2 isolated from *Bothrops brazili* snake venom and synthetic peptides derived from their C-terminal region: Cytotoxic effect on microorganism and tumor cells. *Peptides* **2008**, *29*, 1645–1656. [CrossRef]
17. Mendes, B.; Almeida, J.R.; Vale, N.; Gomes, P.; Gadelha, F.R.; Da Silva, S.L.; Miguel, D.C. Potential use of 13-mer peptides based on phospholipase and oligoarginine as leishmanicidal agents. *Comp. Biochem. Physiol. Pt. C Toxicol. Pharmacol.* **2019**, *226*, 108612. [CrossRef]
18. Santos-Filho, N.A.; Lorenzon, E.N.; Ramos, M.A.; Santos, C.T.; Piccoli, J.P.; Bauab, T.M.; Fusco-Almeida, A.M.; Cilli, E.M. Synthesis and characterization of an antibacterial and non-toxic dimeric peptide derived from the C-terminal region of Bothropstoxin-I. *Toxicon* **2015**, *103*, 160–168. [CrossRef]
19. Almeida, J.R.; Mendes, B.; Lancellotti, M.; Marangoni, S.; Vale, N.; Passos, Ó.; Ramos, M.J.; Fernandes, P.A.; Gomes, P.; Da Silva, S.L. A novel synthetic peptide inspired on Lys49 phospholipase A_2 from *Crotalus oreganus abyssus* snake venom active against multidrug-resistant clinical isolates. *Eur. J. Med. Chem.* **2018**, *149*, 248–256. [CrossRef]
20. Dennison, S.R.; Whittaker, M.; Harris, F.; Phoenix, D.A. Anticancer alpha-helical peptides and structure/function relationships underpinning their interactions with tumor cell membranes. *Curr. Protein Pept. Sci.* **2006**, *7*, 487–499. [CrossRef]
21. Chiangjong, W.; Chutipongtanate, S.; Hongeng, S. Anticancer peptide: Physicochemical property, functional aspect and trend in clinical application. *Int. J. Oncol.* **2020**, *57*, 678–696. [CrossRef]
22. Proaño-Bolanos, C.; Blasco-Zuniga, A.; Almeida, J.R.; Wang, L.; Llumiquinga, M.A.; Rivera, M.; Zhou, M.; Chen, T.; Shaw, C. Unravelling the skin secretion peptides of the gliding leaf frog, *Agalychnis spurrelli* (Hylidae). *Biomolecules* **2019**, *9*, 667. [CrossRef]
23. van Blitterswijk, W.J.; De Veer, G.; Krol, J.H.; Emmelot, P. Comparative lipid analysis of purified plasma membranes and shed extracellular membrane vesicles from normal murine thymocytes and leukemic GRSL cells. *Biochim. Biophys. Acta* **1982**, *14*, 495–504. [CrossRef]
24. Van Der Spoel, D.; Lindahl, E.; Hess, B.; Groenhof, G.; Mark, A.E.; Berendsen, H.J. GROMACS: Fast, flexible, and free. *J. Comput. Chem.* **2005**, *26*, 1701–1718. [CrossRef]
25. Berendsen, H.J.; Postma, J.V.; van Gunsteren, W.F.; DiNola, A.R.H.J.; Haak, J.R. Molecular dynamics with coupling to an external bath. *J. Chem. Phys.* **1984**, *81*, 3684–3690. [CrossRef]
26. Darden, T.; York, D.; Pedersen, L. Particle mesh Ewald: An N·log (N) method for Ewald sums in large systems. *J. Chem. Phys.* **1993**, *98*, 10089–10092. [CrossRef]
27. Vlieghe, P.; Lisowski, V.; Martinez, J.; Khrestchatisky, M. Synthetic therapeutic peptides: Science and market. *Drug Discov. Today* **2010**, *15*, 40–56. [CrossRef]
28. Robles-Loaiza, A.A.; Pinos-Tamayo, E.A.; Mendes, B.; Teixeira, C.; Alves, C.; Gomes, P.; Almeida, J.R. Peptides to tackle leishmaniasis: Current status and future directions. *Int. J. Mol. Sci.* **2021**, *22*, 4400. [CrossRef]
29. Zhu, S.; Aumelas, A.; Gao, B. Convergent evolution-guided design of antimicrobial peptides derived from Influenza A virus hemagglutinin. *J. Med. Chem.* **2011**, *54*, 1091–1095. [CrossRef]
30. Herzig, V.; Cristofori-Armstrong, B.; Israel, M.R.; Nixon, S.A.; Vetter, I.; King, G.F. Animal toxins—Nature's evolutionary-refined toolkit for basic research and drug discovery. *Biochem. Pharmacol.* **2020**, *181*, 114096. [CrossRef]
31. Adhikari, S.; Leissa, J.A.; Karlsson, A.J. Beyond function: Engineering improved peptides for therapeutic applications. *AIChE J.* **2020**, *66*, e16776. [CrossRef]

32. Plisson, F.; Ramírez-Sánchez, O.; Martínez-Hernández, C. Machine learning-guided discovery and design of non-hemolytic peptides. *Sci. Rep.* **2020**, *10*, 16581. [CrossRef]
33. Cardoso, M.H.; Orozco, R.Q.; Rezende, S.B.; Rodrigues, G.; Oshiro, K.G.N.; Cândido, E.S.; Franco, O.L. Computer-aided design of antimicrobial peptides: Are we generating effective drug candidates? *Front. Microbiol.* **2020**, *10*, 3097. [CrossRef]
34. Yan, J.; Bhadra, P.; Li, A.; Sethiya, P.; Qin, L.; Tai, H.K.; Wong, K.H.; Siu, S.W. Deep-AmPEP30: Improve short antimicrobial peptides prediction with deep learning. *Mol. Ther.-Nucleic Acids* **2020**, *20*, 882–894. [CrossRef]
35. Mustafa, K.; Kanwal, J.; Musaddiq, S.; Khakwani, S. Bioactive peptides and their natural sources. In *Functional Foods and Nutraceuticals: Bioactive Components, Formulations and Innovations*; Egbuna, C., Dable Tupas, G., Eds.; Springer International Publishing: Cham, Switzerland, 2020; pp. 75–97.
36. Lomonte, B.; Rangel, J. Snake venom Lys49 myotoxins: From phospholipases A_2 to non-enzymatic membrane disruptors. *Toxicon* **2012**, *60*, 520–530. [CrossRef]
37. Won, A.; Ianoul, A. Interactions of antimicrobial peptide from C-terminus of myotoxin II with phospholipid mono- and bilayers. *Biochim. Biophys. Acta (BBA) Biomembr.* **2009**, *1788*, 2277–2283. [CrossRef] [PubMed]
38. Huang, H.W.; Charron, N.E. Understanding membrane-active antimicrobial peptides. *Q. Rev. Biophys.* **2017**, *50*, e10. [CrossRef] [PubMed]
39. Tan, L.T.-H.; Chan, K.-G.; Pusparajah, P.; Lee, W.-L.; Chuah, L.-H.; Khan, T.M.; Lee, L.-H.; Goh, B.-H. Targeting membrane lipid a potential cancer cure? *Front. Pharmacol.* **2017**, *8*, 12. [CrossRef] [PubMed]
40. Murillo, L.A.; Lan, C.Y.; Agabian, N.M.; Larios, S.; Lomonte, B. Fungicidal activity of a phospholipase-A_2-derived synthetic peptide variant against *Candida albicans*. *Rev. Esp. Quimioter.* **2007**, *20*, 330–333. [PubMed]
41. Khandelia, H.; Ipsen, J.H.; Mouritsen, O.G. The impact of peptides on lipid membranes. *Biochim. Biophys. Acta (BBA) Biomembr.* **2008**, *1778*, 1528–1536. [CrossRef] [PubMed]
42. Da, W.; Tao, L.; Zhu, Y. The inhibitory effect of CTAB on human osteosarcoma through the PI3K/AKT signaling pathway. *Int. J. Oncol.* **2021**, *59*, 42. [CrossRef] [PubMed]
43. Jendrossek, V.; Handrick, R. Membrane targeted anticancer drugs: Potent inducers of apoptosis and putative radiosensitizers. *Curr. Med. Chem. Anticancer Agents* **2003**, *3*, 343–353. [CrossRef]
44. Zalba, S.; Ten Hagen, T.L.M. Cell membrane modulation as adjuvant in cancer therapy. *Cancer Treat. Rev.* **2017**, *52*, 48–57. [CrossRef]
45. Nyström, L.; Malmsten, M. Membrane interactions and cell selectivity of amphiphilic anticancer peptides. *Curr. Opin. Colloid Interface Sci.* **2018**, *38*, 1–17. [CrossRef]
46. Xie, W.; Mondragón, L.; Mauseth, B.; Wang, Y.; Pol, J.; Lévesque, S.; Zhou, H.; Yamazaki, T.; Eksteen, J.J.; Zitvogel, L.; et al. Tumor lysis with LTX-401 creates anticancer immunity. *OncoImmunology* **2019**, *8*, e1594555. [CrossRef]
47. Fleten, K.G.; Eksteen, J.J.; Mauseth, B.; Camilio, K.A.; Vasskog, T.; Sveinbjørnsson, B.; Rekdal, Ø.; Mælandsmo, G.M.; Flatmark, K. Oncolytic peptides DTT-205 and DTT-304 induce complete regression and protective immune response in experimental murine colorectal cancer. *Sci. Rep.* **2021**, *11*, 6731. [CrossRef]
48. Peña-Carrillo, M.S.; Pinos-Tamayo, E.A.; Mendes, B.; Domínguez-Borbor, C.; Proaño-Bolaños, C.; Miguel, D.C.; Almeida, J.R. Dissection of phospholipases A_2 reveals multifaceted peptides targeting cancer cells, Leishmania and bacteria. *Bioorg. Chem.* **2021**, *114*, 105041. [CrossRef]
49. Gebrim, L.C.; Marcussi, S.; Menaldo, D.L.; de Menezes, C.S.; Nomizo, A.; Hamaguchi, A.; Silveira-Lacerda, E.P.; Homsi-Brandeburgo, M.I.; Sampaio, S.V.; Soares, A.M.; et al. Antitumor effects of snake venom chemically modified Lys49 phospholipase A_2-like BthTX-I and a synthetic peptide derived from its C-terminal region. *Biologicals* **2009**, *37*, 222–229. [CrossRef]
50. López Cascales, J.J.; Zenak, S.; García de la Torre, J.; Lezama, O.G.; Garro, A.; Enriz, R.D. Small cationic peptides: Influence of charge on their antimicrobial activity. *ACS Omega* **2018**, *3*, 5390–5398. [CrossRef]
51. O'Day, K.; Gorlick, R. Novel therapeutic agents for osteosarcoma. *Expert Rev. Anticancer Ther.* **2009**, *9*, 511–523. [CrossRef]
52. Czyzewski, K.; Galazka, P.; Zalas-Wiecek, P.; Gryniewicz-Kwiatkowska, O.; Gietka, A.; Semczuk, K.; Chelmecka-Wiktorczyk, L.; Zak, I.; Salamonowicz, M.; Fraczkiewicz, J.; et al. Infectious complications in children with malignant bone tumors: A multicenter nationwide study. *Infect. Drug Resist.* **2019**, *12*, 1471–1480. [CrossRef]
53. van Kan, E.J.; Demel, R.A.; van der Bent, A.; de Kruijff, B. The role of the abundant phenylalanines in the mode of action of the antimicrobial peptide clavanin. *Biochim. Biophys. Acta* **2003**, *1615*, 84–92. [CrossRef]
54. Mishra, V.K.; Palgunachari, M.N.; Krishna, R.; Glushka, J.; Segrest, J.P.; Anantharamaiah, G.M. Effect of leucine to phenylalanine substitution on the nonpolar face of a class A amphipathic helical peptide on its interaction with lipid: High resolution solution NMR studies of 4F-dimyristoylphosphatidylcholine discoidal complex. *J. Biol. Chem.* **2008**, *283*, 34393–34402. [CrossRef]
55. Lee, E.; Shin, A.; Jeong, K.-W.; Jin, B.; Jnawali, H.N.; Shin, S.; Shin, S.Y.; Kim, Y. Role of Phenylalanine and valine10 residues in the antimicrobial activity and cytotoxicity of piscidin-1. *PLoS ONE* **2014**, *9*, e114453. [CrossRef]
56. Sahoo, B.R.; Maruyama, K.; Edula, J.R.; Tougan, T.; Lin, Y.; Lee, Y.-H.; Horii, T.; Fujiwara, T. Mechanistic and structural basis of bioengineered bovine Cathelicidin-5 with optimized therapeutic activity. *Sci. Rep.* **2017**, *7*, 44781. [CrossRef]
57. Azmi, S.; Srivastava, S.; Mishra, N.N.; Tripathi, J.K.; Shukla, P.K.; Ghosh, J.K. Characterization of antimicrobial, cytotoxic, and antiendotoxin properties of short peptides with different hydrophobic amino acids at "a" and "d" positions of a heptad repeat sequence. *J. Med. Chem.* **2013**, *56*, 924–939. [CrossRef]

58. Liscano, Y.; Oñate-Garzón, J.; Delgado, J.P. Peptides with dual antimicrobial–anticancer activity: Strategies to overcome peptide limitations and rational design of anticancer peptides. *Molecules* **2020**, *25*, 4245. [CrossRef]
59. Felício, M.R.; Silva, O.N.; Gonçalves, S.; Santos, N.C.; Franco, O. Peptides with dual antimicrobial and anticancer activities. *Front. Chem.* **2017**, *5*, 5. [CrossRef]
60. Fosgerau, K.; Hoffmann, T. Peptide therapeutics: Current status and future directions. *Drug Discov. Today* **2015**, *20*, 122–128. [CrossRef]
61. Perdikouri, E.I.A.; Arvaniti, K.; Lathyris, D.; Apostolidou Kiouti, F.; Siskou, E.; Haidich, A.B.; Papandreou, C. Infections due to multidrug-resistant bacteria in oncological patients: Insights from a five-year epidemiological and clinical analysis. *Microorganisms* **2019**, *7*, 277. [CrossRef]

MDPI
St. Alban-Anlage 66
4052 Basel
Switzerland
Tel. +41 61 683 77 34
Fax +41 61 302 89 18
www.mdpi.com

Current Issues in Molecular Biology Editorial Office
E-mail: cimb@mdpi.com
www.mdpi.com/journal/cimb

www.ingramcontent.com/pod-product-compliance
Lightning Source LLC
LaVergne TN
LVHW070236100526
838202LV00015B/2134